Transportation Depth Reference Manual
for the PE Civil Exam

Third Edition

Norman R. Voigt, PE, PLS

Register Your Book at ppi2pass.com

- Receive the latest exam news.
- Obtain exclusive exam tips and strategies.
- Receive special discounts.

Report Errors for This Book

PPI is grateful to every reader who notifies us of a possible error. Your feedback allows us to improve the quality and accuracy of our products. Report errata at **ppi2pass.com**.

Digital Book Notice

All digital content, regardless of delivery method, is protected by U.S. copyright laws. Access to digital content is limited to the original user/assignee and is non-transferable. PPI may, at its option, revoke access or pursue damages if a user violates copyright law or PPI's end-user license agreement.

TRANSPORTATION DEPTH REFERENCE MANUAL FOR THE PE CIVIL EXAM
Third Edition

Current release of this edition: 1

Release History

date	edition number	revision number	update
Dec 2017	2	2	Minor corrections.
Mar 2018	2	3	Minor corrections.
Aug 2018	3	1	New edition. Code updates. Copyright update.

© 2019 Kaplan, Inc. All rights reserved.

All content is copyrighted by Kaplan, Inc. No part, either text or image, may be used for any purpose other than personal use. Reproduction, modification, storage in a retrieval system or retransmission, in any form or by any means, electronic, mechanical, or otherwise, for reasons other than personal use, without prior written permission from the publisher is strictly prohibited. For written permission, contact permissions@ppi2pass.com.

Printed in the United States of America.

PPI
1250 Fifth Avenue, Belmont, CA 94002
(650) 593-9119
ppi2pass.com

ISBN: 978-1-59126-623-5

F E D C B

Table of Contents

Preface vii

Acknowledgments ix

Design Standards and References xi

Introduction xiii

Topic I: Transportation Planning
1. Introduction 1-1
2. Traffic Flow 1-2
3. Trip Generation and Attraction 1-2
4. Origin-Destination Studies 1-5
5. Traffic Volume Studies 1-7
6. Capacity Planning Analysis 1-10
7. Optimization and Cost Analysis 1-12
8. Site Impact Analysis 1-15
9. Practice Problems 1-18

Topic II: Traffic and Capacity Analysis
1. Travel Time and Delay Studies 2-2
2. Capacity Analysis Procedures 2-9
3. Capacity Analysis for Uninterrupted Flow ... 2-10
4. Two-Lane Highway Analysis 2-19
5. Traffic Signals 2-26
6. Parking Facilities 2-35
7. Practice Problems 2-48

Topic III: Pedestrian and Mass Transit Analysis
1. Non-Motorized Facilities-Pedestrian 3-3
2. Non-Motorized Facilities-Bicycle 3-25
3. Mass Transit System Studies 3-26
4. Practice Problems 3-37

Topic IV: Geometric Design
1. Introduction 4-2
2. Route and Corridor Alignment 4-2
3. Specifying Line Direction 4-3
4. Horizontal Curves 4-5
5. Degree of Curve 4-6
6. Stationing on a Horizontal Curve 4-7
7. Field Layout and Fitting of Horizontal Curves 4-9
8. Spiral Curves 4-16
9. Vertical Curves 4-18
10. Sources of Error in Route Alignment 4-22
11. Sight Distance 4-23
12. Superelevation 4-30
13. Vertical and Horizontal Clearances 4-39
14. Acceleration and Deceleration 4-42
15. Intersections and Interchanges 4-43
16. Railroad Engineering 4-46
17. Bikeway Geometric Design 4-48
18. Mass Transit Geometric Design 4-50
19. Practice Problems 4-54

Topic V: Transportation Construction
1. Introduction 5-1
2. Excavation and Embankment 5-2
3. Mass Diagrams 5-7
4. Pavement Design 5-9
5. Hot Mix Asphalt 5-11
6. Marshall Mix Design 5-18
7. Hveem Mix Design 5-21
8. Superpave Mix Design Procedures 5-23
9. Value Engineering 5-27
10. Practice Problems 5-28

Topic VI: Traffic Safety
1. Traffic Safety and Roadway Design Analysis 6-2
2. Traffic Data Used for Crash Analysis 6-2
3. Roadway Elements for Safe Design 6-2
4. Crash Analysis 6-4
5. Encroachment Crash Analysis 6-5
6. Roadside Clearance Analysis 6-8
7. Economic Analysis 6-19
8. Driver Behavior and Performance 6-21
9. Work Zone Safety 6-24
10. Temporary Traffic Control Zones 6-26
11. Conflict Analysis 6-27
12. Practice Problems 6-28

Topic VII: Support Material
Appendices A-1

Index I-1

Appendix
Table of Contents

2.A Travel Time, Speed, and Delay Study Field Sheet Using the Moving Vehicle Method..... A-1
2.B Warrant 1, Eight-Hour Vehicular Volume A-2
2.C Warrant 2 .... A-3
2.D Warrant 3 .... A-4
2.E Warrant 4 .... A-5
2.F Warrant 9 .... A-7
4.A Design Superelevation Rates A-9
4.B Minimum Radii for Design Superelevation Rates and Design Speeds A-14
4.C Superelevation, Radius, and Design Speed for Low-Speed Urban Streets* A-23
4.D Minimum Radii and Superelevation for Low-Speed Urban Streets[a,b,c] A-25
4.E Minimum Superelevation Runoff for Horizontal Curves A-29
4.F Acceleration Distances for Passenger Cars, Level Conditions A-31
4.G Deceleration Distances for Passenger Vehicles Approaching Intersections A-32
4.H AASHTO Design Vehicle Dimensions A-33
4.I Minimum Turning Radii of Design Vehicles ... A-35
4.J Minimum Turning Path for Passenger Car (P) Design Vehicle[a,b] A-37
4.K Turning Characteristics of a Typical Tractor/Semitrailer Combination Truck A-38
4.L Minimum Turning Path for Single-Unit Truck (SU-12 [SU-40]) Design Vehicle[a,b] ... A-39
4.M Minimum Turning Path for Intermediate Semitrailer WB-40 (WB-12) Design Vehicle[a,b,c] A-40
4.N Minimum Turning Path for Interstate Semitrailer WB-67 (WB-20) Design Vehicle[a,b,c] . A-41
4.O Minimum Turning Path for Double-Trailer Combination (WB-20D [WB-67D]) Design Vehicle A-42
4.P Minimum Turning Path for City Transit Bus (CITY-BUS) Design Vehicle[a,b] A-43
4.Q Minimum Turning Path for Conventional School Bus (S-BUS36 [S-BUS11]) Design Vehicle[a,b,c] A-44
4.R Minimum Traveled Way, Passenger (P) Design Vehicle Path A-45
4.S Minimum Traveled Way Designs, Single-Unit Trucks and City Transit Buses A-47
4.T Edge-of-Traveled-Way for Turns at Intersections A-49
4.U Effect of Curb Radii on Right-Turning Paths of Various Vehicles A-57
4.V Cross Street Widths Occupied by Turning Vehicles* A-58
4.W Crosswalk Length Variations with Different Curb Radii and Width of Borders A-60
4.X Corner Setbacks with Different Curb Radii and Width of Borders A-61
5.A AASHTO Soil Classification System A-62
5.B Unified Soil Classification System A-63
5.C USDA Soil Triangle A-64
5.D Performance-Graded Asphalt Binder Specification A-65
5.E Superpave Mix Design Procedural Outline A-67

Preface

While there are many quick exam practice problem books on the market whose only purpose is to simply pass the exam, the *Transportation Depth Reference Manual* works to provide a more complete understanding of the material. The organization of this book parallels the planning, development, and construction of transportation system roadway components. Chapter 1 covers the planning process, assessing the desires and needs of a transportation system. Chapter 2 addresses the size and capacity of various vehicular and pedestrian components. Chapter 3 illustrates the processes of evaluating pedestrian, bicycle, and mass transit facilities. Chapter 4 analyzes details of geometry and physical features of a transportation network. Chapter 5 covers construction activities relating to transportation. Chapter 6 details advances in cost-effective analysis of traffic safety features. Taken together, the chapters reflect the almost overwhelming impact transportation has on the daily activities of individual users and communities.

Topics of highway safety, pollution control, construction, and alternative analysis are often considered superior elements of transportation design; however, the basics of geometry cannot be neglected, as those are the foundation of all transportation design. Treating geometry as a minor issue covered by simple computerization can lead to disastrous circumstances if not thoroughly understood and checked by the layout engineer. With this in mind, I focused on areas where there is often confusion, such as circular curve terminology, proper baseline or centerline stationing, and the basics of fitting the alignment to topography. Alignment layout accuracy requires professional field procedures, working hand-in-hand with office procedures, to close the traverse control PI points with minimal error. The understanding is that electronic computations serve well to memorialize poor engineering surveying.

Writing study material for the PE civil exam is challenging, given the frequent updates to exam design standards and references, as well as the continuous changes to the exam specifications by NCEES. This edition of the *Transportation Depth Reference Manual* began as a result of the new edition of the *Highway Capacity Manual* (*HCM*) put out by the Transportation Research Board of the National Academy of Sciences. (*HCM* has changed from using the date of publication in the title to simply calling out the edition number, currently the 6th edition.)

The changes to *HCM* in the 6th edition are fairly extensive. Procedures for studying and analyzing highway traffic that were introduced in *HCM 2010* have been updated to depend heavily on locally generated data, a process the *HCM* calls *calibration*. Broad generalizations and default values of past editions have been expanded to account more precisely for traffic demand, weather events, traffic incidents, and other influences that result in capacity and speed adjustments.

There are many new exhibits that say required data "must be provided" rather than give default values, which of course means more information than before must be given to completely solve related problems. More information is provided for active traffic demand management (ATDM), such as use of high occupancy vehicle lanes or other restricted-use lanes that have become prevalent in urban locations. Freeway and multilane highway analysis has been combined into one chapter, as the information and analysis procedures are much the same for these facilities. (Combining these types of roadways has always been the practice in earlier editions of the *Transportation Depth Reference Manual*.) Two-lane highways remain in a separate analysis section, but there have been changes to truck equivalents, including treating RVs as single-unit trucks. The other modes—transit, bicycle, and pedestrian—use similar analysis procedures as in previous *HCM* editions, with extensions from new and continuing research. Interrupted flow remains basically the same, but again, use of locally provided data from on-site observations is emphasized. This third edition of the *Transportation Depth Reference Manual* has been fully updated to reflect these changes; everything from the text to the figures and tables have been revised, rearranged, and rewritten to the *HCM*'s new specifications.

This manual is an extension of transportation material covered in the *PE Civil Reference Manual*, supplemented by the *Transportation Depth Six-Minute Problems for the PE Civil Exam*. Combined, these three products provide nearly 100% of the knowledge and reference material needed to pass the PE civil transportation exam. It is my hope that the *Transportation Depth Reference Manual* will be your primary transportation reference for the PE exam. Just as the *PE Civil Reference Manual* has been found on many practicing engineers' bookshelves, I hope you will also find the *Transportation Depth Reference Manual* useful as a quick reference for transportation work in your professional career.

I ask that you submit suspected errors to PPI's errata website at **ppi2pass.com**. All errors will be reviewed, and verified mistakes or misprints will be incorporated into future editions of this manual.

Norman R. Voigt, PE, PLS

Acknowledgments

For many pursuing an engineering career, working as a team becomes second nature. Most projects require the knowledge and capabilities of many, and the sense of accomplishment is shared by all who participate. I am privileged to have started my career with one of the finest bridge engineers of the mid-twentieth century, George S. Richardson, PE, at Richardson, Gordon & Associates (RG&A) in Pittsburgh, PA. Mr. Richardson and his partners assembled a group of leading structural transportation engineers to tackle some of the most difficult road and bridge design challenges in the mid-Atlantic region and beyond. With over half the staff being registered engineers, RG&A received more awards for structural and roadway design projects than any firm we knew. This was quite an accomplishment for a medium-sized firm of the day, which has no comparison to today's mega-size firms.

I am particularly indebted to one of my mentors, James K. Arentz, PE, who was one of the designers of the western end of the original Pennsylvania Turnpike. Mr. Arentz was fond of describing how he participated in route layout reconnaissance using a bombsight mounted in the belly of a small plane. The plane would fly the proposed route at low altitude, and Mr. Arentz would view the terrain through the bombsight, envisioning the roadway alignment hugging the sides of hills and following the contours of the valleys. Mr. Arentz taught me how to apply curve geometry and superelevation along challenging routes to make a smooth and comfortable roadway. Our goal was a highway that would ride much like the finest railroad alignments used by first-class passenger trains. The standards set by the Pennsylvania Turnpike were copied in Germany's autobahns and became the basis for the U.S. Interstate Highway System. It took me many years to appreciate what I learned from Mr. Richardson, Mr. Arentz, and their talented colleagues. The kind of mentoring they gave is what true professional development is all about.

Through many years of coordinating and teaching portions of the intensive PE civil review course sponsored by The Pennsylvania State University, I came to know many engineers and engineering candidates. I have shared their experiences and struggles to learn about the practical side of engineering, applying theory to a workable finished product. And, in some cases, defending the basics of engineering design against the business and political nature of large public engineering projects. The emphasis of the Penn State review courses has always been that engineering licensure is about the process of analysis and application of solutions, not simply finding a quick answer to a problem. My colleagues in the program, David Morse, Tom Leech, Kashi Banerjee, Sam Shamsi, and the original energy behind the program, Joe DeSalvo, have all set the tone for proper engineering analysis where the application of realistic solutions is a prerequisite for a commendable finished project.

I would like to thank the individuals who worked to ensure the accuracy of this book, including Joshua T. Frohman, PE, Keith A. Johnson, and Mark Magalotti, PE, MSCE, for technically reviewing this book, and Ralph Arcena, EIT, for performing the calculation check.

I have received outstanding support from PPI on this project. There are many current and former employees (a project of this size doesn't just happen overnight) who helped with the many drafts and my scribbled comments. Thank you to Grace Wong, director of editorial operations; Cathy Schrott, editorial operations manager; Bradley Burch, production editor; Tyler Hayes, senior copy editor; Richard Iriye, typesetter; Beth Christmas, freelance proofreader; Tom Bergstrom, technical illustrator; Ellen Nordman, publishing systems specialist; and Steve Buehler, director of product. Michael R. Lindeburg, PE, graciously provided use of sections, tables, and figures from the *PE Civil Reference Manual*.

The final and most important supporting person has been my wife, Mary Jean, a professional librarian with a background in technology. She has been able to keep me headed in the right direction, teaching me how the written word should be handled so that others can share in the knowledge being presented.

Norman R. Voigt, PE, PLS

Design Standards and References

The *PE Civil Reference Manual* and the *Transportation Depth Reference Manual* are the minimum recommended library for the PE civil transportation depth exam. The exam is based on the following design standards, as noted by the NCEES transportation specifications.

DESIGN STANDARDS AND REFERENCES USED ON THE EXAM

AASHTO: *A Policy on Geometric Design of Highways and Streets*, 6th ed., 2011 (including November 2013 errata), American Association of State Highway and Transportation Officials, Washington, DC

AASHTO: *Guide for Design of Pavement Structures*, (GDPS-4-M) 1993, and 1998 supplement, American Association of State Highway and Transportation Officials, Washington, DC

AASHTO: *Highway Safety Manual*, 1st ed., 2010, vols. 1–3 (including September 2010, February 2012, and March 2016 errata), American Association of State Highway and Transportation Officials, Washington, DC

AASHTO RSDG: *Roadside Design Guide*, 4th ed., 2011 (including February 2012 and July 2015 errata), American Association of State Highway and Transportation Officials, Washington, DC

AASHTO: *Guide for the Planning, Design, and Operation of Pedestrian Facilities*, 1st ed., 2004, American Association of State Highway and Transportation Officials, Washington, DC

AASHTO: *Mechanistic-Empirical Pavement Design Guide: A Manual of Practice*, 2nd ed., July 2015, American Association of State Highway and Transportation Officials, Washington, DC

AI: *The Asphalt Handbook* (MS-4), 7th ed., 2007, Asphalt Institute, Lexington, KY

FHWA: *Hydraulic Design of Highway Culverts*, Hydraulic Design Series Number 5, Publication No. FHWA-HIF-12-026, 3rd ed., April 2012, U.S. Department of Transportation—Federal Highway Administration, Washington, DC

HCM: *Highway Capacity Manual*, 6th ed., Transportation Research Board—National Research Council, Washington, DC

MUTCD: *Manual on Uniform Traffic Control Devices*, 2009, including Revisions 1 and 2, May 2012, U.S. Department of Transportation—Federal Highway Administration, Washington, DC

PCA: *Design and Control of Concrete Mixtures*, 16th ed., 2016, Portland Cement Association, Skokie, IL

PROWAG: *Proposed Accessibility Guidelines for Pedestrian Facilities in the Public Right-of-Way*, July 26, 2011, and supplemental notice of February 13, 2013, United States Access Board, Washington, DC

REFERENCES USED IN THIS BOOK

The following references were used to prepare this book. You may also find them useful references to bring with you to the exam.

Elementary Surveying. Russell C. Brinker and Paul R. Wolf. HarperCollins.

Engineering Economic Analysis. Michael R. Lindeburg, PE. Professional Publications, Inc.

Fundamentals of Traffic Engineering. Wolfgang S. Homburger, Jerome W. Hall, Edward C. Sullivan, and William R. Reilly. University of California, Berkeley.

Highway Engineering. Paul H. Wright and Karen Dixon. John Wiley & Sons.

Manual of Transportation Engineering Studies. Institute of Transportation Engineers.

Pedestrian Planning and Design. John J. Fruin. Metropolitan Association of Urban Designers and Environmental Planners.

Principles of Highway Engineering and Traffic Analysis. Fred L. Mannering, Scott S. Washburn, and Walter P. Kilareski. John Wiley & Sons.

Project Management for Engineering and Construction. Garold D. Oberlender. McGraw-Hill.

Railroad Curves and Earthwork. C. Frank Allen. Norwood Press.

Route Location and Design. Thomas F. Hickerson. McGraw-Hill.

Route Surveying and Design. Carl L. Meyer and David W. Gibson. Harper & Row.

Surveying. Francis H. Moffitt and Harry Bouchard. Harper & Row.

The Surveying Handbook. Russell C. Brinker and Roy Minnick, eds. Kluwer Academic Publishers.

Superpave Mix Design (SP-2). The Asphalt Institute.

Traffic and Highway Engineering. Nicholas J. Garber and Lester A. Hoel. Cengage Learning.

Traffic System Analysis for Engineers and Planners. Martin Wohl and Brian V. Martin. McGraw-Hill.

Transition Curves for Highways. Joseph Barnett. Public Roads Administration.

Transportation Engineering Basics. A. S. Narasimha Murthy and Henry R. Mohle. American Society of Civil Engineers.

Transportation Engineering and Planning. C. S. Papacostas and P. D. Prevedouros. Prentice-Hall.

Trip Generation. The Institute of Transportation Engineers.

Introduction

The National Council of Examiners for Engineering and Surveying (NCEES) develops, administers, and scores the examinations used for engineering and surveying professional licensure in the United States. The PE civil exam provides the qualifying test for candidates seeking registration as civil engineers. The civil transportation exam is intended to assess your knowledge of transportation design principles and techniques of field practice.

This book is written with the exam in mind. Major topics, equations, and example problems are presented and explained, along with additional practice problems and solutions in order to develop a broad understanding of the concept so that a problem that targets specific focus on one or a few selected parts of a concept can be related to the overall process involved. National design standards are referenced throughout the chapters and used in examples and practice problems. Appropriate sections of the standards are explained and analyzed to help gain familiarity with applicable standards before using them during the exam. This book provides a comprehensive guide and reference for self-study of civil transportation engineering.

This book is organized in six chapters that approximate the development of a transportation project from planning, alternative solutions, economic analysis, community impact projection, environmental concerns, safety analysis and predictions, and finally, project construction.

The current NCEES exam specifications cover both the knowledge breadth exam and the discipline depth exam. The breadth portion of the exam covers topics that are generic to all disciplines of civil engineering, while the breadth portion of the exam covers topics specific to the discipline within civil engineering. At the time of publication, the civil transportation specifications are as follows.

I. **Traffic Engineering (Capacity Analysis and Transportation Planning) (11 questions)**

Uninterrupted flow; street segment interrupted flow; intersection capacity; traffic analysis; trip generation and traffic impact studies; accident analysis; nonmotorized vehicle facilities; traffic forecast; highway safety analysis

II. **Horizontal Design (4 questions)**

Basic curve elements; sight distance considerations; superelevation; special horizontal curves

III. **Vertical Design (4 questions)**

Vertical curve geometry; stopping and passing sight distance; vertical clearance

IV. **Intersection Geometry (4 questions)**

Intersection sight distance; interchanges; at-grade intersection layout; including roundabouts

V. **Roadside and Cross-Section Design (4 questions)**

Forgiving roadside concepts; barrier design; cross-section elements; Americans with Disabilities Act (ADA) design considerations

VI. **Signal Design (3 questions)**

Signal timing; signal warrants

VII. **Traffic Control Design (3 questions)**

Signs and pavement markings; temporary traffic control

VIII. **Geotechnical and Pavement (4 questions)**

Sampling and testing; soil stabilization techniques, settlement and compaction, excavation, embankment, and mass balance; design traffic analysis and pavement design procedures; pavement evaluation and maintenance measures

IX. **Drainage (2 questions)**

Hydrology; hydraulics, including culvert and stormwater collection system design, and open-channel flow

X. **Alternatives Analysis (1 question)**

Economic analysis

HOW TO USE THIS BOOK

The *Transportation Depth Reference Manual* provides comprehensive coverage of the major topics on the transportation depth exam and is designed to be used in conjunction with the *PE Civil Reference Manual*, which should be your primary breadth exam review resource. Start by reviewing the exam topics (listed in this introduction) and familiarizing yourself with the content and format of this book. Review the table of contents and the index, and flip through the chapters. Each chapter begins with a nomenclature list of the chapter's variables and ends with practice problems covering the chapter's major topics. Every significant term and concept has been indexed to provide a method of finding topics and data quickly.

Create a study schedule based on your strengths and weaknesses, and on how much time you think you'll need to spend reviewing each chapter. While chapters can be reviewed and referenced independently, each chapter builds on the topics presented previously. As you read each chapter, work the example problems and review the presented solution. At the end of the chapter, assess your understanding by following the example problems and solutions, and the end-of-chapter practice problems. The practice problems are designed to give you experience applying relevant equations, data, and design standards to a given problem. Restrain yourself from reviewing the solutions until after you've solved each problem, then compare your solving approach with that given in the solution. With practice, you will be able to quickly decide which design standards, data, and equations are applicable to the problem at hand.

The PPI Learning Hub (**ppi2pass.com**) offers online versions of this book, a quiz generator with over 900 multiple-choice, exam-like practice problems, over 350 solved problems, full-length practice exams, and learning management tools.

1 Transportation Planning

1. Introduction 1-1
2. Traffic Flow 1-2
3. Trip Generation and Attraction 1-2
4. Origin-Destination Studies 1-5
5. Traffic Volume Studies 1-7
6. Capacity Planning Analysis 1-10
7. Optimization and Cost Analysis 1-12
8. Site Impact Analysis 1-15
9. Practice Problems 1-18

Nomenclature

a	curve adjustment factor	–	–
A	annual amount	$	$
AW	annual worth	$	$
B	benefit	$	$
C	cost	$	$
CF	count expansion factor	–	–
d	delay	min	min
f	factor	–	–
i	effective annual interest rate	%	%
L	sound level	dB	dB
n	number of compounding periods	–	–
N	number	–	–
p	proportion	–	–
P	period	min	min
r	correlation coefficient	–	–
r^2	coefficient of determination	–	–
t	time	min	min
t	user time	p-min	p-min
T	number of trips	–	–
v	flow rate	vph	vph
V	actual volume	veh	veh
V'	adjusted volume	veh	veh
X	quantifiable parameter	various	various

Subscripts

a	adults
A	annual maintenance
B	break
C	counting
e	employed adults
eq	hourly equivalent
h	households
i	initial
l	licensed drivers
m	maintenance
p	peak or person-time
s	school-age children
seg	segment
sig	signal
t	time period, total, ten years, or transit-accessible locations
ts	transit stop
v	vehicles
vtr	vehicle reduction

1. INTRODUCTION

Public transportation planning follows formal steps, such as *coordinating* public involvement, *planning* jurisdictional level responsibilities, *programming* design requirement priorities, and resource *funding*.[1] Important transportation acts include the 1991 Intermodal Surface Transportation Efficiency (ISTEA); the 1998–2003 Transportation Equity Act for the 21st Century (TEA-21); the 2005–2011 Safe, Accountable, Flexible, Efficient Transportation Equity Act: A Legacy for Users (SAFETEA-LU); and the 2015 Fixing America's Surface Transportation Act (FAST). Legislation continues to evolve, setting the standards and compliance criteria necessary for planning and funding of projects. Other requirements include environmental conditions based on the 1970–1990 Clean Air Act.[2] The challenge transportation planners face is devising strategies that meet the capacity needs and livability standards of the future while minimizing negative impacts and decreasing pollution. For more information on federal highway legislation, see www.fhwa.dot.gov/resources/legsregs/.

Some acts, such as the Urban Mass Transportation Assistance Act of 1970 and the Federal-Aid Highway Act of 1956, require considering multiple modes of

---

[1]From *Planning for Transportation in Rural Areas*, Federal Highway Administration, © 2004, Dye Management Group.
[2]The *Clean Air Act* was enacted by U.S. Congress in 1990 and defines the Environmental Protection Agency's (EPA) responsibilities for protecting and improving the nation's air quality. The EPA sets air pollution limits for the United States, and states are prohibited from having less stringent pollution controls than those set for the nation. However, while the EPA sets the national standards, states are responsible for carrying out the act, as combating pollution requires regional knowledge.

transportation. The Federal-Aid Highway Act of 1973 increased metropolitan planning organization (MTO) involvement and the *National Mass Transit Assistance Act* of 1974 extended federal funds for assisting in operating transit systems. In the Highway Capacity Manual (*HCM*), the multimodal trend continues with the inclusion of automobile, transit, pedestrian, and bicycle modes in the comprehensive transportation planning process, instead of treating each mode as separate (and often competing) considerations.

The Federal Highway Administration (FHWA) has a standardized process for all regions with a population of greater than 50,000.[3] The FHWA works with a region's existing plans for land use, environmental, or economic development to establish a transportation system's *goals* (irreducible values) and *objectives* (attainable targets). *Alternative strategies* for managing travel demand are proposed, tested, and evaluated using the modeled projections and the goals and objectives. Changes are made as needed to produce an optimal strategy to implement the proposed transportation system. The *costs* of alternative strategies are compiled, and a draft report is prepared for public comment. Simultaneously, a *draft environmental impact statement* (DEIS) is prepared to determine the projected environmental effects of each alternative. After a period of public input and assessment by the appropriate decision makers, a strategy is selected, the funding stream is secured, and the system's final design begins. Traffic volumes along planned travel paths, which were developed from the travel demand modeling, are used to design the facility to meet the required capacity.

2. TRAFFIC FLOW

Traffic is the flow of people and vehicles on roads. It occurs for many reasons, such as the necessity to transport goods and the human desire to travel. Traffic involves interactions between vehicles, drivers, and the roadway system.

Traffic flow studies are performed when the effort is warranted, such as for estimating the needs of a new trip generator (e.g., a new shopping center or park) or for a planned improvement to an existing location. Broad traffic studies, such as regional or corridor studies, concentrate on traffic volumes, travel times, and capacities. Limited area studies, such as those performed at intersections, tend to focus on crash rates, delays, gaps, and detailed physical conditions. Therefore, depending on their size and purpose, traffic flow studies usually cover

- *volume*, on each leg, link, or segment
- *crash experience*, especially at intersections
- *speed*, primarily in peak periods
- *capacity*, primarily in peak periods
- *physical conditions*, including effect on safety
- *travel time*, in relation to other factors
- *delays*, especially those caused by signals
- *gaps*, or the ability to enter the flow

The data gathered in traffic flow studies are used by traffic engineers who need specific data about study locations in order to create accurate model analyses. General information available in traffic standards and handbooks can then be used to compare data and help validate study conclusions.

3. TRIP GENERATION AND ATTRACTION

The simple form of a trip has an origin and a destination and a link between the two. A complex form may have many incremental segments, with origins, destinations, and purposeful activities along the way. The *trip destination*, commonly called the *trip generator*, determines significant characteristics of the trip. *Trip generation studies* are used to determine the predominant trip generators in the study region, which are then used to estimate the number of trips generated. Examples of trip generators include factories, office buildings, shopping centers, resorts, state parks, airports, transit centers, county fairs, and sporting arenas. Trip attraction studies predict the number of trip ends to a nonhome destination (e.g., to a school or shopping center). Trip generation and attraction studies are closely related and often share methodologies.

Each type of trip purpose, such as for work, shopping, or recreation, has characteristics that transportation planners may use to project trip behavior. The Institute of Transportation Engineers (ITE) publishes the *Trip Generation Manual* (*Trip Generation*), which defines a large number of widely accepted trip generator types and characteristics. Developed using current methodology of land-use-based traffic and transportation engineering, it describes procedures for quantifying the number of trips generated and attracted for various types of land uses. Trip generation studies are often used as *traffic impact studies* to determine the volume generated by a new trip generator and its effect on surrounding roadways. These data are then used to evaluate both the existing and new facilities to determine design improvements to accommodate the increase in trip volume. Because trip generation studies link trip volume and land use, trip generation data are often used in zoning and land use planning. Data determined using trip generation studies are often used in *site-impact studies* to estimate environmental impacts due to new development.

---

[3]Local and state governments set standards for areas with populations of fewer than 50,000.

Trip Generation provides a collection of data for 12 major land use categories comprising more than 4800 traffic studies conducted at over 550 sites in the U.S. and Canada. Figure 1.1 provides an example of data from *Trip Generation* for an office park and shows the data given for each land use. The equations and weighted average trip generation rate given in *Trip Generation* are usually sufficient to determine the number of trip ends, but they can also be determined from local data.

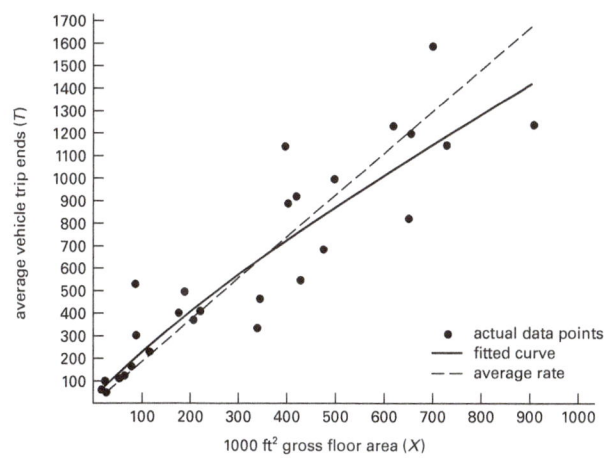

Figure 1.1 Office Park Trip Generation Curve

average vehicle trip ends vs: 1000 ft² gross floor area
on a: weekday,
peak hour of adjacent street traffic
one hour between 7 a.m. and 9 a.m.

number of studies: 27
average 1000 ft² GFA: 331
directional distribution: 89% entering, 11% exiting

trip generator per 1000 ft² gross floor area

average rate	range of rates	standard deviation
1.84	0.98–5.89	1.50

data plot equation

fitted curve equation: $\ln T = 1.679 + 0.818 \ln X$
$r^2 = 0.89$

(Multiply ft² by 0.929 to obtain m².)

From *Trip Generation Manual*, © 2000, Institute of Transportation Engineers. Used by permission.

For trip generation studies, the study area is divided into zones, or tracts, and data are obtained for each zone. Trip generation studies typically use socioeconomic data, such as household income or car ownership, while trip attraction studies most often use employment data, dividing employment by type, such as retail or service. Socioeconomic data can be obtained from sources such as the U.S. Census, while employment information is most often obtained from workplace or household surveys.

Regression Equations

The number of generated trips can be calculated using independent variables, or can be estimated using traffic counts. Independent variables, commonly denoted as X, differ based on the zone and its land use. X must be a quantifiable parameter, such as the number of dwelling units or the number of beds in a hospital. For residential zones, independent variables include household income, car ownership, and family size. For retail zones, examples of independent variables include store type, gross leasable area, customer income level, and proximity to transit. Trip generation studies are often based on several independent variables to develop a zone's total number of trips in order to cover the anticipated behavior of the majority of trip purposes.

Once independent variables are selected and the necessary data are collected, the total number of trips, or total trip ends, can be calculated using either a linear model, Eq. 1.1, or a nonlinear model, Eq. 1.2. T is the total number of trips, bX and $b \ln X$ are the slopes of the plotted curve, and a is the curve's vertical intercept. The units of X depend on the variable being used. For example, if X were the gross leasable area, the units would be square feet.

$$T = a + bX \quad \text{[linear model]} \quad 1.1$$

$$\ln T = a + b \ln X \quad \text{[nonlinear model]} \quad 1.2$$

The ITE refers to both the linear and nonlinear models as *regression equations*, or *fitted curve equations*. A regression equation must be determined for each independent variable used. The choice of the linear model versus the nonlinear model is typically based on the model that has the highest coefficient of determination. The *coefficient of determination*, r^2, is a statistical measure that determines how well the least squares relate to the original data. The value of r^2 ranges from 0 to 1, with 1 indicating that all plotted data fit on the line (i.e., an ideal fit). 0 indicates that the data are scattered and does not indicate an adequate correlation between the data points and the regression equation.

The data given in *Trip Generation* include a regression equation for each land use, although care should be taken when selecting regression equations. Whether the regression equation is calculated or selected from *Trip Generation*, the suitability of the equation to the location should be verified against plotted data points to determine the suitability of fit. The standard deviation and the coefficient of determination for a land use's data are provided as an indication of the variation in trip generation data for the land use. Therefore, proper engineering judgment should be used when using the average rates and regression equations from *Trip Generation*.

Weighted Average Generation Rate

The generated trip ends can also be calculated using known vehicle counts and applying weighted averages. The *weighted average trip generation rate* is the sum of all trip ends during the study period divided by the sum of all independent variables. For example, if the independent variable, X, is the gross floor area, the weighted average rate is the sum of all vehicle trips divided by the sum of the square footage of the gross leasable area in the zone and would be reported as the weighted average rate per 1000 ft^2. The weighted average trip rate is then multiplied by X to find the number of generated trip ends. The weighted average rate can then be plotted as shown in Fig. 1.1 using the assumption that a linear relationship between trip ends and the independent variable exists.

Data Suitability

As outlined by the ITE, the suitability of the regression equation and the weighted average trip generation rate to the location should be verified. *Trip Generation* suggests first comparing the number of trip ends calculated with the regression equation and the weighted average rate. If X is within the range of data points, the curve is not projecting beyond known information. If the difference between the two methods is less than 5%, then either method is acceptable. However, if there is more than a 5% spread, and if the fitted curve shows zero or close to zero trips when $X = 0$, then the equation for the fitted curve should be used. While it is recognized that $\ln 0$ is undefined, a very small number, such as $X = 1$, can be used.

If the previous conditions are not met, then data points should be compared with the independent variable X to see if there are data points near that value to favor one plot over the other.

As a final check, the standard deviation and the coefficient of determination, r^2, the square of the *correlation coefficient*, r, can be compared to the data. As previously outlined, an r^2 value of 1 indicates that the data have a perfect fit, and a value close to 0 indicates that the data are scattered. For trip generation studies, an r^2 value less than 0.75 indicates that the weighted average rate should be used, and an r^2 value greater than or equal to 0.75 indicates that the regression equation should be used. The standard deviation should also be compared to the weighted average rate. If the standard deviation is greater than 110% of the weighted average rate, the regression equation should be used. Likewise, if the standard deviation is less than or equal to 110% of the weighted average rate, the weighted average rate should be used.

In problems where none of the previous conditions are acceptably met, the trip planner must make a selection using acceptable levels of data collected at or near the study site and at least two other sites of similar nature to confirm the data.

Example 1.1

Using Fig. 1.1, estimate the number of vehicle trip ends for the peak one hour between 7:00 a.m. and 9:00 a.m. for an office park with 625,000 ft^2 gross floor area (GFA), and determine which rate is more accurate.

Solution

The example may be solved either graphically or mathematically. To solve graphically, examine the average rate and the fitted curve plots. As shown in the following illustration, $X = 625$ on the horizontal scale intersects the average rate line at 1150 trip ends and the fitted curve at 1050 trip ends.

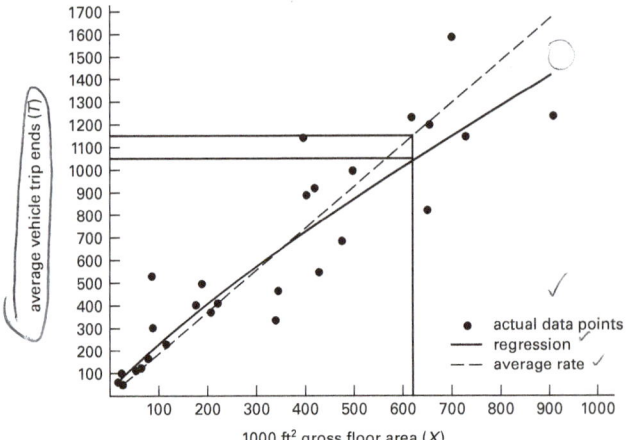

regression equation: $\ln T = 1.679 + 0.818 \ln X$
$r^2 = 0.89$

To solve mathematically, use the weighted average rate and the regression equation. For the weighted average rate,

$$T = (1.84 \text{ trip ends}) \left(\frac{625{,}000 \text{ ft}^2}{1000 \text{ ft}^2} \right) = 1150 \text{ trip ends}$$

X is the gross floor area per 1000 ft^2. Using Eq. 1.2 and the values shown in Fig. 1.1,

$$\ln T = 1.679 + 0.818 \ln X$$
$$= 1.679 + 0.818 \ln \frac{625{,}000 \text{ ft}^2}{1000 \text{ ft}^2}$$
$$= 6.945$$
$$T = e^{\ln T} = e^{6.945}$$
$$= 1038 \text{ trip ends}$$

The graphical procedure shows 1150 trip ends on the average curve and 1050 trip ends on the fitted curve, while the values calculated are 1150 trip ends for the weighted average rate and 1038 trip ends for the regression equation. Although mathematical results are preferable, either graphical or calculated results are acceptable, as 1050 trip ends is within 1% of 1038 trip ends.

Although the difference between the two methods is less than 5% (i.e., either result is acceptable), additional data checks can be performed. There are 27 data points, as shown in the illustration. Substituting $X = 0$ into the fitted curve equation yields 5 trips, which is close to 0 and indicates that the fitted curve is valid.

Perform the final check by evaluating the standard deviation and the correlation coefficient of determination. The standard deviation is (1.84 trip ends/1.50 trip ends) $\times$ 100% = 123%, meaning it is greater than 110% and the regression equation rate should be used. The correlation coefficient of determination, r^2, is greater than 0.75, showing that the regression equation rate is the better choice.

Trip generation data development can be ranked in a general hierarchy, relating reliability (i.e., data quality) to the data source. The list shown is a general hierarchy of data quality, ordered from most to least reliable.

1. field data from actual survey counts at subject study site
2. calculated traffic rate from fitted curve equation
3. average calculated traffic rate (straight line)
4. graphical interpretation between known data points
5. data point near subject study site with similar characteristics of flow (i.e., % T and B, near same V, near same street width, same number of lanes)

4. ORIGIN-DESTINATION STUDIES

Origin-destination studies, often called *O-D studies* or *O-D surveys*, are performed during the preparation of comprehensive transportation plans for a large area. The O-D study report is combined with other information, such as economic and employment data, and is used as baseline information for recommending or justifying transportation development programs and projects 15–25 years forward.[4]

Origins are considered *trip generators* and destinations are considered *trip attractions*.

Figure 1.2 illustrates the features of individual links within the study.

Figure 1.2 Principal Features of Origin-Destination (O-D) Studies Links

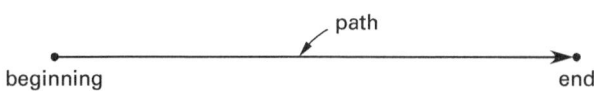

O-D studies classify trips into three distinct categories. *External-external trips* pass through a study area, but have neither their origin nor their destination within the study zone. *Internal-internal trips* have both their origin and destination within the study area. *External trips* (*internal-external* or *external-internal*) have either their origin or their destination outside of the study area. Trip categories are determined from the data collected during an O-D study, which can be performed using a variety of methods.

The results of an O-D study can be applied to traffic volume counts on a segment in order to approximate the type of traffic using that segment, in terms of trip purpose. More precise estimates of trip purpose can be found using trip generation data.

To perform an O-D survey, the study area must first be chosen. A *study area* can be a single zone or can be divided into several zones. For instance, a large study area for a city could have zones designated for major clusters of activity within the city boundary. Examples include a central business district (CBD), a university center, a dense residential community, a park district with low density housing, a manufacturing center, and so on. The boundaries of the zones are usually selected along streets, highways, railroads, waterways, or other easily recognizable features to avoid misunderstandings by interview subjects. For regional attractions, such as shopping districts and industrial parks, O-D data can be aided by customer trip information developed by tenant businesses. Chain stores, restaurants, and other retail and industrial enterprises maintain extensive databases of travel and buying habits. Current data, including future projections, are kept up-to-date so that the business owners can plan investments. Use of these data provides a reliable source of traffic information. For very small areas, such as a signalized traffic intersection, O-D data can be as simple as traffic counts on the approaching roadways.

Boundary locations of the study must ensure all traffic entering and leaving the study area or zone passes through a counting station. Factors are incorporated into the data evaluation to adjust for counting errors, but careful selection of counting station locations and data collection techniques can greatly reduce the amount of error correction needed, improving data reliability.

Study Methods

The *home-interview origin-designation survey* interviews household members to obtain information on the number, purpose, mode, origin, and destination of all trips made on a certain day. However, this method is invasive and time consuming and the data are not often used.

A *license plate study* establishes multiple stations throughout the designated study area to read passing vehicles' license plates. The license plate numbers are then tracked to establish travel paths. Home addresses

---

[4]Because O-D studies involve such a large investment of effort, they are not warranted for shorter-term projects.

can be checked through motor vehicle registration lists to determine where the vehicle most likely came from or was headed. Using electronic remote license plate reading systems, license plate studies blend well with automated data collection techniques.

A *tag-on vehicle study*, also known as the *vehicle intercept method*, involves placing a small tag on a vehicle or handing a colored card to the driver of a vehicle as it enters the study area. The tag or card is color coded to identify the location where it was first assigned. Internal stations can monitor the vehicle's path by observing the color of the tag or card. As the vehicle leaves the study area, the driver is asked to return the card to an observer.

A *cordon survey* encircles a study zone with counting stations located on all travel paths into and out of the zone. Cordon surveys are comprehensive and can include all modes of transportation. They are often used to determine traffic flow from one major activity center, such as a business center or shopping district, to other activity centers. Figure 1.3 is an example of a cordon counting station setup.

Figure 1.3 Cordon Survey Counting Station Locations

Adapted from *Traffic and Highway Engineering*, 5th Ed., by Nicholas J. Garber and Lester A. Hoel, © 2014, CL Engineering. Used by permission.

Because cordon surveys for very large areas involve all vehicles entering or leaving the activity center, counting stations perform simultaneously for a period of 24 hours. These surveys require months of planning and involve extensive notification of the traveling public. To compensate for travelers' tendency to avoid the study area during a count period, adjustment factors are often inserted into the data. Cordon surveys are complex undertakings and are therefore performed infrequently, as many agencies feel reliable results can be obtained instead through a series of carefully planned corridor surveys on each leg of the activity center approach.

A *corridor survey* covers all major routes through a defined traffic corridor, that is, major traffic flows between significant nodes of a network. They are typically used between towns and major industrial, shopping, and residential areas, and can yield significant data, such as vehicle occupancy and origin-destination information.

Roadside interviews require thorough training of interview personnel, careful selection of interview station sites, and significant advance publicity in order to encourage public cooperation. Sample size is often a function of traffic flow and the public's tolerance for traffic delays. The ITE's *Manual of Transportation Engineering Studies* recommends interviewing the drivers of at least 20% of vehicles, although a 50% sample is recommended to obtain reliable results.

Postcard studies stop motorists at survey stations and distribute survey questions on postcards with prepaid postage. The motorists can complete and return the surveys at their leisure, making postcard studies considerably less invasive than roadside or home interviews. The *Manual of Transportation Engineering Studies* suggests a minimum response rate of 20% is needed to obtain reliable data. Typically, postcard studies have shown responses in the range of 25% to 35%.

Onboard transit surveys offer another important interviewing option that features trained interviewers, usually recently retired transit operators. These interviewers ride on transit vehicles, hand survey forms and pencils to passengers as they board, and then ask the passengers to check off applicable boxes while riding to their destination. This type of survey usually results in reliable data, especially if the interviewers are familiar with the transit route system.

In addition to the survey methods previously discussed, O-D studies may use other data sources (often termed *secondary sources*). One of the most-used secondary sources is census data, because they are updated at least once every 10 years and are uniform in quality. Other secondary sources include voter registration records, city directories, property tax records, and land use data.

Once a method's final data are collected, they are reviewed and checked for accuracy. Data checks include comparing the data against selected known high-accuracy samples, which can be obtained from two or three control points within the study area, such as a bridge or other well-known point of traffic restriction. Accuracy can be checked by comparing origin and destination points against census data to ascertain if driver populations closely match survey data, or by checking employment destinations against businesses' employee counts. After the data's accuracy is verified, the data are plotted in a variety of forms to illustrate the intent and purpose of the study. Links between origin and destination nodes are usually bundled along existing travel corridors and presented as a graphic illustration of the relative numbers of trips along each corridor. Histograms are often prepared to illustrate the daily, weekly, or monthly distribution of trips per unit of time.

Cumulative statistics on *person-trips*, or one person taking one trip, are often shown as simple graphics. Person-trip ends can be illustrated as a series of three-dimensional blocks, with the height drawn in proportion to the number of trip ends, or by using a horizontal bar chart. The ITE defines a *trip* as a single movement between two locations with either the origin or the destination inside the study zone. The ITE uses the term *total trip ends* for the total number of trips in and out of a study zone. Therefore, for the purposes of traffic engineering, the terms "total trips" and "trip ends" can be used synonymously. Vehicular traffic volumes are often plotted using line graphs or as histograms.

Line graphs are used to graphically compare different types of data collected. Figure 1.4 illustrates the traffic volume for a typical 24-hour weekday to and from a CBD. It compares the screen line data (i.e., field-collected data at count stations) with survey data (i.e., home interview surveys). The upper plots show the total traffic volumes. The lower plots show only the commercial traffic.

Figure 1.4 Trip Survey Data Compared with Combined Screen Line Counts by Hour

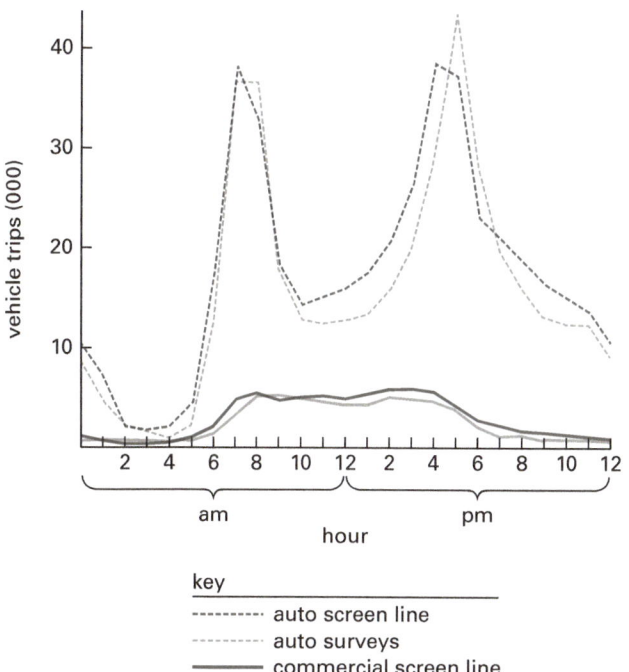

The shape of the curve is common for many mid-size cities with strong commuter and business-oriented traffic. At times, planners want to show the overall proportion of trip types. Planners use traffic volume curves and a variety of graphic displays to illustrate trip purpose, vehicle composition, travel mode, temporal effects, impacts of future changes, and other focus issues.

Figure 1.5 is an example of the representation of trips by auto and trips by transit into and out of a city business center during the morning and afternoon peak periods. The number of trips are also compared to distance, which is illustrated by concentric rings that represent the approximate distance from the city center. In Fig. 1.5, the morning trips entering and leaving the city center are shown on the left, and the afternoon trips are shown on the right. The graphic arrow widths are proportional to the number of trips crossing each concentric distance ring and indicate the direction of trips into and out of the city center. For example, there are 75,000 trips by transit and 15,000 trips by auto crossing ring 1 in the afternoon outbound direction.

Figure 1.5 Internal Trips by Auto and Transit Inbound and Outbound During Peak Periods

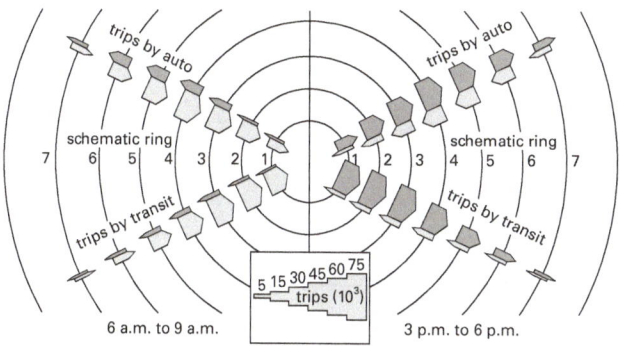

Figure 1.6 is a *traffic flow map*, which illustrates traffic volume using line weight. The line weight is proportional to the relative traffic volume on each segment. Numerical traffic volumes are provided along each segment. Two-directional volumes can be illustrated with separate line weights when the directional traffic volume varies significantly. Unlike the representation shown in Fig. 1.5, traffic flow maps provide the traffic volumes on individual roadway segments and allow comparison between segments. Engineers and statisticians can use these kinds of visualizations to assist in transportation policy decisions, using them to communicate with city officials, planners, and the public, who may not have a background in transportation.

Study formats driven by standardized reporting guidelines and study parameters make it possible to compare interpolated data from region to region and from study period to study period.

5. TRAFFIC VOLUME STUDIES

Traffic volume studies determine the numbers, movements, and classifications of vehicles (and/or bicycles and pedestrians) at specific locations and times. They are essential to define traffic parameters such as annual average daily traffic (AADT, see Chap. 6). The length of a study, or *count period*, depends on the study method and the intended use of the recorded data. Volumes are usually counted over a set time period, which can range from a few minutes to more than one year. The count

Figure 1.6 Traffic Flow Map

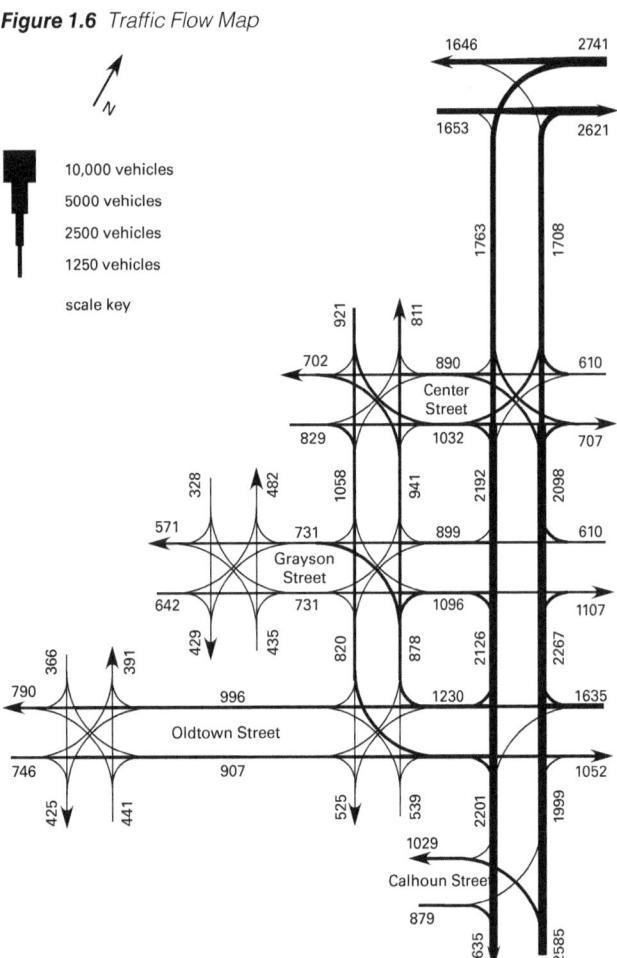

Adapted from *Traffic and Highway Engineering*, 5th Ed., by Nicholas J. Garber and Lester A. Hoel, © 2014, CL Engineering. Used by permission.

period should represent the typical conditions (such as time of day, day of month, or time of year) applicable to the desired analysis. For example, studies at schools should not usually be conducted on weekends. The shortest count periods are used to determine the peak period flow rate. If the peak hour is to be determined, the count period should start one-half hour to one hour before the expected peak hour and extend one-half hour to one hour after the expected peak hour. The peak hour is defined and determined as the four contiguous 15-minute count periods that contribute to the highest one-hour count.

The tools and equipment used to determine the traffic volume can be as simple as tally sheets or manual count boards or as complex as electronic detectors and computer data generation programs. Regardless of the tools used, the purpose for a traffic volume study remains the same: to accurately establish the actual amount of traffic passing a given location during a unit of time.

Counting Methods

There are two counting methods for collecting traffic volume data: *manual* and *automatic*. Manual and automatic counting methods are virtually the same technique for any period being analyzed, except that manual counting methods are usually limited to periods shorter than a few days or hours. Both methods may be used to count vehicles (e.g., to determine the number of multi-axle or large vehicles at a certain location), pedestrians (e.g., to identify shopping, casual, or commuter traffic), and bicycles (e.g., to separate bicycle traffic by recreation or commuter purpose).

Manual Counting Methods

Manual counting methods require observers to go to a site and collect data for a specific time period using stopwatches or timers. They may be used to collect small data samples, or when automatic equipment is not available or is too expensive to use. Examples of data gathered by manual counting include vehicle classifications, direction of travel, vehicle occupancy, and pedestrian movements.

Manual methods are typically used when the count period is limited to a few hours or a few days (also known as *short period sampling*), as longer time periods tend to be fatiguing and generally may be handled more efficiently with automatic methods. In most cases, short period sampling adequately establishes baseline parameters for design purposes. For instance, intersections may be counted at intervals of 5 minutes or 15 minutes to establish peak hour flow rates for a period of 2 hours over the morning or afternoon peak. For roadway segments that lie between intersections, the peak 15-minute and 60-minute flow periods are selected to determine the peak hour flow rate and the peak hour factor. Midday counts at intervals between 9:00 a.m. and 3:00 p.m. may be used to determine base flow rates. Full 12-hour daytime counts (7:00 a.m. to 7:00 p.m.) may be used to determine typical design-day traffic flow. Except for intense evening shopping activity, the remaining 12 hours rarely accounts for more than 15% of the total daily traffic.

The number of observers needed to collect data will depend on the number of lanes or crosswalks being observed, the types of data collected, the traffic volume, and the length of the count period. There should be a sufficient number of observers at count stations to provide frequent intervals of relief. For example, on busy intersections with *full movement activity*,[5] two or more observers may use short-break and/or alternating counting procedures to help fight fatigue. Observers may count for alternating periods of 5 minutes each and use the intervening rest periods to review the count tabulations, reset their counters, and prepare for the next counting period. Rest periods should be alternated

---

[5]Full movement activity is categorized as traffic patterns involving through traffic in four directions, including turning traffic.

among the observers, and there will be little loss in count accuracy if observers carefully coordinate their break periods.

An *adjusted volume*, V', is calculated to account for break periods. To determine the adjusted volume for each counting period, P_C, the actual volume, V, is expanded in proportion to each break period, P_B, by a count expansion factor, CF. When the counting periods and break periods are of constant duration and are equal to each other, the number of cycles can be substituted for P_C and P_B to calculate the count expansion factor.

$$V' = V(\text{CF}) \quad \quad 1.3$$

$$\text{CF} = \frac{P_C + P_B}{P_C} \quad \quad 1.4$$

In order to perform a manual count, observers need portable recording devices. Three devices commonly used are tally sheets, mechanical counting boards, and electronic counting boards. *Tally sheets* are simple, inexpensive tools on which observers record data. The sheets may be prepared in advance, or observers may draw a diagram of the location and use labeled boxes to record tallies of the desired information for each period.

Mechanical counting boards have counters mounted to a board in various configurations to represent typical intersection layouts. Observers carry the board as they move to accurately observe and record vehicle flows. After the data have been collected, the totals are summarized on field forms and then entered into a computer to provide a convenient check on the count accuracy. *Electronic counting boards* are battery-operated, handheld devices that have been developed to bridge the gap from mechanical boards to computer input by providing electronic input of data. Electronic counting boards have counting buttons and an internal clock that automatically separates data by time intervals. End-of-day data are downloaded to a computer. Fully equipped offices may have weatherproof laptops or tablet computers to enter data directly at the field count station. Data reduction, such as accuracy checks and count comparisons, can be performed on the portable computer prior to leaving the count location.

Example 1.2

Observers are stationed at an intersection with a traffic signal that operates on a 90 sec cycle. The observers each take counts for 7 complete cycles, then break for 2 cycles in order to review their count tabulations and prepare for the next series of counts. The total vehicle count averages 160 veh during the 7 cycle counting period for one of the observer's movements. What is the adjusted vehicle volume for this movement?

Solution

The total counting period duration is

$$P_C = \frac{\left(90 \, \frac{\text{sec}}{\text{cycle}}\right)(7 \text{ cycles})}{60 \, \frac{\text{sec}}{\text{min}}} = 10.5 \text{ min}$$

The total break period duration is

$$P_B = \frac{\left(90 \, \frac{\text{sec}}{\text{cycle}}\right)(2 \text{ cycles})}{60 \, \frac{\text{sec}}{\text{min}}} = 3 \text{ min}$$

Using Eq. 1.4, the count expansion factor is

$$\text{CF} = \frac{P_C + P_B}{P_C} = \frac{10.5 \text{ min} + 3 \text{ min}}{10.5 \text{ min}} = 1.29$$

Alternatively, since the counting and break periods are equal (i.e., 90 sec/cycle), the count expansion factor could also have been calculated using the number of cycles. There are 7 counting periods and 2 break periods. Therefore, the count expansion factor is

$$\text{CF} = \frac{N_C + N_B}{N_C} = \frac{7 + 2}{7} = 1.29$$

Using Eq. 1.3, the adjusted volume is

$$V' = V(\text{CF}) = (160 \text{ veh})(1.29)$$
$$= 206 \text{ veh}$$

Automatic Counting Methods

Automatic counting methods gather large quantities of traffic data over an extended, often unattended, period of time. They are generally used to determine traffic patterns and trends, and observers may use permanent or portable automatic counters to collect data.

Permanent automatic counters consist of vehicle detection devices placed along a roadway, such as magnetic loops imbedded into the pavement surface. Placing a single loop in each lane will count the total vehicle flow along a roadway segment. Placing a pair of loops a set distance apart will detect the average traffic speed as well.

Portable automatic counters are generally used to collect data similar to that collected in manual counts, but for longer periods of time. They use pneumatic road tubes that are temporarily attached to the roadway surface. The air in the tubes is compressed each time an axle crosses the tubes, which actuates a switch in the control box. Timers may be set to determine the number of actuations in preset periods. In order to convert axle counts to vehicle counts, a manual observation count is taken over a sample period to determine the typical vehicle mix. Software is available to process data and provide detailed information based on the setting of the pneumatic tubes. The processed data from the sample observations are applied to the axle counts to determine the extracted vehicle counts. Pneumatic tubes may also be placed a set distance apart to determine average traffic speed. This is useful for setting the *amber interval* (i.e., the length of time the yellow light is on) for traffic signals.

Count Locations

Count locations are selected based on the purpose of the study. Observers are positioned to have a clear view of traffic, with two of the most common locations being intersections and roadway or pathway mid-block segments. *Intersection counts* are used to collect approach volumes and turning volumes for traffic signal timing, for high-crash intersection analysis, and for evaluating congestion. The intersection and/or the approach roadways are the most likely locations to perform an intersection study, and the upstream counts (or approach roadway counts) are used to establish arrival type.

Roadway or *pathway mid-block segment counts* are used to determine network volumes and traffic flow or capacity requirements. For mid-block or roadway segment counts, the count can be located anywhere convenient and accessible that has mostly uniform traffic flow. Mid-block count locations are selected at a point where intersection turning movements least affect the traffic flow. Roadway segment count locations are selected far enough from the survey end points that traffic represents uniform flow conditions of the segment.

6. CAPACITY PLANNING ANALYSIS

Capacity planning analysis is used to estimate the traffic carrying ability of transportation network segments at prescribed levels of operation. Criteria for operational levels are characterized by the level of service for each type of facility, which relates to the amount of traffic accommodated within a desired time period (i.e., the flow rate). The *Highway Capacity Manual* (*HCM*) lists six *levels of service* (LOS), which are described in Table 1.1.

The criteria for the various LOS categories are set using base conditions (i.e., ideal conditions), such as 12 ft (3.7 m) wide lanes, near-level grades, and adequate sight distances, for each roadway type. The characteristics of a particular study segment are compared to the base conditions using adjustments to the flow rate made in passenger car equivalents (pce) for capacity and LOS predictions.

Capacity analysis is performed on two major network elements—*intersections* and *uniform segments* of extended roadway length. The analysis begins with volume and speed studies during the planning stage, which deal primarily with base conditions over broader segments of the network. This is in contrast to later operational analyses, which deal with precise definitions of roadway characteristics over shorter segments, spot conditions, and site-specific variations from the base conditions. The concepts of LOS and capacity flow of network segments form the fundamentals of network planning.

Table 1.1 Levels of Service

LOS	description
A	*free flow:* traffic flows at or above the posted speed limit; all motorists have complete mobility between lanes
B	*reasonably free flow:* slightly more congested, with some impingement of maneuverability
C	*stable flow:* roads remain below but efficiently close to capacity; posted speed is maintained, but the ability to pass or change lanes is not always assured
D	*approaching unstable flow:* speeds are somewhat reduced; motorists are hemmed in by other cars and trucks
E	*unstable flow:* flow is irregular and speed varies rapidly, but rarely reaches the posted limit
F	*forced or breakdown flow:* flow is forced, with frequent drops in speed to nearly 0 mph (0 kph)

Volume and Speed Relationships

Once trip paths are defined by a regional study, a series of *desire lines* develop, which may be formed into a connecting link between pairs of origin and destination nodes. The accumulation of O-D nodes and links between nodes may be developed into a *traffic network diagram* to estimate expected volumes. Desire lines are plotted using demand models to apply the trip generation data to the individual links. Link volumes include volumes from the demand model studies, as well as traffic not included in demand model studies. Existing traffic is verified by on-site counts, which are assigned growth factors to the same target dates used in the demand models for planned developments. In order for the planner to determine how much capacity to build into the connecting link, decisions must be made that weigh the link's capacity against a variety of factors, such as economic constraints, the available right-of-way, available funds for improvements, and the type of facility proposed.

Traffic link and node diagrams are developed from traffic network diagrams and traffic volume maps by codifying nodes at major intersections or activity centers. Figure 1.7 shows a traffic link network diagram, with the numbers shown representing the *average annual daily traffic*, or AADT. AADT is the total traffic for one year divided by 365 days. Once the nodes are assigned,

Figure 1.7 Traffic Network Diagram

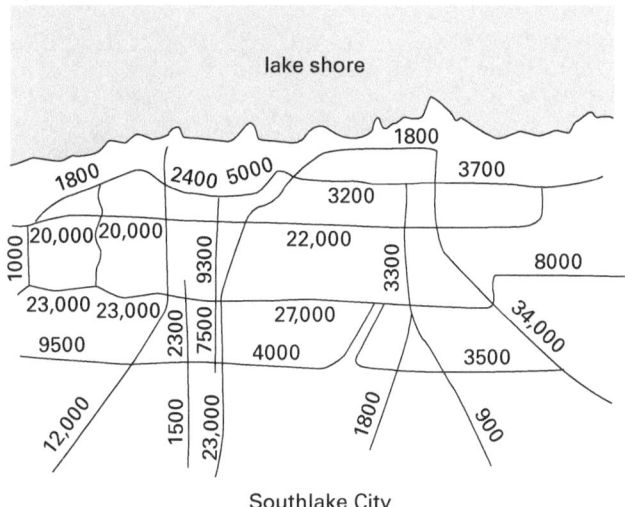

the links between the nodes can be assigned traffic volumes. Figure 1.8 is an example of a typical link and node diagram. Traffic data for various links and nodes are recorded in a spreadsheet format for further computational analysis.

Figure 1.8 Link and Node Diagram

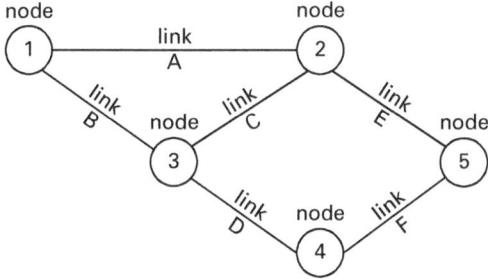

Roadway Types

Each roadway type has a capacity limit based on geometry, access point density, type of traffic control, and number of lanes. Generally, roadways are broken down into the categories of freeways, multilane highways, two-lane highways, arterials, and local streets.

Freeways are designed for safe operating speeds ranging from 50 mph to 70 mph (80 kph to 110 kph) or greater. *Multilane* and *two-lane highways* generally have design speeds of 40 mph to 65 mph (65 kph to 105 kph). *Arterials* are usually designed for at least 30 mph (50 kph). Speed limits posted at or just below the roadway design speed will improve the safety factor. They may also result in greater capacity in dense traffic conditions by encouraging more uniform flow. The posted speed should not be greater than the design speed, but it may be difficult to enforce if the posted speed is unrealistically low.

Typical roadway analysis or design criteria develop a relationship between traffic characteristics such as volume, capacity, trucks, grades, number of lanes, peak hour factors, and speed. Alternatives combine economics with design criteria to make comparisons between selected corridors. The outcome may be different for differing types of economic analysis. For instance, an economic analysis based on delays in shopping center access may show a considerably different solution than an analysis that emphasizes minimal through-traffic delay. Furthermore, opinions on the monetary value of time vary considerably, making it difficult to assess the purely economic value of many incremental improvements, such as intersection widening or the addition of turning lanes.

The *HCM* occasionally refers to a lane as being the "outside lane" without clearly defining which lane is being referred to.[6] Figure 1.9 illustrates typical unadjusted or base lane configurations and terminology used in the *HCM*, as well as the conditions in which they occur.

Figure 1.9 HCM Multilane Highway Lane Designations for Base Conditions

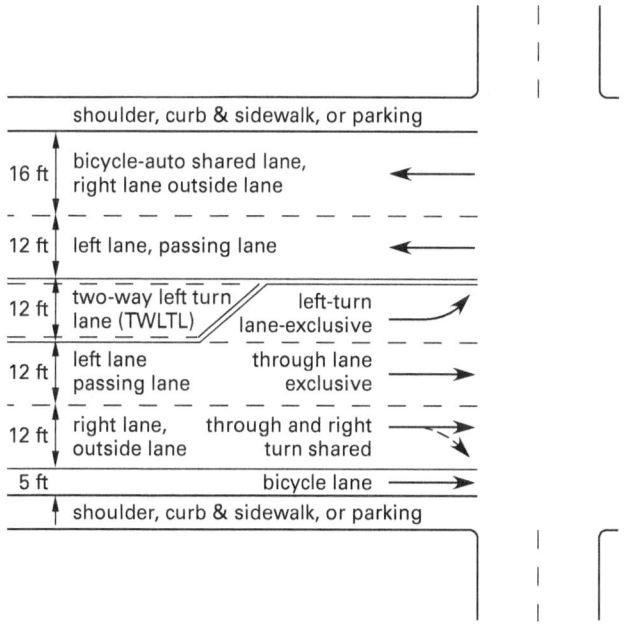

Used with permission and compiled from *Highway Capacity Manual*, 6th Edition: A Guide for Multimodal Mobility Analysis, 2016, by the Transportation Research Board of the National Academies of Sciences, Engineering, and Medicine, Washington DC. DOI: 10.17226/24798

Intersections

An *intersection* is the place where two or more vehicle pathways join, cross, or separate. With pathways that join or cross at unsignalized intersections, each traffic flow must seek gaps in the other flows to avoid collision.

---

[6] *Highway Capacity Manual*, National Academy of Sciences, © 2016, Transportation Research Board, p. 12-43, p. 15-9, and p. 15–43.

Should an arriving vehicle not find a gap of sufficient size to join or cross a path, a delay[7] will occur while that vehicle waits for a gap. To manage traffic gaps, a traffic signaling system may be installed. Gaps become systematized and predictable. However, in order to create gaps, traffic flow delays inevitably occur.

Capacity and Other Types of Analysis

Many preliminary intersection studies address only the capacity and not the details of intersection design. These studies, called *planning analyses*, are broad and give an estimation of an intersection's level of use, but provide little information on delay. *Operational analysis*, on the other hand, is performed to evaluate the full details of four primary components: demand, or service flow; signalization; geometric design; and delay, or level of service. Operational analysis can be used to test various intersection configurations and the effect of detail adjustments on the four components.

7. OPTIMIZATION AND COST ANALYSIS

Optimization is the process of comparing alternative strategies in order to satisfy a project's objectives. Optimization involves choosing among alternative plans based on costs and/or benefits in order to achieve the best possible approach to a project's objective. Optimization is performed in conjunction with other elements of project execution, and often requires weighting objectives when they are in conflict with each other. Optimization can also vary factors to achieve a desired performance of a material, design, or a standard of compliance.

There are three basic characteristics of optimization—*objective functions*, *variables*, and *constraints*. The first step in optimization is to define a project's *objective function*, that is, something to be maximized or minimized. Objective functions are usually attainable values, such as cost, a defined spread of data values, traffic density threshold values, durability functions, safety factors, or other parameters of traffic conditions. For instance, route selection analysis between two points often begins with comparing travel times over alternative route proposals. An assumption can be made that most of the volume between two points travels the route with the shortest travel time.

Example 1.3

There are 10,000 vehicles traveling between two locations over three alternate routes: A, B, or C. Route A takes 15 min, route B takes 18 min, and route C takes 24 min. Determine the proportion volumes on each route according to the inverse ratios of travel time.

Solution

$$\text{total of inverse route times} = \frac{1 \text{ min}}{15 \text{ min}} + \frac{1 \text{ min}}{18 \text{ min}} + \frac{1 \text{ min}}{24 \text{ min}} = 0.164$$

$$\text{route A} = \left(\frac{0.067}{0.164}\right)(10{,}000) \text{ veh} = 4085 \text{ veh}$$

$$\text{route B} = \left(\frac{0.056}{0.164}\right)(10{,}000) \text{ veh} = 3415 \text{ veh}$$

$$\text{route C} = \left(\frac{0.042}{0.164}\right)(10{,}000) \text{ veh} = 2561 \text{ veh}$$

$$4085 \text{ veh} + 3415 \text{ veh} + 2561 \text{ veh} = 10{,}061 \text{ veh}$$

Direct proportion is not always true, such as when increased volume causes enough congestion along the preferred route to increase the travel time to be greater than that of a less preferred route. More complicated modeling is necessary when actual volume is field-measured on each alternate route, or for travelers' preferred conditions, or when constraints are imposed by facility owners.

There can be multiple objective functions, in which a number of different objectives are to be optimized simultaneously. These objectives are often noncomplementary or noncompatible, such as minimizing initial cost while also minimizing operating and maintenance cost. For example, traffic volume is usually expressed as the number of vehicles passing a point. A fixed range of vehicle numbers per unit time (i.e., density) is assigned *level-of-service ratings* (LOS) from A to E. Planning usually targets a particular LOS for the roadway location and the intended use of the roadway. A sublevel objective may be to evaluate a pedestrian-activated crossing signal at an intersection, which can increase or decrease the through-traffic density, depending on the varying delay caused by the varying density rate of pedestrian flow. The vehicular flow competes with the pedestrian flow to attain the objective functions of the intersection. The evaluation process assigns relative values to each delay—vehicular and pedestrian—to optimize the desired objective function.

Most often, multiple objective functions will result in a less-than-optimal use of one or more resources in order to optimize the project as a whole. Having multiple objectives or less-than-optimal usage can lead to simplification by weighting the variables and reforming the functions into a single objective function. However, sublevel objective functions can also be studied in projects with multiple objectives when the sublevel objectives are not fixed in time or value. For example, adding a turning lane at an intersection may improve the flow of automobile traffic but result in a longer green time because of the increased walking distance for crossing

---

[7]A *delay* can be considered the additional time needed to travel a route below a normal free-flow speed. Significant delays create driver discomfort and frustration, lost time, and a less smooth flow of transport.

pedestrians. Usually, the sublevel objectives are optimized before the superior objectives are analyzed. However, the superior objectives can be evaluated using a "what if" scenario to set the performance level of a sublevel objective. In the previous example of increased walking distance, the sublevel objective may be to make the intersection more pedestrian friendly at the additional expense of increased vehicle delay.

The second optimization characteristic, a set of *variables*, or *unknowns*, determines the performance value of the objective function. The variables can include factors such as flow volume, delay time, construction cost, or maintenance responsibility. The previous intersection example has two variables—vehicular flow rate and pedestrian flow rate, which vary according to the time of day and the day of the week. These flow rates may be somewhat consistent, inconsistent, or both may be unknown.

The third characteristic, *constraints*, sets the condition of the variables. However, constraints are not essential to a project's optimization if there are no variables. With variables, the objective function is judged by the level of achievement attained. Usually constraints involve thresholds, such as greater than or less than values, acceptable ranges, and negative eliminators. In transportation projects, constraints are primarily specifications, codes or laws, design guides, standards of practice, costs, and human factors responses. Four modes are evaluated: automobile, transit, pedestrian, and bicycle. Each has its own set of objective functions, variables, and constraints. When the four modes compete for space, time, and enjoyment, optimizing includes multiple objectives, subobjectives, and non-compatible objectives. Community input consulting and jurisdictional policies are necessary to weigh objective values for each mode. For example, recreational areas may seek to optimize pedestrian and bicycle activity. Downtown areas may seek optimization of transit and pedestrian activity. University zones may strongly disoptimize automobile mode. Rural areas may consider automobile mode the highest priority—ignoring or minimizing other modes.

Optimization analysis can be performed as a one-time condition, such as selecting the number of lanes on a roadway. The analysis can also be a continuous process, such as adjusting the timing sequence of a traffic signal as data becomes available. In cases where the data are not accurate or the future predictions are not entirely known, the analysis must take into account these uncertainties. In either case, the output of the optimization analysis requires a *recourse action* to adjust input parameters such as new beginning values of land use or traffic volume data. After the recourse action, the optimization analysis is performed again, and the output is adjusted using the new data.

During the planning stage, engineers employ various models to determine not only construction costs, but also operating and replacement costs that will be necessary to attain certain benefits. The optimal solution may be the one with the least cost, or it may be the one that provides the most benefit to the expected user. It is important that the objectives of the finished product are clearly defined. The degrees of attainment of objectives become qualifying factors for ranking the alternative outcomes against the cost to build each alternative. *Life-cycle cost analysis* (LCCA) involves predicting the expected useful life of a project and the costs of construction, operation and maintenance, and demolition and disposal, especially when environmentally sensitive materials are used in construction.

There are several different types of optimization models, including linear programming, integer programming, and dynamic programming models. *Linear programming* (LP) is used in optimization of a linear objective function. There are two types of LP models: *single objective*, in which there is one objective to optimize, and *multiple objective*, which tries to find solutions satisfying more than one objective. *Integer programming* restricts some or all of the variables in optimization to be integers. *Dynamic programming* (DP) breaks up a problem into smaller, simpler components. It is best used when the issue of optimization is very complex, as it may be easier to find solutions to each separate component, rather than try to solve the problem as a whole.

Example 1.4

An isolated signalized intersection with multiple approaches and movements is being studied for optimum cycle length. After selecting a lane configuration, the criterion for setting the cycle length is to minimize average delay of all approaching vehicles. Several cycle lengths have been tested with the following results.

cycle length (sec)	average delay (sec/veh)
90	95
105	83
120	69
135	68
150	78

When will the minimum delay occur?

Solution

The solution can be found graphically by plotting the data and fitting a curve through the data points.

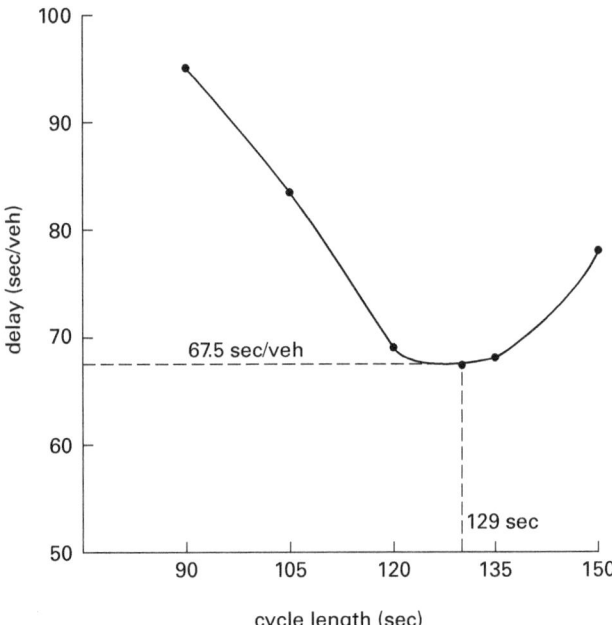

By inspection, the curve shows a minimum delay of 67.5 sec/veh occurring with a cycle length of 129 sec.

Setting Optimization Criteria

Optimization can be performed at many stages in a project. The criterion for optimization depends on the overall goals of the project and the objectives that are to be met at a particular level of execution. In the early stages, the criteria to optimize a project's major components are usually broad. Such criteria would be added to project-cost estimates and funding requirements, and include the following components: community needs, schedule and phasing, material selection, site restrictions, environmental controls, risk assessments, and neighborhood impacts.

Community needs often start with the desire to build or modify a particular transportation facility, such as a bridge, roadway, or intersection. From these needs, *planning studies* are used to determine optimization based on capacity analysis, access improvements, and crash mitigation.

Schedule and phasing are set by both internal and external constraints. External events, such as coordination with transportation network connections, seasonal weather changes, and major community activities should be optimized during the project planning phase. Internal events, such as material production, delivery schedules, staging, and environmental controls, would have more precise criteria to optimize the cost and quality of the completed project.

Material selection focuses on the cost and availability of primary materials. The cost of purchase, delivery, placement, and the impact on the project schedule are criteria for optimization.

Site restriction optimization based on topographic features means selecting the most effective method to build over watercourses, hills, and valleys, in various soil conditions, and to account for climate factors such as ice and snow. Optimizing property usage aims to make efficient use of the space available in relation to the value of the property.

Environmental controls include noise and dust mitigation, VOC emissions, and trash disposal from the jobsite. Compliance with community regulations for dust, noise, and silt runoff is often handled by firms specializing in these tasks. Optimization of productivity, or the dollar value of work performed per unit of time, is compared to the cost of complying with applicable environmental regulations and controls.

Risk assessment is the comparison of the cost of doing or not doing something with the probability that there will be a negative or undesirable result. While risk assessment is constantly changing, there are many established standards (e.g., AASHTO's *A Policy on Design Standards—Interstate System*, AASHTO's *Highway Safety Manual*, or the FHWA's *Manual on Uniform Traffic Control Devices* (*MUTCD*)) that have been thoroughly tested and that represent the level of achievement to be attained during optimization analysis of risk.

Neighborhood impact is the effect of the project on the quality of life. Noise barriers, access controls, school bus safety, and public safety issues are traditional concerns, while a new emphasis is being placed on sustainable living and close-knit neighborhood development. Optimization of these values not only affects the initial cost of a project, but has long-term consequences as well.

Optimization During Later Phases of a Project

As the project design progresses, optimization moves from a broad perspective to a narrower focus on the best solution for the objective at hand. For instance, transportation planners that are optimizing access to a commercial district might initially select a highway as the best solution. Once a highway is selected, the next phase is to select the type of highway—limited access, multilane arterial, or two-lane. This phase involves careful analysis of benefits, user and construction cost, capacity, funding availability, and institutional support. Optimization of each of these elements involves cost and traffic projections and assurances of future use patterns. When the type of highway is selected, designs are tested and optimized using delay models, user-cost models, environmental cost models, and so on. These cost models can be very specific when the base data are highly reliable.

As a project moves into final design and construction, optimization focuses on material selection, source of supply, labor cost, contractor capability, and the specifics of scheduling and phasing the work. Through the progression of a transportation project's stages, it becomes increasingly difficult to return to earlier definitions of broad optimization. For instance, once a concrete bridge has been selected and is in production, changing to a steel superstructure would require including the cost of reconstructing already completed portions in the evaluation of optimizing the change to steel.

Benefit-Cost Analysis

In the simplest terms, cost optimization is choosing among alternatives based on costs (i.e., dollar value) versus benefits. *Benefit-cost analysis* is used on almost all projects when evaluating the design of the project as a whole, or for sub-elements. For instance, pavement designs are usually analyzed for the optimal balance between a predicted service life (the benefit) and the cost of construction.

The benefit-cost ratio can be calculated from Eq. 1.5.

$$B/C = \frac{\Delta \text{user benefits}}{\Delta \text{investment cost} + \Delta \text{maintenance} - \Delta \text{residual value}} \quad 1.5$$

Equation 1.5 is a conventional method of comparing the gross benefits of a project in the numerator to all of the costs in the denominator, which are the initial investment plus the operating and maintenance costs, less the residual value. The *residual value* (also known as the *salvage value*) is the market value at the end of the project life.

In some cases, part of the cost elements may not be known or may not be available. Typically, the residual value or market worth at the end of service life will not be known or readily predicted. Evaluation can proceed, however, recognizing that once the residual value is known, the ratio may change.

One of the biggest drawbacks of benefit-cost analysis is the inability to accurately predict the dollar value of future benefits. This is especially true for projects that have a long expected life (50 years or more), such as major water transportation projects. The closer the benefit-cost ratio moves to unity (1.00), the more the inaccuracies of predictions increase the opportunities for negative benefit conditions. The project then becomes more difficult to sell to the public based on the benefit-cost ratio alone.

Example 1.5

A shipping channel is to be dredged to a new depth at an estimated cost of $248 million. Annual maintenance of the channel at its new depth is expected to cost $15 million for the first year. The port facility served by the shipping channel expects additional revenue of $70 million the first year to occur due to the increased access by larger cargo ships. There is no residual value for the project. Using an effective inflation rate of 5%, what is the benefit-cost ratio of the project for the initial 10 years, assuming both maintenance and additional revenues increase at the same rate of inflation?

Solution

Calculate the present worth of the costs. The total cost for 10 years, C_t, will be the initial construction cost, C_i, plus the present value of the annual maintenance cost, C_m.

$$C_t = C_i + C_m(P/A, 5\%, 10)$$
$$= \$248\text{M} + (\$15\text{M})(7.7217)$$
$$= \$364\text{M}$$

Calculate the present worth of the benefits over the same period of time.

$$B = \$70\text{M}(P/A, 5\%, 10)$$
$$= (\$70\text{M})(7.7217)$$
$$= \$541\text{M}$$

The benefit-cost ratio, B/C, is

$$B/C = \frac{\$541\text{M}}{\$364\text{M}} = 1.49$$

Therefore, the benefits for the first 10 years of life of the project will be 1.49 times more than the cost to install the deeper channel and to maintain it for 10 years.

8. SITE IMPACT ANALYSIS

Site impact analysis involves examining the environmental issues of transportation projects. Site impact analysis can include, but is not limited to, examining a project's potential effect on

- natural resources
- the relocation of residential areas and businesses
- air quality
- sound and noise
- wetlands and coastal zones
- society and the economy
- water quality

- flood hazards, floodplains, and flood liability
- emergency management

Sound, noise, and air quality are especially pertinent to civil engineers, who must monitor and control these factors during project construction.

Sound and Noise

Sound occurs when an ear senses pressure vibrations, while *noise* is unwanted sound. *Noise studies* are conducted to determine the nature and intensity of noise generated by a roadway. They determine existing background noise levels, called *ambient sound* levels, which may also include existing transportation facilities. Sound levels are measured at the L_{10}, L_{50}, and L_{90} levels, which means the sound levels are exceeded 10%, 50%, and 90% of the time, respectively. Traffic noise studies are most often conducted to project the noise levels for a new facility. However, noise levels for an existing facility can also be determined. Data from trip generation studies are often used to estimate the effect of a new facility on existing noise levels.

Sound levels are expressed in levels of pressure and measured in decibels (dB). The decibel scale is logarithmic at base 10, meaning a change of 1 dB reflects a ten-fold change in sound pressure level. The sound level *A-weighted scale*, which suppresses low frequency sounds, is often used to simulate human hearing response, in which case the sound level is given in A-weighted decibels (dBA). The *reference level* for sound in air is usually chosen as 0.02 MPa. This is the lowest level of normal sensitivity for the human ear found in those without hearing damage due to loud noise exposure. The *maximum audible field* (MAF) threshold varies with audible frequency, measured in hertz. The following four levels are often used for reference.

- 70 dB at 2 Hz
- 100 dB at 100 Hz
- 0 dB at 2000 Hz
- 20 dB at 16,000 Hz

The *threshold of discomfort* also varies with audible frequency, but is generally accepted at approximately 120 dB for all cases. The *threshold of pain* is approximately 140 dB, and the *threshold of hearing damage* occurs at approximately 145 dB.

Traffic noise studies are commonly required by the FHWA for federally funded transportation projects. The FHWA classifies transportation projects as either type I or type II. A *type I project* is one that involves constructing a new highway or altering an existing highway such that significant horizontal or vertical alignment changes are made or a new lane is added. The definition of "significant" changes to horizontal and vertical alignment is defined by a state's Department of Transportation. A noise study must be performed for all type I projects. *Type II projects*, also known as *retrofit projects*, must also include a noise study and are defined as projects undertaken on existing roadways to provide noise abatement.

Sound levels are recorded over a period of hours, days, or months, depending on the study objective. Recorded noise levels are then compared to noise abatement criteria to determine if traffic noise impacts exist. A *traffic noise impact* occurs when existing or projected noise levels approach or exceed the noise abatement criteria or when a substantial increase in the existing noise environment occurs. The FHWA defines a substantial increase as 10–15 dBA higher than the current noise levels, but this may vary depending on state policies. The *noise abatement criteria* (NAC) are the absolute noise levels for different land use categories and are outlined in 23 CFR 772 (*Code of Federal Regulations*, Title 23, Part 772) and given in Table 1.2. The FHWA defines *absolute noise levels* as those that are unacceptably high for a given land use. L_{10} is a time-varying traffic noise level that is measured over a one-hour period and is calculated as the percentage of the study period, in this case 10% of the time, that a sound level is exceeded. L_{eq} is the hourly equivalent sound level, which is the equivalent constant sound level which contains the same amount of energy as the time-varying noise level (i.e., L_{10}), both measured for one hour. For most traffic conditions, L_{eq} is about 3 dBA less than L_{10} for the same study period.

If traffic noise impacts are determined to occur, abatement measures must be considered. These measures can include sound barrier walls, berms, dikes, earth mounds, vegetation, or other passive devices to deflect or soften the sound from entering the sensitive sites. Additional measures may be employed as needed, such as changing the grade line of a project, rerouting the configuration ramps, adjusting the location of truck-climbing lanes, and so forth. In most instances, small changes needed to accommodate noise mitigation measures have little effect on a project's cost. However, in some cases, expensive sound barriers may be needed when retrofitting sensitive neighborhoods, such as those near hospitals. This cost must be compared with the cost of relocating the roadway to a less sensitive area.

Air Quality

Transportation, by its very nature of propulsion energy, affects air quality, whether through power plant emissions for electrically powered transit and vehicles or exhaust pipe emissions for combustion systems. Therefore, when a transportation project is proposed, *air quality studies* must be conducted. Like noise studies, air quality studies for new facilities are often based on data obtained through trip generation studies.

The Environmental Protection Agency (EPA) is tasked with developing emission factors for all emission sources. Air quality impact studies must be directly related to the project at hand in order to have valid application. Studies

Table 1.2 Noise Abatement Criteria (NAC), Hourly Sound Level in A-Weighted Decibels (dBA)[a]

activity category	L_{eq}[b](dBA)	L_{10}(dBA)	description of activity category
A	57 (exterior)	60 (exterior)	lands on which serenity and quiet are of extraordinary significance and serve an important public need, and where the preservation of those qualities is essential if the area is to continue to serve its intended purpose
B	67 (exterior)	70 (exterior)	picnic areas, recreation areas, playgrounds, active sports areas, parks, residences, motels, hotels, schools, churches, libraries, and hospitals
C	72 (exterior)	75 (exterior)	developed lands, properties, or activities not included in categories A or B
D	–	–	undeveloped lands
E	52 (interior)	55 (interior)	residences, motels, hotels, public meeting rooms, schools, churches, libraries, hospitals, and auditoriums

[a]Either L_{eq} or L_{10} (but not both) may be used on a project.
[b]L_{eq} represents the hourly equivalent sound level.

Reprinted from *Code of Federal Regulations*, Title 23, Part 772, Table 1, U.S. Department of Transportation, Federal Highway Administration.

that exist whether or not the transportation project is implemented are not valid uses of transportation funds (e.g., studying mitigation methods for the effects of acid rain would not be included in a transportation air quality impact study). Starting with a concise project description, the report should clearly show the nature of the project, the proposed alternatives, unusual topographic features, and ambient conditions.

The data collection and analysis must be complete and sufficient to provide reliable projections based on current industry standards and government regulations or guidelines. The report should contain a complete air quality analysis of each alternative. Projections of total emissions for the entire project and analysis for the year of completion should show the operating emissions environment at the completion of construction.

Transportation is one of the largest energy consumers and is often targeted as a significant contributor of pollutants to *nonattainment areas* (areas that do not meet national primary or secondary air quality standards for a pollutant). The EPA establishes criteria pollutants and the *National Ambient Air Quality Standards* (NAAQS), which can be found at epa.gov. Though not all polluting emissions are applicable to a transportation project, all motor-powered units emit some pollutants into the atmosphere. Depending on the project, air quality studies may test for carbon monoxide (CO), carbon dioxide (CO_2), hydrocarbons (HC), nitrous oxides (NOx), ozone (O_3), lead (Pb, where leaded fuel is still used), sulfurous compounds, mercury (Hg), nuclear radiation (from power plants), and, in some cases, heat energy.

Pollutant abatement from sources not directly related to the transportation project at hand, such as noxious odors from a sewage treatment plant or coke oven, are generally not eligible for transportation project funding.

Air quality studies are either mesoscale or microscale studies. *Mesoscale* studies include the total emissions generated by various project alternatives and involve calculating the impacts of photochemical oxidants as related to local jurisdiction requirements. Mesoscale studies cover contributing sources for several miles outside the project location and include changes in traffic on nearby highways or other transportation segments resulting from the proposed project. Computer modeling is usually employed in these studies.

Microscale studies concentrate on one or a few pollutants, such as carbon monoxide or lead. Dispersion models are used to predict concentrations at critical locations. As with mesoscale studies, computer modeling is used extensively to analyze the data and reduce the outputs into an understandable presentation. Transportation construction projects can significantly affect air quality to such an extent that a separate air quality impact study may need to be specifically performed for construction activities.

Transportation construction activities usually impact air quality in three major ways. First, impact initially occurs when earth is excavated and the protective vegetative cover is removed. This generates a considerable amount of dust that is picked up by the surrounding atmosphere, even in areas where there is little or no vegetation covering the native soil. The U.S. Department of Agriculture's (USDA) Natural Resources Conservation Service (NRCS), formerly the Soil Conservation Service, has determined the susceptibility of different soils to wind erosion. (See NRCS' website at nrcs.usda.gov for more information.) Once a soil type is known, dust suppression methods may be built into a transportation project's construction procedures. Second, transportation construction projects affect air quality through the concentrated collections of exhaust-gas emissions that are caused by traffic congestion. To offset these emissions, traffic routing and congestion relief measures may be specified for a project based on the amount of excess emissions permitted during periods of restricted traffic-flow rates. Third, emissions from the construction equipment affect air quality. In response, equipment manufacturers have steadily improved engine performance and emissions control (for instance, the large clouds of black diesel smoke pouring out from exhaust stacks have been reduced dramatically). EPA standards may also be utilized to prepare project impact reports from construction equipment operations.

9. PRACTICE PROBLEMS

1. An interchange is being evaluated for construction cost versus user cost, with the lowest user cost being preferred. There are two major traffic movements through the interchange, with two alternative ramp configurations being considered. User cost is composed of the vehicle cost and the person cost. Ramp lengths are measured from the same points on the approach roadways for each alternative. The difference in route lengths is due to ramp configuration. Use a 3% annual growth in traffic volume and a 3% annual increase in per-mile and per-hour user costs. A 20 yr projected user benefit-cost analysis is to be used.

(a) The current daily user time (in person-minutes per day) is, most nearly, how much larger for alternative A than alternative B?

(A) 36,500 p-min/day

(B) 80,600 p-min/day

(C) 90,700 p-min/day

(D) 95,300 p-min/day

(b) What is most nearly the current daily total user cost savings for alternative B?

(A) $10,800

(B) $21,500

(C) $25,400

(D) $52,200

	alternative A	alternative B
route 10 traffic volume	42,000 ADT	42,000 ADT
route 14 traffic volume	18,000 ADT	18,000 ADT
ramp length and design speed		
route 10	1.32 mi, 25 mph	0.65 mi, 45 mph
route 14	0.97 mi, 30 mph	0.97 mi, 30 mph
total bridge cost and length	none	$8,500,000; 0.1 mi
time per vehicle in interchange		
route 10	3.16 min	1.0 min
route 14	1.93 min	1.93 min
vehicle cost	$0.45/veh-mi	$0.45/veh-mi
person cost	$8.00/p-hr	$8.00/p-hr
vehicle occupancy	1.05 p/veh	1.05 p/veh
ramp construction cost (excluding bridge costs)	$6,000,000/mi	$6,000,000/mi

2. Travel behavior and population characteristics along a corridor are found to correspond to the following factors derived from surveying a sample population of 1000 people.

proportion of households in the population, p_h	0.35
number of adults per household, N_a	0.74
number of vehicles per household, N_v	1.55
number of licensed drivers per adult, N_l	0.61

The census population of the study zone is 25,000, and there are 14,000 average one-way, weekday trip ends generated from the study zone.

(a) The number of households in the study zone is most nearly

(A) 3500 households

(B) 4400 households

(C) 4900 households

(D) 8800 households

(b) The number of vehicles in the study zone is most nearly

(A) 1550 veh

(B) 13,600 veh

(C) 21,700 veh

(D) 38,800 veh

(c) How many one-way, weekday trip ends are generated within the study zone for each licensed driver?

(A) 1.12 trip ends/licensed driver

(B) 1.40 trip ends/licensed driver

(C) 3.54 trip ends/licensed driver

(D) 7.09 trip ends/licensed driver

3. A residential zone contains 1700 households with proximity to transit. Planners are making a projection of transit ridership potential. A survey of neighborhood characteristics finds the data shown.

proportion of households in the population, p_h	0.35
number of employed adults per household, p_e	0.74
proportion of employed adults working at transit-accessible locations, $p_{a,t}$	0.31

(a) The population of the study zone is most nearly

(A) 600

(B) 1700

(C) 3590

(D) 4860

(b) The number of employed adults living in the study zone is most nearly

(A) 378

(B) 510

(C) 3560

(D) 6570

(c) The number of employed adults living in the study zone who have transit access is most nearly

(A) 440

(B) 1070

(C) 1460

(D) 5670

4. A medium-sized, 20 ac, mixed-use commercial development is planned, consisting of offices, retail stores, and light manufacturing. The existing frontage road is a nearly level, straight, two-lane highway with unlimited sight distance and no median. The posted speed and average speed are both 45 mph. Traffic currently operates at LOS C, and there is no signal planned along the proposed frontage. Analysis of site access is being performed to determine design elements. For vehicular access, five common design groups from AASHTO design guides are being used.

- P: Passenger car
- SU-9 (SU-30): single-unit truck
- WB-12 (WB-40): intermediate semitrailer
- WB-20 (WB-67): interstate semitrailer
- WB-22 (WB-67D): double-trailer combination

(a) What is the minimum design vehicle for this site?

(A) SU-9 (SU-30)

(B) WB-12 (WB-40)

(C) WB-20 (WB-67)

(D) WB-22 (WB-67D)

(b) Proposed alternatives to access the driveway from the existing frontage road include

  I. three driveway entrances: one entrance for delivery trucks and two smaller entrances where delivery trucks would be prohibited

 II. a continuous left-turn lane along the frontage road and multiple access driveways for the retail businesses

III. a single large driveway entrance for all vehicles

 IV. a jug handle to the right for left turns

Which alternative would be the least confusing for unfamiliar drivers?

(A) I

(B) II

(C) III

(D) IV

(c) The local planning board suggests minimizing the traffic impact on the frontage road by designing the entrance driveways for quick entry into the site and restricting the exit flow rate with narrow exit lanes and speed control humps. What is the main reason restricting the exit flow rate is NOT a good idea?

(A) Drivers leaving the site may become frustrated with the exit restrictions and attempt to leave by traveling the wrong way on the entrance driveways.

(B) Designing exit driveways to restrict both truck and automobile movement is difficult, as trucks are much larger than automobiles.

(C) Drivers wanting to leave the site in a hurry may become discouraged and not return, opting to visit businesses at more accessible locations.

(D) Restricting egress from a site limits the speed of and capacity for emergency evacuation.

5. Two zones, A and B, are connected by three alternative highway travel routes. There are 10,000 vehicles traveling from zone A to zone B during a weekday. The distances and average travel speeds are shown. Route 1 has 2 signals, each with a 70 sec delay, route 2 has 15 signals with a 30 sec delay each, and route 3 has 2 signals with a 30 sec delay each.

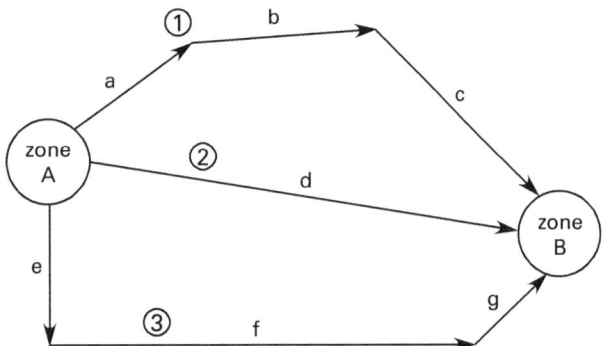

segment	segment length (mi)	average travel speed (mph)
a	4	25
b	5	35
c	7	30
d	15	40
e	4	30
f	8	65
g	3	30

Trips are to be assigned to each route based on the inverse of the proportion of the total travel time delay.

(a) What is most nearly the trip delay per vehicle on the quickest route?

(A) 22.4 min on route 3
(B) 23.0 min on route 2
(C) 30.0 min on route 2
(D) 34.5 min on route 1

(b) What percentage of vehicles uses the route with the least delay?

(A) 27.1%
(B) 31.2%
(C) 42.1%
(D) 72.9%

(c) The number of vehicles on the route with the least delay is most nearly

(A) 2320 veh on route 1
(B) 2710 veh on route 2
(C) 3120 veh on route 2
(D) 4210 veh on route 3

SOLUTIONS

1. (a) The daily user time for alternative A is

$$t_A = t_{A,10} + t_{A,14} = (42{,}000 \text{ veh})\left(1.05 \ \frac{\text{p}}{\text{veh}}\right)(3.16 \text{ min})$$
$$+ (18{,}000 \text{ veh})\left(1.05 \ \frac{\text{p}}{\text{veh}}\right)(1.93 \text{ min})$$
$$= 175{,}833 \text{ p-min/day}$$

The daily user time for alternative B is

$$t_B = t_{B,10} + t_{B,14} = (42{,}000 \text{ veh})\left(1.05 \ \frac{\text{p}}{\text{veh}}\right)(1.0 \text{ min})$$
$$+ (18{,}000 \text{ veh})\left(1.05 \ \frac{\text{p}}{\text{veh}}\right)(1.93 \text{ min})$$
$$= 80{,}577 \text{ p-min/day}$$

The time difference is

$$t_A - t_B = 175{,}833 \ \frac{\text{p-min}}{\text{day}} - 80{,}577 \ \frac{\text{p-min}}{\text{day}}$$
$$= 95{,}256 \text{ p-min/day} \quad (95{,}300 \text{ p-min/day})$$

The daily user time for alternative A is 95,256 p-min/day (95,300 p-min/day) more than the daily user time for alternative B.

The answer is (D).

(b) The daily total user cost is the per vehicle cost plus the per person cost.

The per vehicle cost for alternative A, $C_{A,v}$, is

$$C_{A,v} = \left(0.45 \ \frac{\$}{\text{veh-mi}}\right)\binom{(42{,}000 \text{ veh})(1.32 \text{ mi})}{+(18{,}000 \text{ veh})(0.97 \text{ mi})}$$
$$= \$32{,}805$$

The per person cost for alternative A, $C_{A,p}$, is

$$C_{A,p} = \frac{(175{,}833 \text{ p-min})\left(8.00 \ \frac{\$}{\text{p-hr}}\right)}{60 \ \frac{\text{min}}{\text{hr}}}$$
$$= \$23{,}444$$

The daily total user cost for alternative A, $C_{A,t}$, is

$$C_{A,t} = C_{A,v} + C_{A,p}$$
$$= \$32{,}805 + \$23{,}444$$
$$= \$56{,}249$$

The per vehicle cost for alternative B, $C_{B,v}$, is

$$C_{B,v} = \left(0.45 \ \frac{\$}{\text{veh-mi}}\right)\binom{(42{,}000 \text{ veh})(0.65 \text{ mi})}{+(18{,}000 \text{ veh})(0.97 \text{ mi})}$$
$$= \$20{,}142$$

The per person cost for alternative B, $C_{B,p}$, is

$$C_{B,p} = \frac{(80{,}577 \text{ p-min})\left(8.00 \ \frac{\$}{\text{p-hr}}\right)}{60 \ \frac{\text{min}}{\text{hr}}} = \$10{,}744$$

The daily total user cost for alternative B, $C_{B,t}$, is

$$C_{B,t} = C_{B,v} + C_{B,p}$$
$$= \$20{,}142 + \$10{,}744$$
$$= \$30{,}886$$

The current daily user benefit for alternative B is

$$B = C_{A,t} - C_{B,t}$$
$$= \$56{,}249 - \$30{,}886$$
$$= \$25{,}363 \quad (\$25{,}400)$$

The answer is (C).

2. (a) The number of households is found by multiplying the proportion of households in the population by the population.

$$N_h = p_h(\text{population})$$
$$= (0.35 \text{ households})(25{,}000)$$
$$= 8750 \text{ households} \quad (8800 \text{ households})$$

The answer is (D).

(b) The number of vehicles is found by multiplying the number of vehicles per household by the number of households.

$$\text{no. of vehicles} = N_v N_h$$
$$= \left(1.55 \ \frac{\text{veh}}{\text{household}}\right)(8750 \text{ households})$$
$$= 13{,}563 \text{ veh} \quad (13{,}600 \text{ veh})$$

The answer is (B).

(c) The amount of one-way weekday trip ends generated within the study zone for each licensed driver is

$$T = \frac{T_t}{N_l N_a N_h}$$

$$= \frac{14{,}000 \text{ trip ends}}{\left(0.61\, \frac{\text{licensed driver}}{\text{adult}}\right)\left(0.74\, \frac{\text{adults}}{\text{household}}\right)}$$
$$\times (8750 \text{ households})$$

$$= 3.54 \text{ trip ends/licensed driver}$$

The answer is (C).

3. (a) The population is found by dividing the number of households, N_h, by the proportion of households in the population, p_h.

$$\text{population} = \frac{N_h}{p_h}$$
$$= \frac{1700 \text{ households}}{0.35 \text{ households}}$$
$$= 4857 \quad (4860)$$

The answer is (D).

(b) The number of employed adults is found by multiplying the total population of the study zone by the proportion of employed adults per household.

$$N_a = (\text{population}) p_e$$
$$= (4860)(0.74)$$
$$= 3596 \quad (3560)$$

The answer is (C).

(c) To find the number of employed adults that have transit access, multiply the number of employed adults by the proportion of employed adults working at transit-accessible locations.

$$N_{a,t} = N_a p_{a,t}$$
$$= (3560)(0.31)$$
$$= 1068 \quad (1070)$$

The answer is (B).

4. (a) While retail commercial stores can usually be served by SU minimum design, light manufacturing more often needs access by WB-12 (WB-40) intermediate semitrailer. This design can accommodate occasional WB-20 (WB-67) interstate semitrailer if there are few obstructions close to the driveway.

The answer is (B).

(b) The most comprehensible access involves a single combined access point with adequate capacity for all vehicle types and the expected traffic volume. Option III provides a single access point for all drivers. Multiple access points with restrictions may cause confusion to unfamiliar drivers and lead to potential violations of the restrictions, even with careful signage or other control devices. A continuous left-turn lane should be avoided for the same reason and will become less effective under increased traffic volumes. A jug handle can be confusing because of the need to go to the right to turn left. A jug handle also increases the amount of traffic crossing the mainline flow.

The answer is (C).

(c) Fire safety and life should be paramount over all other concerns when designing site access. The ability to exit the site quickly is important if there is a need to evacuate people in an emergency situation. Restricting exiting ability can be dangerous and increase the potential for panic.

The answer is (D).

5. (a) Trip delay times are found for each route by adding the travel delay for each segment together with the signal delay.

The trip delay time for route 1 is

$$d_{R1} = \sum_{i=1}^{m} d_{\text{seg},i} + \sum_{i=1}^{n} d_{\text{sig},i}$$
$$= d_{\text{seg a}} + d_{\text{seg b}} + d_{\text{seg c}} + N_{\text{sig}} d_{\text{sig}}$$
$$= \left(\frac{4 \text{ mi}}{25\, \frac{\text{mi}}{\text{hr}}} + \frac{5 \text{ mi}}{35\, \frac{\text{mi}}{\text{hr}}} + \frac{7 \text{ mi}}{30\, \frac{\text{mi}}{\text{hr}}}\right)\left(60\, \frac{\text{min}}{\text{hr}}\right)$$
$$+ \frac{(2 \text{ signals})\left(70\, \frac{\text{sec}}{\text{signal}}\right)}{60\, \frac{\text{sec}}{\text{min}}}$$
$$= 34.5 \text{ min}$$

The trip delay time for route 2 is

$$d_{R2} = \sum_{i=1}^{m} d_{\text{seg},i} + \sum_{i=1}^{n} d_{\text{sig},i} = d_{\text{seg d}} + N_{\text{sig}} d_{\text{sig}}$$
$$= \left(\frac{15 \text{ mi}}{40\, \frac{\text{mi}}{\text{hr}}}\right)\left(60\, \frac{\text{min}}{\text{hr}}\right)$$
$$+ \frac{(15 \text{ signals})\left(30\, \frac{\text{sec}}{\text{signal}}\right)}{60\, \frac{\text{sec}}{\text{min}}}$$
$$= 30.0 \text{ min}$$

The trip delay time for route 3 is

$$d_{R3} = \sum_{i=1}^{m} d_{seg,i} + \sum_{i=1}^{n} d_{sig,i} = d_{sege} + d_{segf} + d_{segg} + N_{sig}d_{sig}$$

$$= \left(\frac{4 \text{ mi}}{30 \frac{\text{mi}}{\text{hr}}} + \frac{8 \text{ mi}}{65 \frac{\text{mi}}{\text{hr}}} + \frac{3 \text{ mi}}{30 \frac{\text{mi}}{\text{hr}}} \right) \left(60 \frac{\text{min}}{\text{hr}} \right)$$

$$+ \frac{(2 \text{ signals})\left(30 \frac{\text{sec}}{\text{signal}} \right)}{60 \frac{\text{sec}}{\text{min}}}$$

$$= 22.4 \text{ min}$$

The smallest delay is 22.4 min on route 3.

The answer is (A).

(b) Proportioning the traffic according to the inverse of the time delay means the route with the least delay receives the largest proportion of traffic. To find the proportions, the inverse of delay for each route is divided by the total of inverse delays for all routes. The total of inverse delays is

$$\sum_{1}^{3} d = \frac{1}{d_{R1}} + \frac{1}{d_{R2}} + \frac{1}{d_{R3}}$$

$$= \frac{1}{34.5 \text{ min}} + \frac{1}{30.0 \text{ min}} + \frac{1}{22.4 \text{ min}}$$

$$= 0.029 \frac{1}{\text{min}} + 0.033 \frac{1}{\text{min}} + 0.045 \frac{1}{\text{min}}$$

$$= 0.107 \text{ 1/min}$$

The proportion for route 1 is

$$p_1 = \frac{0.029 \frac{1}{\text{min}}}{0.107 \frac{1}{\text{min}}} = 0.271$$

The proportion for route 2 is

$$p_2 = \frac{0.033 \frac{1}{\text{min}}}{0.107 \frac{1}{\text{min}}} = 0.308$$

The proportion for route 3 is

$$p_3 = \frac{0.045 \frac{1}{\text{min}}}{0.107 \frac{1}{\text{min}}} = 0.421$$

Check that the proportions total 1.00.

$$p_t = 0.271 + 0.308 + 0.421 = 1.00 \quad [\text{OK}]$$

The proportion of traffic for route 3 is 0.421 (42.1%).

The answer is (C).

(c) The vehicle volumes are proportioned by

$$V_i = p_i V_t$$

Route 3 has the least delay. The vehicle volume on route 3 is

$$V_3 = p_3 V_t = (0.421)(10{,}000 \text{ veh}) = 4210 \text{ veh}$$

The answer is (D).

Traffic and Capacity Analysis

1. Travel Time and Delay Studies 2-2
2. Capacity Analysis Procedures 2-9
3. Capacity Analysis for Uninterrupted Flow 2-10
4. Two-Lane Highway Analysis 2-19
5. Traffic Signals 2-26
6. Parking Facilities 2-35
7. Practice Problems 2-48

Nomenclature

A	area	ft²	m²
A_{pbT}	permitted phase adjustment	–	–
ATS	average travel speed	mph	kph
BFFS	base free-flow speed	mph	kph
BP	breakpoint	pc/hr/ln	pc/h/ln
BPTSF	base percent time spent following	%	%
c	capacity	vph	vph
c_i	capacity of lane group	–	–
C	confidence level	%	%
C	cycle length	sec	s
d	control delay	sec/veh	s/veh
d	distance	mi	km
d_1	uniform control delay	sec/veh	s/veh
d_2	incremental delay	sec/veh	s/veh
d_3	initial queue delay	sec/veh	s/veh
df	degrees of freedom	–	–
D	density	various	various
e	effective green time	sec	s
E_R	extended upgrades	–	–
E_T	extended downgrades	–	–
f	adjustment factor	various	various
f_p	driver population factor	–	–
f_M	median adjustment factor	–	–
f_w	lane width adjustment factor	–	–
F_c	grade adjustment factor	–	–
FFS	free-flow speed	mph	kph
g	effective green time	sec	s
g_s	time to clear waiting queue	sec	s
HV	percentage of heavy vehicles	%	%
k	incremental delay calibration factor	–	–
l_1	start up lost time	sec	s
l_2	clearance lost time	sec	s
L	limit of acceptable error	–	–
L	lost time	sec	s
L	study segment length	mi	km
M	vehicles met	veh	veh
MSF	maximum service flow	–	–
N	number, number of lanes, or sample size	–	–
O	overtaking vehicles	veh	veh
P	proportion	–	–
P	vehicles passed	veh	veh
P_{TC}	flow rate of trucks at crawl	–	–
PHF	peak hour factor	–	–
PTSF	percent time spent following	%	%
Q_b	initial queue	veh	veh
R_c	red clearance interval	sec	s
R_p	platoon ratio	–	–
$\overline{R}$	average range in running speed	mph	kph
s	saturation flow rate	various	various
s	sample standard deviation	mph	kph
S	travel speed	mph	kph
t	Student's t-distribution parameter	–	–
t	time	various	various
$\bar{t}$	adjusted average travel time	min	min
t_e	time extension for random arrivals	sec	s
T	analysis time period duration	hr	h
TLC	total lateral clearance	ft	m
v	actual or projected demand flow rate	vph	vph
v	flow rate, observed flow rate	various	various
v_i	demand flow rate	–	–
v_g	unadjusted demand flow rate for lane group	vph	vph
v_{gl}	unadjusted demand flow rate for single lane with highest volume	vph	vph
v_p	demand or peak flow rate	–	–
v/c	volume-capacity ratio	–	–
V	volume	vph	vph
V_i	demand volume	–	–
W	width	ft	m
X	lane group v/c ratio or degree of saturation	–	–
X_i	degree of saturation	–	–

Y	yellow change interval	sec	s
z	confidence limit	–	–

Symbols

α	significance level	%	%
σ	standard deviation	mph	kph

Subscripts

a	area type
adj	adjusted
ave	average
A	access-point density or approach A
b	bus
bb	bus blockage
B	bus
c	clearing, conflicting, or critical
cc	curb-to-curb
d	directional traffic distribution
EB	eastbound
f	field-observed, free-flow, or percent time spent following
FM	field-measured
g, G	grade
HV	heavy vehicle
i	ith vehicle or lane group i
I	intersection control delay
ID	interchange density
L	lanes, left, or lost
LC	lateral clearance
Lpb	left-turning pedestrian group
LS	lane and shoulder
LT	left turn
LU	lane utilization
LW	lane width
m	parking maneuvers or number of group lanes
min	minimum
ms	downstream lane blockage
M	median or multilane
n	northbound
np	no-passing zone
o	base or optimal
p	driver population, parking, passenger, passenger car, peak, or pedestrian
pb	pedestrian and bicycle conflicts
pr	perception
pce	passenger car equivalent
ped	pedestrian
r	receiving
R	recreational vehicle or right
Rpb	right-turning pedestrian group
RT	right turn
s	average travel speed, saturated, southbound, or speed
SB	southbound
sp	spillback
t	test run, time period, total, travel time, or turn
T	truck
TLC	total lateral clearance
wz	work zone

1. TRAVEL TIME AND DELAY STUDIES

Travel time and delay studies on highway networks are dominated by *automobile mode* transportation. In addition, *pedestrian*, *bicycle*, and *transit modes* are evaluated where these modes are a significant part of the travel activity, and where these modes cause delay to travel patterns. The data collected by travel time and delay studies are used to identify a route's problem locations and to determine design or operational improvements that will better facilitate traffic flow.

Highways are either *rural* or *urban* types of traffic systems. They are subdivided into *principal arterials*, *minor arterials*, *collector systems*, and *urban local streets* or *rural local roads*, based on the amount of traffic carried and the function of the type of roadway. *Principal arterial systems* carry the majority of travel between cities or major regional activity centers and involve the longest travel movements. These arterials include *freeways*, *multilane highways*, and *two-lane highways*.

Minor arterials extend access to the principal arterials from larger towns and other traffic generators, integrate interstate networks, and provide corridor movement for intraregional trips. Minor arterials can include freeways, but generally include a larger proportion of multilane and two-lane highways.

Collector systems provide access service and circulation within residential, commercial, and industrial sites for urban and developed areas, and link intracounty activities with arterials in rural areas. *Major collector roads* and *minor collector roads* accommodate traffic to and from local roads. Collectors use multilane highways and two-lane highways, with frequent interruptions caused by driveway access points and interrupted flow intersections.

Urban local streets and *rural local roads* are the primary access to abutting property and activity, connecting individual properties to the collector systems. Vehicles include automobiles, light trucks, and municipal utility service vehicles. Local streets may have bus service, and usually accommodate bicycles and pedestrian traffic. Automobiles and light trucks predominate on rural local roads.

Urban areas have closer-spaced housing of many styles, ranging from large city to small town environments. Urban-like environments can also appear in rural areas as village-type housing clusters with populations as low as 5000. *Rural areas* are considered to be any place outside of the boundaries of urban areas.

Terminology Typically Reported in Travel Time and Delay Studies

Travel time is the time taken by a vehicle to travel over a given segment of roadway, which includes delay.

Travel speed (running speed) is the distance traveled divided by the running time of a single vehicle.

Running time is the time that a vehicle is actually in motion.

Space-mean speed, *mean travel speed*, or *mean running speed* is the segment distance divided by the mean travel time over the segment of several vehicles, or of a single vehicle making several trips.

Delay is the travel time lost by a vehicle due to causes beyond the control of the driver.

Operational delay is caused by the presence and interference of other traffic.

Fixed delay is the component of delay caused by traffic control devices; it is independent of traffic volume and operational delay.

Stopped-time delay (or *stopped delay*) is when the vehicle is not moving (or is moving slower than a predetermined speed).

Travel-time delay is the difference between the actual travel time at an average speed in uncongested traffic flow on a segment compared with unrestricted speed on the same segment. For intersections, travel-time delay is the difference between the time a vehicle passes a point downstream of the intersection where it has regained normal speed and the time it would have passed that point had it been able to continue through the intersection at its approach speed.

Time-in-queue delay is the difference between the time a vehicle joins the rear of a queue and the time it clears the intersection or another queue point.

Total delay is the difference between the time that it takes a vehicle to traverse through a segment or an intersection and the time it would have taken if allowed to travel at its desired speed.

The essential measurements of a traffic system's performance are travel time, delay, and vehicle speed. Traffic studies are most frequently conducted on major arterials leading to and from a region's central business district (CBD), but a study segment can include any type of roadway.

Average travel speed (ATS) is a *space mean speed* because it is measured with respect to the segment length. Equation 2.1 is used to calculate the average travel speed, S_{ave}, where L is the study segment length (in miles or kilometers), N_t is the number of test runs observed, and t_i is the travel time (in hours) of the ith vehicle to travel the segment.

$$S_{\text{ave}} = \frac{N_t L}{\sum_{i=1}^{N_t} t_{i,\text{hr}}} = \frac{L}{\frac{1}{N_t}\sum_{i=1}^{N_t} t_{i,\text{hr}}} \quad 2.1$$

If the average travel time, t_{ave} (in hours), for the study segment length is known, Eq. 2.1 can be simplified to Eq. 2.2.

$$S_{\text{ave}} = \frac{L}{t_{\text{ave,hr}}} \quad 2.2$$

Equation 2.3 is used to calculate the running speed, S, for a single test run. L is the study segment length, and t is the travel time.

$$S = \frac{L}{t} \quad 2.3$$

Confidence Level

Conclusions and predictions based on sampling studies are not 100% accurate 100% of the time, and travel time and delay studies are no exception. Studies should be repeated several times, and test runs with unusual measurements should be removed from the final data, in order to increase confidence in the results.

Confidence in the results of a study is measured in terms of the *confidence level*, C, a percentage expressing the likelihood that the results from a given study are accurate and correct. For example, results with a 95% confidence level are considered to have a 5% chance of being erroneous in some way.

The complement of the confidence level is the *significance level*, α. The significance level is a percentage expressing the chance that given results are incorrect. In other words, the significance level for given results is always equal to 100% minus the confidence level for those results; and conversely, the confidence level for given data is always equal to 100% minus the significance level for that data. The 95% confidence level, or 5% significance level, is the lowest level generally accepted as being statistically significant. Results with a confidence level of 99% are said to be *highly significant*.

Sample Size

A study's minimum sample size can be determined using the procedure outlined in the Institute of Transportation Engineers' (ITE's) *Manual of Transportation Engineering Studies*. The *minimum sample size* is the number of test runs required to achieve a certain permitted error. The *permitted error* is determined by the level of accuracy required in the study and is defined as the tolerance for error in a specific parameter. The ITE procedure applies the permitted error to the average range in running speed, $\bar{R}$, and has units of miles (kilometers) per hour. A transportation engineer typically determines the permitted error before data collection begins, and the value of permitted error may be increased or decreased based on an agency's reliability requirements, the availability of accurate data, and staff, time, and financial resources.

The ITE suggests the following ranges of permitted error for various types of studies. However, engineering judgment should be used when selecting a value of permitted error.

- transportation planning and highway capacity studies: ±3–5 mph (±5–8 kph)
- traffic operations, trend analysis, and economic evaluations: ±2–4 mph (±4–7 kph)
- before-and-after studies: ±1–3 mph (±2–5 kph)

Once four initial test runs are completed, the average range in running speed, $\bar{R}$, is determined using Eq. 2.4.

$$\bar{R} = \frac{|S_1 - S_2| + |S_2 - S_3| + \cdots + |S_{N_t-1} - S_{N_t}|}{N_t - 1} \qquad 2.4$$

The permitted error and the average range in running speed are compared to Table 2.1 to establish the minimum sample size necessary to achieve a 95% confidence level. Table values should not be interpolated. If the average range in running speed falls between two listed ranges, the higher range is always used. Additional test runs are completed as necessary to satisfy the minimum sample size requirement.

Table 2.1 Minimum Sample Size Requirements for Travel Time and Delay Studies with a Confidence Level of 95%

average range in running speed, $\bar{R}$, (mph)	minimum number of test runs, N_t, for a permitted error of				
	1.0 mph	2.0 mph	3.0 mph	4.0 mph	5.0 mph
2.5	4	2	2	2	2
5.0	8	4	3	2	2
10.0	21	8	5	4	3
15.0	38	14	8	6	5
20.0	59	21	12	8	6

(Multiply mph by 1.609 to obtain kph.)

Reprinted with permission from the *Manual of Transportation Engineering Studies*, 2nd Ed., © 2010. Institute of Transportation Engineers, Washington, DC.

The sample size can also be calculated using either the normal distribution or the *t*-distribution. The *normal distribution*, also referred to as the *Gaussian distribution*, is a symmetrical distribution that is used extensively in statistical analysis. The normal distribution is based on known values of the standard deviation and mean. A *t-distribution* is used when the true population standard deviation and mean are unknown. *Student's t-distribution* is a standardized *t*-distribution that is commonly used in travel time and delay studies. The Federal Highway Administration (FHWA) recommends that a *t*-distribution be used if the estimated sample size, N, is less than 30, which applies to the majority of travel time and delay studies.

Equation 2.5 calculates a study's minimum sample size, N, using Student's *t*-distribution. L is the limit of acceptable error, t_α is the *t*-distribution parameter for a desired confidence level, and s is the sample's standard deviation in mph (kph).

$$N = \left(\frac{t_\alpha s}{L}\right)^2 \qquad 2.5$$

The *t*-distribution parameter is typically determined from Student's *t*-distribution tables using the degrees of freedom, df, which is one less than the sample size. However, the sample size is determined using the *t*-distribution parameter. Therefore, the one- and two-tail limits for a normal distribution, z, are substituted for t_α to calculate an estimated sample size. Values of z for various confidences levels on a standard normal curve are given in Table 2.2. The estimated sample size calculated using the values from Table 2.2 should be checked by calculating the degrees of freedom (i.e., the estimated sample size minus 1), determining the *t*-distribution parameter from Student's *t*-distribution tables, and recalculating the sample size. Both calculated sample sizes should be relatively close in value.

Table 2.2 Confidence Levels for One- and Two-Tail Confidence Limits

confidence level	one-tail limit, ±z	two-tail limit, ±z
90%	1.28	1.645
95%	1.645	1.96
97.5%	1.96	2.17
99%	2.33	2.575
99.5%	2.575	2.81
99.75%	2.81	3.00

Example 2.1

A major arterial through a shopping district is being evaluated for signal upgrades. The arterial is 2.3 mi long and has 14 intersections with varying spacing. Four test runs have been completed with the results shown in the table.

run number	elapsed time (sec)	speed (mph)
1	420	17.3
2	370	19.7
3	340	21.3
4	520	14.0

The permitted error is 1.0 mph with a confidence level of 95%. How many additional test runs are needed to complete the study?

Solution

Use Eq. 2.4 to find the average range in running speed.

$$\overline{R} = \frac{|S_1 - S_2| + |S_2 - S_3| + |S_3 - S_4|}{N_t - 1}$$

$$= \frac{\left|17.3 \frac{\text{mi}}{\text{hr}} - 19.7 \frac{\text{mi}}{\text{hr}}\right| + \left|19.7 \frac{\text{mi}}{\text{hr}} - 21.3 \frac{\text{mi}}{\text{hr}}\right| + \left|21.3 \frac{\text{mi}}{\text{hr}} - 14.0 \frac{\text{mi}}{\text{hr}}\right|}{4 - 1}$$

$$= 3.77 \text{ mi/hr}$$

Use Table 2.1 to find the minimum sample size requirement for the study. 3.77 mph is between 2.5 mph and 5.0 mph, so use the values for an average range in running speed of 5.0 mph. For a 5.0 mph average range and a permitted error of ±1.0 mph, a total of eight test runs are needed. Four test runs have already been completed, so four more test runs are needed to complete the study.

Data Collection

Data collection for travel time and delay studies consists of measuring the travel time and delay time. Multiple test runs are performed until the sample size is large enough to determine ATS, running speed, and/or space mean speed.

Manual data collection requires observers. In most tests, observers stay out of sight on the roadside. If a test vehicle is to be used, an observer can serve as the driver of the test vehicle or ride in the test vehicle as a passenger. Observers record data at predetermined points along the study segment using two stopwatches, one tracking total elapsed time and one tracking delay time.

Automatic data collection uses a variety of devices to measure total time and delay time. One common method of automatic data collection is installing sensors connected to a laptop computer in a test vehicle. An observer riding in or driving the test vehicle keys into the computer when the test vehicle passes the starting point, and again when the test vehicle passes the ending point. A computer program then uses the data collected from the sensors and the information keyed in by the observer to calculate the travel time and delay time.

Another common method of automatic data collection is the use of sensors or other devices installed at various locations along a study segment, providing continuous data on an observed vehicle's speed and position. This method typically uses sensors installed for other primary purposes, such as real-time traffic monitoring or vehicles' global positioning systems (GPS). Ultimately, the advantages of automatic data collection are the reduction of the number of field personnel required for data collection and the decreased possibility of human error.

Regardless of the type of data collection, personnel involved in the study should take note of the type, location, duration, and cause of any traffic delays occurring during a given test run. This information is necessary for proper data analysis.

Test Vehicle Methods

Test vehicle methods require a test vehicle, a driver, and/or an observer. Depending on the data collection method used, the observer may be a passenger in the test vehicle, or the driver may serve as the observer. Test vehicle methods typically require study segments of no less than 1 mi (1.6 km), with end points distant enough from intersections to obtain accurate approach and recovery speeds. An intersection can be used as the starting point or end point for a study segment when using a test vehicle method, as long as the test vehicle can proceed uninhibited through the intersection and the speed at which vehicles travel through the intersection matches the travel flow speed.

Three common methods requiring a test vehicle are the floating car, average speed, and moving vehicle test methods. When using the *floating car method*, the driver "floats" with the traffic flow, attempting to safely pass as many vehicles as pass the test vehicle. The total elapsed time is recorded from the beginning to the end of the study segment. Trips are recorded several times in each direction to establish a reliable average travel time.

For the *average speed method*, the driver makes several practice runs, then drives the length of the study segment going a speed that, in the opinion of the driver, is the average speed of the traffic flow.

The *moving vehicle method* is used on two-directional roadways where opposing traffic is visible at all times, usually over a study segment with a convenient turn-around just beyond the segment. The moving vehicle method requires several round trips, and it produces results and data that can be used to calculate the average travel time, space mean speed, and hourly volume for the study segment. Observers count vehicles traveling in the opposite direction of the test vehicle (i.e., *vehicles met*), vehicles that pass the test vehicle (i.e., *vehicles overtaking*) and vehicles moving in the same direction that are passed by the test vehicle (i.e., *vehicles passed*). While simply recording the total travel time for a segment may be sufficient for a study's purpose, data gathered at intermediate points along the study segment, such as intersections, can be recorded to study speed variations along the route. Additional points should be included only if the observers are able to safely and accurately record the data. The additional data can be used to determine the location of the most severe delay or locations where speed improvements would have the greatest effect on the overall average travel time.

An example of a travel time and delay study field sheet for the moving vehicle method can be found in App. 2.A. Rows and columns can be arranged to meet the needs of the study and location so data can be readily entered by observers, allowing for faster recording and reducing errors caused by missing critical information.

Figure 2.1 shows a typical study segment for a study using the moving vehicle method, with end points at Haysville St. and Emsworth Ave. Table 2.3 shows the data for one round trip on the study segment. Data for each intersection include the arrival time, which is the time the vehicle enters the intersection, and the departure time, which is the time the vehicle leaves the intersection. The delay is found by subtracting the arrival time from the departure time. Because the only typical cause of delay along the study segment is delay from time spent waiting at traffic signals, all the delay for this study is classified as stopped delay. The travel time is determined by subtracting the departure time of the last intersection from the arrival time of the current intersection. For example, the travel time for Chateau Ave. for the southbound run is calculated as the time the test vehicle departed Belleview Ave., 92 sec, subtracted from the time the test vehicle arrived at Chateau Ave., 139 sec. Therefore, the travel time for Chateau Ave. is 47 sec.

The *hourly volume* is the number of vehicles that will pass the beginning of the study segment in the time it takes the test vehicle to make a round trip through the study segment. The calculations assume a number of vehicles met, M, will pass the starting point between the time the test vehicle passes the starting point and the time the test vehicle returns to the starting point. Vehicles overtaking the test vehicle, O, minus those passed by the test vehicle, P, compensate for the test vehicle not traveling at the exact speed of traffic. The following moving vehicle method equations can be rewritten to determine the directional volume, average travel time, and space mean speed for eastbound and westbound segments. Equation 2.6 and Eq. 2.7 are used to determine directional volumes.

$$V_s = \frac{M_n + O_s - P_s}{t_n + t_s} \quad \text{[southbound]} \quad 2.6$$

$$V_n = \frac{M_s + O_n - P_n}{t_n + t_s} \quad \text{[northbound]} \quad 2.7$$

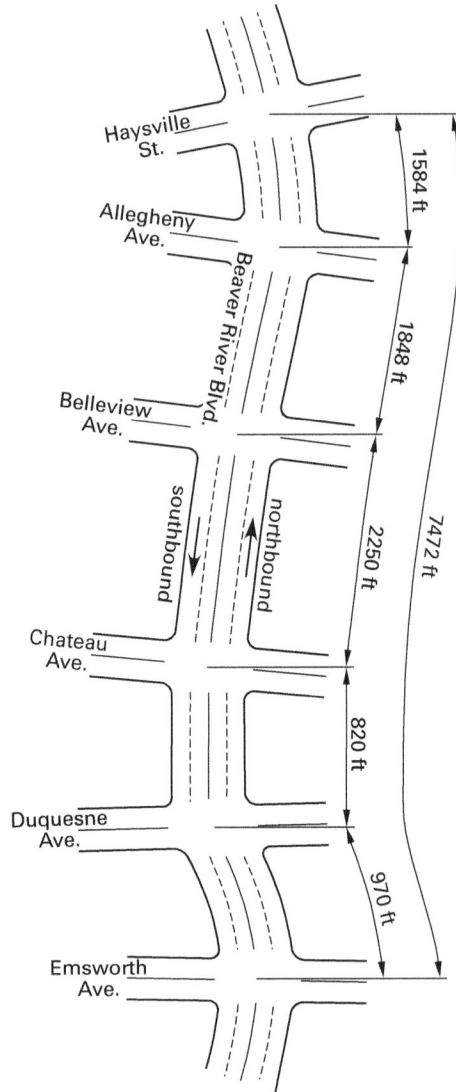

Figure 2.1 Beaver River Blvd. Study Segment for the Moving Vehicle Method

Volume in vehicles per minute is multiplied by 60 min/hr to obtain the volume in vehicles per hour, which is how volumes are usually reported.

Volumes are estimates, as errors in counting are inevitable. For example, some vehicles will not be counted during the time it takes to turn the test vehicle around at the end of the segment, opposing vehicles that exit or enter at cross streets are not accounted for, and so on.

The adjusted average travel time is determined from the average of all test run travel times adjusted by the vehicles passed and the overtaking vehicles, as shown in Eq. 2.8 and Eq. 2.9.

$$\bar{t}_s = t_s - \frac{O_s - P_s}{V_s} \quad \text{[southbound]} \quad 2.8$$

$$\bar{t}_n = t_n - \frac{O_n - P_n}{V_n} \quad \text{[northbound]} \quad 2.9$$

TRAFFIC AND CAPACITY ANALYSIS

Table 2.3 Moving Vehicle Method Data for Beaver River Blvd. Study Segment

Study segment: Beaver River Blvd. Date: May 12
Weather: 75°F, clear Time: 4:30–6:00 p.m. End points: Emsworth Ave. and Haysville St.

way point	distance (ft)	arrival time (sec)	departure time (sec)	vehicles met	vehicles passed	overtaking vehicles	travel time (sec)	stopped delay (sec)	delay reason	segment speed (mph)
run 1 southbound										
Haysville St.	0	–	0	–	–	–	–	–	–	–
Allegheny Ave.	1584	30	30	21	3	1	30	0	–	35.9
Belleview Ave.	1848	70	92	20	2	2	40	22	signal	31.4
Chateau Ave.	2250	139	142	23	3	1	47	3	signal	32.6
Duquesne Ave.	820	159	159	9	1	0	17	0	–	32.8
Emsworth Ave.	970	176	–	11	1	3	17	–	–	38.8
total	7472	176	–	84	10	7	151	25	–	–
average										28.8
run 1 northbound										
Emsworth Ave.	0	–	0	–	–	–	–	–	–	–
Duquesne Ave.	970	17	17	18	1	0	17	0	–	38.8
Chateau Ave.	820	31	39	16	0	1	14	8	signal	39.8
Belleview Ave.	2250	81	96	41	3	2	42	415	signal	36.4
Allegheny Ave.	1848	132	132	31	2	3	36	0	–	34.9
Haysville St.	1584	164	–	29	1	0	32	–	–	33.7
total	7472	164	–	135	7	6	141	23	–	–
average										31.0

(Multiply ft by 0.305 to obtain m.)
(Multiply mph by 1.609 to obtain kph.)

The space mean speed for the moving vehicle method is calculated using Eq. 2.10 or Eq. 2.11.

$$S_s = \frac{d}{\bar{t}_s} \quad \text{[southbound]} \quad 2.10$$

$$S_n = \frac{d}{\bar{t}_n} \quad \text{[northbound]} \quad 2.11$$

Example 2.2

Seven test trips over the Monongahela Blvd. study segment were conducted using the moving vehicle method. The total study segment distance is 8150 ft, and only the total travel time from end point to end point was recorded during the test. Using the test trip data in the following table, calculate (a) the directional volumes, (b) the directional adjusted average travel times, and (c) the directional space mean speeds using customary U.S. units.

test trip summary, Monongahela Blvd.

southbound trips	t_s (sec)	M_s (veh)	O_s (veh)	P_s (veh)
1	214	84	7	10
2	220	90	6	9
3	223	82	7	8
4	210	75	8	11
5	214	81	8	10
6	235	85	4	3
7	212	90	6	7
total	1528	587	46	58
average	218.3	83.9	6.6	8.3

test trip summary, Monongahela Blvd.

northbound trips	t_n (sec)	M_n (veh)	O_n (veh)	P_n (veh)
1	208	135	6	7
2	206	131	5	5
3	211	142	6	4
4	196	120	4	5
5	210	128	3	2
6	215	143	3	1
7	208	143	1	2
total	1454	942	28	26
average	207.7	134.6	4.0	3.7

Solution

(a) The directional volumes are calculated using Eq. 2.6 and Eq. 2.7 and the averages of all runs. For the southbound direction, use Eq. 2.6.

$$V_s = \frac{M_n + O_s - P_s}{t_n + t_s} = \left(\frac{134.6 \text{ veh} + 6.6 \text{ veh} - 8.3 \text{ veh}}{207.7 \text{ sec} + 218.3 \text{ sec}}\right)$$
$$\times \left(60 \; \frac{\text{min}}{\text{hr}}\right)\left(60 \; \frac{\text{sec}}{\text{min}}\right)$$
$$= 1123 \text{ vph}$$

For the northbound direction, use Eq. 2.7.

$$V_n = \frac{M_s + O_n - P_n}{t_n + t_s} = \left(\frac{83.9 \text{ veh} + 4.0 \text{ veh} - 3.7 \text{ veh}}{207.7 \text{ sec} + 218.3 \text{ sec}}\right)$$
$$\times \left(60 \; \frac{\text{min}}{\text{hr}}\right)\left(60 \; \frac{\text{sec}}{\text{min}}\right)$$
$$= 712 \text{ vph}$$

(b) For the adjusted average travel time in the southbound direction, use Eq. 2.8.

$$\bar{t}_s = t_s - \frac{O_s - P_s}{V_s} = \frac{218.3 \text{ sec}}{60 \; \frac{\text{sec}}{\text{min}}} - \frac{(6.6 \text{ veh} - 8.3 \text{ veh}) \times \left(60 \; \frac{\text{min}}{\text{hr}}\right)}{1123 \; \frac{\text{veh}}{\text{hr}}}$$
$$= 3.73 \text{ min}$$

For the northbound direction, use Eq. 2.9.

$$\bar{t}_n = t_n - \frac{O_n - P_n}{V_n}$$
$$= \frac{207.7 \text{ sec}}{60 \; \frac{\text{sec}}{\text{min}}} - \frac{(4.0 \text{ veh} - 3.7 \text{ veh})\left(60 \; \frac{\text{min}}{\text{hr}}\right)}{712 \; \frac{\text{veh}}{\text{hr}}}$$
$$= 3.44 \text{ min}$$

(c) The space mean speed in the southbound direction can be found using Eq. 2.10.

$$S_s = \frac{d}{\bar{t}_s} = \frac{\left(\frac{8150 \text{ ft}}{5280 \; \frac{\text{ft}}{\text{mi}}}\right)\left(60 \; \frac{\text{min}}{\text{hr}}\right)}{3.73 \text{ min}} = 24.8 \text{ mph}$$

For the northbound direction, use Eq. 2.11.

$$S_n = \frac{d}{\bar{t}_n} = \frac{\left(\frac{8150 \text{ ft}}{5280 \; \frac{\text{ft}}{\text{mi}}}\right)\left(60 \; \frac{\text{min}}{\text{hr}}\right)}{3.44 \text{ min}} = 26.9 \text{ mph}$$

Methods Without Test Vehicles

Many transportation studies do not require test vehicles to gather data. For instance, the *direct observation method* involves observers placed at an elevated vantage point who measure travel time between two points a known distance apart. This method requires good visibility and is suitable for study segments of less than 0.5 mi (0.8 km).

The *license plate method* requires observers, placed at checkpoints along the test segment, to record the last two, three, or four digits of vehicle license plates and the time at which the vehicle passes each checkpoint. The spacing between checkpoints will vary depending on the roadway's characteristics, the configuration of nearby roads, and the purpose of the study. The FHWA guidelines for checkpoint spacing are 1–3 mi (1.6–4.8 km) between checkpoints on freeways with high levels of access; 3–5 mi (4.8–8.0 km) on freeways with low levels of access; 0.5–1 mi (0.8–1.6 km) on arterials with high levels of cross street and driveway access; and 1–2 mi (1.6–3.2 km) on arterials with low levels of cross street and driveway access. Clocks are coordinated between checkpoints to ensure accuracy. Tape recorders can be used to record the precise time of day at periodic intervals and the reading of license plates at each checkpoint in order to reduce the number of missed vehicles. The plate readings are matched along with the observed times to obtain travel time between the checkpoints. According to the *Manual of Transportation Engineering Studies*, a sample size of 50 matched plate readings usually provides sufficient accuracy.

A variation of the license plate method uses *toll road cards* or automatic toll collection system data. The time a vehicle crosses the toll gateway is compared with the time the vehicle crosses the exit recorder, and the ATS is calculated using the distance between the two locations. This method is useful for comparing travel speeds at various times of the day, as toll information is recorded continuously. Toll road cards can only be used to obtain data on travel speed, not actual running speed, as delay cannot be obtained from toll collection data.

An *interview method* can be set up using selected individuals, such as truck, taxi, or delivery drivers. The drivers are asked to record their start and end times for designated routes or test segments. Volunteer participants need a short training session, a stop watch, and a log sheet to record their time. This method provides a lot of data in a short amount of time with little expense, although the reliability of the data may be less than data collected by fully trained observers.

Data Analysis

Data must be analyzed once it is collected. After the speeds, delays, and travel and running times have been tabulated, the data can be analyzed to determine whether delays exist. Once the types and severity of delays are established, operational improvements that may reduce delay and improve overall flow can be determined.

Data can be reported in tabular form but are often shown using graphical plots for enhanced visualization. This kind of plot is particularly useful when the study zone covers a very large area, such as an entire CBD. Figure 2.2 shows a typical travel time, speed, and delay diagram along a study route. Study results can also be displayed graphically as speed contours laid over a map of the study zone.

Figure 2.2 Travel Time, Speed, and Delay Diagram Along a Route

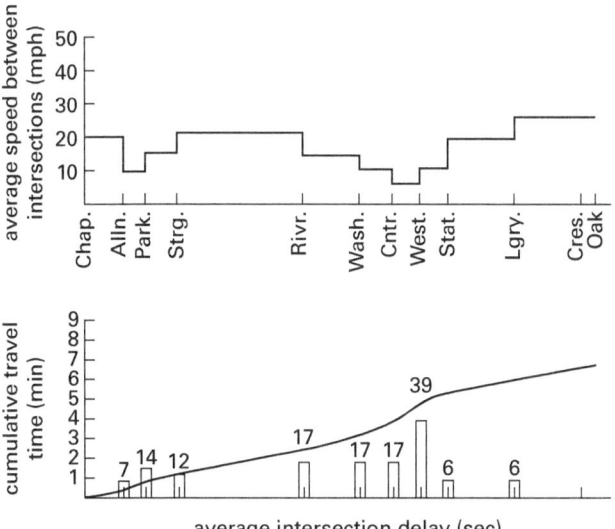

Reprinted with permission from the Institute of Transportation Engineers, *Manual of Transportation Engineering Studies*.

2. CAPACITY ANALYSIS PROCEDURES

Traffic capacity analysis estimates the ability of transportation facilities to sustain traffic under a range of defined conditions. The *HCM* is the primary authority on capacity analysis. The methods presented in the *HCM* are used to analyze capacity and predict traffic flow by accounting for a facility's specific characteristics, such as roadway type, traffic flow, and control systems. The *HCM* procedures use data accumulated over several decades. Due to improvements in computer-assisted analysis, the mathematical models the procedures are based on now closely represent actual traffic conditions. While there is no substitute for using field data for the specific location being evaluated, the *HCM* procedures give a highly probable, mathematical approximation of what may occur under actual field conditions.

Analytical Procedures and Applications

Traffic capacity analysis in the *HCM* divides a given roadway into facilities, segments, and points. A *facility* is a length of roadway composed of contiguous segments. A *segment* is a one-directional length of roadway, marked by two *points* that indicate where a given segment begins and ends. Points are usually locations where traffic enters, exits, or crosses a facility. Two-lane highway segments are unique in that segments can be analyzed as either two-way segments or as one-way segments with a passing lane.

The *HCM* divides transportation facilities into several categories. The two major facility categories are uninterrupted flow facilities and interrupted flow facilities. *Uninterrupted flow facilities* include freeways, multilane highways, and two-lane highways on segments that do not have intersections that affect the traffic flow. *Interrupted flow facilities* include unsignalized and signalized intersections, urban streets, roundabouts (sometimes called traffic circles), and transit, pedestrian, and bicycle facilities.

The *HCM* also classifies transportation facilities by mode of transportation. Transit, pedestrian, and bicycle modes are self-explanatory. *Highway mode* includes freeways, multilane highways, two-lane highways, and urban streets. A *freeway* is a divided highway with two or more lanes in each direction, fully controlled access, and no fixed interruption points (e.g., signalized intersections). *Multilane highways* have at least two lanes in each direction, no or partial control of access, and signalized intersections spaced 2 mi (3 km) or more apart. *Two-lane highways* have one lane in each direction and may have signalized intersections and no or partial controlled access. *Urban streets* typically have a high density of access points and signalized intersections less than 2 mi (3 km) apart.

Capacity analysis is performed on segments where *demand* (expressed as vehicles or passenger cars per hour) and *capacity* (the maximum sustainable flow rate over a segment during a specific time period under set conditions) are relatively constant. Capacity analysis methods compare a facility's demand to its capacity. The procedure used for capacity analysis depends on whether or not the segment being analyzed includes signalized intersections. Capacity analysis for unsignalized intersections is covered in Sec. 2.3. Section 2.5 covers the fundamentals of traffic signals.

Analytical methods are designed to be sensitive to roadway and traffic characteristics. However, the methods cannot predict the effects of posted speed limits, police enforcement, safety features, driver education, or unusual vehicle performance, as these are isolated conditions that do not represent the majority of traffic flow.

3. CAPACITY ANALYSIS FOR UNINTERRUPTED FLOW

Unsignalized intersection capacity analysis follows a step-by-step procedure: (1) adjust the traffic volume to determine the demand flow rate, (2) calculate the *free-flow speed* (FFS) (either from field measurement or by applying adjustment factors to a *base free-flow speed* (BFFS)), and (3) determine the level of service (LOS). While methods for determining the flow rate, speed, and LOS vary based on facility type and roadway characteristics, using this core procedure simplifies analysis.

Uninterrupted-flow facilities have no fixed causes of delay or interruption caused by events other than actions within the traffic stream itself. The primary analysis is applied to *unsaturated* conditions, where demand is less than a roadway's capacity, and occasionally to *oversaturated* conditions, where demand exceeds capacity.

The methodologies cover freeways and major arterials in a nearly parallel process, and two-lane highways under three types of operating conditions. The methodologies (procedures, practices, and processes) are applied in a step-wise fashion, covering *planning analysis*, *preliminary engineering*, and *design analysis* levels. Preliminary engineering and planning analysis pass over operational refinements using default values in order to provide general overall assessments, and concentrate on future projections or trends. The most intensive level of analysis is operational design, which concentrates on near-term conditions, using actual field data as much as possible to refine detail designs.

Flow Rates

Flow rate is the number of vehicles (or bicycles or pedestrians) passing a given point in an hour. If the vehicle count (or bicycle or pedestrian count) applies to a time period other than one hour, the rate is converted to and expressed as vehicles (or bicycles or pedestrians) per hour.

Flow rate is closely related to density and speed, as shown in Eq. 2.12, and knowing any two values can determine the third. *Density* is the number of vehicles averaged over the roadway segment space and is typically given in units of vehicles per mile or vehicles per kilometer (veh/mi or veh/km), vehicles per mile per lane or vehicles per kilometer per lane (veh/mi-ln or veh/km·ln), or passenger cars per mile per lane or passenger cars per kilometer per lane (pc/mi-ln or pc/km·ln). Density is difficult to measure in the field and is typically calculated using the following speed-flow rate relationship.

$$D_{\text{veh/mi}} = \frac{v_{\text{vph}}}{S} \quad [HCM \text{ Eq. 4-4}] \quad 2.12$$

Density is particularly applicable to uninterrupted flow since it is the quality of traffic operations from the perspective of freedom to maneuver within the traffic stream. However, increases in flow rate and speed under low density conditions are limited by geometry (see Chap. 6) or driver comfort, whichever is lower. For design and analysis purposes, the *HCM* uses a maximum speed of 75 mph. Flow rates at greater speeds are not considered for analysis.

Flow rate is also affected by *flow friction*, which is caused by conditions deviating from the default (i.e., ideal) conditions. For example, the *HCM* sets the base lane width at 12 ft (3.7 m). Lane widths narrower than the base width cause drivers to decrease speed, which decreases the flow rate. The *HCM* uses a series of adjustment factors to correct the flow rate when roadway conditions fall outside the default values.

For unsaturated flow conditions, the terms flow rate, demand, demand flow rate, volume, and flow volume tend to be used interchangeably, which can be confusing. In general, the *demand* is the number of drivers desiring service on a given roadway, expressed in vehicles or passenger cars per hour. Demand is determined by actual vehicle counts and includes a mix of vehicle types (e.g., automobiles, trucks, and buses). For oversaturated conditions (i.e., when demand exceeds capacity), it is appropriate to refer to demand rather than flow rate, as the flow is often near zero. *Volume* is similar to flow rate and is based on the number of vehicles passing a point during a given interval, often one hour. Volume is typically given in vehicles per hour. Unlike flow rate, volume is not divided by the observation time. For example, a volume of 20 vph counted over a 15 min period corresponds to a flow rate of 80 vph (i.e., the volume divided by the counting period length).

Because freeway and highway traffic are composed mainly of passenger cars, flow rate uses passenger cars as its unit of measure. Other types of vehicles, such as trucks, buses, or recreational vehicles (RVs), must be converted to *passenger car equivalents* (PCEs, or the number of passenger cars displaced by a heavy vehicle under set roadway, traffic, and control conditions) when calculating the flow rate.

Levels of Service (LOS) Concepts

The *level of service* (LOS) rating system outlined in the *HCM* was created to measure the quality of service on a roadway as viewed by traffic participants and traffic analysts. LOS designations are arranged into five levels of traffic flow accommodations, ranging from A to E, with a sixth level, F, designating when demand is such that the capacity is overwhelmed and flow comes to a standstill. The *HCM* flow levels are

- A: free flow
- B: reasonable free flow
- C: stable flow
- D: approaching unstable flow

- E: fully saturated flow, considered capacity flow, subject to minor interruptions creating unstable flow
- F: forced or breakdown flow, oversaturation, jam density

LOS A, the lowest traffic density and highest freedom to maneuver condition, and LOS E, the maximum capacity with least freedom to maneuver condition, represent the upper and lower limits of LOS rating. LOS B, C, and D are relative gradations between the two traffic extremes. The absolute threshold ranges for each LOS level vary according to mode and type of facility, using LOS C as an approximate mid-level of operational acceptability. Capacity calculations provide quantitative measures of the demand that a roadway can handle. The LOS methodology expands these quantitative measures to illustrate the operational conditions of the roadway. Levels of service can be based on flow speed, flow rate, and measures of delay, as well as from a perceptive viewpoint of the facility user, depending on the mode being evaluated. While the other factors can be quantified from traffic data, the user's perception is related to the user's expectations. Interpretation of user perception of service quality can be estimated by traffic engineers and planners; however, user expectations can vary widely from location to location and from mode to mode, such that user perception of LOS often requires adjustment to local conditions. *HCM* suggests user expectations be developed from qualitative methods, such as interviews of users, when suggesting thresholds of value to determine breakpoints between levels B, C, and D, for modes of bicycle, pedestrian, and transit.

The freedom to maneuver is inversely related to density, as there is greater maneuverability space at lower density conditions. Roadway density, which is calculated using Eq. 2.12, is most often used to determine LOS levels for highways. The LOS can also be based on the service flow rate. The *service flow rate*, in pc/hr-ln, is the hourly flow rate at which vehicles can reasonably be expected to cross a given point during a period of time (typically 15 min) under typical roadway conditions, while maintaining a designated LOS. Because the service flow rate is the maximum flow rate for a given LOS, the service flow rate is the effective boundary between levels of service. The flow rates and densities are therefore considered threshold values. The maximum service flow rate for LOS E is considered the maximum sustainable flow rate and is referred to as the *limit of capacity*.

When comparing LOS for individual modes on shared-mode facilities, conflict occurs when one mode is restricted to improve the LOS of another mode. For example, at urban intersections, pedestrian crossing delays compete with traffic movement delays, such that reducing delays for pedestrian crossings increases delays for vehicle traffic.

Freeway and Multilane Highway Analysis

HCM has organized the computations for freeways and multilane highways into a series of six steps to determine the LOS of a freeway or multilane highway segment. The steps are shown in Fig. 2.3, with all references from the *HCM*.

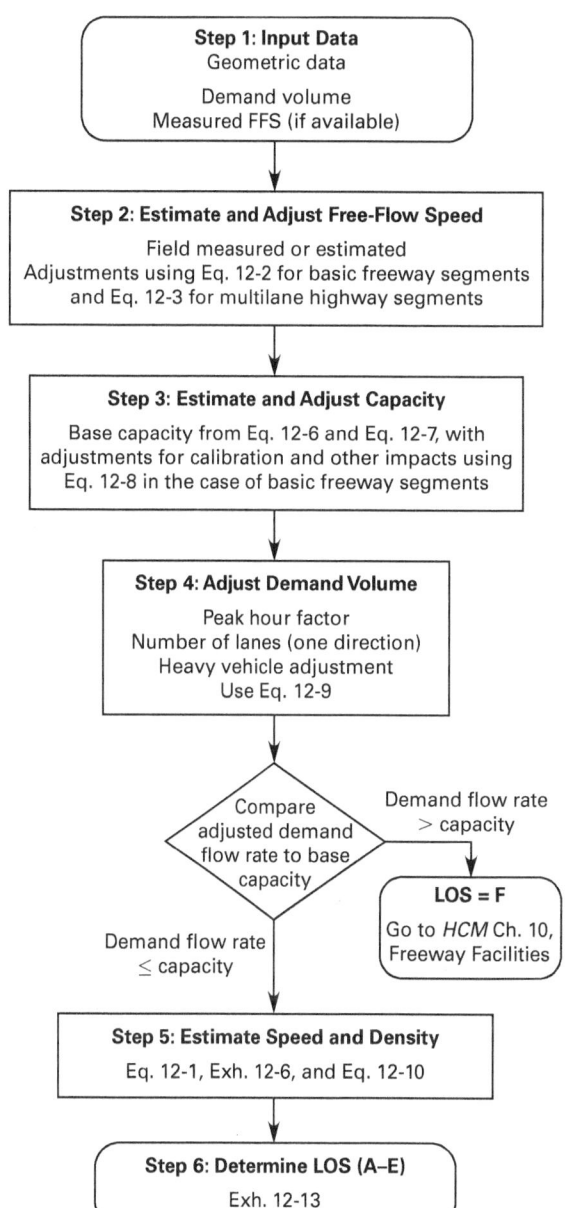

Figure 2.3 Flowchart of Freeway and Multilane Highway Methodology

Used with permission from *Highway Capacity Manual*, 6th Edition: A Guide for Multimodal Mobility Analysis, 2016, Exhibit 12-19, by the Transportation Research Board of the National Academies of Sciences, Engineering, and Medicine, Washington DC. DOI: 10.17226/24798

The procedures are similar and use the same equations and many of the same tables. Differences are found in a few conditions unique to each type of highway. Note that step 2 is skipped if measured FFS is available. Also, if demand is greater than base capacity, LOS = F, no further analysis is necessary, except in special freeway situations.

Base Conditions

Traffic analysis uses a set of base conditions for comparison of a particular segment in order to establish the full capacity. These conditions include good weather, good visibility, no incidents or crashes, no work zone activity, and no pavement deterioration that affects operations. In addition, the traffic stream has a low number of heavy vehicles, the driver population is primarily regular users who are familiar with the facility, and the lanes are 12 ft wide with at least 6 ft right-side clearance (on freeways). The traffic volume is considered *undersaturated*, except in special circumstances of *queue discharge*, *queue backup*, and *oversaturated flow*.

Default values can be substituted for data that is not available in *design analysis* and *planning and preliminary engineering*, and for system-wide *service flow rate and service volume* analysis. However, default values will affect the accuracy of the output. For *operational analysis*, actual field observed or field derived values should be used to assure that the analysis results appropriately fit the segment condition being described.

Default values for freeways and multilane highways are listed in Table 2.4.

Step 1: Check the Available Input Data. The primary inputs are *geometric data*, *demand volume* (which is the vehicle count not adjusted for PCEs for trucks, buses, and RVs), and *field measured free-flow speed* (if available). Other inputs are *lane width*, *right-side lateral clearance* (freeways), *median type* (multilane highways), *roadside clearance* (multilane highways), *number of lanes*, *traffic composition* (%T&B and %RV), *access points per mile*, *terrain* (length and % of grades), base free-flow speed, *type of traffic* (familiar or tourist), *peak hour factor* (PHF), and *peak-hour volume* (PHV).

Step 2: Determine Free-Flow Travel Speed on Multilane Highways and Freeways. Free-flow speed (FFS) is the theoretical average, field-measured speed of traffic on a study segment or a nearby segment. The FFS assumes ideal conditions (12 ft lanes, adequate lateral clearance). Field counts of at least 1000 data points should be used in order to ensure reliability. If necessary, FFS values for nearby areas can be used as a base free-flow speed (BFFS) for the study segment, then adjusted for conditions. According to the *HCM*, the most prominent adjustments for FFS are lane width, lateral clearance, ramp or access point density, and for multilane highways, median type. Speed variations at lower flow rates (500–800 pcphpl) are not drastically affected by these adjustments, so long as the attributes in question are consistent throughout the segment. Although default values of BFFS have been suggested, driver behavior varies enough that reliance on a default value can result in misleading conclusions. Speed limits are not considered a reliable estimate of FFS, since drivers tend to move with the speed of traffic flow regardless of posted speed limits. In the absence of field data, a very rough estimate of FFS can be found using the design speed for the study segment or similar nearby segments, so long as the design speed is at or near the minimum standard for the study location. A reasonable expected speed for similar nearby segments can also be used as a rough estimate.

FFS for a freeway can be estimated from the BFFS using Eq. 2.13 [*HCM* Eq. 12-2].

$$\text{FFS} = \text{BFFS} - f_{\text{LW}} - f_{\text{RLC}} - 3.22\text{TRD}^{0.84} \quad 2.13$$

FFS for multilane highways can be estimated with the following formula [*HCM* Eq. 12-3].

$$\text{FFS} = \text{BFFS} - f_{\text{LW}} - f_{\text{TLC}} - f_M - f_A \quad 2.14$$

Lane widths of 12 ft are considered an ideal standard, which provides the perception of a clear path for drivers of all skill levels without requiring undue attention to the precise positioning of the vehicle within the lane itself. Lane widths narrower than 10 ft are not practical for traffic with a mixed proportion of trucks and buses, as there is not enough room for comfortable passing. Table 2.5 gives adjustments for lane width on freeways and multilane highways.

Reduced clearance from the edge of a lane reduces a driver's feeling of safety. Drivers tend to reduce speed when right-side shoulders are narrower than 6 ft. On freeways, the left shoulder next to the median is rarely less than 2 ft, so adjustments for left-shoulder clearance are not normally applied. Table 2.6 shows right-side clearance on freeways.

On freeways, the free-flow speed is further reduced by total ramp density (TRD). Figure 2.4 illustrates the effect of ramp density on free-flow speed.

The *total ramp density* is equal to the number of on- and off-ramps in one direction that are located from 3 mi upstream to 3 mi downstream of the study segment, divided by the length of the study segment plus 6 mi.

For multilane highways, the effect of reduced lateral clearance on free-flow speed is based on the lateral clearance to fixed objects along both sides of the roadway. Fixed objects include signs, trees, bridge rails, traffic barriers, and retaining walls. Standard raised curbs are not considered fixed objects for determining lateral clearance. For divided multilane highways, the lateral clearances for both sides of the roadway are added together. For undivided highways, the left-shoulder lateral clearance is always taken as 6 ft when *two-way left-turn lanes* (TWLTLs) are present. Lateral clearances on a multilane

Table 2.4 Potential Data Sources and Default Values for Freeway and Multilane Highway Analysis

data category	potential data source(s)	suggested default value
time periods	user-defined study period, representative data of base scenario, and RRP	must be provided
demand multipliers	field data or modeling to generate day-of-week by month-of-year demand factors	urban and rural defaults provided in *HCM* Sec. 5
weather	online database for probabilities of various intensities of rain, snow, cold, and low visibility by month	defaults for 101 largest U.S. metropolitan areas provided in *HCM* Chap. 25
incidents	field data estimates of frequencies of occurrence of shoulder and lane closures per study period for each month, incident severity distribution, and average incident durations; alternatively, crash rate and incident-to-crash ratio for the facility, in combination with defaulted incident type probability and duration data	estimated from segment AADT and lengths as described in *HCM* Chap. 25
work zones and special events	user input on changes to base conditions and their schedule	must be provided
nearest city	select from the list of metropolitan areas provided in the *HCM* Vol. 4 technical reference library	must be provided when default weather data are used
geometrics	no details beyond core methodology needed; obtained from road inventory or aerial photo	must be provided
traffic counts	demand multiplier represented in base dataset; base scenario data from field data or modeling	must be provided

Used with permission from *Highway Capacity Manual*, 6th Edition: A Guide for Multimodal Mobility Analysis, 2016, Exhibit 11-10, by the Transportation Research Board of the National Academies of Sciences, Engineering, and Medicine, Washington DC. DOI: 10.17226/24798

highway are shown in Table 2.7 [*HCM* Exhibit 12-22]. Note that total lateral clearance greater than 12 ft results in no reduction of free-flow speed.

On undivided multilane highways, drivers often shy away from the median stripe, but cross the median stripe if no traffic is coming the other way and there is a need to provide more room for passing a vehicle. The combined effect results in a small reduction in free-flow speed, as shown in Table 2.8.

Table 2.5 FFS Adjustment for Average Lane Width for Freeways and Multilane Highways

lane width (ft)	reduction in FFS (mph)
≥12	0.0
≥11–12	1.9
≥10–11	6.6

(Multiply ft by 0.305 to obtain m.)
(Multiply mph by 1.609 to obtain kph.)

Used with permission from *Highway Capacity Manual*, 6th Edition: A Guide for Multimodal Mobility Analysis, 2016, Exhibit 12-20, by the Transportation Research Board of the National Academies of Sciences, Engineering, and Medicine, Washington DC. DOI: 10.17226/24798

Table 2.6 FFS Adjustment for Right-Side Lateral Clearance on Freeways, f_{LC}

right-side lateral clearance (ft)	lanes in one direction			
	2	3	4	≥5
≥6	0.0	0.0	0.0	0.0
5	0.6	0.4	0.2	0.1
4	1.2	0.8	0.4	0.2
3	1.8	1.2	0.6	0.3
2	2.4	1.6	0.8	0.4
1	3.0	2.0	1.0	0.5
0	3.6	2.4	1.2	0.6

Note: Interpolate for noninteger values of right-side lateral clearance.

Used with permission from *Highway Capacity Manual*, 6th Edition: A Guide for Multimodal Mobility Analysis, 2016, Exhibit 12-21, by the Transportation Research Board of the National Academies of Sciences, Engineering, and Medicine, Washington DC. DOI: 10.17226/24798

Table 2.7 FFS Adjustment for Total Lateral Clearance (TLC) on Multilane Highways, f_{LC}

four-lane highway		six-lane highway	
total lateral clearance (ft)	reduction in FFS (mph)	total lateral clearance (ft)	reduction in FFS (mph)
12	0	12	0
10	0.4	10	0.4
8	0.9	8	0.9
6	1.3	6	1.3
4	1.8	4	1.7
2	3.6	2	2.8
0	5.4	0	3.9

(Multiply ft by 0.305 to obtain m.)
(Multiply mph by 1.609 to obtain kph.)
Note: Interpolation to the nearest 0.1 is recommended.

Used with permission from *Highway Capacity Manual*, 6th Edition: A Guide for Multimodal Mobility Analysis, 2016, Exhibit 12-22, by the Transportation Research Board of the National Academies of Sciences, Engineering, and Medicine, Washington DC. DOI: 10.17226/24798

Figure 2.4 FFS Effect Due to Ramp Density

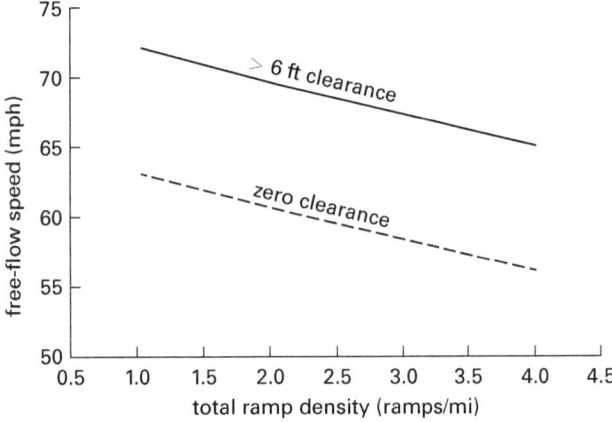

Note: Calculated by using this chapter's methods. Fixed values include BFFS = 75.4 mph for a basic freeway segment and f_{LW} = 6.6 for 10-ft lanes.

Used with permission from *Highway Capacity Manual*, 6th Edition: A Guide for Multimodal Mobility Analysis, 2016, Exhibit 12.32, by the Transportation Research Board of the National Academies of Sciences, Engineering, and Medicine, Washington DC. DOI: 10.17226/24798

Table 2.8 FFS Adjustment, f_M, for Median Type on Multilane Highways

median type	reduction in FFS (mph)
undivided highways	1.6
divided highways (including TWLTLs*)	0.0

(Multiply mph by 1.609 to obtain kph.)

*two-way left-turn lanes

Used with permission from *Highway Capacity Manual*, 6th Edition: A Guide for Multimodal Mobility Analysis, 2016, Exhibit 12-23, by the Transportation Research Board of the National Academies of Sciences, Engineering, and Medicine, Washington DC. DOI: 10.17226/24798

Some multilane highways have *uncontrolled access points*, such as intersections or driveways, that can interrupt the flow of traffic. BFFS assumes there are no access points along the study segment. The total number of right-side access points in the direction of travel is divided by the length of the study segment to determine the *access-point density* per mile. Access points that do not noticeably affect free-flow speed should not be counted. Access-point density on multilane highways is found in Table 2.9.

Table 2.9 FFS Adjustment for Access-Point Density on Multilane Highways, f_A

access points per mile	reduction in FFS (mph)
0	0.0
10	2.5
20	5.0
30	7.5
≥ 40	10.0

(Multiply mph by 1.609 to obtain kph.)

Used with permission from *Highway Capacity Manual*, 6th Edition: A Guide for Multimodal Mobility Analysis, 2016, Exhibit 12-24, by the Transportation Research Board of the National Academies of Sciences, Engineering, and Medicine, Washington DC. DOI:10.17226/24798

Step 3: Select FFS Curve. Free-flow speed curves for *freeways* are plotted in Fig. 2.5.

$$S = \text{FFS}_{\text{adj}} \quad [v_p \leq \text{BP}] \quad \text{[SI]} \quad 2.15(a)$$

$$S = \text{FFS}_{\text{adj}} - \frac{\left(\text{FFS}_{\text{adj}} - \dfrac{c_{\text{adj}}}{D_c}\right)(v_p - \text{BP})}{(c_{\text{adj}} - \text{BP})} \quad \text{[U.S.]} \quad 2.15(b)$$

$$[\text{BP} \leq v_p \leq c]$$

Figure 2.5 Speed-Flow Curves for Basic Freeway Segments Under Base Conditions

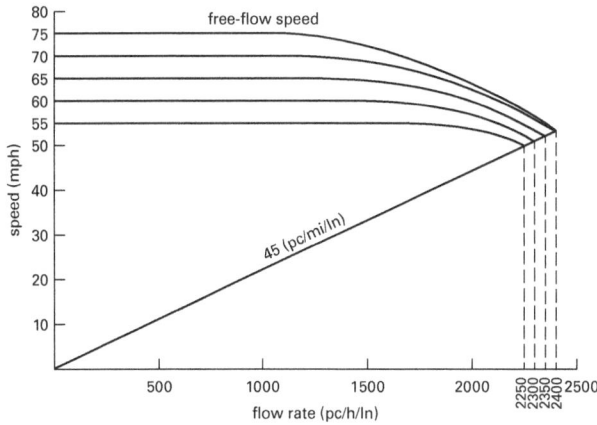

Used with permission from *Highway Capacity Manual*, 6th Edition: A Guide for Multimodal Mobility Analysis, 2016, Exhibit 12-7, by the Transportation Research Board of the National Academies of Sciences, Engineering, and Medicine, Washington DC. DOI:10.17226/24798

There are two ranges on the curves. In the first range, the FFS remains constant for each curve from 0 pcplph to the breakpoint. After the breakpoint, the second range (from 1000 pcphpl at 75 mph to 1800 pcphpl at 55 mph on freeways) shows the density increasing enough to interfere with FFS, and speeds decreasing with density until capacity is reached.

On multilane highways, there are more opportunities for traffic flow interference than on freeways, such as uncontrolled driveways and intersections, isolated signalized intersections, more severe geometry restrictions, and undivided medians separating opposing traffic. The result is lower speed expectations than on freeways. For each of the curves, the FFS remains constant up to the breakpoint of 1450 pcphpl. After the breakpoint, the density increases to interfere with FFS, and speeds decrease until capacity is reached. When solving for flow rate, the FFS should be rounded to the nearest 5 mph without interpolation due to variations in both observed and predicted values. Speed flow curves for freeways are shown in Fig. 2.5, and for multilane highways in Fig. 2.6.

Figure 2.6 Speed-Flow Curves for Multilane Highways Under Basic Conditions

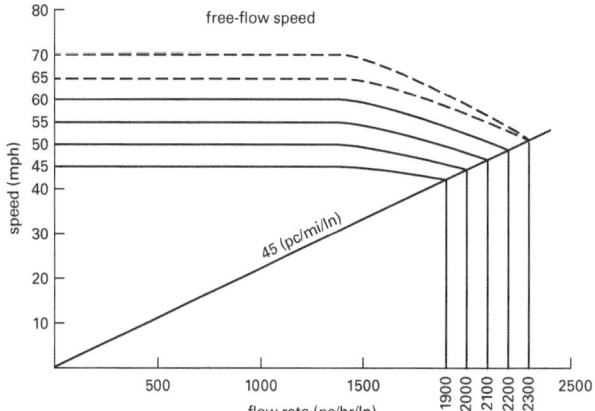

Note: Dashed curves are extrapolated and not based on field data.

Used with permission from *Highway Capacity Manual*, 6th Edition: A Guide for Multimodal Mobility Analysis, 2016, Exhibit 12-8, by the Transportation Research Board of the National Academies of Sciences, Engineering, and Medicine, Washington DC. DOI: 10.17226/24798

Speed curves and other traffic data values are general approximations of average conditions suitable for planning and preliminary design. Examples of more reliable data sources used for final roadway conditions are shown in Table 2.4.

Although direct, field-measured conditions are the most reliable data source, certain variables can use default values for preliminary analysis when field data is not yet available. Table 2.10 shows some of the possible data sources and the related default values.

Step 4: Adjust the Demand Volume. Freeways and multilane highways use the following equation to determine the flow rate [*HCM* Eq. 12-9].

$$v_p = \frac{V}{(\text{PHF})N f_{\text{HV}}} \quad 2.16$$

The units of v_p are passenger cars per hour per lane (pc/hr/ln).

Table 2.10 Default Values and Data Sources for Freeways and Multilane Highways

required data and units	potential data source(s)	suggested default value
geometric data—basic freeway segments		
free-flow speed[a] (mi/hr)	direct speed measurements, estimate from design speed or speed limit	base free-flow speed: speed limit + 5 mi/hr (range 55–75 mi/hr)
number of mainline freeway lanes in one direction[b] (ln)	road inventory, aerial photo	at least 2
lane width (ft)	road inventory, aerial photo	12 ft (range 10–12 ft)
right-side lateral clearance (ft)	road inventory, aerial photo	10 ft (range 0–10 ft)
total ramp density (ramps/mi)	road inventory, aerial photo	must be provided (range 0–6 ramps/mi)
terrain type (level, rolling, specific grade)	design plans, analyst judgment	must be provided
geometric data—multilane highway segments		
free-flow speed[a] (mi/hr)	direct speed measurements, estimate from design speed or speed limit	base free-flow speed: speed limit + 5 mi/hr (range 45–70 mi/hr)
number of mainline freeway lanes (one direction)[b]	road inventory, aerial photo	at least 2
lane width (ft)	road inventory, aerial photo	12 ft (range 10–12 ft)
right-side lateral clearance (ft)	road inventory, aerial photo	6 ft (range 0–6 ft)
median (left-side) lateral clearance (ft)	road inventory, aerial photo	6 ft (range 0–6 ft)
access point density (points/mi)	road inventory, aerial photo	8 access points/mi (rural) 16 access points/mi (low-density suburban) 25 access points/mi (high-density suburban)
terrain type (level, rolling, specific grade)	design plans, analyst judgment	must be provided
median type (divided, undivided, TWLTL[c])	road inventory, aerial photo	must be provided
demand data—basic freeway and multilane highway segments		
hourly demand volume (veh/hr)	field data, modeling	must be provided
heavy vehicle percentage (%)	field data	5% (urban) 12% (rural)[d]
peak hour factor[e] (decimal)	field data	basic freeway segments 0.94 multilane highways 0.95 (urban) or 0.88 (rural)
driver population, capacity, and free-flow speed adjustment factors	field data	1.0 (see also *HCM* Chap. 26)

[a]Moderate sensitivity (10%–20% change) of service measure to the choice of default value
[b]High sensitivity (>20% change) of service measure to the choice of default value
[c]TWLTL = two-way left-turn lane
[d]See *HCM* Chap. 26 in Volume 4 for state-specific default heavy vehicle percentages and driver population adjustment factors.
[e]Moderate to high sensitivity of service measures for very low PHF values. See the discussion in the text. PHF is not required when peak 15 min demand volumes are provided.

Used with permission from *Highway Capacity Manual*, 6th Edition: A Guide for Multimodal Mobility Analysis, 2016, Exhibit 12-18, by the Transportation Research Board of the National Academies of Sciences, Engineering, and Medicine, Washington DC. DOI: 10.17226/24798

Heavy vehicle adjustments are determined by this formula [*HCM* Eq. 12-10]. P_T and P_R are decimal ratios.

$$f_{\text{HV}} = \frac{1}{1 + P_T(E_T - 1)} \quad 2.17$$

Truck and bus counts can be combined because of similar performance characteristics. *HCM* calls for separate counts of single-unit trucks (SUT) and trailer trucks (TT) due to differences in weight-to-horsepower ratios.

Roadway terrain is classified into three general types. *Level terrain* (*general terrain*) is any combination of grades and vertical or horizontal alignment that permits heavy vehicles to operate at nearly the same speed as passenger cars. Grades are limited to 1% or 2%, or short duration.

Rolling terrain (*specific upgrades* or *specific downgrades*) is any combination of grades and horizontal or vertical alignment that causes heavy vehicles to reduce their operating speed substantially below that of passenger cars, but does not reduce heavy vehicles to operate at crawl speeds for a significant length of time or extended distances.

Mountainous terrain consists of any combination of grades and vertical or horizontal alignment that causes heavy vehicles to operate at crawl speeds for significant distances or at frequent intervals.

Extended freeway or highway segments can involve a series of upgrades and downgrades that do not have a significant influence on traffic flow speed. This can include grades of up to 3% for distances of 0.25 mi to 1 mi. Grades greater than 3% for distances longer than 0.25 mi should be a specific grade to be analyzed as a separate segment.

Table 2.11 shows PCEs for general terrain segments on freeways and for multilane highways.

Table 2.11 Heavy Vehicle PCEs for General Terrain Segments on Freeways and Multilane Highways

factor	type of terrain	
	level	rolling
trucks and buses, E_T	2.0	3.0

Used with permission from *Highway Capacity Manual*, 6th Edition: A Guide for Multimodal Mobility Analysis, 2016, Exhibit 12-25, by the Transportation Research Board of the National Academies of Sciences, Engineering, and Medicine, Washington DC. DOI: 10.17226/24798

On extended grades, trucks and buses gradually lose momentum and their speed drops below the normal travel speed of automobiles. Isolated or few trucks in the traffic flow have more effect on automobile traffic than with larger proportions of trucks. Increased proportions of trucks tend to add stability to the traffic flow, reducing the delay effects of lane changing and passing maneuvers.[1] Check for LOS E before proceeding.

Table 2.12 (at the end of Sec. 2.6) shows PCEs for trucks and buses on upgrades for freeways and multilane highways.

Step 5: Estimate Speed and Density. Using the adjusted demand volume and the appropriate FFS curve, the speed can be estimated from the curve selected in step 3. Note that the FFS curve and demand adjustments are needed only when field-measured FFS is not available.

When the demand does not exceed capacity, this speed is divided into the adjusted demand flow rate to determine *density* [*HCM* Eq. 12-11].

$$D = \frac{v_p}{S} \qquad 2.18$$

Step 6: Determine Level of Service (LOS). Table 2.13 is used to determine LOS on freeways and multilane highways.

Maximum flow rate (pc/hr/ln) can also be used for LOS designations on basic freeway sections, although *HCM* emphasizes density as the preferred measure. LOS maximum service flow rates for freeways, listed in Table 2.14, are interpreted from plotted curves in Fig. 2.7.

LOS densities at breakpoints from LOS A through D have been set as consensus limits of quality of flow ranges.

Table 2.13 LOS Criteria for Basic Freeway and Multilane Highway Segments

LOS	density (pc/mi/ln)
A	≤11
B	>11–18
C	>18–26
D	>26–35
E	>35–45
F	demand exceeds capacity >45

Used with permission from *Highway Capacity Manual*, 6th Edition: A Guide for Multimodal Mobility Analysis, 2016, Exhibit 12-15, by the Transportation Research Board of the National Academies of Sciences, Engineering, and Medicine, Washington DC. DOI: 10.17226/24798

Table 2.14 Level of Service for Basic Freeway Sections, Based on Maximum Service Flow Rate

	maximum service flow (pc/hr/ln)				
FFS	LOS A	LOS B	LOS C	LOS D	LOS E
75 mph	820	1310	1830	2110	2400
70 mph	770	1250	1690	2070	2400
65 mph	710	1170	1630	2030	2350
60 mph	660	1080	1560	2010	2300
55 mph	600	990	1430	1900	2250

Used with permission from *Highway Capacity Manual*, 6th Edition: A Guide for Multimodal Mobility Analysis, 2016, Exhibit 12-37, by the Transportation Research Board of the National Academies of Sciences, Engineering, and Medicine, Washington DC. DOI: 10.17226/24798

---

[1]Additional details about analysis of truck PCEs in mountainous terrain are covered in *HCM*.

Figure 2.7 *LOS for Basic Freeway Segments Shown on Speed-Flow Curves*

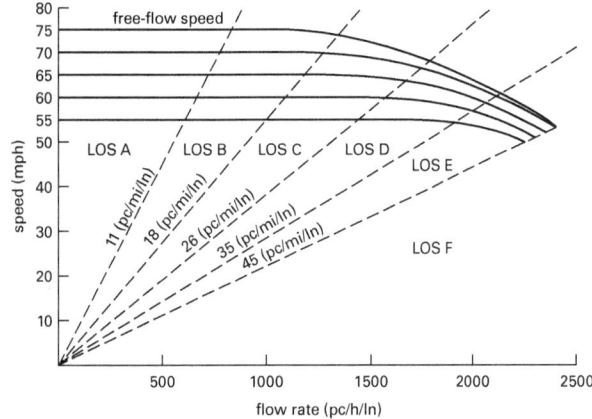

Used with permission from *Highway Capacity Manual*, 6th Edition: A Guide for Multimodal Mobility Analysis, 2016, Exhibit 12-16, by the Transportation Research Board of the National Academies of Sciences, Engineering, and Medicine, Washington DC. DOI:10.17226/24798

Figure 2.8 shows multilane highway LOS plotted on speed vs. flow-rate curves.

Multilane highway service flow rates (which are interpreted from plotted curves in Fig. 2.8) under base conditions are shown in Table 2.15.

Figure 2.8 *LOS Ranges Plotted on Multilane Highway Speed-Flow Curves*

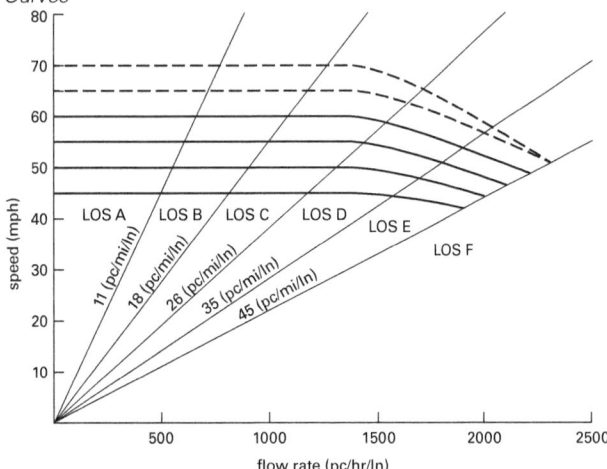

Note: Dashed curves are extrapolated and not based on field data.

Used with permission from *Highway Capacity Manual*, 6th Edition: A Guide for Multimodal Mobility Analysis, 2016, Exhibit 12-17, by the Transportation Research Board of the National Academies of Sciences, Engineering, and Medicine, Washington DC. DOI:10.17226/24798

Table 2.15 *Multilane Highway Maximum Service Flow Rates Under Base Conditions*

FFS (mph)	target LOS				
	A	B	C	D	E
60	660	1080	1550	1980	2200
55	600	990	1430	1850	2100
50	550	900	1300	1710	2000
45	490	810	1170	1550	1900

Used with permission from *Highway Capacity Manual*, 6th Edition: A Guide for Multimodal Mobility Analysis, 2016, Exhibit 12-38, by the Transportation Research Board of the National Academies of Sciences, Engineering, and Medicine, Washington DC. DOI: 10.17226/24798

An additional step can be performed to determine the number of lanes required for a target LOS. For multilane highways, the minimum number of lanes necessary to accommodate the target LOS is found by dividing the demand service flow by the *maximum service flow* (MSF) by using Eq. 2.19 [*HCM* Eq. 12-23].

$$N = \frac{V}{\text{MSF}_i (\text{PHF}) f_{\text{HV}}} \quad 2.19$$

While N should be rounded to the next highest integer, there may be cases in which a target LOS can be shifted one level lower in order to avoid the expense of over-constructing a facility. For instance, if LOS B is targeted, and N is 2.1 from the equation, LOS C may result if two lanes are provided. Some judgment is often necessary when borderline LOS thresholds are approached.

Using the K factor (the proportion of AADT during the peak hour) and D (the directional distribution) for typical urban and rural multilane highways in level terrain and rolling terrain, LOS target volumes can be selected from Table 2.16 and Table 2.17 at the end of Sec. 2.6.

Generalized Daily Service Volumes for Multilane Highways

For *planning* and *preliminary engineering* purposes, and using a set of typical conditions for rural and urban multilane highways, daily service volume tables have been developed targeting several operational LOS thresholds. The generalized conditions used in these tables are

- 12% HV for rural highways and 8% HV for urban highways
- FFS = 60 mph for both locations
- PHF = 0.88 for rural highways, PHF = 0.95 for urban highways

The generalized service volumes shown in Table 2.16 and Table 2.17 (at the end of Sec. 2.6) are for level or rolling terrain. The tables are very approximate, and the user should make appropriate adjustments for local and regional factors before using the listed values for design recommendations.

To summarize LOS conditions, a density of 11 pc/mi/ln or less is considered LOS A for both freeways and multilane highways over all speed ranges. On the other end of the scale, maximum flow capacity, which is represented by LOS E, is 45 pc/mi/ln for freeways, and from 40 to 45 pc/mi/ln for multilane highways, depending on free-flow speed. The maximum flow rate—ranging from 2250 pc/hr/ln to 2400 pc/hr/ln for freeways, and from 1900 pc/hr/ln to 2200 pc/hr/ln for multilane highways—is considered the limit of capacity for each highway type.

Example 2.3

An undivided four-lane highway has a total lateral clearance of 4 ft. The lanes are 11 ft wide, and there are 10 access points per mile. The flow rate is 1150 pcphpl. Assuming a base free-flow speed of 55 mph, what are the free-flow speed and LOS?

Solution

The BFFS is given as 55 mph. From Table 2.5, the lane width adjustment factor, f_{LW}, for 11 ft lanes is 1.9 mph. From Table 2.7, the lateral clearance adjustment factor, f_{LC}, for a four-lane highway with a total lateral clearance of 4 ft is 1.8 mph. From Table 2.8, the adjustment for median type, f_M, for an undivided roadway is 1.6 mph. From Table 2.9, the adjustment for access-point density of 10 access points per mile is 2.5 mph.

$$\text{FFS} = \text{BFFS} - f_{LW} - f_{LC} - f_M - f_A$$
$$= 55\,\frac{\text{mi}}{\text{hr}} - 1.9\,\frac{\text{mi}}{\text{hr}} - 1.8\,\frac{\text{mi}}{\text{hr}} - 1.6\,\frac{\text{mi}}{\text{hr}} - 2.5\,\frac{\text{mi}}{\text{hr}}$$
$$= 47.2\;\text{mi/hr}\quad(47\text{ mph})$$

For a flow rate of 1150 pcphpl,

$$D = \frac{1150\,\frac{\text{pc}}{\text{hr-ln}}}{47\,\frac{\text{mi}}{\text{hr}}} = 24\;\text{pcpmpl}$$

From Fig. 2.8, the LOS is C.

4. TWO-LANE HIGHWAY ANALYSIS

A two-lane highway segment with no signals for at least 2 mi (or for 3 mi or more) can use *uninterrupted flow* methodologies for analysis, planning, design, and operations—for both automobile and bicycle modes. Two-lane highway uninterrupted flow methodology includes procedures for predicting the effect of passing and truck climbing lanes. These methodologies are presented in *HCM* Chap. 15. On segments where signalized intersections are less than 2.0 mi apart, the facility should be classified as an urban street, and *interrupted flow* methodologies covered in *HCM* Chap. 16 or Chap. 17 should be used. On urban streets, it is assumed that no passing occurs in the opposing lanes.

Two-lane highway conditions cover a very broad range of functions in the transportation network. The range of functions can overlap, and often be in conflict with each other. In order to more clearly define functional capability, two-lane highways are classified under three groups.

Class I two-lane highways. These highways are major intercity routes—connecting major traffic generators—and carry traffic consisting of daily commuters, or they are major links in the national highway network. Trips are mostly long distance and provide connections between other facilities that serve long-distance trips. Travel speed is often relatively high, similar to multilane highways, with average travel up to or near freeway speeds.

Class II two-lane highways. These highways function as access to class I facilities, can serve as scenic or recreational routes, or pass through rugged terrain. Trip lengths are relatively short, serving the beginning or ending of longer trips; or sightseeing or recreational activity may occur on them. Travel speed is not expected to be as high as class I facilities, but may be greater than urban streets for extended segments.

Class III two-lane highways. These highways serve moderately developed areas and may be portions of class I or class II highways. Class III highway traffic has a mixture of local and through traffic, involving unsignalized roadside access that is noticeably denser than in a purely rural area. Class III segments often have reduced speed limits that reflect higher activity levels and roadside interference.

Figure 2.9 is a flowchart of the two-lane highway *HCM* methodology, which includes references to appropriate *HCM* exhibits.

Step 1: Input Data. Table 2.18 shows input data and default values used for basic two-lane highway segment analysis. Default values are used when site-specific data are not available. *Planning* and *preliminary engineering* studies often use default data for establishing general road construction and traffic conditions; however, default data is far less accurate than locally derived and field verified data needed for *operational analysis*.

Step 2: Estimate FFS on Two-Lane Highways. Table 2.19 summarizes *HCM* exhibits for speed-determining methodologies. PTSF is the percent time spent following.

The BFFS is generally specified or estimated by the agency, and usually is somewhat higher (e.g., 10 mph) than the posted speed. The design speed, if available, can also be used for BFFS. In locations of low flow rates (less than 200 vph), it may be difficult to observe an accurate field-measured speed. In these cases, a speed sample may be taken at higher flow rates and adjusted for the reduced volume using Eq. 2.20 [*HCM* Eq. 15-1]. In this equation, v is the observed flow rate for the period corresponding to S_{FM}.

$$\text{FFS} = S_{FM} + 0.00776\frac{v}{f_{HV,ATS}} \quad\quad 2.20$$

[field measured S]

When data are not available, an estimate of BFFS can be assumed as the design speed if the design speed is somewhat consistent over the study segment. If the design speed is unknown, the posted speed can be used as a rough estimate of BFFS, assuming the posted speed seems realistic. After determining BFFS, FFS can be estimated from Eq. 2.21 [*HCM* Eq. 15-2]. For this equation, f_{LS} is the *lane and shoulder width adjustment factor*. f_A is the *adjustment for access point density* (intersections and driveways), or number of access points per mile.

$$\text{FFS} = \text{BFFS} - f_{LS} - f_A \quad\quad 2.21$$

Table 2.20 shows adjustment factors for land and shoulder width. Table 2.21 shows adjustment factors for access point density.

Figure 2.9 *Flowchart of Two-Lane Highway Methodology*

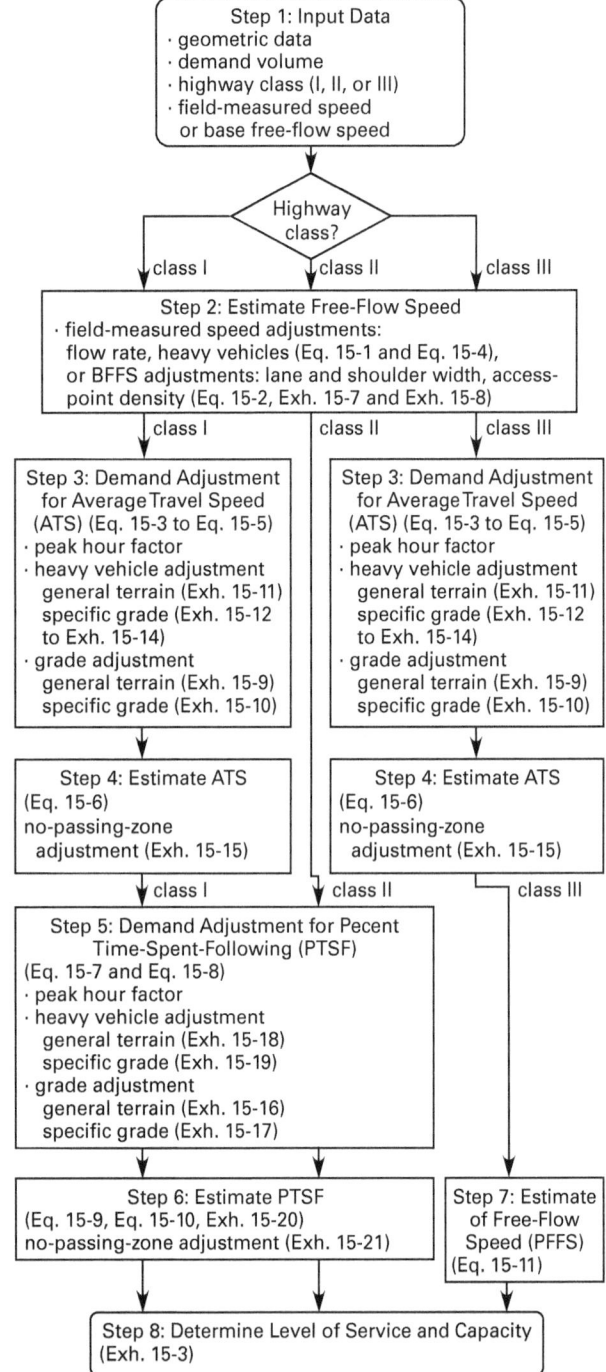

Used with permission from *Highway Capacity Manual*, 6th Edition: A Guide for Multimodal Mobility Analysis, 2016, Exhibit 15-6, by the Transportation Research Board of the National Academies of Sciences, Engineering, and Medicine, Washington DC. DOI:10.17226/24798

Table 2.18 *Required Input Data and Default Values for Two-Lane Highways*

required data	suggested default value	potential data source(s)
geometric data		
highway class	must be provided	determine from functional class, land use, motorist expectation
lane width	12 ft	road inventory, aerial photo
shoulder width	6 ft	road inventory, aerial photo
access-point density (both sides)	classes I and II: 8/mi, class III: 16/mi	field data, aerial photo
terrain type	must be provided	design plans, analyst judgment
percent no-passing zone[a]	level: 20%, rolling: 40%, more extreme: 80%	road inventory, aerial photo
free-flow speed	speed limit + 10 mph	direct speed measurements, estimates from design speed or speed limit
length of passing lane (if present)	must be provided	field data, road inventory, aerial photo
demand data		
hourly demand volume	must be provided	field data, modeling
length of analysis period	15 min (0.25 h)	must be provided
peak hour factor	0.88	field data
directional volume split	must be provided	field data, modeling
heavy vehicle percentage[b]	6%	field data

[a]Percent no-passing zone may be different in each direction.
[b]See Chap. 26 in Vol. 4 for state-specific default heavy vehicle percentages.
Used with permission from *Highway Capacity Manual*, 6th Edition: A Guide for Multimodal Mobility Analysis, 2016, Exhibit 15-5, by the Transportation Research Board of the National Academies of Sciences, Engineering, and Medicine, Washington DC. DOI: 10.17226/24798

Table 2.19 HCM Exhibits for Speed-Determining Methodologies

		ATS	PTSF
general sections			
	grade adjustment, f_G	15-9	15-16
	truck equivalent, E_T	15-11	15-18
	recreational vehicle, E_R	15-11	15-18
specific upgrades			
	grade adjustment, f_G	15-10	15-17
	truck equivalent, E_T	15-12	15-19
	recreational vehicle, E_R	15-13	15-19

Table 2.20 Adjustment Factors for Lane and Shoulder Width on Two-Lane Highways, f_{LS}

	shoulder width (ft)			
lane width (ft)	≥ 0 and < 2	≥ 2 and < 4	≥ 4 and < 6	≥ 6
≥ 9 and < 10	6.4	4.8	3.5	2.2
≥ 10 and < 11	5.3	3.7	2.4	1.1
≥ 11 and < 12	4.7	3.0	1.7	0.4
≥ 12	4.2	2.6	1.3	0.0

(Multiply ft by 0.305 to obtain m.)
(Multiply mph by 1.609 to obtain kph.)

Used with permission from *Highway Capacity Manual*, 6th Edition: A Guide for Multimodal Mobility Analysis, 2016, Exhibit 15-7, by the Transportation Research Board of the National Academies of Sciences, Engineering, and Medicine, Washington DC. DOI: 10.17226/24798

Table 2.21 Adjustment for Access-Point Density, f_A, on Two-Lane Highways

access points per mile	reduction in FFS (mph)
0	0.0
10	2.5
20	5.0
30	7.5
40	10.0

(Multiply mph by 1.609 to obtain kph.)

Used with permission from *Highway Capacity Manual*, 6th Edition: A Guide for Multimodal Mobility Analysis, 2016, Exhibit 15-8, by the Transportation Research Board of the National Academies of Sciences, Engineering, and Medicine, Washington DC. DOI: 10.17226/24798

To determine access point density, divide the total number of unsignalized intersections and driveways on both sides of the roadway segment by the length of the segment (in miles). The FFS is the same in both directions of a two-lane segment using the same estimated BFFS; however, field-measured FFS could be different in each direction.

Step 3: Demand Adjustment for ATS. This step applies to class I and class III two-lane highways only.

The demand flow rate, v, is the adjusted hourly demand passenger car equivalents for the peak 15 min, calculated from the hourly volume, V. To make the demand adjustment for ATS, the demand flow rate, $v_{i,\text{ATS}}$, is found by adjusting the demand volume, V_i, by the heavy vehicle adjustment factor, f_{HV}, the grade adjustment factor, f_G, and the PHF. The demand flow rate and the demand volume are determined for the analysis direction ($i = d$), or the opposing direction ($i = o$), as separate calculations using Eq. 2.22 [*HCM* Eq. 15-3].

$$v_{i,\text{ATS}} = \frac{V_i}{\text{PHF} f_{g,\text{ATS}} f_{HV,\text{ATS}}} \quad 2.22$$

The PHF considers variation in traffic flow within the hour. Two-lane highway analysis is based on the demand flow rates for a peak 15 min period within the analysis hour, usually (but not necessarily) the peak hour. If flow rates for the peak 15 min have been directly measured, the PHF is set to 1.00.

The ATS grade adjustment factor for segments 2 mi or longer on level terrain, rolling terrain, and specific downgrades where trucks do not operate at a crawl speed is found using Table 2.22. Interpolation between values to the nearest 0.01 is suggested.

Table 2.22 ATS Grade Adjustment Factor for Level Terrain, Rolling Terrain, and Specific Downgrades, $f_{g,\text{ATS}}$

	adjustment factor	
one-direction demand flow rate, v_{vph} (vph)	level terrain and specific downgrades	rolling terrain
≤ 100	1.00	0.67
200	1.00	0.75
300	1.00	0.83
400	1.00	0.90
500	1.00	0.95
600	1.00	0.97
700	1.00	0.98
800	1.00	0.99
≥ 900	1.00	1.00

Note: Interpolation to the nearest 0.01 is recommended.

Used with permission from *Highway Capacity Manual*, 6th Edition: A Guide for Multimodal Mobility Analysis, 2016, Exhibit 15-9, by the Transportation Research Board of the National Academies of Sciences, Engineering, and Medicine, Washington DC. DOI: 10.17226/24798

The *ATS heavy vehicle adjustment factor* for specific upgrades is related to steepness of grade, grade length, and directional flow rate, as shown in Table 2.23 at the end of Sec. 2.6. Again, interpolation between values to the nearest 0.01 is suggested.

ATS PCEs for trucks and RVs for level terrain, rolling terrain, and specific downgrades are found in Table 2.24. Passenger car equivalents are the number of passenger cars displaced from the vehicle stream by one truck or RV, or any combination of cars with a trailer, or a truck with a trailer, that displaces more than the equivalent of a single car.

Table 2.24 ATS Passenger Car Equivalents for Trucks, E_T, and RVs, E_R, for Level Terrain, Rolling Terrain, and Specific Downgrades

vehicle type	directional demand flow rate, v_{vph} (vph)	level terrain and specific downgrades	rolling terrain
trucks, E_T	≤100	1.9	2.7
	200	1.5	2.3
	300	1.4	2.1
	400	1.3	2.0
	500	1.2	1.8
	600	1.1	1.7
	700	1.1	1.6
	800	1.1	1.4
	≥900	1.0	1.3
RVs, E_R	all flows	1.0	1.1

Note: Interpolation to the nearest 0.1 is recommended.
Used with permission from *Highway Capacity Manual*, 6th Edition: A Guide for Multimodal Mobility Analysis, 2016, Exhibit 15-11, by the Transportation Research Board of the National Academies of Sciences, Engineering, and Medicine, Washington DC. DOI: 10.17226/24798

The heavy vehicle adjustment factor, $f_{HV,ATS}$, is found using Eq. 2.23 [*HCM* Eq. 15-4].

$$f_{HV,ATS} = \frac{1}{1 + P_T(E_T - 1) + P_R(E_R - 1)} \quad 2.23$$

PCEs for trucks on specific upgrades are found in Table 2.25 at the end of Sec. 2.6. RCEs for RVs for specific upgrades are found in Table 2.26 at the end of Sec. 2.6.

On specific downgrades where trucks travel at crawl speed, PCEs are much greater than on level or rolling terrain. When trucks operate at crawl speed while moving downhill, the driver is operating the vehicle at a greatly reduced speed to reduce momentum so that the normal brake system is sufficient to slow or stop the vehicle in a safe manner. For these conditions, Eq. 2.24 [*HCM* Eq. 15-5] applies.

$$f_{HV,ATS} = \frac{1}{1 + P_{TC}P_T(E_{TC} - 1) + (1 - P_{TC})} \quad 2.24$$
$$\times P_T(E_T - 1) + P_R(E_R - 1)$$

The variable P_{TC} is the flow rate of trucks traveling at crawl speed, divided by the flow rate of all trucks. E_{TC} is found using Table 2.27 at the end of Sec. 2.6.

Step 4: Estimate ATS_d. This step applies only to class I and class III two-lane highways, since class II two-lane highways do not use ATS as an LOS measure.

Using Eq. 2.25 [*HCM* Eq. 15-6], the ATS is estimated from the FFS, the demand flow rate, the opposing flow rate, and the percentage of no-passing zones in the analysis direction.

$$ATS_d = FFS - 0.00776(v_{d,ATS} + v_{o,ATS}) - f_{np,ATS} \quad 2.25$$

Table 2.28 is used to find the ATS adjustment factor for no-passing zones.

Table 2.28 ATS Adjustment Factor for No-Passing Zones, $f_{np,ATS}$

opposing demand flow rate, v_o (pc/hr)	percent no-passing zones				
	≤ 20	40	60	80	100
FFS ≥ 65 mph					
≤100	1.1	2.2	2.8	3.0	3.1
200	2.2	3.3	3.9	4.0	4.2
400	1.6	2.3	2.7	2.8	2.9
600	1.4	1.5	1.7	1.9	2.0
800	0.7	1.0	1.2	1.4	1.5
1000	0.6	0.8	1.1	1.1	1.2
1200	0.6	0.8	0.9	1.0	1.1
1400	0.6	0.7	0.9	0.9	0.9
≥1600	0.6	0.7	0.7	0.7	0.8
FFS = 60 mph					
≤100	0.7	1.7	2.5	2.8	2.9
200	1.9	2.9	3.7	4.0	4.2
400	1.4	2.0	2.5	2.7	3.9
600	1.1	1.3	1.6	1.9	2.0
800	0.6	0.9	1.1	1.3	1.4
1000	0.6	0.7	0.9	1.1	1.2
1200	0.5	0.7	0.9	0.9	1.1
1400	0.5	0.6	0.8	0.8	0.9
≥1600	0.5	0.6	0.7	0.7	0.7
FFS = 55 mph					
≤100	0.5	1.2	2.2	2.6	2.7
200	1.5	2.4	3.5	3.9	4.1
400	1.3	1.9	2.4	2.7	2.8
600	0.9	1.1	1.6	1.8	1.9
800	0.5	0.7	1.1	1.2	1.4
1000	0.5	0.6	0.8	0.9	1.1
1200	0.5	0.6	0.7	0.9	1.0
1400	0.5	0.6	0.7	0.7	0.9
≥1600	0.5	0.6	0.6	0.6	0.7
FFS = 50 mph					
≤100	0.2	0.7	1.9	2.4	2.5
200	1.2	2.0	3.3	3.9	4.0
400	1.1	1.6	2.2	2.6	2.7
600	0.6	0.9	1.4	1.7	1.9
800	0.4	0.6	0.9	1.2	1.3
1000	0.4	0.4	0.7	0.9	1.1
1200	0.4	0.4	0.7	0.8	1.0
1400	0.4	0.4	0.6	0.7	0.8
≥1600	0.4	0.4	0.5	0.5	0.5
FFS ≤ 45 mph					
≤100	0.1	0.4	1.7	2.2	2.4
200	0.9	1.6	3.1	3.8	4.0
400	0.9	0.5	2.0	2.5	2.7
600	0.4	0.3	1.3	1.7	1.8
800	0.3	0.3	0.8	1.1	1.2
1000	0.3	0.3	0.6	0.8	1.1
1200	0.3	0.3	0.6	0.7	1.0
1400	0.3	0.3	0.6	0.6	0.7
≥1600	0.3	0.3	0.4	0.4	0.6

Note: Interpolation of $f_{np,ATS}$ for percent no-passing zones, demand flow rate, and FFS to the nearest 0.1 is recommended.
Used with permission from *Highway Capacity Manual*, 6th Edition: A Guide for Multimodal Mobility Analysis, 2016, Exhibit 15-15, by the Transportation Research Board of the National Academies of Sciences, Engineering, and Medicine, Washington DC. DOI: 10.17226/24798

Step 5: Demand Adjustment for PTSF. This step applies only to class I and class II two-lane highways, since LOS on class III highways is not based on PTSF.

The demand adjustment is calculated using Eq. 2.26 [*HCM* Eq. 15-7] and Eq. 2.27 [*HCM* Eq. 15-8].

$$v_{i,\text{PTSF}} = \frac{V_i}{\text{PHF} f_{g,\text{PTSF}} f_{\text{HV,PTSF}}} \quad 2.26$$

$$f_{\text{HV,PTSF}} = \frac{1}{1 + P_T(E_T - 1) + P_R(E_R - 1)} \quad 2.27$$

Table 2.29 to Table 2.31 (after Sec. 2.6) provide PTSF grade adjustment factors and passenger car equivalents.

Table 2.32 shows PTSF equivalents for trucks and buses based on terrain.

Table 2.32 *PTSF Passenger Car Equivalents for Trucks and RVs for Level Terrain, Rolling Terrain, and Specific Downgrades*

vehicle type	directional demand flow rate (veh/h)	level and specific downgrade	rolling
trucks, E_T	≤100	1.1	1.9
	200	1.1	1.8
	300	1.1	1.7
	400	1.1	1.6
	500	1.0	1.4
	600	1.0	1.2
	700	1.0	1.0
	800	1.0	1.0
	≥900	1.0	1.0
RVs, E_R	all	1.0	1.0

Note: Interpolation is not recommended.

Used with permission from *Highway Capacity Manual*, 6th Edition: A Guide for Multimodal Mobility Analysis, 2016, Exhibit 15-18, by the Transportation Research Board of the National Academies of Sciences, Engineering, and Medicine, Washington DC. DOI: 10.17226/24798

Step 6: Estimate PTSF. This step applies only to class I and class II two-lane highways.

The PTSF in the analysis direction is estimated using Eq. 2.28 [*HCM* Eq. 15-9].

$$\text{PTSF}_d = \text{BPTSF}_d + f_{\text{np,PTSF}} \left(\frac{v_{d,\text{PTSF}}}{v_{d,\text{PTSF}} + v_{o,\text{PTSF}}} \right) \quad 2.28$$

The fraction in this equation is the demand volume in the analysis direction divided by the total of the demand volume in the analysis direction and the opposing direction.

BPTSF is estimated by using Eq. 2.29 [*HCM* Eq. 15-10] and Table 2.33.

$$\text{BPTSF}_d = 100\left(1 - \exp(av_d^b)\right) \quad 2.29$$

Table 2.33 *PTSF Coefficients for Use in Eq. 2.29 for Estimating BPTSF*

opposing demand flow rate, v_o (pc/hr)	coefficient a	coefficient b
≤ 200	−0.0014	0.973
400	−0.0022	0.923
600	−0.0033	0.870
800	−0.0045	0.833
1000	−0.0049	0.829
1200	−0.0054	0.825
1400	−0.0058	0.821
≥ 1600	−0.0062	0.817

Note: Straight-line interpolation of a to the nearest 0.0001 and b to the nearest 0.001 is recommended.

Used with permission from *Highway Capacity Manual*, 6th Edition: A Guide for Multimodal Mobility Analysis, 2016, Exhibit 15-20, by the Transportation Research Board of the National Academies of Sciences, Engineering, and Medicine, Washington DC. DOI: 10.17226/24798

Table 2.34 lists the no-passing zone adjustment factors.

Step 7: Estimate the PFFS. This step applies only to class III two-lane highways. Percent of free-flow speed (PFFS) is estimated from Eq. 2.30 [*HCM* Eq. 15-11].

$$\text{PFFS} = \frac{\text{ATS}_d}{\text{FFS}} \quad 2.30$$

Step 8: Determine LOS and Capacity. For the automobile mode, *HCM* procedures use three measures, or expectations of quality of service: ATS, PTSF, and PFFS. For class I two-lane highways, both ATS and PTSF are considered. ATS is a measure of mobility and determines the average travel time to traverse a segment length. PTSF is a measure of hindrance to maneuvering freedom, due to the inability to pass slower-moving vehicles. When selecting LOS from Table 2.35, should ATS LOS and PTSF LOS differ, the lower of the two levels prevails, with LOS A being the highest level and LOS E being the lowest level. A surrogate measure uses a headway of less than 3.0 sec as the rate at which PTSF becomes the dominant measure of LOS, as well as the approximate percentage of vehicles traveling in platoons.

For class II two-lane highways, ATS is not significant, as PTSF is the predominant measure of maneuvering hindrance. On class III two-lane highways, passing rarely occurs and travel speed is reduced greatly due to the many flow frictions present. For these types of two-lane highways, PFFS is used as the gauge by which service quality is related to driver expectations; this is defined

Table 2.34 No-Passing Zone Adjustment Factor for Determination of PTSF, $f_{np,PTSF}$

total two-way flow rate, $v = v_d + v_o$ (pc/hr)	\multicolumn{6}{c}{percent no-passing zones}
	0	20	40	60	80	100
\multicolumn{7}{c}{directional split = 50/50}
≤ 200	9.0	29.2	43.4	49.4	51.0	52.6
400	16.2	41.0	54.2	61.6	63.8	65.8
600	15.8	38.2	47.8	53.2	55.2	56.8
800	15.8	33.8	40.4	44.0	44.8	46.6
1400	12.8	20.0	23.8	26.2	27.4	28.6
2000	10.0	13.6	15.8	17.4	18.2	18.8
2600	5.5	7.7	8.7	9.5	10.1	10.3
3200	3.3	4.7	5.1	5.5	5.7	6.1
\multicolumn{7}{c}{directional split = 60/40}
≤ 200	11.0	30.6	41.0	51.2	52.3	53.5
400	14.6	36.1	44.8	53.4	55.0	56.3
600	14.8	36.9	44.0	51.1	52.8	54.6
800	13.6	28.2	33.4	38.6	39.9	41.3
1400	11.8	18.9	22.1	25.4	26.4	27.3
2000	9.1	13.5	15.6	16.0	16.8	17.3
2600	5.9	7.7	8.6	9.6	10.0	10.2
\multicolumn{7}{c}{directional split = 70/30}
≤ 200	9.9	28.1	38.0	47.8	48.5	49.0
400	10.6	30.3	38.6	46.7	47.7	48.8
600	10.9	30.9	37.5	43.9	45.4	47.0
800	10.3	23.6	28.4	33.3	34.5	35.5
1400	8.0	14.6	17.7	20.8	21.6	22.3
2000	7.3	9.7	11.7	13.3	14.0	14.5
\multicolumn{7}{c}{directional split = 80/20}
≤ 200	8.9	27.1	37.1	47.0	47.4	47.9
400	6.6	26.1	34.5	42.7	43.5	44.1
600	4.0	24.5	31.3	38.1	39.1	40.0
800	3.8	18.5	23.5	28.4	29.1	29.9
1400	3.5	10.3	13.3	16.3	16.9	32.2
2000	3.5	7.0	8.5	10.1	10.4	10.7
\multicolumn{7}{c}{directional split = 90/10}
≤ 200	4.6	24.1	33.6	43.1	43.4	43.6
400	0.0	20.2	28.3	36.3	36.7	37.0
600	−3.1	16.8	23.5	30.1	30.6	31.1
800	−2.8	10.5	15.2	19.9	20.3	20.8
1400	−1.2	5.5	8.3	11.0	11.5	11.9

Note: Straight-line interpolation of $f_{np,PTSF}$ for percent no-passing zones, demand flow rate, and directional split is recommended to the nearest 0.1.

Used with permission from *Highway Capacity Manual*, 6th Edition: A Guide for Multimodal Mobility Analysis, 2016, Exhibit 15-21, by the Transportation Research Board of the National Academies of Sciences, Engineering, and Medicine, Washington DC. DOI: 10.17226/24798

by the LOS scores shown in Table 2.35. For class II and class III two-lane highways, only one measure of LOS is utilized.

Table 2.35 Automobile LOS for Two-Lane Highways

	\multicolumn{2}{c}{class I highways}	class II highways	class III highways
LOS	ATS (mph)	PTSF (%)	PTSF (%)	PFFS (%)
A	> 55	≤ 35	≤ 40	> 91.7
B	> 50–55	> 35–50	> 40–55	> 83.3–91.7
C	> 45–50	> 50–65	> 55–70	> 75.0–83.3
D	> 40–45	> 65–80	> 70–85	> 66.7–75.0
E	≤ 40	> 80	> 85	≤ 66.7
F	\multicolumn{4}{c}{demand exceeds capacity}

Used with permission from *Highway Capacity Manual*, 6th Edition: A Guide for Multimodal Mobility Analysis, 2016, Exhibit 15-3, by the Transportation Research Board of the National Academies of Sciences, Engineering, and Medicine, Washington DC. DOI: 10.17226/24798

When considering passing conditions, driver perceptions of service are affected by *passing capacity* and *passing demand*. As the opposing service volume increases, the ability to pass is diminished. As passing demand increases, platoons form behind slower vehicles. The interplay of passing demand and passing capacity is the primary influence on PTSF.

Capacity

For base conditions, the capacity of a two-lane highway is 1700 pc/hr in one direction. This base capacity is adjusted using Eq. 2.31 [HCM Eq. 15-12] and Eq. 2.32 [HCM Eq. 15-13] for the peak 15 min flow, PHF = 1.00.

$$c_{dATS} = 1700 f_{g,ATS} f_{HV,ATS} \quad 2.31$$

$$c_{dPTSF} = 1700 f_{g,PTSF} f_{HV,PTSF} \quad 2.32$$

For class I highways, the lower of the two capacities prevails. For class II highways, only PTSF capacity is determined. For class III highways, only ATS capacity is determined.

The demand-flow rate base of 1700 pc/hr should be adjusted for grade and heavy vehicle components; however, these adjustments can result in iterative solutions, and the adjustments are unlikely to reduce the demand flow rate to less than 900 vph. Also, from Table 2.35, the two-way flow rate with an ideal 50/50 distribution is considered 3200 pc/hr, which means the highest opposing flow rate reaches an upper limit of 1500 pc/hr.

Heavy Vehicle Factor

Heavy vehicles (defined by the *HCM* as vehicles with more than four tires touching the pavement) in the traffic stream can change the free-flow speed because trucks, buses, and RVs occupy more space and have different performance characteristics than automobiles. A heavy vehicle adjustment factor, f_{HV}, is needed so that the equivalent flow rate can be expressed in pc/hr-ln (pc/h·ln). The heavy vehicle adjustment factor is found from Eq. 2.33.

$$f_{HV} = \frac{1}{1 + P_T(E_T - 1) + P_R(E_R - 1)} \quad 2.33$$
$$[HCM \text{ Eq. 15-4}]$$

E_T and E_R are the passenger car equivalents for trucks and buses and for RVs, respectively. Truck and bus counts can be combined because trucks and buses share similar performance characteristics. When the number of trucks is more than five times the number of RVs for multilane highways and freeways, the RV count can be included with trucks rather than evaluated separately. The type of terrain dictates the E_T and E_R values, as shown in the following section. P_T and P_R are the proportion of trucks and buses and of RVs in the vehicle mix, respectively.

Example 2.4

A two-lane class II highway is used as a scenic and recreational route. The following data apply.

- 950 vph two-way volume
- 4% trucks and buses
- 0% RVs
- 0.85 PHF
- level terrain
- 3 ft shoulder width
- 10 ft lane width
- 60/40 directional split
- 50 mph base speed
- 5.0 mi road length
- 12 access points per mile
- 20% no-passing zones

What is the LOS and capacity in the prevailing direction for this highway?

Solution

Use the procedure from *HCM*.

Step 1: Review the input data. For a class II two-lane highway, average travel speed (ATS) is not used; therefore, the analysis proceeds with percent time-spent-following (PTSF). Field measured FFS is not given; therefore, FFS must be estimated from the given BFFS.

Step 2: Estimate FFS using *HCM* Eq. 15-2, Exhibit 15-7, and Table 2.21 [*HCM* Exhibit 15-8].

$$\begin{aligned} \text{FFS} &= \text{BFFS} - f_{LS} - f_A \\ &= 50 \text{ mph} - 3.7 \text{ mph} - \left(2.5 + \left(\frac{2}{10}\right)(2.5)\right) \\ &= 43.3 \text{ mph} \end{aligned}$$

Find access point density by proportioning speed reduction values shown in Table 2.21.

Steps 3 and 4 apply to class I and class III two-lane highways, but not class II two-lane highways.

Step 5: Adjust for PTSF.

The directional demand adjusted for PHF is

$$v_d = (0.6)\left(\frac{950 \text{ vph}}{0.85}\right) = 671 \text{ vph}$$

The opposing peak directional demand adjusted for PHF is

$$v_o = (0.4)\left(\frac{950 \text{ vph}}{0.85}\right) = 447 \text{ vph}$$

Calculate the heavy vehicle factor, $f_{HV,PTSF}$, using *HCM* Eq. 15-8.

$$\begin{aligned} f_{HV,PTSF} &= \frac{1}{1 + P_T(E_t - 1) + P_R(E_r - 1)} \\ &= \frac{1}{1 + (0.04)(1.0 - 1)} \\ &= 1.0 \end{aligned}$$

For level terrain, $f_{g,PTSF}$, the grade adjustment factor is 1.0.

Calculate the demand flow rate adjusted for grade and heavy vehicles using *HCM* Eq. 15-7.

$$\begin{aligned} v_{i,PTSF} &= \frac{v}{\text{PHF}(f_{g,PTSF})(f_{HV,PTSF})} \\ &= \frac{(0.6)(950 \text{ vph})}{(0.85)(1.0)(1.0)} \\ &= 671 \text{ vph} \end{aligned}$$

$$\begin{aligned} v_{o,PTSF} &= \frac{v}{\text{PHF}(f_{g,PTSF})(f_{HV,PTSF})} \\ &= \frac{(0.4)(950 \text{ vph})}{(0.85)(1.0)(1.0)} \\ &= 447 \text{ vph} \end{aligned}$$

Step 6: Estimate PTSF.

First, estimate $BPTSF_d$ using *HCM* Eq. 15-10. Select coefficient values for a and b from *HCM* Exh. 15-20. Determine no-passing zone adjustment factors from *HCM* Exh. 15-21. Factors are proportioned between values shown in exhibits.

$$BPTSF_d = 100\left(1 - \exp(av_d^b)\right)$$
$$= (100)\left(1 - \exp(-0.0025)(671^{0.9105})\right)$$
$$= 60.7$$

Use *HCM* Eq. 15-9 and Exh. 15-21.

$$PTSF_d = BPTSF_d + f_{np,PTSF}\left(\frac{v_{d,PTSF}}{v_{d,PTSF} + v_{o,PTSF}}\right)$$
$$= 60.7\% + (25.9\%)\left(\frac{671 \text{ vph}}{671 \text{ vph} + 447 \text{ vph}}\right)$$
$$= 76.2\%$$

Step 7: Estimate PFFS. This step applies only to class III two-lane highways.

Step 8: Determine LOS and capacity.

From *HCM* Exh. 15-3, for class II highways with PTSF >70–85%, the LOS is D.

Determine capacity by using *HCM* Eq. 15-13.

$$c_{d,PTSF} = 1700 f_{g,PTSF} f_{HV,PTSF}$$
$$= (1700)(1.0)(1.0) \text{ vph}$$
$$= 1700 \text{ vph}$$

Example 2.5

A two-lane highway has a 70/30 directional split with 40% no-passing zones. The adjusted two-way flow rate is 800 pc/hr. What is the PTSF in the prevailing direction?

Solution

This procedure applies only to class I and class II two-lane highways. PTSF is not used to determine LOS for class III two-lane highways.

$$v_d = 0.70(800 \text{ vph}) = 560 \text{ vph}$$
$$v_o = 0.30(800 \text{ vph}) = 240 \text{ vph}$$

Find PTSF using *HCM* Eq. 15-9 and Eq. 15-10.

Find $BPTSF_d$ using Eq. 15-10. The values for coefficient a and coefficient b are found by interpolating between values in *HCM* Exh. 15-20.

$$BPTSF_d = 100[1 - \exp(av_d^b)]$$
$$= 100[1 - \exp(-0.0016(560)^{0.963})]$$
$$= 50.8\%$$

Use *HCM* Eq 15-9 and *HCM* Exh. 15-21 to find the value for $f_{n,PTSF}$.

$$PTSF_d = BPTSF_d + f_{np,PTSF}\left(\frac{v_{n,PTSF}}{v_{d,PTSF} + v_{o,PTSF}}\right)$$
$$= 50.8\% + 28.4\%\left(\frac{560 \text{ vph}}{560 \text{ vph} + 240 \text{ vph}}\right)$$
$$= 70.7\%$$

PTSF is 70.7% in the prevailing direction.

5. TRAFFIC SIGNALS

The *Manual on Uniform Traffic Control Devices* (*MUTCD*) defines a *traffic signal* as any device that alternately directs traffic, including pedestrians, bicyclists, and vehicles, to stop and then proceed. Traffic signals are used to reduce traffic conflict, crashes, and delays; to reduce police involvement in traffic management; and to improve the overall safety of the road system.

All traffic signal systems have at least three round signals arranged in a vertical configuration. Each signal (also called an *aspect*) is a different color. The top signal is always red, the center signal is always amber (also called yellow), and the bottom signal is always green.

In special, low-clearance cases, a traffic signal system may be arranged horizontally instead of vertically. In these instances, the leftmost signal is red, the center signal is amber, and the rightmost signal is green. A green arrow is often added to the bottom of a signal system to indicate permitted left or right movements (see Fig. 2.10). Arrows may also be added at complex intersections to indicate movements at various angles other than straight ahead.

Figure 2.10 *Common Signal Arrangement*

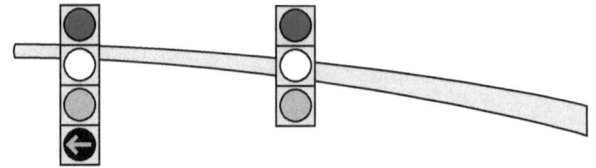

The red aspect indicates all approaching traffic must stop. The green aspect is turned on (and the red aspect is turned off) when one or more approaching traffic streams is permitted to enter the intersection. At the end of the permitted approach time, the amber aspect turns on and the green aspect turns off to indicate that the green interval is about to end. After the red aspect turns on again, the approaching traffic must stop and allow a different traffic stream to enter the intersection.

Traffic Signal Warrants

Traffic signals are a significant tool for managing traffic systems. The design, installation, operation, and maintenance of a signal system involves considerable expense and places substantial liability for proper operation on those having responsibility for the signal. The need for a signal is determined by a traffic engineering study showing that installing a traffic control signal will improve the overall safety and/or operation of the intersection. In addition to meeting the various prescribed warrants, a traffic signal should be installed only when it does not disrupt the progression of the flow of traffic. The study should also describe intersection improvements that would reduce the impact of the signal, such as realigning the horizontal and vertical geometry of the intersection or reconfiguring the approach lanes. The study should also evaluate improvements that could possibly reduce the complexity of the signal, or eliminate the need for a signal.

When a signal is warranted, the usual goal is to minimize delay through the intersection for the total traffic flow. This includes pedestrian, bicycle, and transit flows and the necessary safety factors, such as red-red overlap and right-turn-on-red controls.

A detailed explanation of traffic signal warrants is covered by the *Manual on Uniform Traffic Control Devices* (*MUTCD*), Chap. 4C. To view the complete *MUTCD*, go to http://mutcd.fhwa.dot.gov. This site also lists the known errata.

The following is a summary of the warrants.

- *Warrant 1, Eight-Hour Vehicular Volume:* Satisfying either condition A or condition B, or a combination of both A and B, will satisfy Warrant 1. During the highest eight-hour period of the day, condition A is when the volume on a major street (in both directions) is at least 500 vph for one-lane approaches, or 600 vph for two-lane approaches, and the minor street has at least 150 vph on the highest one-lane approach, or 200 vph on a two-lane approach, during the same eight hours. Condition B is called for when there is interruption to continuous traffic, which is when the major street flow is so heavy that the minor street flow cannot enter without significant delay. For this condition, the major street two-way volume is at least 750 vph for one-lane approaches, or 900 vph for two-lane approaches, and the minor street highest entering flow is at least 75 vph for one-lane approaches, or 100 vph for two-lane approaches. For combinations of condition A and B 80% of the minimum warrants can be used. When the major street approach speed is greater than 40 mph (70 kph) in isolated communities of 10,000 or less, 70% of the minimum warrants can be used for A or B, or 56% for combinations of A and B.

- *Warrant 2, Four-Hour Vehicular Volume:* The same minimum volumes and conditions apply as in Warrant 1, except the volumes occur during the highest four-hour time period. The highest volumes for the major street and the highest volumes for the minor street do not have to occur during the same four-hour period.

- *Warrant 3, Peak Hour:* This warrant applies at certain intersections where traffic on the minor approach experiences undue delay for at least one hour during the average day. These intersections are near facilities—such as office complexes, industrial parks, manufacturing plants, or other vehicle-intensive sites— that have a high volume of vehicle traffic over a short time. Three conditions must be met: vehicle delay on a minor approach is a minimum of four vehicle-hours for one lane and five vehicle-hours for two lanes; four-hour volume must be at least 100 vehicles per hour for one lane and 150 vehicles per hour for two-lane approaches; and the total volume must be at least 650 vehicles per hour for a three-approach intersection, or 800 vehicles per hour for intersections with four or more approaches. Alternative criteria are given in graphical form (*MUTCD* Fig. 4C-3 and Fig. 4C-4), which may be used to satisfy Warrant 3 as well.

- *Warrant 4, Pedestrian Volume:* This warrant applies to both mid-block and pedestrian crossings at intersections, so long as the nearest existing signal is at least 300 ft (90 m) away. In order to apply this warrant, there must be at least 100 pedestrians during each of any four hours, or at least 190 pedestrians during any one hour, and fewer than 60 gaps in the traffic stream large enough for a pedestrian to safely cross the street.

- *Warrant 5, School Crossing:* During the time period that children are using the crossing, this warrant applies if there are fewer adequate crossing gaps in the traffic stream than the minutes required for the children to cross, and there are at least 20 students during the highest crossing hour. The signal is not warranted within 300 ft (90 m) of another signal, unless the school crossing signal is coordinated not to restrict the progressive movement of traffic.

- *Warrant 6, Coordinated Signal System:* On one-way streets, two-way streets with predominantly directional flow, or two-way streets, additional signals can be warranted if the existing signals are too far apart to maintain efficient platooning, as long as the new signals are spaced at least 1000 ft (300 m) from existing signals.

- *Warrant 7, Crash Experience:* This warrant requires three conditions, all of which must be met: (a) trial periods of lesser measures have not reduced crash frequency; (b) five or more reportable crashes that were susceptible to correction by a traffic control signal, and that involved property damage or personal injury, have occurred within a 12-month period; and (c) the intersection experiences 80% of the traffic volumes shown in Warrant 1.

- *Warrant 8, Roadway Network:* Generally, this warrant calls for at least 1000 vehicles per hour during the peak hour of a typical workday, using current traffic or the projected volume within the next five years, and at least one of the following: (a) the intersection is part of a principal network for through traffic; (b) the intersection includes rural or suburban highways near or through a city; and (c) the intersection appears as a major route on an official transportation plan.

The *MUTCD* provides details on the application of these warrants. Simply meeting these warrants' criteria does not mean that the installation (or removal) of a traffic signal is mandatory. Traffic signals, even when they are warranted, can be ineffectively placed, incorrectly operated, or improperly maintained. Such signals create excessive delays, signal disobediences, and increased use of alternate roads to avoid the signal, and/or escalate collision rates. Before a signal is installed or removed, the engineering study must determine improvements that will reduce the impact of the signal. The justification for a signal installation should be the overall effect on the orderly progression and safety of traffic flow. As with any traffic control device, the installation and enforcement has to be backed up by proper legislation, ordinances, and directives.

In general, the idea of traffic signals is to reduce traffic conflicts and delay, to reduce crashes, and to minimize the use of police time, as well as to improve the overall safety of the road system.

Signal Design

Many commercial computer programs are available to organize intersection data and prepare inputs for signal system design. Signal design procedures can also be performed using hand calculations on less complex intersections. *HCM* illustrates a complete set of worksheet input forms that are designed to reduce much of the necessary raw calculation effort. These automated calculation programs are of great help, particularly when working with large intersections and complicated phasing systems. The basics of data analysis remain the same, regardless of the size or intersection complexity. *HCM* shows a step-wise framework for applying the automobile methodology to signalized intersections, as reproduced in Fig. 2.11.

Signal Computational Steps

Step 1: Determine Movement Groups and Lane Groups. On each approach, the general rules are to select the lane groups by

- exclusive through movements with one or more lanes
- shared left-turn and through lanes
- shared right-turn and through lanes
- exclusive left-turn movements with one or more lanes
- exclusive right-turn movements with one or more lanes
- bus, pedestrian, bike delays
- parking activity
- work zones
- downstream blockages

Lane groups are then placed in movement groups in a logical fashion. The rules result in the designation of one to three movement groups for each approach. Figure 2.12 illustrates the typical movement groups and lane groups.

Step 2: Determine Movement Group Flow Rate. The total movement group flow rate is first determined, and then right-turn-on-red (RTOR) and left-turn-on-red (LTOR) flow are subtracted from the movement group flow rates.

Step 3: Determine Lane Group Flow Rate. Lane groups within a movement group may have significant flow rate differences, such that an individual lane group becomes the critical flow, which may not be apparent when working with the total movement group.

When there are two or more lanes on the approach, and one or more shared lanes, the lane group is computed using the procedure described in Chap. 31 of *HCM*. The presumption is that drivers select the approach lane that minimizes their delay at the intersection. The results can vary if drivers preselect an approach lane to anticipate a turn at a downstream intersection. In these conditions, the analysis requires the demand flow rate for each lane on the approach, and the individual lane approach demand can be aggregated for the lane group flow rate. Separating approach lanes into individual lane groups may be particularly desirable if the approach demand for each lane significantly differs because of this prepositioning.

Step 4: Determine the Adjusted Saturation Flow Rate. In automobile mode, for each lane, the base equation used for interrupted flow through a signalized intersection is shown in Eq. 2.34 [see *HCM* Eq. 19-8].

$$s = s_O f_w f_{HVg} f_p f_{bb} f_a f_{LU} f_{LT} f_{RT} f_{Lfb} f_{Rfb} f_{wz} f_{ms} f_{sp} \quad 2.34$$

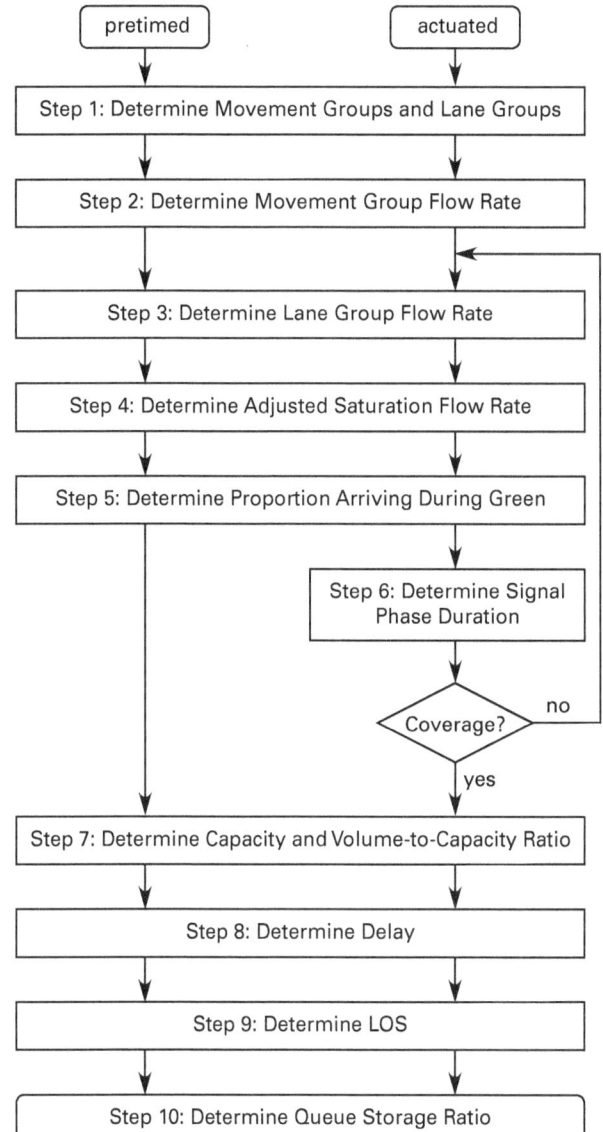

Figure 2.11 Step-by-Step Process of Automobile Signalized Intersection Analysis

Used with permission from *Highway Capacity Manual*, 6th Edition: A Guide for Multimodal Mobility Analysis, 2016, Exhibit 19-18, by the Transportation Research Board of the National Academies of Sciences, Engineering, and Medicine, Washington DC. DOI: 10.17226/24798

s is saturation flow rate, s_0 is base saturation flow rate, and f_w is lane width in feet. For populations greater than or equal to 250,000 use 1900 pcphpl; otherwise use 1750 pcphpl.

Use Table 2.36 to find the lane width factor.

Table 2.36 Lane Width Adjustment Factor, f_w

average lane width (ft)	adjustment factor f_w
<10.0*	0.96
≥10.0–12.9	1.00
>12.9	1.04

*Factors apply to average lane widths of 8.0 ft or more.

Used with permission from *Highway Capacity Manual*, 6th Edition: A Guide for Multimodal Mobility Analysis, 2016, Exhibit 19-20, by the Transportation Research Board of the National Academies of Sciences, Engineering, and Medicine, Washington DC. DOI: 10.17226/24798

From *HCM* Eq. 19-9, for a downgrade approach,

$$f_{\text{HVg}} = \frac{100}{100 + P_{\text{HVg}}} = \frac{100 - 0.79 P_{\text{HV}} - 2.07 P_g}{100} \quad 2.35$$

P_{HV} is the decimal proportion of heavy vehicles in the traffic flow.

From *HCM* Eq. 19-10, for zero approach or an upgrade approach,

$$f_{\text{HVg}} = \frac{100 - 0.78 P_{\text{HV}} - 0.31 P_g^2}{100} \quad 2.36$$

f_p is the curb parking manuevers along the approach lanes. N is the number of lanes in the lane group. N_m is the number of parking maneuvers per hour.

$$N_{m,\max} = 180$$

Equation 2.37 [*HCM* Eq. 19-11] is

$$f_p = \frac{N - 0.1 - \dfrac{18 N_m}{3600}}{N} \geq 0.050 \quad 2.37$$

Table 2.37 lists default parking maneuver rates.

Table 2.37 Default Parking Maneuver Rates

street type	number of spaces in 250 ft	parking time limit (hr)	turnover rate (vph)	maneuver rate (maneuvers/hr)
two-way	10	1	1.0	16
		2	0.5	8
one-way	20	1	1.0	32
		2	0.5	16

Used with permission from *Highway Capacity Manual*, 6th Edition: A Guide for Multimodal Mobility Analysis, 2016, Exhibit 19-16, by the Transportation Research Board of the National Academies of Sciences, Engineering, and Medicine, Washington DC. DOI: 10.17226/24798

Figure 2.12 *Typical Movement Groups and Lane Groups*

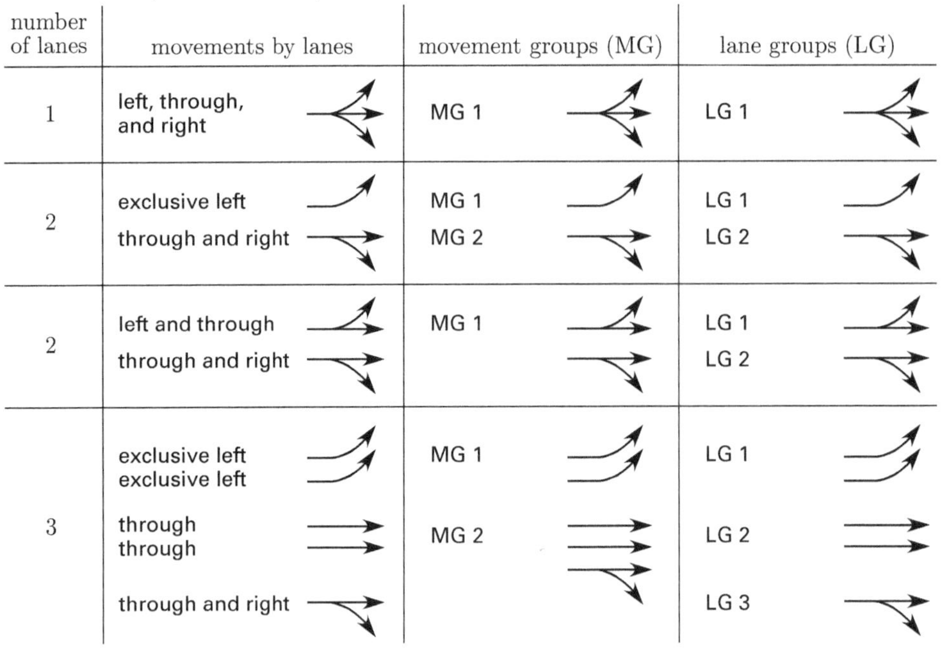

number of lanes	movements by lanes	movement groups (MG)	lane groups (LG)
1	left, through, and right	MG 1	LG 1
2	exclusive left	MG 1	LG 1
	through and right	MG 2	LG 2
2	left and through	MG 1	LG 1
	through and right		LG 2
3	exclusive left, exclusive left	MG 1	LG 1
	through, through	MG 2	LG 2
	through and right		LG 3

Used with permission from *Highway Capacity Manual*, 6th Edition: A Guide for Multimodal Mobility Analysis, 2016, Exhibit 19-19, by the Transportation Research Board of the National Academies of Sciences, Engineering, and Medicine, Washington DC. DOI: 10.17226/24798

f_{bb} is bus blockage, N is the number of lanes in the lane group, and N_b is the number of buses stopping per hour on the approach. The minimum value from the equation is 0.050 when buses are present. The practical upper limit is 250 buses per hour.

Equation 2.38 [*HCM* Eq. 19-12] is

$$f_{bb} = \frac{N - \frac{14.4 N_b}{3600}}{N} \geq 0.050 \quad \quad 2.38$$

f_a is the type. For urban intersections use 0.90; use 1.0 for other locations.

f_{LU} is the lane utilization. For values, refer to Table 2.38.

f_{LT} is the left-turn adjustment factor. For left-turn movements in protected-permitted mode with an exclusive left-turn lane, reduce the left-turn volume, V_{LT}, by two vehicles to account for sneakers. The minimum phase length should allow no fewer than four vehicles per cycle. For a single exclusive left-turn lane use $f_{LT} = 0.95$. For double exclusive protected left-turn lanes use $f_{LT} = 0.92$. For left turn movements unopposed because of one-way street or T-intersection configuration, allow for pedestrians by using corresponding values of 0.85 for one-lane and 0.75 for two lanes. Left-turn volume is entered as zero for exclusive left turns that operate in permitted mode.

Table 2.38 *Default Lane Utilization Adjustment Factor*

movement group	number of lanes in movement group (ln)	traffic in most heavily traveled lane (%)	lane utilization adjustment factor, f_{LU}
exclusive through	1	100.0	1.000
	2	52.5	0.952
	3*	36.7	0.908
exclusive left turn	1	100.0	1.000
	2*	51.5	0.971
exclusive right turn	1	100.0	1.000
	2*	56.5	0.885

*If a movement group has more lanes than shown in this exhibit, it is recommended that field surveys be conducted or that the smallest f_{LU} value shown for that type of movement group be used.

Used with permission from *Highway Capacity Manual*, 6th Edition: A Guide for Multimodal Mobility Analysis, 2016, Exhibit 19-15, by the Transportation Research Board of the National Academies of Sciences, Engineering, and Medicine, Washington DC. DOI: 10.17226/24798

HCM has limited information on how to handle shared-lane volume, since the service volume calculation is applied to each lane or lane group, rather than the total approach. However, the previous method of applying a left-turn factor, f_{LT}, or right-turn factor, f_{RT}, to a shared lane may apply in some cases to take into account the flow friction caused by left or right-turn movements in a shared lane. For these conditions, using Eq. 2.39(a) and 1.05 for the equivalent number of through cars for a protected left-turning vehicle, E_L, or using Eq. 2.39(b) and 1.18 for the equivalent number of through cars for a protected right-turning vehicle, E_R,

$$f_{LT} = \frac{1}{1.05 P_{LT}} \quad \text{2.39(a)}$$

$$f_{RT} = \frac{1}{1.18 P_{RT}} = \frac{1}{E_{RT}} \quad \text{2.39(b)}$$

f_{LT} is then 0.95 and f_{RT} is 0.85.

For pedestrian and bicycle conflicts, use the equations for right turns and left turns, respectively.

$$f_{Rpb} = 1.0 - P_{RT}(1 - A_{pbT})(1 - P_{RTA})$$

$$f_{Lpb} = 1.0 - P_{LT}(1 - A_{pbT})(1 - P_{LTA})$$

In the case of pedestrian-only conflicts,

$$A_{prT} = 1 - \frac{f V_{\text{ped}}}{2000}$$

The value for f is either 0.6 or 1. If the number of turning lanes is the same as the number of receiving lanes, then $f = 0.6$. If the number of turning lanes is smaller than the number of receiving lanes, then $f = 1$.

Each lane is considered individually, as one or more of the flow factors may vary across the approach lane group or movement group. The default values shown are set by the practical limits of the movement conditions. When field observations are not available, other default values are shown, which are found in Table 2.39 (located after Sec. 2.6).

For f_{Lpb}, f_{Rpb}, f_{wz}, f_{ms}, f_{sp}, see *HCM* Chap. 31. If a given condition is not present, the factor is equal to 1.0.

Step 5: Determine Proportion Arriving During Green. Calculate Eq. 2.40 [*HCM* Eq. 19-15].

$$P = R_P \frac{g}{C} \quad \text{2.40}$$

R_P is the platoon ratio, which is the progression quality during the permitted green indication for the indicated phase. Values typically range from 0.3 to 2.0. Use $R_P = 1.0$ if no other information is given. Table 2.40 lists progression quality arrival types. Table 2.41 shows the quality of progression associated with selected platoon ratio values.

For pretimed operation the effective green, g, and the cycle length, C, are known. For actuated phase control, the effective green is determined in the next step.

Table 2.40 Progression Quality Arrival Type

arrival type	signal spacing (ft)	conditions under which arrival type is likely to occur
1	≤1600	coordinated operation on a two-way street where the subject direction does not receive good progression
2	>1600–3200	a less extreme version of arrival type 1
3	>3200	isolated signals or widely spaced coordinated signals
4	>1600–3200	coordinated operation on a two-way street where the subject direction receives good progression
5	≤1600	coordinated operation on a two-way street where the subject direction receives good progression
6	≤800	coordinated operation on a one-way street in dense networks and central business districts

Used with permission from *Highway Capacity Manual*, 6th Edition: A Guide for Multimodal Mobility Analysis, 2016, Exhibit 19-14, by the Transportation Research Board of the National Academies of Sciences, Engineering, and Medicine, Washington DC. DOI: 10.17226/24798

Table 2.41 Selected Values of Platoon Ratio Showing Arrival Type and Progression Quality

arrival type	platoon ratio	progression quality
1	0.33	very poor
2	0.67	unfavorable
3	1.00	random arrivals
4	1.33	favorable
5	1.67	highly favorable
6	2.00	exceptionally favorable

Used with permission from *Highway Capacity Manual*, 6th Edition: A Guide for Multimodal Mobility Analysis, 2016, Exhibit 19-13, by the Transportation Research Board of the National Academies of Sciences, Engineering, and Medicine, Washington DC. DOI: 10.17226/24798

Step 6: Determine Signal Phase Duration. For pretimed signals this step is skipped. For an actuated phase, the duration or the phase is ideally defined in Eq. 2.41 [*HCM* Eq. 19-2]. This procedure can also be used to estimate the length of a pretimed phase.

$$D_p = l_1 + g_s + g_e + Y + R_c \quad 2.41$$

The first period, l_1, is the time lost when the green first appears. This is often called the start-up lost time. The second period, g_s, is the time required to clear the waiting queue. The third period, g_e, is the time extension for random arrivals. The fourth period, Y, is the yellow change interval, and the fifth period, R_c, is the red clearance interval. The effective green becomes Eq. 2.42 [*HCM* Eq. 19-3]

$$g = D_p - l_1 - l_2 = g_s + g_e + e \quad 2.42$$

l_2 is the clearance lost time, $iY + R_c - e$. The extension of effective green, e, is usually 2 sec.

For default values, see *HCM* Exh. 19-17.

Step 7: Determine Capacity and v/c Ratio. The capacity of any lane group, c_i, is determined by Eq. 2.43 [see *HCM* Eq. 19-16].

$$c_i = Ns_i \frac{g_i}{C} \quad 2.43$$

The lane group does not include shared lanes or a permitted left-turn lane group.

The degree of saturation, X_i, is estimated using Eq. 2.44 [*HCM* Eq. 19-17].

$$X_i = \left(\frac{v}{c}\right)_i = \frac{v_i}{s_i\left(\frac{g_i}{C}\right)} = \frac{v_i C}{s_i g_i} \quad 2.44$$

The value of X_i ranges from 1.0 to 0. Values greater than 1.0 indicate an excess of demand over capacity.

For peak periods, evaluating X_c can help in assessing timing problems.

If all $X_i \leq 1$ and $X_c \leq 1$, then the signal timings are adequate, but not necessarily good or optimal. If all $X_i > 1$ and $X_c \leq 1$, then the cycle length is adequate, but the green allocation is incorrect. If $X_i =$ any value and $X_c > 1$, then the cycle length is too short. If several $X_i > 1$ and $X_c > 1$, then another, possibly simpler, phasing may yield improvement. If this fails, and $C > 150$ sec, then the ability of the signal to handle the traffic demand has been exhausted. Measures such as lane addition and left-turn movement deletion are needed.

For intersections without overlapping phases, the intersection can be evaluated by considering critical lane groups for each phase. The critical v/c ratio can be determined by Eq. 2.45 [*HCM* Eq. 19-30] and Eq. 2.46 [*HCM* 19-31].

$$X_c = \sum \left(\frac{v}{s}\right)_{ci}\left(\frac{C}{C-L}\right) = \left(\frac{C}{C-L}\right)\sum_{i \in ci} y_{c,i} \quad 2.45$$

$$L = \sum_{i \in ci} l_{t,i} \quad 2.46$$

The second form is as shown in *HCM*. Usually the lost time is the change period, which is the yellow change interval plus the red clearance interval, or 4 sec for each phase. L is lost time per cycle, computed as lost time, t_L, for the critical path of movements.

For intersections with *pedestrian crossings*, the minimum green interval may be selected by

$$G_{p,\min} = t_{\text{pr}} + \frac{L_{\text{cc}}}{S_p} - Y - R_c \quad 2.47$$

Pedestrian perception time, t_{pr}, is set at 7.0 sec.

The curb-to-curb length of the crosswalk, L_{cc}, is divided by the pedestrian walking speed, S_p, average of 3.5 fps, to obtain the time required for a pedestrian to cross. Procedures for crossing large platoons of pedestrians are covered under pedestrian mode analysis.

Step 8: Determine Delay. The average control delay experienced by all vehicles that arrive in the analysis period determines the LOS for the intersection. The following formula from *HCM* is used to determine the average control delay [*HCM* Eq. 19-18].

$$d = d_1 + d_2 + d_3 \quad 2.48$$

d_2 is the incremental delay to account for effect of random arrivals and oversaturation queues, adjusted for the duration of the analysis period and the type of signal control. This delay component assumes that there is an initial queue for the lane group at the start of the analysis period. d_3 is the initial queue delay, which accounts for delay to all vehicles in the analysis period due to the initial queue at the start of the analysis period (sec/veh).

Uniform Delay, d_1

The *uniform delay*, d_1, is determined assuming uniform arrivals, stable flow, and no initial queue at the beginning of green [*HCM* Eq. 19-19].

$$d_1 = \frac{0.5C\left(1 - \frac{g}{C}\right)^2}{1 - (\min(1, X))\left(\frac{g}{C}\right)} \quad 2.49$$

The value min (1,X) is equal to 1 or X, whichever is less. Cycle length is used in pretimed signal control, or average cycle length for actuated control. Green time is used in pretimed signal control, or average lane group effective green time for actuated control.

For actuated signal control see *HCM*, Chap. 31, for signal timing estimation of parameters for C and g.

Incremental Delay, d_2

The *incremental delay*, d_2, is used to estimate delay due to non-uniform arrivals [see *HCM* Eq. 19-26]. The assumption is that there is no unmet demand that causes initial queues at the start of the analysis period.

For pretimed signals, $k = 0.50$, which is based on random arrivals and uniform service time. For actuated controllers, k can vary from 0.04 to 0.50, depending on the degree of saturation and the unit extension. (See *HCM*, Chap. 19, pages 50-54, for further information.) PT is the passage time, which is the maximum amount of time one vehicle actuation can extend the green interval while the green is displayed.

$$d_2 = 900\,T\left[(X_A-1) + \sqrt{(X_A-1)^2 + \frac{8kIX_A}{cT}}\,\right] \quad 2.50$$

$$k = (1 - 2k_{\min})\left(\frac{v}{c_a} - 0.5\right) + k_{\min} \le 0.50 \quad 2.51$$

$$k_{\min} = -0.375 + 0.354(\text{PT}) - 0.0910(\text{PT})^2 \\ + 0.00889(\text{PT})^3 \quad 2.52$$
$$\ge 0.04$$

$$c_a = 3600\frac{g_a s N}{C} \quad 2.53$$

$$g_a = G_{\max} + Y + R_c - l_1 - l_2 \quad 2.54$$

Initial Queue Delay, q_3

The *initial queue delay*, q_3, is equal to 0.0 for lane groups that have no initial queue. When an initial queue is present, the following equations apply [*HCM* Eq. 19-44]

$$d_3 = \frac{3600}{vT}\left(t_A\frac{Q_b + Q_e - Q_{e0}}{2} + \frac{Q_e^2 - Q_{e0}^2}{2c_A} - \frac{Q_b^2}{2c_A}\right) \quad 2.55$$

$$Q_e = Q_b + t_A(v - c_A)$$

If $v \ge c_A$, then

$$Q_{e0} = T(v - c_A) \\ t_A = T \quad 2.56$$

If $v < c_A$, then

$$Q_{e0} = 0.0 \text{ veh}$$
$$t_A = \frac{Q_b}{c_A - v} \le T \quad 2.57$$

Aggregated Delay Estimates

Lane groups for an approach can be aggregated to provide an estimation of control delay for the entire approach. This aggregation computes weighted averages, where the lane group delays are weighted by the adjusted flows in the lane groups. For an individual approach, use Eq. 2.58 [*HCM* Eq. 19-28].

$$d_{A,j} = \frac{\sum_{i=1}^{m_j} d_i v_i}{\sum_{i=1}^{m_j} v_i} \quad 2.58$$

Control delays for the approaches can be further aggregated to provide an estimate of the average delay for the entire intersection using Eq. 2.59 [*HCM* Eq. 19-29].

$$d_I = \frac{\sum d_i v_i}{\sum v_i} \quad 2.59$$

Step 9: Determine Intersection LOS. Intersection LOS is determined by the average control delay per vehicle. The LOS designations are found in Table 2.42.

Table 2.42 LOS Designation for Automobile Mode Based on Signal Control Delay

LOS ≤ 1.0	control delay (sec/veh)
A	≤ 10
B	>10–20
C	>20–35
D	>35–55
E	>55–80
F	>80

Used with permission from *Highway Capacity Manual*, 6th Edition: A Guide for Multimodal Mobility Analysis, 2016, Exhibit 19-8, by the Transportation Research Board of the National Academies of Sciences, Engineering, and Medicine, Washington DC. DOI: 10.17226/24798

Step 10: Determine Queue Storage Ratio. *HCM* Chap. 31 describes a procedure to evaluate queue storage ratio. The position of the back of queue for arrivals during the previous red signal, indicating the number of vehicles that did not clear the intersection during the previous cycle, plus those that arrived during the previous red signal, is determined. This back of queue position may be the limit of storage. If upon subsequent signal this required back of queue storage space is exceeded, that is the ratio exceeds

1.0, the storage space will overflow and the queued vehicles may block other vehicles from moving forward. The queue storage ratio will exceed 1.0 until the arrival rate reduces to below capacity.

Multiple Time Period Evaluations

The total time period of evaluation, T, should cover several cycles such that the initial queue, Q_b, for the second and subsequent time period is equal to the final queue, Q_e, from the previous period. Ideally, the beginning of the 10-step sequence of evaluation should begin with an undersaturated signal cycle, that is a cycle in which the demand flow rate does not have an initial queue, and the flow is sufficiently low that the entire queue arriving on red can clear the subsequent green phase. The evaluation time period should be extended until the final queue is reduced to 0 or near zero.

It should be noted that v/c on one or more approaches can be > 1.0 while the overall intersection may be determined to have acceptable delay. Also, the intersection may be determined to have unacceptable delay even though v/c on one or more approaches < 1.0. These conditions can result from poor progression, or from flow variations considerably different from the analysis period. When these occur, it may be necessary to evaluate several analysis periods in order to obtain more acceptable timing procedures.

Cycle Length

For coordinated signal systems, compromises are made based on intersection capacity, queue size, phase sequence, segment length, speed, and progression quality. Table 2.43 shows suggested default cycle lengths for several of these conditions.

A more detailed evaluation of phase length and total cycle length is performed in step 7 of the 10-step signalized methodology.

Traffic Circles and Roundabouts

Traffic circles reduce or eliminate the number of conflict points in an intersection. A conventional traffic circle is an intersection with a raised circular island in the center, directing traffic in a one-way (counterclockwise) flow around the circle. All traffic enters and leaves the circle using right-turn movements. Straight through and left-turn movements are accomplished by entering and traveling around the circle, and then turning right at the selected exit. Typical large traffic circles are found in Washington DC, Paris, Rome, and several other major cities.

A *roundabout* is a variation of a traffic circle that is used in smaller business areas and residential neighborhoods. Roundabouts have a smaller radius, resulting in lower operating speeds, and their physical geometry includes yield-at-entry street connections, which encourage "traffic calming." Most often, a roundabout intersection eliminates traffic signals. An added advantage for residential neighborhoods is to discourage large trucks from using residential streets as shortcuts to bypass major street congestion. There is usually a raised paved area with a mountable curb surrounding a central island to allow occasional larger vehicles to travel around without damaging the central island. Figure 2.13 illustrates a typical modern roundabout.

Figure 2.13 A Typical Modern Roundabout

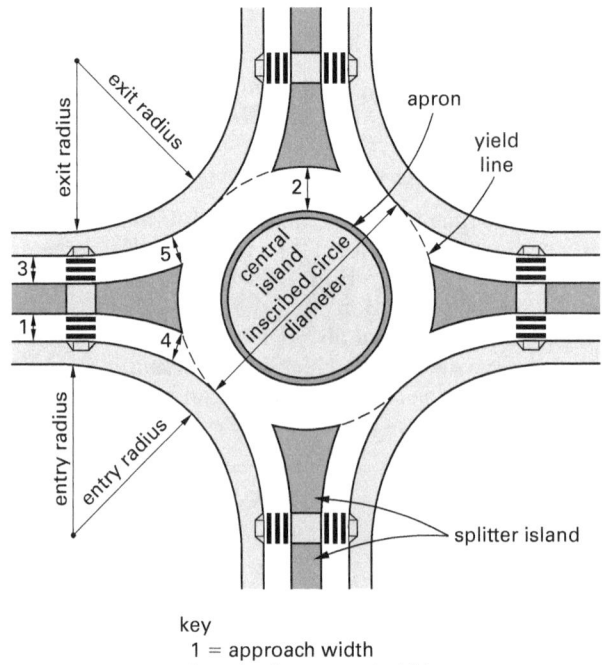

key
1 = approach width
2 = circulatory road width
3 = departure width
4 = entry width
5 = exit width

Source: Wright, Paul H. *Highway Engineering*. John Wiley & Sons, Inc. 2004. This figure is from FHA publications.

Despite their technical differences, the terms "traffic circle," "roundabout," and "rotary" are functionally synonymous.

Roundabout capacity is controlled by the number of conflicting movements at street entrances and exits. The capacity of a single-lane roundabout is limited to 1800 vph entry flow rate, simultaneous with a conflicting flow rate (those vehicles already on the roundabout) of 1800 vph. This is analogous to a saturation flow rate of 3600 vph. Roundabouts typically have less vehicle delay than traffic signals. $c_{e,pce}$, the lane capacity adjusted for heavy vehicles (pc/h), and $v_{c,pce}$, the conflicting flow rate, are related by Eq. 2.60 [*HCM* Eq. 22-1].

$$c_{e,pce} = 1130 e^{(-1.0 \times 10^{-3}) v_{c,pce}} \quad 2.60$$

An example of one-lane entry conflict with one circulating lane is shown in Fig. 2.14.

Table 2.43 Default System Cycle Length

	cycle length by street class and left-turn phasing (sec)[b]					
	major arterial street			minor arterial street or grid network		
average segment length (ft)[a]	no left-turn phases	left-turn phases on one street	left-turn phases on both streets	no left-turn phases	left-turn phases on one street	left-turn phases on both streets
1300	90	120	150	60	80	120
2600	90	120	150	100	100	120
3900	110	120	150			

[a] Average length based on all street segments in the signal system.
[b] Selected left-turn phasing column should describe the phase sequence at the high-volume intersections in the system.
Used with permission from *Highway Capacity Manual*, 6th Edition: A Guide for Multimodal Mobility Analysis, 2016, Exhibit 19-17, by the Transportation Research Board of the National Academies of Sciences, Engineering, and Medicine, Washington DC. DOI: 10.17226/24798

Figure 2.14 One-Lane Entry Conflict with One Circulating Lane

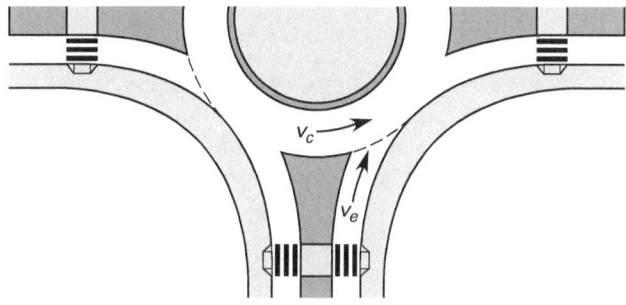

Used with permission from *Highway Capacity Manual*, 6th Edition: A Guide for Multimodal Mobility Analysis, 2016, Exhibit 22-2, by the Transportation Research Board of the National Academies of Sciences, Engineering, and Medicine, Washington DC. DOI: 10.17226/24798.

6. PARKING FACILITIES

Introduction

Parking is an integral part of any transportation system. Parking provides storage for vehicles, and it serves as a place to change transportation modes (e.g., park and ride lots). *Parking facilities* are structures containing multiple parking spaces (e.g., parking lots or parking garages). Parking facilities are often difficult to design and costly to construct, and design of parking facilities is affected by many factors unique to these facilities.

Requirements for parking design are dictated both by local codes and ordinances and by more universal design standards, such as the *International Building Code* (IBC). The Americans with Disabilities Act (ADA) *Standards for Accessible Design* governs design of *accessible parking spaces*, which are an essential design consideration for any parking facility. Other common design requirements are maneuvering room, total capacity, stall size, vertical clearance, and site dimensions.

Users have varying parking needs that must be considered and accommodated when designing a parking facility. For example, commuters who frequently use a given facility will adapt to the facility's design constraints better than tourists or shoppers who use the facility infrequently or only once. If a facility will primarily be used by drivers who are not familiar with the facility, measures such as additional signage may be needed to better accommodate the needs of unfamiliar users. Parking is sometimes segregated by user type (e.g., separate employee parking at a shopping mall) so the design can be tailored more easily to meet the needs of different types of users.

Because parking facilities often have large capital and maintenance costs, user patterns must be studied to ensure that the facility will be adequately used. Like roadway design, parking facility design is concerned with the volume of drivers and often uses the same peak hour analysis method. Trip generation for parking facilities is covered in more detail in the ITE's *Trip Generation*.

Considerations unique to parking facilities include peak arrival rates, peak departure rates, and turnover rates. Because the exact volume of parking facilities is highly variable, peak rates are measured in terms of the percentage of the capacity of a parking facility that arrives or departs over the course of a given hour (e.g., a peak arrival rate of 70 %/hr for a facility with a capacity of 100 vehicles means that 70 vehicles arrive at the facility and park during the peak hour). The *peak arrival rate* is the percentage of the total facility capacity that arrives and parks during the peak hour, and the *peak departure rate* is the percentage of the capacity that leaves the facility during the peak hour. Peak arrival and departure rates vary according to the facility type, as shown in Table 2.44.

Table 2.44 Typical Peak Hour Parking Rates

	percentage of total facility capacity arriving or departing in one hour[a]			
	peak a.m. hour		peak p.m. hour	
land use	arrival (%/hr)	departure (%/hr)	arrival (%/hr)	departure (%/hr)
residential	5–10	30–50	30–50	10–30
hotel/motel	30–50	50–80	30–60	10–30
office	40–70	5–15	5–20	40–70
general retail/restaurant	20–50	30–60	30–60	30–60
convenience retail/ banking	80–150	80–150	80–150	80–150
central business district[b]	20–60	10–60	10–50	20–60
medical office	40–60	50–80	60–80	60–90
hospital				
visitor spaces	30–40	40–50	40–60	50–75
employee spaces	60–75	5–10	10–15	60–75
airport				
short-term (0–3 hr)	50–75	80–100	90–100	90–100
mid-term (3–24 hr)	10–30	5–10	10–30	10–30
long-term (24+ hr)	5–10	5–10	5–10	5–10
special event[c]	80–100	85–200		

[a]As a general rule, the larger the facility and/or the more diverse the tenants of the generating land uses, the lower the peak hour volume as a percentage of the capacity.
[b]It is generally more accurate to determine what fraction of the spaces are allotted to retail, office, and other uses than to use these typical rates.
[c]Assuming 100% of the capacity leaves over the course of 30 min. The equivalent volume in a full hour is 200% of capacity.

Reprinted with permission from Anthony P. Chrest, Mary S. Smith, and Sam Bhuyan, *Parking Structures: Planning Design, Construction, Maintenance, and Repair*, copyright © 2001, by Springer.

Turnover is based on the duration for which vehicles park and varies by the user types and trip purposes associated with the facility. Parking at residential or office parking facilities tends to be long-term (i.e., 6–8 hr), and retail parking tends to be short-term (i.e., less than 3 hr). A high turnover rate means that the parking space can be shared among multiple users throughout the day. Areas with high parking demand, such as busy downtown areas, often promote high turnover by limiting the parking duration allowed at nearby parking facilities.

Capacity of Internal Layouts

The capacity of a parking facility is based on the dimensions of the *parking stalls* (commonly referred to as *parking spaces*). Parking spaces are typically between 8.5 ft and 9.5 ft (2.6 m and 2.9 m) wide and between 17 ft and 18 ft (5.2 m and 5.5 m) long. These dimensions are based on *design vehicles*, which are determined from typical vehicle specifications for a wide range of U.S. vehicles. The American Association of State Highway and Transportation Officials (AASHTO) uses a passenger car design vehicle 7 ft (2.1 m) wide and 19 ft (5.8 m) long. (Design vehicles are covered in more detail in Chap. 4.) However, the selection of stall dimensions should consider land use, turnover, vehicle size, and user needs. For example, the use of shopping carts in a parking facility with high turnover may necessitate greater stall widths to allow quick loading and unloading of packages.

Designs for a parking facility may include the dimensions of a *parking module*, a section of a parking facility that includes one or two rows of parking spaces, as well as an aisle where vehicles can maneuver into and out of spaces. When module dimensions are given instead of parking dimensions, the area required for two facing spaces (the width of a parking space multiplied by the length of a parking module) can be used to calculate capacity; this will ensure that the capacity includes an allowance for the aisles between spaces.

Once the parking stall and module dimensions are selected, the capacity of the parking facility can be determined from the total available area. Before the capacity is estimated, a portion of the total area of the facility must be subtracted to allow for end row turning, exit and entrance ramps, turning loops between floors of a multi-story facility, ticket booths and gates at ticketed facilities, and any other similar components of a facility that subtract from the total space available to park. This allowance can either be calculated as an estimate of the area required for all nonparking components of the facility (e.g., if a ticket booth is known to have an area of 1500 ft^2 (1350 m^2), 1500 ft^2 (1350 m^2) can be subtracted directly from the available area) or estimated as a percentage of the total available area (e.g., 5% of the total facility area will be required for nonparking construction). Once the available area is adjusted to account for these parts of the facility, the remaining available area is divided by the area of a single parking space to find the capacity of the parking facility. This procedure only produces an estimate, but is much faster than preparing a detailed layout.

Example 2.6

A parking facility is to be built on a 120 ft wide × 180 ft long rectangular site. The dimensions of a parking stall are designed as shown in the following illustration. Allowing 5% of the available area for maneuvering space beyond the aisles and parking spaces, how many parking spaces can the parking facility accommodate?

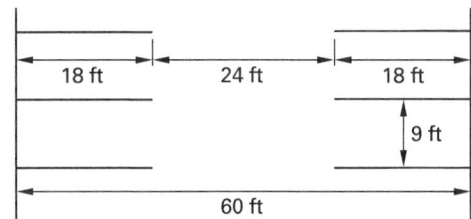

Solution

Parking spaces are 9 ft wide, and one parking module is 60 ft long, so the area required for two facing parking spaces and an aisle is

$$A = bW = (60 \text{ ft})(9 \text{ ft}) = 540 \text{ ft}^2$$

Determine the area of the site.

$$A = bW = (180 \text{ ft})(120 \text{ ft}) = 21{,}600 \text{ ft}^2$$

Subtract the maneuvering areas from the site area to determine the total area available for parking.

$$A_{\text{parking}} = 21{,}600 \text{ ft}^2 - (0.05)(21{,}600 \text{ ft}^2) = 20{,}520 \text{ ft}^2$$

The number of parking spaces available is

$$N_{\text{parking spaces}} = (2)\left(\frac{20{,}520 \text{ ft}^2}{540 \text{ ft}^2}\right) = 76$$

LOS Determination

LOS for parking facilities is based on the delay when attempting to use a given facility, measured in seconds per vehicle. User delay can be caused by a variety of factors that are dependent on actual site conditions and cannot be easily modeled. When a parking facility is designed, LOS is often approached in terms of the delays users will tolerate in the facility rather than designing for an acceptable level of delay.

For simplicity, delay tolerance is examined from the standpoint of users who are visitors to a given facility and users who regularly use the facility (e.g., employees). Table 2.45 shows the levels of service different kinds of users will accept for different design considerations. In general, visitors are more tolerant of longer travel distances and larger numbers of turns, while employees and other frequent users are more tolerant of lower turning radii and lower flow capacities.

Table 2.45 Level of Service Tolerance for Parking Facilities

		acceptable LOS			
design consideration	chief factor	A	B	C	D
turning radii, ramp slopes, etc.	freedom to maneuver		visitors		employees
travel distance, number of turns, etc.	travel time		employees		visitors
geometrics	freedom to maneuver		visitors		employees
flow capacity	v/c ratio		visitors		employees
entries/exits	average wait		employees		visitors

Reprinted with permission from Anthony P. Chrest, Mary S. Smith, and Sam Bhuyan, *Parking Structures: Planning Design, Construction, Maintenance, and Repair*, copyright © 2001, by Springer.

Table 2.46 (located after Sec. 2.6) shows sample design parameters that are commonly found in facilities meeting LOS A through LOS D. LOS E is generally considered unacceptable for parking facilities, as near-maximum flows will not typically be tolerated by users. LOS D is only minimally acceptable but will tend to be tolerated in high-traffic areas like CBDs or parking facilities for major events. Designers tend to design for LOS B or LOS C when designing parking facilities for suburban locations.

Table 2.12 Specific Segment PCEs for Mix of Single Unit Trucks and Tractor Trailers

grade (%)	length (mi)	percentage of trucks (%)								
		2%	4%	5%	6%	8%	10%	15%	20%	>25%
−2	0.125	2.62	2.37	2.30	2.24	2.17	2.12	2.04	1.99	1.97
	0.375	2.62	2.37	2.30	2.24	2.17	2.12	2.04	1.99	1.97
	0.625	2.62	2.37	2.30	2.24	2.17	2.12	2.04	1.99	1.97
	0.875	2.62	2.37	2.30	2.24	2.17	2.12	2.04	1.99	1.97
	1.25	2.62	2.37	2.30	2.24	2.17	2.12	2.04	1.99	1.97
	1.5	2.62	2.37	2.30	2.24	2.17	2.12	2.04	1.99	1.97
0	0.125	2.62	2.37	2.30	2.24	2.17	2.12	2.04	1.99	1.97
	0.375	2.62	2.37	2.30	2.24	2.17	2.12	2.04	1.99	1.97
	0.625	2.62	2.37	2.30	2.24	2.17	2.12	2.04	1.99	1.97
	0.875	2.62	2.37	2.30	2.24	2.17	2.12	2.04	1.99	1.97
	1.25	2.62	2.37	2.30	2.24	2.17	2.12	2.04	1.99	1.97
	1.5	2.62	2.37	2.30	2.24	2.17	2.12	2.04	1.99	1.97
2	0.125	2.62	2.37	2.24	2.24	2.17	2.12	2.04	1.99	1.97
	0.375	3.76	2.96	2.78	2.65	2.48	2.38	2.22	2.14	2.09
	0.375	3.76	2.96	2.78	2.65	2.48	2.38	2.22	2.14	2.09
	0.875	4.80	3.50	3.22	3.03	2.77	2.61	2.39	2.28	2.21
	1.25	5.00	3.60	3.30	3.09	2.83	2.66	2.42	2.30	2.23
	1.5	5.04	3.62	3.32	3.11	2.84	2.67	2.43	2.31	2.23
2.5	0.125	2.62	2.37	2.30	2.24	2.17	2.12	2.04	1.99	1.97
	0.375	4.11	3.14	2.93	2.78	2.58	2.46	2.28	2.19	2.13
	0.625	5.04	3.62	3.32	3.11	2.84	2.67	2.43	2.31	2.23
	0.875	5.48	3.85	3.51	3.27	2.96	2.77	2.50	2.36	2.28
	1.25	5.73	3.98	3.61	3.36	3.03	2.83	2.54	2.40	2.31
	1.5	5.80	4.02	3.64	3.38	3.05	2.84	2.55	2.41	2.32
3.5	0.125	2.62	2.37	2.30	2.24	2.17	2.12	2.04	1.99	1.97
	0.375	4.88	3.54	3.25	3.05	2.80	2.63	2.41	2.29	2.22
	0.625	6.34	4.30	3.87	3.58	3.20	2.97	2.64	2.48	2.38
	0.875	7.03	4.66	4.16	3.83	3.39	3.12	2.76	2.57	2.46
	1.25	7.44	4.87	4.33	3.97	3.50	3.22	2.82	2.62	2.50
	1.5	7.53	4.92	4.38	4.01	3.53	3.24	2.84	2.63	2.51
4.5	0.125	2.62	2.37	2.30	2.24	2.17	2.12	2.04	1.99	1.97
	0.375	5.80	4.02	3.64	3.38	3.05	2.84	2.55	2.41	2.32
	0.625	7.90	5.11	4.53	4.14	3.63	3.32	2.90	2.68	2.55
	0.875	8.91	5.64	4.96	4.50	3.92	3.56	3.07	2.82	2.67
	1	9.19	5.78	5.08	4.60	3.99	3.62	3.11	2.85	2.70
5.5	0.125	2.62	2.37	2.30	2.24	2.17	2.12	2.04	1.99	1.97
	0.375	6.87	4.58	4.10	3.77	3.35	3.09	2.73	2.55	2.44
	0.625	9.78	6.09	5.33	4.82	4.16	3.76	3.21	2.93	2.77
	0.875	11.20	6.83	5.94	5.33	4.56	4.09	3.45	3.12	2.93
	1	11.60	7.04	6.11	5.47	4.67	4.18	3.51	3.17	2.97
6	0.125	2.62	2.37	2.30	2.24	2.17	2.12	2.04	1.99	1.97
	0.375	7.48	4.90	4.36	3.99	3.52	3.23	2.83	2.63	2.51
	0.625	10.87	6.66	5.79	5.21	4.46	4.01	3.39	3.08	2.89
	0.875	12.54	7.54	6.51	5.81	4.94	4.40	3.67	3.30	3.08
	1	13.02	7.78	6.71	5.99	5.07	4.51	3.75	3.37	3.14

Note: Interpolation is permitted.

Used with permission from *Highway Capacity Manual*, 6th Edition: A Guide for Multimodal Mobility Analysis, 2016, Exhibits 12-26, 12-27, 12-28, by the Transportation Research Board of the National Academies of Sciences, Engineering, and Medicine, Washington DC. DOI: 10.17226/24798

Table 2.16 Generalized DSV for Rural Multilane Highways, Based on 1000 VPD Volume

K	D	four-lane highways				six-lane highways				eight-lane highways			
		LOS B	LOS C	LOS D	LOS E	LOS B	LOS C	LOS D	LOS E	LOS B	LOS C	LOS D	LOS E
						level terrain							
0.08	0.50	42.4	60.9	75.8	86.4	63.6	91.3	113.7	129.6	84.9	121.8	151.6	172.9
	0.55	38.6	55.4	68.9	78.6	57.9	83.0	103.4	117.9	77.1	110.7	137.8	157.1
	0.60	35.4	50.7	63.2	72.0	53.0	76.1	94.8	108.0	70.7	101.5	126.3	144.0
	0.65	32.6	46.8	58.3	66.5	49.0	70.3	87.5	99.7	65.3	93.7	116.6	133.0
0.09	0.50	37.7	54.1	67.4	76.8	56.6	81.2	101.1	115.2	75.4	108.2	134.8	153.7
	0.55	34.3	49.2	61.3	69.8	51.4	73.8	91.9	104.8	68.6	98.4	122.5	139.7
	0.60	31.4	45.1	56.2	64.0	47.1	67.7	84.2	96.0	62.9	90.2	112.3	128.0
	0.65	29.0	41.6	51.8	59.1	43.5	62.4	77.7	88.6	58.0	83.3	103.7	118.2
0.10	0.50	33.9	48.7	60.6	69.1	50.9	73.1	91.0	103.7	67.9	97.4	121.3	138.3
	0.55	30.9	44.3	55.1	62.9	46.3	66.4	82.7	94.3	61.7	88.6	110.3	125.7
	0.60	28.3	40.6	50.5	57.6	42.4	60.9	75.8	86.4	56.6	81.2	101.1	115.2
	0.65	26.1	37.5	46.6	53.2	39.2	56.2	70.0	79.8	52.2	74.9	93.3	106.4
0.11	0.50	30.9	44.3	55.1	62.9	46.3	66.4	82.7	94.3	61.7	88.6	110.3	125.7
	0.55	28.1	40.3	50.1	57.1	42.1	60.4	75.2	85.7	56.1	80.5	100.2	114.3
	0.60	25.7	36.9	45.9	52.4	38.6	55.4	68.9	78.6	51.4	73.8	91.9	104.8
	0.65	23.7	34.1	42.4	48.4	35.6	51.1	63.6	72.5	47.5	68.1	84.8	96.7
0.12	0.50	28.3	40.6	50.5	57.6	42.4	60.9	75.8	86.4	56.6	81.2	101.1	115.2
	0.55	25.7	36.9	45.9	52.4	38.6	55.4	68.9	78.6	51.4	73.8	91.9	104.8
	0.60	23.6	33.8	42.1	48.0	35.4	50.7	63.2	72.0	47.1	67.7	84.2	96.0
	0.65	21.8	31.2	38.9	44.3	32.6	46.8	58.3	66.5	43.5	62.4	77.7	88.6
						rolling terrain							
0.08	0.50	38.3	55.0	68.5	78.1	57.5	82.5	102.7	117.1	76.6	110.0	136.9	156.1
	0.55	34.8	50.0	62.2	71.0	52.3	75.0	93.4	106.5	69.7	100.0	124.5	141.9
	0.60	31.9	45.8	57.1	65.1	47.9	68.7	85.6	97.6	63.9	91.7	114.1	130.1
	0.65	29.5	42.3	52.7	60.0	44.2	63.5	79.0	90.1	59.0	84.6	105.3	120.1
0.09	0.50	34.1	48.9	60.9	69.4	51.1	73.3	91.3	104.1	68.1	97.8	121.7	138.8
	0.55	31.0	44.4	55.3	63.1	46.5	66.7	83.0	94.6	61.9	88.9	110.7	126.2
	0.60	28.4	40.7	50.7	57.8	42.6	61.1	76.1	86.7	56.8	81.5	101.4	115.7
	0.65	26.2	37.6	46.8	53.4	39.3	56.4	70.2	80.1	52.4	75.2	93.6	106.8
0.10	0.50	30.7	44.0	54.8	62.5	46.0	66.0	82.2	93.7	61.3	88.0	109.6	124.9
	0.55	27.9	40.0	49.8	56.8	41.8	60.0	74.7	85.2	55.7	80.0	99.6	113.5
	0.60	25.5	36.7	45.6	52.0	38.3	55.0	68.5	78.1	51.1	73.3	91.3	104.1
	0.65	23.6	33.8	42.1	48.0	35.4	50.8	63.2	72.1	47.2	67.7	84.3	96.1
0.11	0.50	27.9	40.0	49.8	56.8	41.8	60.0	74.7	85.2	55.7	80.0	99.6	113.5
	0.55	25.3	36.4	45.3	51.6	38.0	54.5	67.9	77.4	50.7	72.7	90.5	103.2
	0.60	23.2	33.3	41.5	47.3	34.8	50.0	62.2	71.0	46.5	66.7	83.0	94.6
	0.65	21.4	30.8	38.3	43.7	32.2	46.1	57.5	65.5	42.9	61.5	76.6	87.3
0.12	0.50	25.5	36.7	45.6	52.0	38.3	55.0	68.5	78.1	51.1	73.3	91.3	104.1
	0.55	23.2	33.3	41.5	47.3	34.8	50.0	62.2	71.0	46.5	66.7	83.0	94.6
	0.60	21.3	30.6	38.0	43.4	31.9	45.8	57.1	65.1	42.6	61.1	76.1	86.7
	0.65	19.7	28.2	35.1	40.0	29.5	42.3	52.7	60.0	39.3	56.4	70.2	80.1

Note: Key assumptions: 12% trucks, PHF = 0.88, FFS = 60 mi/h.

Used with permission from *Highway Capacity Manual*, 6th Edition: A Guide for Multimodal Mobility Analysis, 2016, Exhibit 12-42, by the Transportation Research Board of the National Academies of Sciences, Engineering, and Medicine, Washington DC. DOI: 10.17226/24798

Table 2.17 Generalized DSV for Urban Multilane Highways, Based on 1000 VPD Volume

K	D	four-lane highways				six-lane highways				eight-lane highways			
		LOS B	LOS C	LOS D	LOS E	LOS B	LOS C	LOS D	LOS E	LOS B	LOS C	LOS D	LOS E
						level terrain							
0.08	0.50	47.5	68.2	84.9	96.8	71.3	102.2	127.3	145.1	95.0	136.3	169.7	193.5
	0.55	43.2	62.0	77.2	88.0	64.8	93.0	115.7	131.9	86.4	123.9	154.3	175.9
	0.60	39.6	56.8	70.7	80.6	59.4	85.2	106.1	120.9	79.2	113.6	141.4	161.3
	0.65	36.5	52.4	65.3	74.4	54.8	78.7	97.9	111.6	73.1	104.9	130.6	148.9
0.09	0.50	42.2	60.6	75.4	86.0	63.3	90.9	113.2	129.0	84.4	121.2	150.9	172.0
	0.55	38.4	55.1	68.6	78.2	57.6	82.6	102.9	117.3	76.8	110.2	137.2	156.4
	0.60	35.2	50.5	62.9	71.7	52.8	75.7	94.3	107.5	70.4	101.0	125.7	143.3
	0.65	32.5	46.6	58.0	66.2	48.7	69.9	87.0	99.2	65.0	93.2	116.1	132.3
0.10	0.50	38.0	54.5	67.9	77.4	57.0	81.8	101.8	116.1	76.0	109.1	135.8	154.8
	0.55	34.5	49.6	61.7	70.4	51.8	74.4	92.6	105.6	69.1	99.1	123.4	140.7
	0.60	31.7	45.4	56.6	64.5	47.5	68.2	84.9	96.8	63.3	90.9	113.2	129.0
	0.65	29.2	41.9	52.2	59.5	43.8	62.9	78.3	89.3	58.5	83.9	104.5	119.1
0.11	0.50	34.5	49.6	61.7	70.4	51.8	74.4	92.6	105.6	69.1	99.1	123.4	140.7
	0.55	31.4	45.1	56.1	64.0	47.1	67.6	84.2	96.0	62.8	90.1	112.2	127.9
	0.60	28.8	41.3	51.4	58.6	43.2	62.0	77.2	88.0	57.6	82.6	102.9	117.3
	0.65	26.6	38.1	47.5	54.1	39.9	57.2	71.2	81.2	53.1	76.3	95.0	108.3
0.12	0.50	31.7	45.4	56.6	64.5	47.5	68.2	84.9	96.8	63.3	90.9	113.2	129.0
	0.55	28.8	41.3	51.4	58.6	43.2	62.0	77.2	88.0	57.6	82.6	102.9	117.3
	0.60	26.4	37.9	47.1	53.8	39.6	56.8	70.7	80.6	52.8	75.7	94.3	107.5
	0.65	24.4	35.0	43.5	49.6	36.5	52.4	65.3	74.4	48.7	69.9	87.0	99.2
						rolling terrain							
0.08	0.50	44.2	63.5	79.0	90.1	66.3	95.2	118.5	135.1	88.4	126.9	158.0	180.2
	0.55	40.2	57.7	71.8	81.9	60.3	86.5	107.7	122.8	80.4	115.4	143.7	163.8
	0.60	36.9	52.9	65.8	75.1	55.3	79.3	98.8	112.6	73.7	105.8	131.7	150.1
	0.65	34.0	48.8	60.8	69.3	51.0	73.2	91.2	103.9	68.0	97.6	121.6	138.6
0.09	0.50	39.3	56.4	70.2	80.1	59.0	84.6	105.4	120.1	78.6	112.8	140.5	160.2
	0.55	35.7	51.3	63.8	72.8	53.6	76.9	95.8	109.2	71.5	102.6	127.7	145.6
	0.60	32.8	47.0	58.5	66.7	49.1	70.5	87.8	100.1	65.5	94.0	117.1	133.5
	0.65	30.2	43.4	54.0	61.6	45.4	65.1	81.0	92.4	60.5	86.8	108.1	123.2
0.10	0.50	35.4	50.8	63.2	72.1	53.1	76.2	94.8	108.1	70.8	101.5	126.4	144.1
	0.55	32.2	46.2	57.5	65.5	48.2	69.2	86.2	98.3	64.3	92.3	114.9	131.0
	0.60	29.5	42.3	52.7	60.1	44.2	63.5	79.0	90.1	59.0	84.6	105.4	120.1
	0.65	27.2	39.1	48.6	55.4	40.8	58.6	72.9	83.2	54.4	78.1	97.2	110.9
0.11	0.50	32.2	46.2	57.5	65.5	48.2	69.2	86.2	98.3	64.3	92.3	114.9	131.0
	0.55	29.2	42.0	52.2	59.6	43.9	62.9	78.4	89.3	58.5	83.9	104.5	119.1
	0.60	26.8	38.5	47.9	54.6	40.2	57.7	71.8	81.9	53.6	76.9	95.8	109.2
	0.65	24.7	35.5	44.2	50.4	37.1	53.3	66.3	75.6	49.5	71.0	88.4	100.8
0.12	0.50	29.5	42.3	52.7	60.1	44.2	63.5	79.0	90.1	59.0	84.6	105.4	120.1
	0.55	26.8	38.5	47.9	54.6	40.2	57.7	71.8	81.9	53.6	76.9	95.8	109.2
	0.60	24.6	35.3	43.9	50.0	36.9	52.9	65.8	75.1	49.1	70.5	87.8	100.1
	0.65	22.7	32.5	40.5	46.2	34.0	48.8	60.8	69.3	45.4	65.1	81.0	92.4

Note: Key assumptions: 8% trucks, PHF = 0.95, FFS = 60 mi/h.

Used with permission from *Highway Capacity Manual*, 6th Edition: A Guide for Multimodal Mobility Analysis, 2016, Exhibit 12-41, by the Transportation Research Board of the National Academies of Sciences, Engineering, and Medicine, Washington DC. DOI: 10.17226/24798

Table 2.23 ATS Grade Adjustment Factor for Specific Upgrades, $f_{g,\text{ATS}}$

grade (%)	grade length (mi)	directional demand flow rate, v_{vph} (vph)								
		≤100	200	300	400	500	600	700	800	≥900
≥3 <3.5	0.25	0.78	0.84	0.87	0.91	1.00	1.00	1.00	1.00	1.00
	0.50	0.75	0.83	0.86	0.90	1.00	1.00	1.00	1.00	1.00
	0.75	0.73	0.81	0.85	0.89	1.00	1.00	1.00	1.00	1.00
	1.00	0.73	0.79	0.83	0.88	1.00	1.00	1.00	1.00	1.00
	1.50	0.73	0.79	0.83	0.87	0.99	0.99	1.00	1.00	1.00
	2.00	0.73	0.79	0.82	0.86	0.98	0.98	0.99	1.00	1.00
	3.00	0.73	0.78	0.82	0.85	0.95	0.96	0.96	0.97	0.98
	≥4.00	0.73	0.78	0.81	0.85	0.94	0.94	0.95	0.95	0.96
≥3.5 <4.5	0.25	0.75	0.83	0.86	0.90	1.00	1.00	1.00	1.00	1.00
	0.50	0.72	0.80	0.84	0.88	1.00	1.00	1.00	1.00	1.00
	0.75	0.67	0.77	0.81	0.86	1.00	1.00	1.00	1.00	1.00
	1.00	0.65	0.73	0.77	0.81	0.94	0.95	0.97	1.00	1.00
	1.50	0.63	0.72	0.76	0.80	0.93	0.95	0.96	1.00	1.00
	2.00	0.62	0.70	0.74	0.79	0.93	0.94	0.96	1.00	1.00
	3.00	0.61	0.69	0.74	0.78	0.92	0.93	0.94	0.98	1.00
	≥4.00	0.61	0.69	0.73	0.78	0.91	0.91	0.92	0.96	1.00
≥4.5 <5.5	0.25	0.71	0.79	0.83	0.88	1.00	1.00	1.00	1.00	1.00
	0.50	0.60	0.70	0.74	0.79	0.94	0.95	0.97	1.00	1.00
	0.75	0.55	0.65	0.70	0.75	0.91	0.93	0.95	1.00	1.00
	1.00	0.54	0.64	0.69	0.74	0.91	0.93	0.95	1.00	1.00
	1.50	0.52	0.62	0.67	0.72	0.88	0.90	0.93	1.00	1.00
	2.00	0.51	0.61	0.66	0.71	0.87	0.89	0.92	0.99	1.00
	3.00	0.51	0.61	0.65	0.70	0.86	0.88	0.91	0.98	0.99
	≥4.00	0.51	0.60	0.65	0.69	0.84	0.86	0.88	0.95	0.97
≥5.5 <6.5	0.25	0.57	0.68	0.72	0.77	0.93	0.94	0.96	1.00	1.00
	0.50	0.52	0.62	0.66	0.71	0.87	0.90	0.92	1.00	1.00
	0.75	0.49	0.57	0.62	0.68	0.85	0.88	0.90	1.00	1.00
	1.00	0.46	0.56	0.60	0.65	0.82	0.85	0.88	1.00	1.00
	1.50	0.44	0.54	0.59	0.64	0.81	0.84	0.87	0.98	1.00
	2.00	0.43	0.53	0.58	0.63	0.81	0.83	0.86	0.97	0.99
	3.00	0.41	0.51	0.56	0.61	0.79	0.82	0.85	0.97	0.99
	≥4.00	0.40	0.50	0.55	0.61	0.79	0.82	0.85	0.97	0.99
≥6.5	0.25	0.54	0.64	0.68	0.73	0.88	0.90	0.92	1.00	1.00
	0.50	0.43	0.53	0.57	0.62	0.79	0.82	0.85	0.98	1.00
	0.75	0.39	0.49	0.54	0.59	0.77	0.80	0.83	0.96	1.00
	1.00	0.37	0.45	0.50	0.54	0.74	0.77	0.81	0.96	1.00
	1.50	0.35	0.45	0.49	0.54	0.71	0.75	0.79	0.96	1.00
	2.00	0.34	0.44	0.48	0.53	0.71	0.74	0.78	0.94	0.99
	3.00	0.34	0.44	0.48	0.53	0.70	0.73	0.77	0.93	0.98
	≥4.00	0.33	0.43	0.47	0.52	0.70	0.73	0.77	0.91	0.95

Note: Straight-line interpolation of $f_{g,\text{ATS}}$ for length of grade and demand flow is permitted to the nearest 0.01.

Used with permission from *Highway Capacity Manual*, 6th Edition: A Guide for Multimodal Mobility Analysis, 2016, Exhibit 15-10, by the Transportation Research Board of the National Academies of Sciences, Engineering, and Medicine, Washington DC. DOI: 10.17226/24798

Table 2.25 ATS Passenger Car Equivalents for Trucks on Specific Upgrades, E_T

grade (%)	grade length (mi)	\multicolumn{9}{c}{directional demand flow rate, v_{vph} (vph)}								
		≤ 100	200	300	400	500	600	700	800	≥ 900
≥ 3 < 3.5	0.25	2.6	2.4	2.3	2.2	1.8	1.8	1.7	1.3	1.1
	0.50	3.7	3.4	3.3	3.2	2.7	2.6	2.6	1.9	1.6
	0.75	4.6	4.4	4.3	4.2	3.7	3.6	3.4	2.4	1.9
	1.00	5.2	5.0	4.9	4.9	4.4	4.2	4.1	3.0	2.3
	1.50	6.2	6.0	5.9	5.8	5.3	5.0	4.8	3.6	2.9
	2.00	7.3	6.9	6.7	6.5	5.7	5.5	5.3	4.1	3.5
	3.00	8.4	8.0	7.7	7.5	6.5	6.2	6.0	4.6	3.9
	≥ 4.00	9.4	8.8	8.6	8.3	7.2	6.9	6.6	4.8	3.7
≥ 3.5 < 4.5	0.25	3.8	3.4	3.2	3.0	2.3	2.2	2.2	1.7	1.5
	0.50	5.5	5.3	5.1	5.0	4.4	4.2	4.0	2.8	2.2
	0.75	6.5	6.4	6.5	6.5	6.3	5.9	5.6	3.6	2.6
	1.00	7.9	7.6	7.4	7.3	6.7	6.6	6.4	5.3	4.7
	1.50	9.6	9.2	9.0	8.9	8.1	7.9	7.7	6.5	5.9
	2.00	10.3	10.1	10.0	9.9	9.4	9.1	8.9	7.4	6.7
	3.00	11.4	11.3	11.2	11.2	10.7	10.3	10.0	8.0	7.0
	≥ 4.00	12.4	12.2	12.2	12.1	11.5	11.2	10.8	8.6	7.5
≥ 4.5 < 5.5	0.25	4.4	4.0	3.7	3.5	2.7	2.7	2.7	2.6	2.5
	0.50	6.0	6.0	6.0	6.0	5.9	5.7	5.6	4.6	4.2
	0.75	7.5	7.5	7.5	7.5	7.5	7.5	7.5	7.5	7.5
	1.00	9.2	9.2	9.1	9.1	9.0	9.0	9.0	8.9	8.8
	1.50	10.6	10.6	10.6	10.6	10.5	10.4	10.4	10.2	10.1
	2.00	11.8	11.8	11.8	11.8	11.6	11.6	11.5	11.1	10.9
	3.00	13.7	13.7	13.6	13.6	13.3	13.1	13.0	11.9	11.3
	≥ 4.00	15.3	15.3	15.2	15.2	14.6	14.2	13.8	11.3	10.0
≥ 5.5 < 6.5	0.25	4.8	4.6	4.5	4.4	4.0	3.9	3.8	3.2	2.9
	0.50	7.2	7.2	7.2	7.2	7.2	7.2	7.2	7.2	7.2
	0.75	9.1	9.1	9.1	9.1	9.1	9.1	9.1	9.1	9.1
	1.00	10.3	10.3	10.3	10.3	10.3	10.3	10.3	10.2	10.1
	1.50	11.9	11.9	11.9	11.9	11.8	11.8	11.8	11.7	11.6
	2.00	12.8	12.8	12.8	12.8	12.7	12.7	12.7	12.6	12.5
	3.00	14.4	14.4	14.4	14.4	14.3	14.3	14.3	14.2	14.1
	≥ 4.00	15.4	15.4	15.3	15.3	15.2	15.1	15.1	14.9	14.8
≥ 6.5	0.25	5.1	5.1	5.0	5.0	4.8	4.7	4.7	4.5	4.4
	0.50	7.8	7.8	7.8	7.8	7.8	7.8	7.8	7.8	7.8
	0.75	9.8	9.8	9.8	9.8	9.8	9.8	9.8	9.8	9.8
	1.00	10.4	10.4	10.4	10.4	10.4	10.4	10.4	10.3	10.2
	1.50	12.0	12.0	12.0	12.0	11.9	11.9	11.9	11.8	11.7
	2.00	12.9	12.9	12.9	12.9	12.8	12.8	12.8	12.7	12.6
	3.00	14.5	14.5	14.5	14.5	14.4	14.4	14.4	14.3	14.2
	≥ 4.00	15.4	15.4	15.4	15.4	15.3	15.3	15.3	15.2	15.1

Note: Interpolation for length of grade and demand flow rate to the nearest 0.1 is recommended.

Used with permission from *Highway Capacity Manual*, 6th Edition: A Guide for Multimodal Mobility Analysis, 2016, Exhibit 15-12, by the Transportation Research Board of the National Academies of Sciences, Engineering, and Medicine, Washington DC. DOI: 10.17226/24798

Table 2.26 ATS Passenger Car Equivalents for RVs on Specific Upgrades, E_R

grade (%)	grade length (mi)	≤100	200	300	400	500	600	700	800	≥900
≥3 <3.5	≤0.25	1.1	1.1	1.1	1.0	1.0	1.0	1.0	1.0	1.0
	>0.25 ≤0.75	1.2	1.2	1.1	1.1	1.0	1.0	1.0	1.0	1.0
	>0.75 ≤1.25	1.3	1.2	1.2	1.1	1.0	1.0	1.0	1.0	1.0
	>1.25 ≤2.25	1.4	1.3	1.2	1.1	1.0	1.0	1.0	1.0	1.0
	>2.25	1.5	1.4	1.3	1.2	1.0	1.0	1.0	1.0	1.0
≥3.5 <4.5	≤0.75	1.3	1.2	1.2	1.1	1.0	1.0	1.0	1.0	1.0
	>0.75 ≤3.50	1.4	1.3	1.2	1.1	1.0	1.0	1.0	1.0	1.0
	>3.50	1.5	1.4	1.3	1.2	1.0	1.0	1.0	1.0	1.0
≥4.5 <5.5	≤2.50	1.5	1.4	1.3	1.2	1.0	1.0	1.0	1.0	1.0
	>2.50	1.6	1.5	1.4	1.2	1.0	1.0	1.0	1.0	1.0
≥5.5 <6.5	≤0.75	1.5	1.4	1.3	1.1	1.0	1.0	1.0	1.0	1.0
	>0.75 ≤2.50	1.6	1.5	1.4	1.2	1.0	1.0	1.0	1.0	1.0
	>2.50 ≤3.50	1.6	1.5	1.4	1.3	1.2	1.1	1.0	1.0	1.0
	>3.50	1.6	1.6	1.6	1.5	1.5	1.4	1.3	1.2	1.1
≥6.5	≤2.50	1.6	1.5	1.4	1.2	1.0	1.0	1.0	1.0	1.0
	>2.50 ≤3.50	1.6	1.5	1.4	1.2	1.3	1.3	1.3	1.3	1.3
	>3.50	1.6	1.6	1.6	1.5	1.5	1.5	1.4	1.4	1.4

Column header above the data: directional demand flow rate, v_{vph} (vph)

Note: Interpolation in this exhibit is not recommended.

Used with permission from *Highway Capacity Manual*, 6th Edition: A Guide for Multimodal Mobility Analysis, 2016, Exhibit 15-13, by the Transportation Research Board of the National Academies of Sciences, Engineering, and Medicine, Washington DC. DOI: 10.17226/24798

Table 2.27 ATS Passenger Car Equivalents for Trucks on Downgrades Traveling at Crawl Speed, E_{TC}

difference between FFS and truck crawl speed (mph)	≤100	200	300	400	500	600	700	800	≥900
≤15	4.7	4.1	3.6	3.1	2.6	2.1	1.6	1.0	1.0
20	9.9	8.7	7.8	6.7	5.8	4.9	4.0	2.7	1.0
25	15.1	13.5	12.0	10.4	9.0	7.7	6.4	5.1	3.8
30	22.0	19.8	17.5	15.6	13.1	11.6	9.2	6.1	4.1
35	29.0	26.0	23.1	20.1	17.3	14.6	11.9	9.2	6.5
≥40	35.9	32.3	28.6	24.9	21.4	18.1	14.7	11.3	7.9

Column header above the data: directional demand flow rate, v_{vph} (vph)

Note: Interpolation against both speed difference and demand flow rate to the nearest 0.1 is recommended.

Used with permission from *Highway Capacity Manual*, 6th Edition: A Guide for Multimodal Mobility Analysis, 2016, Exhibit 15-14, by the Transportation Research Board of the National Academies of Sciences, Engineering, and Medicine, Washington DC. DOI: 10.17226/24798

Table 2.29 PTSF Grade Adjustment Factor for Level Terrain, Rolling Terrain, and Specific Downgrades, $f_{g,PTSF}$

directional demand flow rate, v_{vph} (vph)	level terrain and specific downgrades	rolling terrain
≤ 100	1.00	0.73
200	1.00	0.80
300	1.00	0.85
400	1.00	0.90
500	1.00	0.96
600	1.00	0.97
700	1.00	0.99
800	1.00	1.00
≥ 900	1.00	1.00

Note: Interpolation to the nearest 0.01 is recommended.

Used with permission from *Highway Capacity Manual*, 6th Edition: A Guide for Multimodal Mobility Analysis, 2016, Exhibit 15-16, by the Transportation Research Board of the National Academies of Sciences, Engineering, and Medicine, Washington DC. DOI: 10.17226/24798

Table 2.30 PTSF Grade Adjustment Factor for Specific Upgrades, $f_{g,PTSF}$

grade (%)	grade length (mi)	directional demand flow rate, v_{vph} (vph)								
		≤ 100	200	300	400	500	600	700	800	≥ 900
$\geq 3 < 3.5$	0.25	1.00	0.99	0.97	0.96	0.92	0.92	0.92	0.92	0.92
	0.50	1.00	0.99	0.98	0.97	0.93	0.93	0.93	0.93	0.93
	0.75	1.00	0.99	0.98	0.97	0.93	0.93	0.93	0.93	0.93
	1.00	1.00	0.99	0.98	0.97	0.93	0.93	0.93	0.93	0.93
	1.50	1.00	0.99	0.98	0.97	0.94	0.94	0.94	0.94	0.94
	2.00	1.00	0.99	0.98	0.98	0.95	0.95	0.95	0.95	0.95
	3.00	1.00	1.00	0.99	0.99	0.97	0.97	0.97	0.96	0.96
	≥ 4.00	1.00	1.00	1.00	1.00	1.00	0.99	0.99	0.97	0.97
$\geq 3.5 < 4.5$	0.25	1.00	0.99	0.98	0.97	0.94	0.93	0.93	0.92	0.92
	0.50	1.00	1.00	0.99	0.99	0.97	0.97	0.97	0.96	0.95
	0.75	1.00	1.00	0.99	0.99	0.97	0.97	0.97	0.96	0.96
	1.00	1.00	1.00	0.99	0.99	0.97	0.97	0.97	0.97	0.97
	1.50	1.00	1.00	0.99	0.99	0.97	0.97	0.97	0.97	0.97
	2.00	1.00	1.00	0.99	0.99	0.98	0.98	0.98	0.98	0.98
	3.00	1.00	1.00	1.00	1.00	1.00	1.00	1.00	1.00	1.00
	≥ 4.00	1.00	1.00	1.00	1.00	1.00	1.00	1.00	1.00	1.00
$\geq 4.5 < 5.5$	0.25	1.00	1.00	1.00	1.00	1.00	0.99	0.99	0.97	0.97
	≥ 0.50	1.00	1.00	1.00	1.00	1.00	1.00	1.00	1.00	1.00
≥ 5.5	all	1.00	1.00	1.00	1.00	1.00	1.00	1.00	1.00	1.00

Note: Interpolation for length of grade and demand flow rate to the nearest 0.01 is recommended.

Used with permission from *Highway Capacity Manual*, 6th Edition: A Guide for Multimodal Mobility Analysis, 2016, Exhibit 15-17, by the Transportation Research Board of the National Academies of Sciences, Engineering, and Medicine, Washington DC. DOI: 10.17226/24798

Table 2.31 PTSF Passenger Car Equivalents for Trucks, E_T, and RVs, E_R, on Specific Upgrades

grade (%)	grade length (mi)	≤ 100	200	300	400	500	600	700	800	≥ 900
		\multicolumn{9}{c}{directional demand flow rate, v_{vph} (vph)}								
\multicolumn{11}{c}{passenger car equivalents for trucks (E_T)}										
≥ 3 < 3.5	≤ 2.00	1.0	1.0	1.0	1.0	1.0	1.0	1.0	1.0	1.0
	3.00	1.5	1.3	1.3	1.2	1.0	1.0	1.0	1.0	1.0
	≥ 4.00	1.6	1.4	1.3	1.3	1.0	1.0	1.0	1.0	1.0
≥ 3.5 < 4.5	≤ 1.00	1.0	1.0	1.0	1.0	1.0	1.0	1.0	1.0	1.0
	1.50	1.1	1.1	1.0	1.0	1.0	1.0	1.0	1.0	1.0
	2.00	1.6	1.3	1.0	1.0	1.0	1.0	1.0	1.0	1.0
	3.00	1.8	1.4	1.1	1.2	1.2	1.2	1.2	1.2	1.2
	≥ 4.00	2.1	1.9	1.8	1.7	1.4	1.4	1.4	1.4	1.4
≥ 4.5 < 5.5	≤ 1.00	1.0	1.0	1.0	1.0	1.0	1.0	1.0	1.0	1.0
	1.50	1.1	1.1	1.1	1.2	1.2	1.2	1.2	1.2	1.2
	2.00	1.7	1.6	1.6	1.6	1.5	1.4	1.4	1.3	1.3
	3.00	2.4	2.2	2.2	2.1	1.9	1.8	1.8	1.7	1.7
	≥ 4.00	3.5	3.1	2.9	2.7	2.1	2.0	2.0	1.8	1.8
≥ 5.5 < 6.5	≤ 0.75	1.0	1.0	1.0	1.0	1.0	1.0	1.0	1.0	1.0
	1.00	1.0	1.0	1.1	1.1	1.2	1.2	1.2	1.2	1.2
	1.50	1.5	1.5	1.5	1.6	1.6	1.6	1.6	1.6	1.6
	2.00	1.9	1.9	1.9	1.9	1.9	1.9	1.9	1.8	1.8
	3.00	3.4	3.2	3.0	2.9	2.4	2.3	2.3	1.9	1.9
	≥ 4.00	4.5	4.1	3.9	3.7	2.9	2.7	2.6	2.0	2.0
≥ 6.5	≤ 0.50	1.0	1.0	1.0	1.0	1.0	1.0	1.0	1.0	1.0
	0.75	1.0	1.0	1.0	1.0	1.1	1.1	1.1	1.0	1.0
	1.00	1.3	1.3	1.3	1.4	1.4	1.5	1.5	1.4	1.4
	1.50	2.1	2.1	2.1	2.1	2.0	2.0	2.0	2.0	2.0
	2.00	2.9	2.8	2.7	2.7	2.4	2.4	2.3	2.3	2.3
	3.00	4.2	3.9	3.7	3.6	3.0	2.8	2.7	2.2	2.2
	≥ 4.00	5.0	4.6	4.4	4.2	3.3	3.1	2.9	2.7	2.5
\multicolumn{11}{c}{passenger car equivalents for RVs (E_R)}										
all	all	1.0	1.0	1.0	1.0	1.0	1.0	1.0	1.0	1.0

Note: Interpolation for length of grade and demand flow rate to the nearest .01 is recommended.

Used with permission from *Highway Capacity Manual*, 6th Edition: A Guide for Multimodal Mobility Analysis, 2016, Exhibit 15-19, by the Transportation Research Board of the National Academies of Sciences, Engineering, and Medicine, Washington DC. DOI: 10.17226/24798

Table 2.39 Default Values for Automobile Mode with Fully or Semi-Actuated Signal Control

data category	input data element	default values
traffic characteristics	demand flow rate	must be provided
	right-turn-on-red flow rate	0.0 vph
	percent heavy vehicles	3%
	intersection peak hour factor	if analysis period is 0.25 hr and hourly data are used: • total entering volume $\geq$ 1000 vph: 0.92 • total entering volume $<$ 1000 vph: 0.90 otherwise: 1.00
	platoon ratio	see discussion
	upstream filtering adjustment factor	1.0
	initial queue	must be provided
	base saturation flow rate	• metropolitan area with population $\geq$ 250,000: 1900 pc/hr/ln • otherwise: 1750 pc/hr/ln
	lane utilization adjustment factor	see discussion
	on-street parking maneuver rate	see discussion
	local bus stopping rate	• central business district: 12 buses/hr • non–central business district: 2 buses/hr
	unsignalized movement delay	see discussion
geometric design	number of lanes	must be provided
	average lane width	12 ft
	number of receiving lanes	must be provided
	turn bay length	must be provided
	presence of on-street parking	must be provided
	approach grade	flat approach: 0%
	(negative for downhill conditions)	moderate grade on approach: 3% steep grade on approach: 6%
signal control	type of signal control	must be provided
	phase sequence	must be provided
	left-turn operational mode	must be provided
	Dallas left-turn phasing option	dictated by local use
	passage time (if actuated)	2.0 sec (presence detection)
	maximum green (if actuated)	• major-street through movement: 50 sec • minor-street through movement: 30 sec • left-turn movement: 20 sec
	green duration (if pretimed)	• major-street through movement: 50 sec • minor-street through movement: 30 sec • left-turn movement: 20 sec
	minimum green	• major-street through movement: 10 sec • minor-street through movement: 8 sec • left-turn movement: 6 sec
	yellow change + red clearance*	4.0 sec
	walk	• actuated: 7.0 sec • pretimed: green interval minus pedestrian clear
	pedestrian clear	based on 3.5 ft/sec walking speed
	phase recall (if actuated)	no recall
	dual entry (if actuated)	not enabled (i.e., use single entry)
	simultaneous gap-out (if actuated)	enabled
	cycle length (if pretimed)	see discussion
other	analysis period duration	0.25 hr
	stop-line detector length	40 ft (presence detection)

*Specific values of yellow change and red clearance should be determined by local guidelines or practice.

Adapted and used with permission from *Highway Capacity Manual*, 6th Edition: A Guide for Multimodal Mobility Analysis, 2016, Exhibit 19-11, by the Transportation Research Board of the National Academies of Sciences, Engineering, and Medicine, Washington DC. DOI: 10.17226/24798

Table 2.46 Typical Recommended Design Parameters to Meet Level of Service Criteria

design standard	LOS A	LOS B	LOS C	LOS D
turning radius, ft[a]	42	36	30	24
driving lane width[b]				
straight				
single lane, ft[c]	12.5	12	11.5	11
multiple lane, ft	11.5	11	10.5	10
24 ft radius				
single lane, ft[d]	14.5	14	13.5	13
multiple lane, ft[e]	14	13.5	13	12.5
clearance from lane, ft[f]	3	2.5	2	1.5
turning bay[g]				
one-way, ft	20.5	19	17.5	16
two-way, concentric, ft[h]	33	31	29	27
two-way, nonconcentric, ft	36	34	32	30
circular helix				
single-threaded[i]				
outside diameter, ft[j]	99	86	73	60
inside diameter, ft[k]	52	42	32	22
slope[l]	4.2%	5.0%	6.1%	7.8%
double-threaded				
outside diameter, ft[j]	112	99	86	73
inside diameter, ft[k]	65	55	45	35
	7.2%	8.3%	9.7%	11.8%
parking ramp slope	3.0%	4.0%	5.0%	6.0%
express ramp slope				
covered	8%	10%	12%	14%
uncovered	6%	8%	10%	12%
speed ramp slope	10%	12%	14%	16%
transition length, ft	13	12	11	10
360° turns to top	2.5	4.0	5.5	7.0
short circuit in long run, ft	250	300	350	400
travel "up" to crossover, ft	300	450	600	750
spaces passed				
angled	400	750	1100	N.R.[m]
perpendicular	250	500	750	N.R.[m]
volume-capacity ratio, v/c[n]	0.6	0.7	0.8	N.R.[m]
walking distance to elevator, ft	150	200	250	300
overhead clearance, ft[o]	10	9	8	7

(Multiply ft by 0.305 to obtain m.)

[a]measured from center to edge of circle traced by outside front wheel of design vehicle
[b]based on AASHTO 1984
[c]add 5 ft (1.5 m) to provide enough space to pass a broken down vehicle
[d]at less than 10 mph (16 kph); add 2 ft (0.6 m) for higher speeds such as in circular helix; add 10 ft (3 m) to allow vehicle to pass breakdown
[e]at less than 10 mph (16 kph); add 2 ft (0.6 m) to each lane for higher speeds
[f]to wall, column, or other obstruction per AASHTO 1984
[g]clear between face of columns; check clearance at parked vehicles with turning template
[h]to be adequate for two-way traffic. If flow is predominately one-way, can reduce 3 ft (0.9 m)
[i]for left-hand turns. Add 5 ft (1.5 m) to diameter for right-hand turns (not applicable to double-threaded designs)
[j]outside-to-outside outer bumper wall
[k]inside-to-inside inner bumper wall using recommended lane widths and clearances above and 6 ft (2 m) walls. Reduce 10 ft (3 m) to provide space to pass breakdown
[l]at centerline of driving lane with 10 ft (3 m) floor-to-floor height
[m]not recommended
[n]volume-capacity ratio; based on the *HCM*.
[o]straight vertical clearance from top of floor surface to underside of overhead obstruction

Reprinted with permission from Anthony P. Chrest, Mary S. Smith, and Sam Bhuyan, *Parking Structures: Planning Design, Construction, Maintenance, and Repair*, copyright © 2001, by Springer.

7. PRACTICE PROBLEMS

1. A six-lane major arterial through a city's central business district is being studied for possible development and improvement. If the development takes place, the anticipated increase in commercial traffic may require additional traffic management measures to minimize delay along the arterial. Therefore, a speed study is being performed along a 2.5 mi long segment to evaluate current speeds and delays along the arterial. If development takes place, the study data will be used as the "before" condition in a before-and-after study. The study segment operates near capacity during peak commuter times and has 14 signalized intersections to allow access to the businesses in the district. Four initial test runs are completed with the results shown.

run no.	elapsed time
1	613 sec
2	484 sec
3	662 sec
4	570 sec

(a) What is most nearly the average running speed of the initial runs?

(A) 15 mph

(B) 16 mph

(C) 19 mph

(D) 23 mph

(b) What is most nearly the average range in running speed?

(A) 3.7 mph

(B) 5.1 mph

(C) 6.5 mph

(D) 7.4 mph

(c) How many additional runs are required for a confidence level of 95%?

(A) 0 runs

(B) 2 runs

(C) 3 runs

(D) 4 runs

2. Interchanges along a 10 mi segment of freeway are spaced at average intervals of 2.5 mi. Each interchange allows full movement between intersecting roadways, with one entrance and one exit ramp on the freeway for each interchange. (See the illustration.) A new interchange is to be built midway between two of the existing interchanges to accommodate a new development project. What is the expected average speed reduction on the freeway caused by the new interchange?

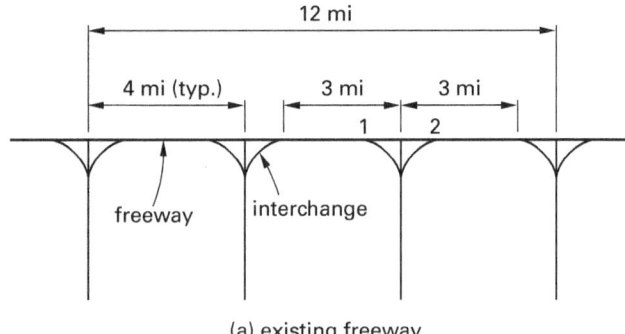

(a) existing freeway

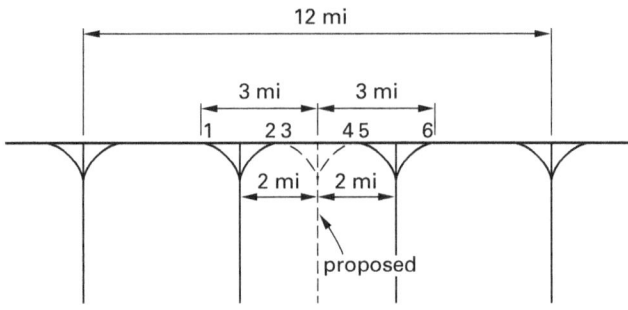

(b) proposed highway interchange

(A) 0.5 mph

(B) 1.4 mph

(C) 3.22 mph

(D) 5.1 mph

3. A 1.7 mi class I two-lane highway segment is on a 3.2% grade. The directional demand flow volume vehicle count is 450 vph in the upgrade direction. There are 20% trucks and no RVs in the traffic flow, and the PHF is 0.85. What is the effective demand flow rate on the upgrade segment?

(A) 390 vph

(B) 775 vph

(C) 855 vph

(D) 1120 vph

4. Which of the following statements are true regarding two-lane highways?

1. Class I and class II two-lane highways serve less developed areas, while class III highways serve areas with increased density of roadside development.

2. Class III two-lane highways occur only in fringe urban areas.

3. Class I two-lane highways serve as major intercity routes or primary connectors of major traffic generators, with travel at relatively high speeds for long-distance trips.

4. Class II two-lane highways do not have passing zones.

5. Class II and class III two-lane highways have all at-grade intersections, while class I two-lane highways have most or all grade-separated intersections.

6. Class III two-lane highways serve moderately developed areas along class I or class II highways, such as through small towns or recreational areas, with a mix of local traffic and with reduced speed limits.

(A) 1, 2, 5

(B) 1, 3, 6

(C) 2, 3, 6

(D) 3, 4, 6

5. A 3 mi section of undivided four-lane highway in level terrain has a 3200 ft long grade of 2%. Lanes are 11 ft wide, with a 4 ft right-side shoulder in both directions. Two-direction traffic volume is 1900 vph, primarily users that are familiar with the highway, with 10% trucks and 5% RVs in the traffic flow. There are 20 access points per mile, and PHF is 0.90. The free-flow speed has been observed at 56 mph. What is the LOS on the 3200 ft long grade?

SOLUTIONS

1. (a) Determine the running speed of each test run using Eq. 2.3.

$$S_1 = \frac{L}{t_{\text{sec},1}} = \frac{(2.5 \text{ mi})\left(3600 \frac{\text{sec}}{\text{hr}}\right)}{613 \text{ sec}}$$
$$= 14.7 \text{ mph}$$

$$S_2 = \frac{L}{t_{\text{sec},2}} = \frac{(2.5 \text{ mi})\left(3600 \frac{\text{sec}}{\text{hr}}\right)}{484 \text{ sec}}$$
$$= 18.6 \text{ mph}$$

$$S_3 = \frac{L}{t_{\text{sec},3}} = \frac{(2.5 \text{ mi})\left(3600 \frac{\text{sec}}{\text{hr}}\right)}{662 \text{ sec}}$$
$$= 13.6 \text{ mph}$$

$$S_4 = \frac{L}{t_{\text{sec},4}} = \frac{(2.5 \text{ mi})\left(3600 \frac{\text{sec}}{\text{hr}}\right)}{570 \text{ sec}}$$
$$= 15.8 \text{ mph}$$

Determine the average speed of all test runs.

$$S_{\text{ave}} = \frac{14.7 \frac{\text{mi}}{\text{hr}} + 18.6 \frac{\text{mi}}{\text{hr}} + 13.6 \frac{\text{mi}}{\text{hr}} + 15.8 \frac{\text{mi}}{\text{hr}}}{4}$$
$$= 15.7 \text{ mi/hr} \quad (16 \text{ mph})$$

The answer is (B).

(b) Determine the average range in running speed using Eq. 2.4.

$$\overline{R} = \frac{|S_1 - S_2| + |S_2 - S_3| + |S_3 - S_4|}{N_t - 1}$$

$$= \frac{\left|14.7 \frac{\text{mi}}{\text{hr}} - 18.6 \frac{\text{mi}}{\text{hr}}\right| + \left|18.6 \frac{\text{mi}}{\text{hr}} - 13.6 \frac{\text{mi}}{\text{hr}}\right| + \left|13.6 \frac{\text{mi}}{\text{hr}} - 15.8 \frac{\text{mi}}{\text{hr}}\right|}{4 - 1}$$

$$= 3.70 \text{ mi/hr} \quad (3.7 \text{ mph})$$

The answer is (A).

(c) Refer to Table 2.1. Since the average range in running speed is 3.7 mph, choose the next highest average range in running speed from Table 2.1, which is 5.0 mph. For economic evaluations, a permitted error of 2.0–4.0 mph is recommended, with 2.0 mph providing greater accuracy. From Table 2.1, for an average range in running speed of 5.0 mph and a permitted error of

2.0 mph, a sample size of four is required. However, should the development take place, the speed study will be used as a "before" condition for any before-and-after study. Therefore, a permitted error of ±1.0–3.0 mph will be required based on the ITE's recommendations. Using a permitted error of 1.0 mph will yield the most accurate results. A total of eight runs are needed to establish a 95% confidence level with a 1.0 mph permitted error. Therefore, four additional test runs are required.

The answer is (D).

2. From Eq. 2.13 [*HCM* Eq. 12-2], the average speed reduction based on ramp density can be found.

$$\text{FFS} = \text{BFFS} - f_{\text{LW}} - f_{\text{LC}} - 3.22\text{TRD}^{0.84}$$

The lane width and lateral clearance are unknown, but are assumed not to change. Total ramp density (TRD) is the number of merge and diverge ramps within a distance of 3 mi upstream and 3 mi downstream of the midpoint of the segment at the new interchange. For the existing interchange spacing, there are two ramps at 4.0 mi spacing.

$$\text{TRD}_{\text{exist}} = \left(\frac{1 \text{ int}}{4.0 \text{ mi}}\right)\left(\frac{2 \text{ ramp}}{\text{int}}\right) = 0.5 \text{ ramp/mi}$$

As proposed, inserting an interchange midway between two of the existing interchanges reduces the spacing to one-half of the existing distance, or 2.0 mi spacing. Included in the distance of 3 mi upstream will be 3 ramps, and downstream will be 3 ramps. Therefore, there will be a total of 6 ramps in the total 6 mi segment.

$$\text{TRD}_{\text{pro}} = \frac{6 \text{ ramps}}{6 \text{ mi}} = 1.0 \text{ ramp/mi}$$

With all other terms remaining constant, the reduction in free-flow speed will be

$$\Delta\text{FFS} = (3.22)(0.5)^{0.84} - (3.22)(1.0)^{0.84}$$
$$= -1.4 \text{ mph} \quad (1.4 \text{ mph})$$

The answer is (B).

3. The presence of trucks in the upgrade vehicle flow increases the passenger car equivalent to an increased demand flow rate. Class I two-lane highways are relatively free of intersection delays; therefore, they are evaluated primarily using average travel speed (ATS) characteristics. Determine the grade adjustment factor ($f_{g,\text{ATS}}$) using Table 2.23 (at the end of Sec. 2.6) [*HCM* Exhibit 15-10].

$f_g = 0.926$ by interpolation.

Determine the heavy vehicle factor, $f_{\text{HV,ATS}}$, using *HCM* Eq. 15-4 and proportioning values shown in *HCM* Exhibit 15-12.

$$f_{\text{HV,ATS}} = \frac{1}{1 + P_T(E_T - 1) + P_R(E_R - 1)}$$
$$= \frac{1}{1 + (0.20)(5.8 - 1) + 0}$$
$$= 0.51$$

Find the estimated demand flow rate using *HCM* Eq. 15-3.

$$v_{i,\text{ATS}} = \frac{V_i}{\text{PHF}(f_{g,\text{ATS}})(f_{\text{HV,ATS}})}$$
$$= \frac{450 \text{ vph}}{0.85(0.926)(0.51)}$$
$$= 1120 \text{ vph}$$

The answer is (D).

4. Statement 2 is incorrect because Class III two-lane highways can occur through small towns in rural areas. Statement 4 is also incorrect since Class II two-lane highways can have passing zones. Finally, statement 5 cannot be true because Class I, II, and III two-lane highways can have a mixture of at-grade or grade-separated intersections; the majority of the intersections are at grade for all three types of highways.

The answer is (B).

5. Convert the demand volume to the demand flow rate in passenger cars per hour using Eq. 2.16 [*HCM* Eq. 12-9].

$$v_p = \frac{V}{\text{PHF} \, N f_{\text{HV}}}$$

f_{HV} is found using Eq. 2.17 [*HCM* Eq. 12-10].

$$f_{\text{HV}} = \frac{1}{1 + P_T(E_T - 1)}$$

The length of grade is

$$\frac{3200 \text{ ft}}{5280 \, \frac{\text{ft}}{\text{mi}}} = 0.61 \text{ mi}$$

It is necessary to find PCEs for both trucks and RVs on the upgrade and on the downgrade.

RVs operate similarly as single-unit trucks (SUT) and are included in the SUT proportion. The total truck volume is 10% + 5% = 15%. The SUT/TT proportion is not given, so assume a base condition of 30% SUT and 70% TT.

Check PCEs for the upgrade and for the downgrade section.

For the upgrade side, interpolating from Table 2.12 [HCM Exh. 12-26], $E_T = 2.30$.

$$f_{HV,\text{upgrade}} = \frac{1}{1 + (0.15)(2.30 - 1)}$$
$$= 0.84$$

For the downgrade side, from Table 2.12 [HCM Exh. 12-26], $E_T = 2.04$.

$$f_{HV,\text{downgrade}} = \frac{1}{1 + (0.15)(2.04 - 1)}$$
$$= 0.86$$

Use the given value of 1900 vph for the demand volume.

$$v_{p,\text{upgrade}} = \frac{1900 \text{ vph}}{(0.90)(2)(0.84)}$$
$$= 1257 \text{ pcpmpl}$$

$$v_{p,\text{downgrade}} = \frac{1900 \text{ vph}}{(0.90)(2)(0.86)}$$
$$= 1227 \text{ pcpmpl}$$

From Table 2.13 [HCM Exh. 12-15], the LOS for both directions is C.

Pedestrian and Mass Transit Analysis

1. Non-Motorized Facilities: Pedestrian 3-3
2. Non-Motorized Facilities-Bicycle 3-25
3. Mass Transit System Studies 3-26
4. Practice Problems 3-37

Nomenclature

a	acceleration	ft/sec²	–
a_1	passenger load weighing factor	–	–
A_p	pedestrian space	ft²/ped	–
c	capacity	–	–
C	cycle length	–	–
CL	1 if trail has a centerline, 0 if no centerline	–	–
d	deceleration, delay	–	–
d_p	average delay	sec	–
d_t	rough vehicle delay	sec/veh	–
d_{pc}	pedestrian delay when crossing segment nearest signalized crossing	sec/p	–
d_{pd}	pedestrian diversion delay	sec/ped	–
d_{pp}	pedestrian delay when walking parallel to segment centerline	sec/ped	–
d_{pw}	pedestrian waiting delay	sec/ped	–
d_{px}	pedestrian crossing delay	sec/ped	–
D_c	diversion distance	ft	–
D_{ped}	density of pedestrian flow	p/ft²	–
DP	minimum of the delayed passings per minute or 1.5	–	–
e	ridership elasticity	–	–
E	weighted events per minute = meetings per minute + 10 × (active passings)	–	–
f	coefficient of side friction (use 0.25)	–	–
F_h	headway factor	–	–
F_l	average passenger load factor	passengers/seat	–
F_m	number of meeting events	events/hr	–
F_p	number of passing events	events/hr	–
F_s	motorized vehicle speed adjustment factor	–	–
F_v	motorized vehicle volume adjustment factor	–	–
F_w	cross-section adjustment factor	–	–
F_{cd}	roadway crossing difficulty factor	–	–
f_{sw}	sidewalk width coefficient	–	–
F_{tt}	perceived travel time factor	–	–
g	green time	–	–
h	headway	–	–
h_2	height of object	ft	m
$I_{p,\text{int}}$	pedestrian LOS score for intersection	–	–
$I_{p,\text{link}}$	pedestrian LOS score for link	–	–
$I_{p,\text{seg}}$	pedestrian LOS score for segment	–	–
L	length of path segment, line of sight (sight distance)	mi	–
L	minimum length of vertical curve	ft	m
L	segment length	ft	–
L_t	total length of urban street segment under analysis	mi	km
M	offset distance to obstruction from center of path to inside of curve	–	–
M_1	meetings per minute of users on the path segment	–	–
N	number	–	–
N_{th}	number of through lanes on segment in the subject direction of travel	–	–
p	proportion	–	–
PHF	peak hour factor	–	–
Q_{ob}	bicycle demand in opposing direction	bicycles/hr	–
Q_{sb}	bicycle demand in same direction	bicycles/hr	–
R	minimum radius of curve, radius of curvature of the center of the path	ft	m
R	red phase time	sec	s
RW	reciprocal of path width = 1/path width	ft	–
S_b	mean bicycle speed	mi/hr	–
S_p	mean pedestrian speed	mi/hr	–
S_{pf}	free-flow pedestrian walking speed	ft/sec	–
S_R	vehicle running speed	mph	–
S_T	travel speed	ft/sec or mi/hr	–
$S_{Tp,\text{seg}}$	travel speed of through pedestrians for segment	ft/sec	–
t, T	time	sec	s
t_c	critical gap	–	–
T	total crosswalk occupancy time	–	–
TS	time-space	–	–
v	pedestrian flow per unit width	ped/ft/min, ped/hr	–
v	normal travel speed	–	–
v_0	initial and final speeds before leaving one station and arriving at next	–	–

Symbol	Description	Units	
v_i	speed of path user of mode i	mph	–
v_I	volume of inbound pedestrians crossing intersection	–	–
v_m	midsegment demand flow rate (direction nearest to subject sidewalk)	vph	–
v_o	volume of outbound pedestrians crossing intersection	–	–
v_p	passenger flow rate	pc/h/ln	–
v_s	transit frequency for segment	veh/hr	–
v_t	total circulating pedestrian flow	–	–
V	design speed or velocity	mph	kph
V_{po}	flow rate of pedestrians in the opposing direction	ped/hr	–
V_{ps}	flow rate of pedestrians in the subject direction	ped/hr	–
v/c	volume-to-capacity ratio	–	–
w_O	effective width of fixed objects	ft	–
W	width of sidewalk, crosswalk, or pathway	ft	–
W_1	total width of shoulder, bicycle lane, and parking lane (see *HCM* Exh. 18-19)	ft	–
W_{aA}	adjusted available sidewalk width	ft	–
W_{buf}	buffer width between roadway and sidewalk	ft	–
W_E	effective sidewalk or crosswalk width	ft	–
W_i	width of intersection (same as length of crosswalk)	ft	–
W_O	adjusted fixed-object effective width	ft	–
W_s	shy distance	ft	–
W_T	total walkway width	ft	–
W_v	effective total width of outside (curbside) through lane, bicycle lane, and shoulder as a function of traffic volume (see *HCM* Exh. 18-19)	ft	–
X	distance of user beyond end of path segment	mi	–

Subscripts

0	at rest
a	acceleration
A	available
at	amenity time or average travel
ave	average
b	bicycle or bike lane
be	benches
btt	base travel time
c	corner, circulating pedestrian, critical gap, or crossing platoon
co	minor street
d	crossing major street or deceleration
do	major street
dt	dwell time
dw	dwell
e	equivalent or exclusive facility
E	effective
ex	excess wait
g	gap
G	group
i	inbound direction, inside (curbside) of crosswalk, or inside (curbside) of sidewalk
j	intersection j
l	loss
LT	left-turn
m	meeting
max	maximum
mi	minor street
min	minimum
mj	major street
N	pedestrian count
o	base condition, opposing, outbound direction, or outside (building side) of sidewalk
occ	occupancy
p	passenger, peak, pedestrian, or platoon
pk	parking
ps	passenger service
pt	passenger trip
ped	pedestrian
ptt	perceived travel time
r	reaction
ret	reentry
Rt	running time
RT	right-turn
s	shared facility, start-up and end clearance, or subject direction
sh	shelters
ST	straight
t	total train, or transit
tco	minor street
tdo	major street
tv	turning vehicle
T	urban street
v	vehicle

1. NON-MOTORIZED FACILITIES: PEDESTRIAN

Pedestrian facilities are areas or structures designed to accommodate pedestrians, either exclusively or in conjunction with nonmotorized and motorized vehicles. Pedestrian facilities include sidewalks, stairways, queuing, off-street paths, and crosswalks. The *Highway Capacity Manual* (*HCM*) provides more information on types of pedestrian facilities.[1]

Pedestrian facilities are evaluated for several types of environments.

- urban walkways along streets (sidewalks)
- interrupted flow at non-signalized crosswalks (considered part of urban street analysis)
- interrupted flow crosswalks at signalized intersections, including platoon-adjusted flow (covered under traffic signals)
- stairways
- off-street walkways (usually 35 ft (10.7 m) or more from a nearby roadway)
- shared-use paths with bicycles
- pedestrians with impaired mobility
- pedestrians using mobility aids

Design of pedestrian facilities is based on optimizing *pedestrian flow rate*, the number of pedestrians that can pass through a given point in a given amount of time (typically measured as either pedestrians per minute or pedestrians per 15 minutes). *Pedestrian flow* is similar to vehicle flow, but is subject to more random and difficult-to-measure variables. This section discusses methods of analyzing pedestrian facilities and determining pedestrian flow for those facilities, as well as those aspects of facility design that must be accounted for by transportation engineers.

Analysis Input Data and Default Values

When analyzing a pedestrian facility, the most reliable data are field-measured values of the pedestrian facility and of the pedestrian flow at the study location. If the facility is not yet constructed, a nearby location that has similar characteristics to the proposed facility can be used for field data. The field study must take into account the flow traffic demographics, regional behavior of pedestrian traffic, geometric conditions, walkway surface, ambient weather and lighting, and other factors that influence traffic flow characteristics. In the absence of field data, default values can be considered. Table 3.1 shows default values that may apply to general situations without known field data.

Walking Speed

Evaluations of pedestrian flow and derivative calculations are in part a function of the average walking speed of pedestrians in the flow. *Average walking speed* (also referred to as *design speed*) is normally measured in feet per minute or feet per second (meters per minute or meters per second) and can be affected by a variety of factors, such as the grade of the walkway or the number of elderly pedestrians or slowly walking children in the pedestrian flow. *Pedestrian trip purpose* can also affect walking speed; studies have found that commuters, students, and people who frequently travel to and from a given location have greater walking speeds than shoppers, children, and older pedestrians.

On normal, level, uncongested sidewalks, average walking speed can be assumed to be 5.0 ft/sec (1.5 m/s). When flow traffic includes elderly persons, up to 20% elderly populations can use 4.0 ft/sec (1.2 m/s), and more than 20% elderly can use 3.5 ft/sec (1.0 m/s). For crosswalks, flow populations of up to 20% elderly can use 3.5 ft/sec (1.0 m/s), and populations with more than 20% elderly should use 3.0 ft/sec (0.9 m/s) average speed. When the walkway has an upgrade of 10% or more, the average speed should be reduced by 0.5 ft/sec (0.2 m/s). Other, less predictable conditions can also affect the average walking speed of pedestrian flow, such as an abundance of items in the path of flow or poor walking surface conditions. Good engineering judgment must be used when determining the average speed of a flow.

For pedestrian facilities where pedestrians will form a line, such as on a mass transit platform or at a signalized intersection, average walking speed must take into account the pedestrian start-up time. *Pedestrian start-up time* is different for each pedestrian, as the pedestrians in the back cannot step forward without interference until the front pedestrians begin to walk. Queues move forward in a *wave pattern* so that as each pedestrian walks, the entire queue moves forward in a fairly ordered progression. The time it takes for the wave progression to move from the front to the back of the queue depends largely on the length of the queue, though an average 3 sec start-up time is used for signal analyses at crosswalks.

The average walking speed is greater in terminals than for crosswalks or street sidewalks. A *terminal* is a transportation facility that serves as a loading and unloading area for passengers, such as a bus depot, train station, or airport. It is often used to transfer between transit modes and vehicles. Terminals are subject to more directed and purposeful flow with little crosswalk interference. Terminal average walking speeds can be increased by 0.5 ft/sec (0.2 m/s) over similar conditions on general sidewalk connections.

The illustrated walking speed, shown in Fig. 3.1, does not include pedestrian start-up time.

---

[1]The Americans with Disabilities Act (ADA) requirements for sidewalks and curb ramps apply to all intersection designs.

Table 3.1 Required Input Data, Potential Data Sources, and Default Values for Off-Street Facility Analysis

required data and units	mode and facility type	potential data sources	suggested default value
facility width	pedestrian (exclusively), bike (all)	field data, aerial photo	must be provided
effective facility width	pedestrian (exclusively)	field data, aerial photo	same as facility width
pedestrian volume	pedestrian (exclusively)	field data	must be provided
bicycle volume	pedestrian (shared)	field data	must be provided
total path volume	bike (all)	field data	must be provided
bicycle mode split	bike (all)	field data	0.55 (i.e., 55% of total path volume)
pedestrian mode split	bike (all)	field data	0.20
runner mode split	bike (all)	field data	0.10
inline skater mode split	bike (all)	field data	0.10
child bicyclist mode split	bike (all)	field data	0.05
peak hour factor[a]	pedestrian (exclusively), pedestrian (shared), bike (all)	field data	0.85
directional volume split[b]	pedestrian (shared), bike (all)	field data	0.50
average pedestrian speed (ft/min)	pedestrian (exclusively)	field data	300 ft/min
average pedestrian speed (mph)	pedestrian (shared), bike (all)	field data	3.4 mph
pedestrian speed standard deviation	bike (all)	field data	0.6 mph
average bicycle speed	pedestrian (shared), bike (all)	field data	12.8 mph
bicycle speed standard deviation	bike (all)	field data	3.4 mph
average runner speed	bike (all)	field data	6.5 mph
runner speed standard deviation	bike (all)	field data	1.2 mph
average inline skater speed	bike (all)	field data	10.1 mph
inline skater speed standard deviation	bike (all)	field data	2.7 mph
average child bicyclist speed	bike (all)	field data	7.9 mph
child bicyclist speed standard deviation	bike (all)	field data	1.9 mph
segment length	bike (all)	field data, aerial photo	must be provided
walkway grade ≤ 5% (yes or no)[c]	pedestrian (exclusively)	field data	must be provided
pedestrian flow type (random or platooned)[d]	pedestrian (exclusively)	field data	must be provided
centerline stripe (yes or no)	bike (all)	field data	must be provided

Notes: ***Bold italic*** indicates high sensitivity (>20% change) of service measure to the choice of value; **bold** indicates moderate sensitivity (10% to 20% change) of service measure to the choice of value.

[a]Not required for pedestrian analysis when peak 15 min demand volumes are provided.
[b]Not required when directional demand volumes are provided.
[c]Pedestrian speeds reduce when grades exceed 5%; the service measure is highly sensitive to average pedestrian speed.
[d]LOS letter result is highly sensitive to the selection of pedestrian flow type.

Used with permission from *Highway Capacity Manual*, 6th Edition: A Guide for Multimodal Mobility Analysis, 2016, Exhibit 24-6, by the Transportation Research Board of the National Academies of Sciences, Engineering, and Medicine, Washington DC. DOI: 10.17226/24798

Figure 3.1 Cumulative Distributions of Walking Speeds at Unsignalized Intersections

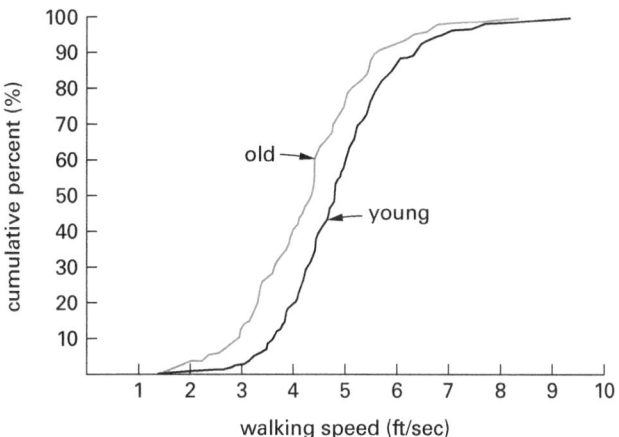

Used with permission from *Highway Capacity Manual*, 6th Edition: A Guide for Multimodal Mobility Analysis, 2016, Exhibit 4-13, by the Transportation Research Board of the National Academies of Sciences, Engineering, and Medicine, Washington DC. DOI: 10.17226/24798

Pedestrian Space Requirements

Average walking speed of pedestrians is directly affected by the amount of space available for each pedestrian to walk in. Design of pedestrian facilities must account for a certain minimum amount of space for each pedestrian.

For the sake of simplicity, a pedestrian is assumed to occupy an oval-shaped body ellipse that represents the minimum space required for the pedestrian to stand and move around without impediment. The elliptical space occupied by a pedestrian standing upright is assumed to be 24 in (0.6 m) wide at the shoulders and 20 in (0.5 m) deep front to back. (See Fig. 3.2(a).) This represents a minimum standard for standing pedestrians.

When walking, normal arm and leg movements will increase the required space both in front and to the sides of a pedestrian. Increasing the width of the ellipse to 2.5 ft (0.8 m) allows for 3 in (8 cm) of extra distance on each side to account for bodily sway, extra berth from carrying packages or avoiding strangers, and other modifications to the pedestrian flow. Pedestrian walkways must be designed to accommodate this 2.5 ft (0.8 m) *minimum walking path*, so that pedestrians can use the space without significant risk of running into obstacles or being forced to slow their walking speed. The minimum walking path is different from the *effective walkway width*, the portion of the walkway that can be used effectively for pedestrian movement. Walkway design must also take the minimum clear width into account, ensuring that proper space is allotted for buffer zones between pedestrians and vehicular traffic or similar environmental concerns.

Design for forward space in a pedestrian facility is not as straightforward as design for lateral space. The amount of forward space a pedestrian requires can be divided

Figure 3.2 Pedestrian Ellipse for Standing Areas and Pedestrian Walking Space Requirements

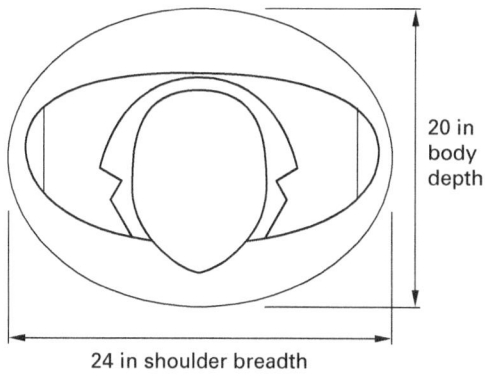

(a) pedestrian body ellipse

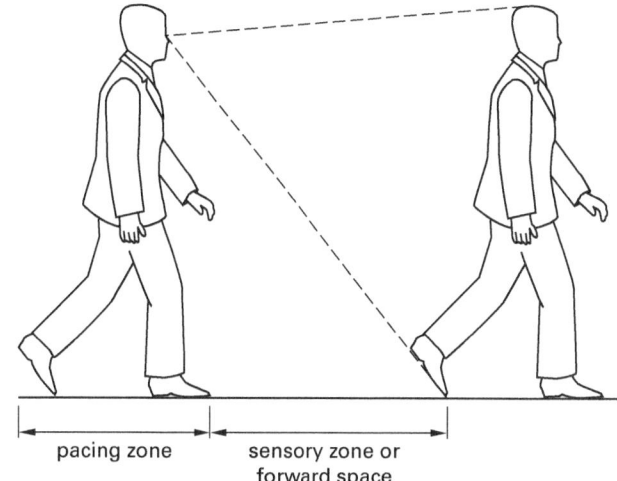

(b) pedestrian walking space requirements

(Multiply in by 2.54 to obtain cm.)

Used with permission from *Highway Capacity Manual*, 6th Edition: A Guide for Multimodal Mobility Analysis, 2016, Exhibit 4-12, by the Transportation Research Board of the National Academies of Sciences, Engineering, and Medicine, Washington DC. DOI: 10.17226/24798

into the *pacing zone*, the area required for a pedestrian to place his or her feet during forward movement, and the *sensory zone*, the area required for a pedestrian to comfortably perceive and react to obstacles and changes in the environment while walking. (See Fig. 3.2(b).) The length of a pacing zone is directly related to the pedestrian's travel speed, but is also dependent on the pedestrian's age, sex, physical condition, and other factors. The sensory zone varies considerably from person to person, depending on a given pedestrian's familiarity with the walkway, the amount of clutter or obstacles in the walkway, the level of lighting, the number of other pedestrians in the flow, and a variety of other factors that could cause a pedestrian to require a greater amount of space in which to perceive and react to changes in the walking environment. For this reason, pedestrian flow must be carefully observed to ensure that the body ellipse used for design does not require modification.

Pedestrian Unit Flow Rate

The *pedestrian unit flow rate* is the number of pedestrians passing a point per unit of time, typically expressed as pedestrians per minute. The pedestrian unit flow rate, v_p, is given by Eq. 3.1. S_p is the average walking speed of pedestrians (*pedestrian speed*) in a given area in feet or meters per minute (found from empirical observation or estimated as explained in the subsection "Walking Speed"), and D_p is *pedestrian density* in the area in pedestrians per square foot or square meter.

$$v_p = S_p D_p \quad [HCM \text{ Eq. 4-11}] \quad 3.1$$

While an increase in pedestrian density will mean an increase in the flow rate, an increase in pedestrian density will also mean a decrease in the physical space available to each pedestrian in the flow, referred to as the *pedestrian space*. As available pedestrian space decreases, pedestrian speed also decreases, due to the decreased room to maneuver around obstacles or slower-moving pedestrians. As pedestrian density increases, average speed decreases, until eventually the average speed of the pedestrian flow is equal to the speed of the slowest person in the flow. Figure 3.3 shows the relationship between pedestrian flow rate and pedestrian space. Figure 3.4 shows the relationship between pedestrian speed and pedestrian space. Figure 3.5 shows the relationship between pedestrian speed and pedestrian density. As shown in Fig. 3.3, the *maximum flow rate* occurs when the available pedestrian space is between 5 ft²/ped and 9 ft²/ped (0.5 m²/ped and 0.8 m²/ped). When available pedestrian space is 5 ft²/ped (0.5 m²/ped) or less, the flow rate drops dramatically, and when the pedestrian space is between 2 ft²/ped and 4 ft²/ped (0.2 m²/ped and 0.4 m²/ped), all flow stops.

Given these correlations, the *HCM* recommends Eq. 3.2 [*HCM* Eq. 4-12] be used to find the pedestrian flow rate rather than Eq. 3.1. A_p is the available pedestrian space, the reciprocal of the pedestrian density, as shown in Eq. 3.3 [*HCM* Eq. 24-4].

$$v_p = \frac{S_p}{M} \quad 3.2$$

$$M = A_p = \frac{1}{D_p} = \frac{S_p}{v_p} \quad 3.3$$

As with the space-flow relationship, *speed-flow* relationships for pedestrian flow are similar to vehicular flow. Under low flow density, the total flow volume increases as speed increases. Increased density causes increased interaction between pedestrians, which in turn causes a reduction in speed. At some point, the total flow volume

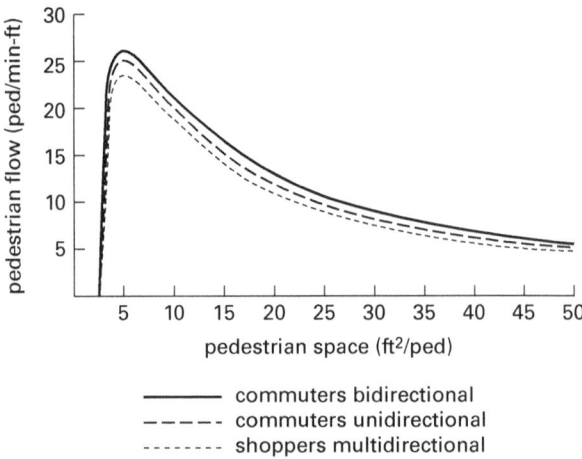

Figure 3.3 Relationships Between Pedestrian Flow and Space

Used with permission from *Highway Capacity Manual*, 6th Edition: A Guide for Multimodal Mobility Analysis, 2016, Exhibit 4-15, by the Transportation Research Board of the National Academies of Sciences, Engineering, and Medicine, Washington DC. DOI: 10.17226/24798

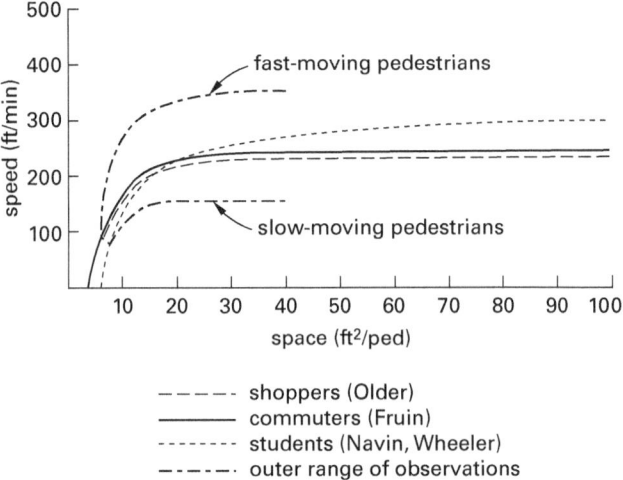

Figure 3.4 Relationships Between Pedestrian Speed and Space per Pedestrian

Used with permission from *Highway Capacity Manual*, 6th Edition: A Guide for Multimodal Mobility Analysis, 2016, Exhibit 4-17, by the Transportation Research Board of the National Academies of Sciences, Engineering, and Medicine, Washington DC. DOI: 10.17226/24798

reaches a peak. As flow volume continues to increase, speeds decline until a critical level of crowding occurs, during which movement becomes very difficult or may stop entirely for brief periods of time. Figure 3.6 illustrates this relationship for several types of pedestrian flow.

Completing the relationships between speed, space, density, and flow rate are pedestrian speed and space. Figure 3.4 illustrates the reduction in speed when space per

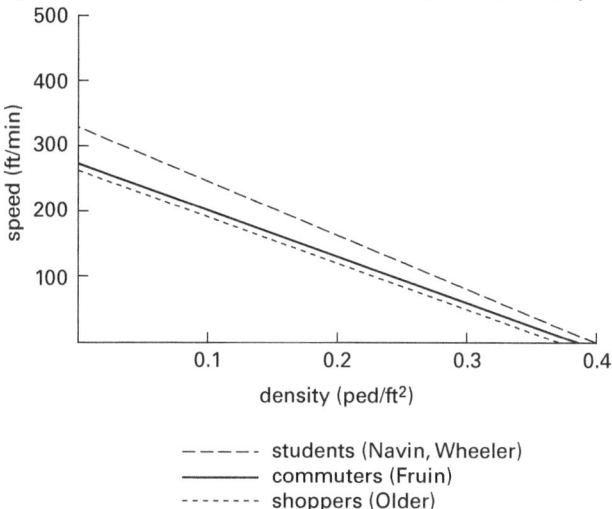

Figure 3.5 Relationship Between Pedestrian Speed and Density

(Multiply ped/min-ft by 0.305 to obtain ped/min·m.)
(Multiply ft²/ped by 0.093 to obtain m²/ped.)

Used with permission from *Highway Capacity Manual*, 6th Edition: A Guide for Multimodal Mobility Analysis, 2016, Exhibit 4-14, by the Transportation Research Board of the National Academies of Sciences, Engineering, and Medicine, Washington DC. DOI: 10.17226/24798

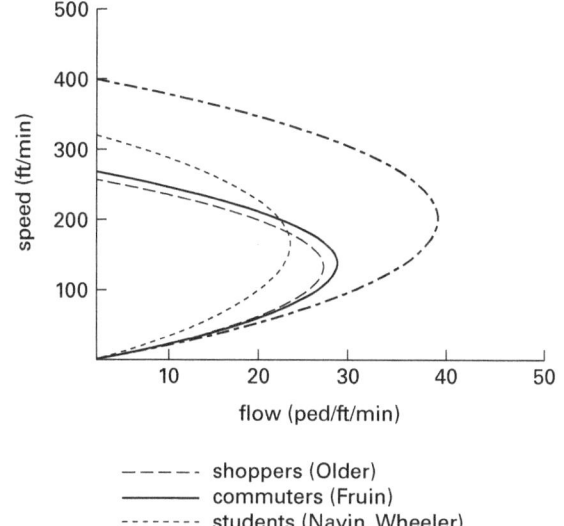

Figure 3.6 Relationships Between Pedestrian Speed and Flow

Used with permission from *Highway Capacity Manual*, 6th Edition: A Guide for Multimodal Mobility Analysis, 2016, Exhibit 4-16, by the Transportation Research Board of the National Academies of Sciences, Engineering, and Medicine, Washington DC. DOI: 10.17226/24798

pedestrian drops below 20 ft²/ped (1.9 m²/ped). Pedestrian speeds tend to converge when space becomes less than 5 ft²/ped (0.5 m²/ped). Figure 3.4 shows speed and space for several categories of pedestrians.

Effective Walkway Width

Effective walkway width is the portion of the walkway that can be used effectively for pedestrian travel. Portions of walkway that cannot be used effectively must be deducted when determining if a walkway meets minimum walking path requirements. For instance, curb edges and vertical surfaces adjacent to walkways, such as building faces, storefronts, walls, and tall vegetation, are not generally used by pedestrians. Sidewalk fixtures and landscaping, such as trash containers, utility poles, signs, fire hydrants, mailboxes, parking meters, or trees, can impede the free movement of pedestrians. Storefront window recesses and column pillars along the sidewalk edge are often used as standing spaces by pedestrians.

Although a single pole or obstruction has little effect on walkway capacity under light to moderate flow conditions, for flows near the capacity limit, the presence of a series of obstructions has an effect similar to that of a *continuous obstruction*. The effective length of an obstruction is assumed to be five times its effective width. Therefore, *occasional obstructions* can be evaluated by multiplying their effective width by the ratio of their effective length to the average distance between them. Equation 3.4 is used to calculate the effective walkway width, W_E. W_t is the total walkway width, and W_o is the sum of obstruction widths and shy distances.

$$W_E = W_t - W_o \quad \quad 3.4$$

Effective width reductions are called *shy distances* (i.e., the distance pedestrians "shy away from" an obstacle), or *preemptive widths*. The shy distances are summed using the most prevalent feature on each side of the walkway. Figure 3.7 illustrates typical walkway width adjustments caused by objects placed in the total walkway width, and Table 3.2 provides a list of shy distances. *HCM* does not provide a clear definition of which side of a walkway is the outside and which is the inside, but it can be inferred from the text that the curbside edge of the walkway is the outside and the edge farthest from the curb is the inside. The applicable shy distances are shown in Fig. 3.7.

Flow-Conflict Performance

A walkway's greatest capacity occurs when 100% of the flow travels in the same direction (*unidirectional flow*). When *two-directional flow* occurs, the capacity is nearly the same as unidirectional flow, provided the flow is balanced (i.e., the directional split is 50/50). When the directional split is considerably unbalanced under high flows, such as 90/10, the minor directional flow will have

Figure 3.7 Width Adjustments for Fixed Objects Along a Walkway

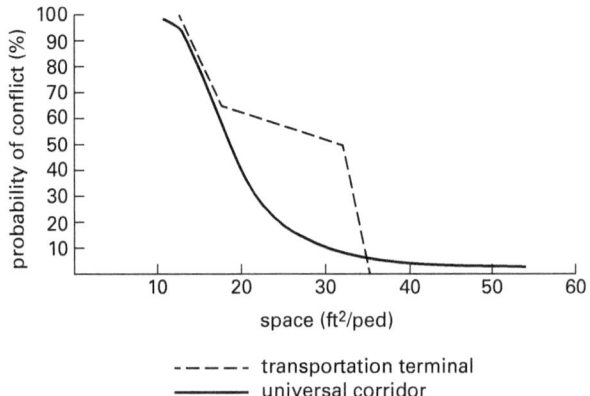

(not to scale)

▢ = shy distance
▇ = fixed-object effective width

Used with permission from *Highway Capacity Manual*, 6th Edition: A Guide for Multimodal Mobility Analysis, 2016, Exhibit 18-18, by the Transportation Research Board of the National Academies of Sciences, Engineering, and Medicine, Washington DC. DOI: 10.17226/24798

considerable difficulty using the walkway, as the major flow often spills over into the minor flow direction. This is called *cross-flow walkway interference* and reduces the capacity of both walkways. The probability of contact between two pedestrians increases when the space per person becomes less than 35 ft²/ped (3.3 m²/ped). When there is 35 ft²/ped (3.3 m²/ped) or greater, the probability of conflict is effectively zero, and as the space approaches 18 ft²/ped (1.7 m²/ped) or less, passing becomes virtually impossible. Between these extremes, conflict probability follows a nonlinear relationship, as shown in Fig. 3.8.

Figure 3.8 Probability of Conflict Within Pedestrian Cross Flows

Used with permission from *Highway Capacity Manual*, 6th Edition: A Guide for Multimodal Mobility Analysis, 2016, Exhibit 4-18, by the Transportation Research Board of the National Academies of Sciences, Engineering, and Medicine, Washington DC. DOI: 10.17226/24798

Table 3.2 Shy Distances*

obstacle	approximate width preempted (ft)
street furniture	
light poles	2.5–3.5
traffic signal poles and boxes	3.0–4.0
fire alarm boxes	2.5–3.5
fire hydrants	2.5–3.0
traffic signs	2.0–2.5
parking meters	2.0
mailboxes (1.7 ft × 1.7 ft)	3.2–3.7
telephone booths (2.7 ft × 2.7 ft)	4.0
trash cans (1.8 ft diameter)	3.0
benches	5.0
bus shelters (on sidewalk)	6.0–7.0
public underground access	
subway stairs	5.5–7.0
subway ventilation gratings (raised)	6.0+
transformer vault ventilation gratings (raised)	6.0+
landscaping	
trees	3.0–4.0
planter boxes	5.0
commercial uses	
newsstands	4.0–13.0
vending stands	variable
advertising displays	variable
store displays	variable
sidewalk cafes (two rows of tables)	7.0
building protrusions	
columns	2.5–3.0
stoops	2.0–6.0
cellar doors	5.0–7.0
standpipe connections	1.0
awning poles	2.5
truck docks (trucks protruding)	variable
garage entrances/exits	variable
driveways	variable

(Multiply ft by 0.305 to obtain m.)

*To account for the avoidance distance between pedestrians and obstacles, 1.0 ft to 1.5 ft (0.3 m to 0.5 m) must be added to the preemption width for individual obstacles. Widths are from curb to edge of object or from building face to edge of object.

Used with permission from *Highway Capacity Manual*, 6th Edition: A Guide for Multimodal Mobility Analysis, 2016, Exhibit 24–9, by the Transportation Research Board of the National Academies of Sciences, Engineering, and Medicine, Washington DC. DOI: 10.17226/24798

When there is a cross flow or an opposite direction flow, an average of at least 60 ft² to 100 ft² (5.6 m² to 9.3 m²) of space per pedestrian is necessary to avoid interference.

Level of Service for Uninterrupted Flow Facilities

Uninterrupted flow pedestrian facilities are established either exclusively for pedestrian use or for shared use with nonmotorized forms of transportation. Because the facilities have no disruptions except for interactions with other pedestrians (or with nonmotorized forms of transportation on shared facilities), pedestrian flow after start-up is relatively uniform over extended distances. Therefore, it can be analyzed using the pedestrian walking speed and pedestrian space requirements given in Table 3.3. Note that Table 3.3 does not consider platoon flow, but assumes average flow throughout the effective walkway width.

Table 3.3 LOS Criteria for Average Walkways and Sidewalks

LOS	average space (ft²/ped)	related measures flow rate (ped/min/ft)[a]	average speed (ft/sec)	v/c ratio[b]	comments
A	>60	≤5	>4.25	≤0.21	ability to move in desired path, no need to alter movements
B	>40–60	>5–7	>4.17–4.25	>0.21–0.31	occasional need to adjust path to avoid conflicts
C	>24–40	>7–10	>4.00–4.17	>0.31–0.44	frequent need to adjust path to avoid conflicts
D	>15–24	>10–15	>3.75–4.00	>0.44–0.65	speed and ability to pass slower pedestrians restricted
E	>8–15[c]	>15–23	>2.50–3.75	>0.65–1.00	speed restricted, very limited ability to pass slower pedestrians
F	≤8[c]	variable	≤2.50	variable	speeds severely restricted, frequent contact with other users

(Multiply ft²/ped by 0.093 to obtain m²/ped.)
(Multiply ped/min-ft by 0.305 to obtain ped/min·m.)
(Multiply ft/sec by 0.305 to obtain m/s.)

[a]Pedestrians per minute per foot of walkway width.
[b]v/c ratio = flow rate/23. LOS is based on average space per pedestrian.
[c]In cross-flow situations, the LOS E–F threshold is 13 ft²/ped.

Used with permission from *Highway Capacity Manual*, 6th Edition: A Guide for Multimodal Mobility Analysis, 2016, Exhibit 24-1, by the Transportation Research Board of the National Academies of Sciences, Engineering, and Medicine, Washington DC. DOI: 10.17226/24798 (Exhibit 24-1 does not apply to walkways with steep grades (>5%). See the Special Cases section for further discussion.)

Sidewalks, concourses, and walkways that separate pedestrians from vehicular traffic and on which bicycles or other nonpedestrian uses are prohibited are often established along wide city streets, through park-like settings, and in airport, bus, or train terminals. These facilities are generally constructed to accommodate high volumes of pedestrian traffic and are designed along straight or gently curving routes with minimal cross-flow interference.[2]

Performance measures can be determined by available space per person along the facility, M, found from Eq. 3.3. Field observations of space are made by counting the number of pedestrians in a given area per unit of time. When calculating the *pedestrian unit flow rate*, v_p, a 15 min unit of time is often used to compensate for normal flow variations, as shown in Eq. 3.5.

$$v_p = \frac{v_{15}}{15\, W_E} \quad [HCM \text{ Eq. 24-3}] \quad 3.5$$

Level of service (LOS) criteria measure the quality of pedestrian flow by comparing flow to defined ranges of available space per pedestrian, flow rates, and speeds. LOS for pedestrians is similar to LOS for highway traffic, using relative freedom to maneuver in traffic flow as a guideline. LOS is divided into five levels of flow, with LOS A describing the least dense flow and LOS F describing the most dense flow that is still moving. LOS E is considered the *nominal walkway capacity*, with flow that is tightly compressed, allowing for very little ability to pass slower pedestrians. As for vehicular flow, LOS F for pedestrians is a condition when all flow is severely restricted and flow is sporadic and unstable.

Assigning a quality of flow index based on pedestrian space establishes a common parameter for flow analysis. Table 3.3 shows LOS criteria for pedestrian walkways.

In addition to space, travel perception can influence LOS ratings. Using "A" for "best" quality of service, and "F" for the worst quality of service, travelers were interviewed for walkway analysis, resulting in a traveler perception score. LOS criteria is based on both space per pedestrian and traveler perception. Table 3.4 shows this relationship.

Service Volumes at Capacity Flow

The *HCM* defines *service volume* as the maximum hourly rate at which pedestrians can be reasonably expected to traverse a point along a walkway while maintaining a designated LOS during an hour where assumed conditions exist. The service volume takes into account the physical conditions of the walkway, such as the grade, effective width, average pedestrian speed, and other features that have an effect on the total rate of traffic flow.

For average flow conditions, LOS based on average space per pedestrian and the accompanying flow rates are shown in Table 3.3.

---

[2]The Americans with Disabilities Act (ADA) requirements for sidewalks and curb ramps apply to all intersection designs.

Table 3.4 Pedestrian-Mode LOS Criteria Using Traveler Perception and Space per Pedestrian

pedestrian LOS score	LOS by average pedestrian space (ft²/ped)					
	>60	>40–60	>24–40	>15–24	>8.0–15*	≤ 8.0
≤2.00	A	B	C	D	E	F
>2.00–2.75	B	B	C	D	E	F
>2.75–3.50	C	C	C	D	E	F
>3.50–4.25	D	D	D	D	E	F
>4.25–5.00	E	E	E	E	E	F
>5.00	F	F	F	F	F	F

*In cross-flow situations, the LOS E/F threshold is 13 ft²/ped.

Used with permission from *Highway Capacity Manual*, 6th Edition: A Guide for Multimodal Mobility Analysis, 2016, Exhibit 18-2, by the Transportation Research Board of the National Academies of Sciences, Engineering, and Medicine, Washington DC. DOI: 10.17226/24798

The usual method of analysis is to consider a time period of, for example, 5 min, 10 min, or 15 min, during which the flow tends to average and is more representative of continuous flow. The selection of a time period is a matter of pedestrian convenience, available space, economics, policy, code requirements, and the type of design conditions being considered. For instance, a fire or panic escape walkway has a different function than a terminal concourse or a street sidewalk. The fire walkway would require higher short-term capacity in order to evacuate a space quickly. On the other hand, travel on an open concourse or street sidewalk has more random purpose, a wider range of speeds, and perhaps a lower overall average speed.

For queuing areas, pedestrians require less area per person than along a walkway. The queues that occur at ticket windows or entrance gates tend to be slow moving and generally orderly. Pedestrians usually tolerate some delay while standing in queues, as long as the delay is not long term.

Walkers vary in the distance between persons when walking two or more abreast. Walkers that know each other will tend to walk more closely together. Pedestrians carrying parcels or baggage take up more width than pedestrians with nothing in their hands. Pedestrian flow along terminal concourses is less affected by pedestrians carrying suitcases and baggage due to the more uniform nature of walking patterns than pedestrians carrying packages along a walkway in a shopping district.

Walkways that allow pedestrians to pass each other in the same or opposing directions should have an effective width of at least 2.5 ft (0.8 m) per person. Using the concept of effective lane width is not necessarily favored by designers; however, the analogy can be useful for establishing basic parameters of flow. Using a minimum width of 5.0 ft (1.5 m) for two-direction flow, this procedure divides the effective walkway width by 2.5 ft (0.8 m) to determine the number of 2.5 ft (0.8 m) equivalent lanes. The number of lanes is rounded to the next lower whole number, as a lane width of less than 2.5 ft (0.8 m) is effectively nonexistent. The equivalent lanes are multiplied by the capacity of one lane of flow, such as 58 persons per minute of flow for a walking speed of 3 ft (0.9 m) per second.

Lane widths of 2.5 ft (0.8 m) are usually not adequate for flows with a significant number of wheelchairs. Providing 3.5 ft (1.1 m) for each lane between handrails or other obstructions (or 7 ft (2.1 m) for two-direction flow) is necessary for wheelchair users to reach over the side of the wheels and to prevent collisions with wheelchair traffic in the opposing direction. Pathways with wheelchair traffic should not be constructed on a longitudinal grade greater than 5%. A grade straighter than 5% but no more than 8.33% is to be designed as a ramp with level platforms spaced at every 30 in (0.8 m) change in elevation. All pathways are to be designed with a cross slope no greater than 2% and without a sharp drop-off at the edge of the pavement. A wheelchair encountering a sharp drop-off can easily overturn if the wheelchair user loses control of the chair. The intersections of walkways used by wheelchairs should have an inside corner radius of 6 ft (4 ft min) (1.8 m (1.2 m min)) to allow the wheelchair to turn without swinging into opposing traffic. Although motorized bicycles, scooters, and skateboards are often banned from walkways, motorized wheelchairs are usually permitted.

While service volumes for pedestrian flow may vary from one location to another, the capacity of a walkway occurs when the maximum flow rate is achieved for a particular facility. Capacity is expressed by *space*, which is square feet per pedestrian, or by *unit flow*, which is pedestrians per minute per foot of walkway width. The following are general conditions for maximum comfortable flow.

- walkways with random flow, 23 ped/min/ft
- walkways with platoon flow, 18 ped/min/ft average over 5 min
- cross-flow areas, 23 ped/min/ft sum of both flows
- stairway up-direction, 15 ped/min/ft

Figure 3.3 shows the relationship between flow rate and space.

The average walking speed at capacity is about half of the average speed under conditions of low congestion. Figure 3.6 shows the relationship between pedestrian speed and flow rate.

Platoon Flow

Pedestrian flow tends to occur in a series of *platoons*, a group of pedestrians walking together. Platooning is generally involuntary and is often caused by an accumulation of faster walkers behind slower ones or a time-dependent event, such as a traffic signal or a bus arrival at a terminal. In transportation terminals, platooning usually occurs along walkways and concourses when large volumes of pedestrians emerge from arriving vehicles.

Short-term flow in periods of 1 min intervals can be as much as double the average rate in 5 min, 10 min, or 15 min average rates. The space per pedestrian within a platoon can approach full capacity (8 ft²/ped (0.7 m²)) or less for short periods of time, and then suddenly expand to relatively uncongested space during the next minute. This platooning occurs because of the unregulated and random arrivals or travel rate of pedestrians. Interruptions of flow at signals, or by walkway width variations caused by obstacles, tend to intensify platooning. Until platoons disperse, pedestrians often scatter to adjacent areas, such as stepping off of the curb and walking in the street near signalized intersections, or walking on the grassed or unpaved landscape paralleling an open walkway.

Platoon flow occurs on narrow passageways of transportation terminals when large groups of people arrive on an airplane, train, or ferryboat, and discharge along a boarding ramp or passageway. Faster walkers have less opportunity to pass slower walkers and form queues along the passageway, which eventually dissipates after the queue flow moves to a wider concourse or walkway. The relationship between platoon flow and average flow is plotted in Fig. 3.9. The LOS for platoon flow is shown in Table 3.5.

Figure 3.9 Relationship Between Platoon Flow and Average Flow

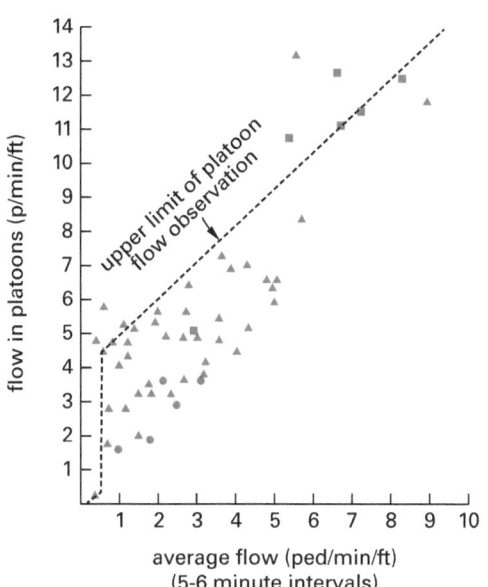

- ■ observations at subway entrances
- ● observations on very wide sidewalks
- ▲ other observations

Used with permission from *Highway Capacity Manual*, 6th Edition: A Guide for Multimodal Mobility Analysis, 2016, Exhibit 4-21, by the Transportation Research Board of the National Academies of Sciences, Engineering, and Medicine, Washington DC. DOI: 10.17226/24798

Table 3.5 LOS Criteria for Platoon-Adjusted Walkways

LOS	average space (ft²/ped)	related measure flow rate[a] (ped/min/ft)[b]	comments
A	>530	≤0.5	ability to move in desired path, no need to alter movements
B	>90–530	>0.5–3	occasional need to adjust path to avoid conflicts
C	>40–90	>3–6	frequent need to adjust path to avoid conflicts
D	>23–40	>6–11	speed and ability to pass slower pedestrians restricted
E	>11–23[c]	>11–18	speed restricted, very limited ability to pass slower pedestrians
F	≤11	>18	speeds severely restricted, frequent contact with other users

[a]Rates in the table represent average flow rates over a 5 min period. Flow rate is directly related to space; however, LOS is based on average space per pedestrian.
[b]Pedestrians per minute per foot of walkway width.
[c]In cross-flow situations, the LOS E–F threshold is 13 ft²/ped.

Used with permission from *Highway Capacity Manual*, 6th Edition: A Guide for Multimodal Mobility Analysis, 2016, Exhibit 24-2, by the Transportation Research Board of the National Academies of Sciences, Engineering, and Medicine, Washington DC. DOI: 10.17226/24798

Interrupted Flow Pedestrian Facilities

Pedestrians experience interrupted flow at signalized intersections. The delay, d_p, experienced at the intersection is calculated from Eq. 3.6.

$$d_p = \frac{(C-g)^2}{2C} \quad [HCM \text{ Eq. 19-70}] \quad 3.6$$

C is the cycle length in seconds, and g is the effective green time in seconds. Pedestrians use both the *"WALK" interval* and the first few seconds of the *"Flashing DON'T WALK"* (FDW) *interval* to enter the intersection. For delay calculations, the *effective green time interval* is equal to the walk interval plus the first 4 sec of the FDW interval.

When considering pedestrian flow as platoon flow, the delay and the flow rate can be used to approximate the LOS, as shown in Table 3.5. *HCM* gives a further refinement of this method using LOS score rates.

The LOS score is derived from the approximate comfort levels of the delay factors, reflecting the roadway cross-section, motorized vehicle volume, motorized vehicle speed, and pedestrian delay. LOS criteria for pedestrians at signalized intersections are shown in Table 3.6. When delay is greater than 30 sec, pedestrians become impatient and will take more risks, such as disregarding signal indicators (i.e., "WALK" and/or "DON'T WALK" intervals). *Pedestrian noncompliance* with signal indicators is reduced by high traffic volumes, as pedestrians have little choice but to wait for the walk signal.

Table 3.6 LOS Criteria for Pedestrians at Signalized Intersections

LOS	LOS Score
A	1.50
B	>1.50–2.50
C	>2.50–3.50
D	>3.50–4.50
E	>4.50–5.50
F	>5.50

Used with permission from *Highway Capacity Manual*, 6th Edition: A Guide for Multimodal Mobility Analysis, 2016, Exhibit 19-9, by the Transportation Research Board of the National Academies of Sciences, Engineering, and Medicine, Washington DC. DOI: 10.17226/24798

Both LOS scores and delay tolerances are locally derived, subjective values, established using statistics from pedestrian interviews. The *HCM* values are meant as general overall guidance.

The general procedure for determining an LOS score shown in Fig. 3.10 is focused on signalized intersections. Table 3.7 shows potential sources for various data, including default values that can be used as benchmarks for analysis.

Figure 3.10 LOS Procedure for Pedestrians at Signalized Intersections

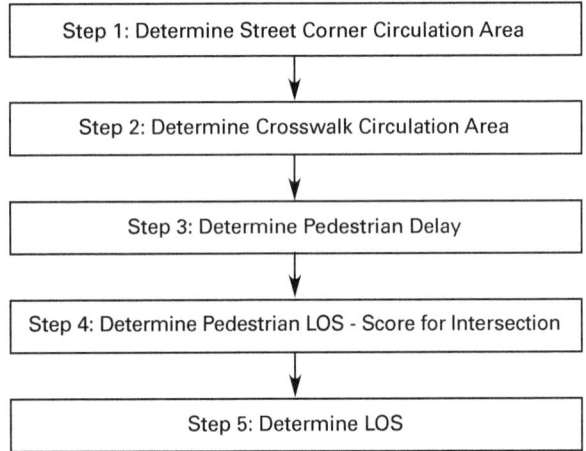

Used with permission from *Highway Capacity Manual*, 6th Edition: A Guide for Multimodal Mobility Analysis, 2016, Exhibit 19-33, by the Transportation Research Board of the National Academies of Sciences, Engineering, and Medicine, Washington DC. DOI: 10.17226/24798

Pedestrian Area Requirements at Street Corners

A certain amount of *pedestrian area* is required at all signalized street corners. A *circulation area* is needed, providing space for pedestrians crossing the street during the green signal phase, pedestrians joining the queue of people waiting to cross during the red signal phase (the *red-phase queue*), and pedestrians moving between the sidewalks adjoining the corner who are not crossing the street. A *temporary holding area* is also required to accommodate the red-phase queue while those pedestrians wait to cross.

Figure 3.11 and Fig. 3.12 show the two *signal phase conditions* assessed when calculating values for crosswalks and street corners. *Condition one*, shown in Fig. 3.11, applies when pedestrians are crossing the minor street and the red-phase queue is waiting to cross the major street. *Condition two*, shown in Fig. 3.12, applies when pedestrians are crossing the major street and the red-phase queue is waiting to cross the minor street.

Figure 3.11 Condition One: Minor Street Crossing

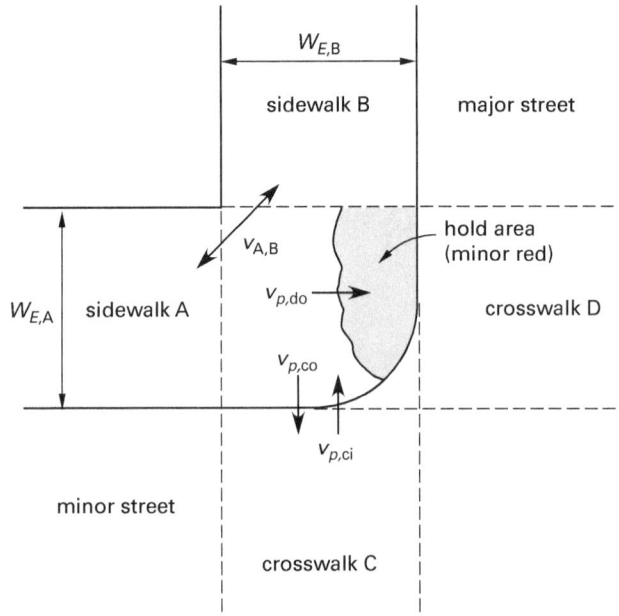

key condition 1

$v_{A,B}$ = sidewalk flow
$v_{p,do}$ = peds joining queue
$v_{p,co}$ = outbound crossing peds
$v_{p,ci}$ = inbound crossing platoon
$W_{E,A \text{ or } B}$ = effective width of sidewalks

Used with permission from *Highway Capacity Manual*, 6th Edition: A Guide for Multimodal Mobility Analysis, 2016, Exhibit 19-29, by the Transportation Research Board of the National Academies of Sciences, Engineering, and Medicine, Washington DC. DOI: 10.17226/24798

Table 3.7 Required Input Data, Potential Data Sources, and Default Values for Off-Street Facility Analysis

required data and units	basis	potential data sources	suggested default value
traffic characteristics			
motorized vehicle demand flow rate (vph)	movement	field data, past counts	must be provided
right-turn-on-red flow rate (vph)	approach	field data, past counts	must be provided
permitted left-turn flow rate (vph)	movement	field data, past counts	see discussion in *HCM*
midsegment 85th percentile speed (mph)	approach	field data	speed limit
pedestrian flow rate (vph)	movement	field data, past counts	must be provided
geometric design			
number of lanes (ln)	leg	field data, aerial photo	must be provided
number of right-turn islands (0, 1, 2)	leg	field data, aerial photo	0
total walkway width (ft)	approach	field data, aerial photo	business or office land use: 9.0 ft; residential or industrial land use: 11.0 ft
crosswalk width (ft)	leg	field data, aerial photo	12 ft
corner radius (ft)	approach	field data, aerial photo	trucks and buses in turn volume: 45 ft; no trucks or buses in turn volume: 25 ft
signal control			
walk (sec)	phase	field data	actuated: 7 sec; pretimed: green interval minus pedestrial clear
pedestrian clear (sec)	phase	field data	based on 3.5 ft/sec walking speed
rest in walk (yes or no)	phase	field data	not enabled
cycle length (sec)	intersection	field data	same as motorized vehicle mode
yellow change + red clearance (sec)[a]	phase	field data	4 sec
duration of phase serving pedestrians (sec)	phase	field data	same as motorized vehicle mode
pedestrian signal head presence (yes or no)	phase	field data	must be provided
other data			
analysis period duration (hr)[b]	intersection	set by analyst	0.25 hr

Notes: movement = one value for each left-turn, through, and right-turn movement
approach = one value for the intersection approach
leg = one value for the intersection leg (approach plus departure sides)
phase = one value or condition for each signal phase
intersection = one value for the intersection

[a]Specific values of yellow change and red clearance should be determined by local guidelines or practice.
[b]Analysis period duration is as defined for HCM Exh. 19-31.
Used with permission from *Highway Capacity Manual*, 6th Edition: A Guide for Multimodal Mobility Analysis, 2016, Exhibit 19-31, by the Transportation Research Board of the National Academies of Sciences, Engineering, and Medicine, Washington DC. DOI: 10.17226/24798

Assessment of pedestrian area requirements at a given street corner compares the time and space available on the corner to the pedestrian arrival demand at that corner. The space available is constrained by sidewalk geometry at the corner, and the time available is determined by the signal timing at the corner.

In order to calculate the circulation area, the total number of circulating pedestrians and the net corner time-space available for circulating pedestrians must be known. The total number of circulating pedestrians, v_t, can be found from Eq. 3.7. The variables required for this equation are illustrated in Fig. 3.11 and Fig. 3.12.

$$v_t = v_{p,\text{ci}} + v_{p,\text{co}} + v_{p,\text{di}} + v_{p,\text{do}} + v_{A,B} \quad 3.7$$

If the number of pedestrians waiting to cross in any direction on either street during one cycle is not known, it can be found from Eq. 3.8.

$$v = \left(\frac{v_p}{15 \text{ min}}\right) C \quad 3.8$$

Finding the net corner time-space available requires the total time-space available and the time-space occupied by pedestrians waiting to cross. The *total time-space available*, TS, at a given corner can be found from

Figure 3.12 Condition Two: Major Street Crossing

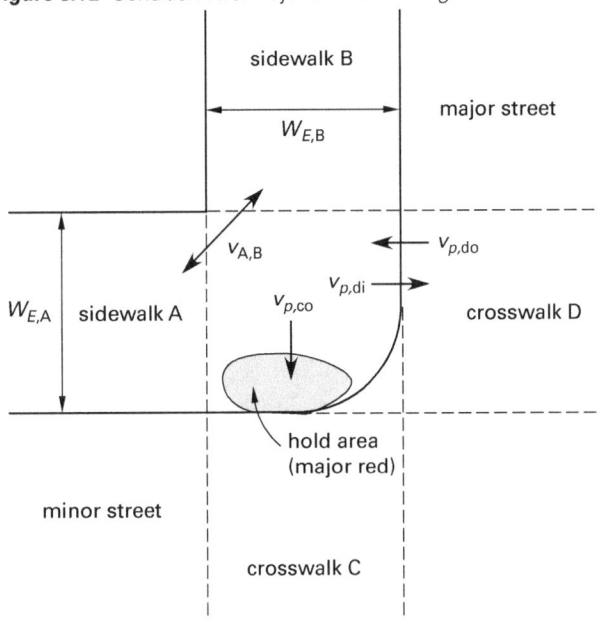

key condition 2

$v_{A,B}$ = sidewalk flow
$v_{p,co}$ = peds joining queue
$v_{p,do}$ = outbound crossing peds
$v_{p,di}$ = inbound crossing platoon
$W_{E,A\,or\,B}$ = effective width of sidewalks

Used with permission from *Highway Capacity Manual*, 6th Edition: A Guide for Multimodal Mobility Analysis, 2016, Exhibit 19-30, by the Transportation Research Board of the National Academies of Sciences, Engineering, and Medicine, Washington DC. DOI: 10.17226/24798

Eq. 3.9. R is the radius of the curb, and $W_{E,A}$ and $W_{E,B}$ are the effective widths of sidewalk A and sidewalk B, respectively, as shown in Fig. 3.11 and Fig. 3.12.

$$\text{TS} = C(W_{E,A} W_{E,B} - 0.215 R^2) \quad 3.9$$
$$[HCM \text{ Eq. } 19\text{-}51]$$

The time-space occupied by pedestrians waiting to cross is a function of the *holding area waiting time* under condition one, Q_{tdo}, given by Eq. 3.10, and the holding area waiting time under condition two, Q_{tco}, given by Eq. 3.11. N_{do} and N_{co} are the number of pedestrians waiting to cross the major and minor streets, respectively, during one cycle. R_{mi} is the length of the minor-street red phase, and R_{mj} is the length of the major-street red phase. If pedestrian signals are installed, the red phase of both streets is equal to the length of the "DON'T WALK" phase. The number of pedestrians waiting to cross either the major or minor streets can be found from Eq. 3.12.

$$Q_{tdo} = \frac{N_{do} R_{mi}^2}{2C} \quad 3.10$$

$$Q_{tco} = \frac{N_{co} R_{mj}^2}{2C} \quad 3.11$$

$$N_{do} = \left(\frac{v_{do}}{3600}\right) C \quad 3.12$$

When all these values are known, the *net corner time-space available*, TS_c, can be found using Eq. 3.13.

$$TS_{c,\text{ft}^2\cdot\text{sec}} = TS_{\text{ft}^2\cdot\text{sec}} - 5(Q_{tdo} + Q_{tco}) \quad 3.13$$
$$[HCM \text{ Eq. } 19\text{-}57]$$

Circulating space for each pedestrian, M, can be found from Eq. 3.14. The total circulation volume, v_t, is multiplied by 4 sec, the assumed average circulation time. The calculated space for each pedestrian can be used with Table 3.3 to determine the LOS for the walkway being studied.

$$M = \frac{TS_c}{4N_{tot}} \quad [HCM \text{ Eq. } 19\text{-}58] \quad 3.14$$

In addition to adequate circulating area at a street corner, adequate circulating area must also be provided in a crosswalk to ensure pedestrians are able to clear the crossing during the green phase. Required circulating area for a crosswalk is determined similarly to required circulating area for a street corner, with some variations to address the safety concerns and time constraints placed on pedestrians crossing a street.

To find the circulation space available per pedestrian in the crosswalk, the crosswalk time-space available and the total crosswalk occupancy time must be known. The *crosswalk time-space available* can be found from Eq. 3.15. L is the length of the crosswalk, W_E is the effective width of the crosswalk, WALK + FDW is the effective green time used by a crossing pedestrian, S_p is the average speed of pedestrians in the crosswalk, and g is the green time when a pedestrian signal is not installed. These variations of the same equation reflect the fact that where pedestrian signals are installed, pedestrians use both the "WALK" interval and the first few seconds of the FDW interval to cross an intersection, whereas when no pedestrian signal is installed, pedestrians will tend to use only the green time available.

$$TS = LW_E\left(\text{WALK} + \text{FDW} - \frac{L}{2S_p}\right) \quad 3.15(a)$$
[when signal is installed]

$$TS_{cw} = L_d W_d g_{\text{walk,mi}} \quad 3.15(b)$$
[when signal is not installed]

Equation 3.16 can be used to calculate the number of pedestrians crossing during a single cycle. v is the volume of pedestrians on the walkway that feeds into the crossing.

$$N_p = N_{do}\frac{C-g}{C} \quad [HCM \text{ Eq. } 19\text{-}66] \quad 3.16$$

To find the total crosswalk occupancy time, the *total crossing time* (also called the *effective green time*), the time it takes for a crosswalk to clear once pedestrians enter the crosswalk, must be known. For calculations of total crossing time, pedestrian start-up time is assumed to be 3.2 sec. Equation 3.17 gives the total crossing time for crosswalks more than 10 ft (3 m) wide, and Eq. 3.18 gives the total crossing time for crosswalks 10 ft (3 m) wide or less.

$$t = 3.2 + \frac{L_{ft}}{S_p} + 2.7\left(\frac{N_p}{W_{ft}}\right) \quad 3.17$$
$$[HCM \text{ Eq. } 19\text{-}64]$$

$$t = 3.2 + \frac{L}{S_p} + 0.27 N_p \quad [HCM \text{ Eq. } 19\text{-}65] \quad 3.18$$

Pedestrian Area Requirements in Crosswalks

The *total crosswalk occupancy time*, T_{occ}, is given by Eq. 3.19. $t_{ps,do}$ is the total crossing time for pedestrians crossing the major street; N_{do} is the number of pedestrians waiting to cross the major street during each cycle, found from Eq. 3.12; $t_{ps,di}$ is the crossing time for pedestrians crossing the minor street; and N_{di} is the number of pedestrians arriving at each corner after crossing the major street.

$$T_{occ} = t_{ps,do} N_{do} + t_{ps,di} N_{di} \quad [HCM \text{ Eq. } 19\text{-}67] \quad 3.19$$

The circulation space per pedestrian in the crosswalk is found from Eq. 3.20. Like the circulation space at a street corner, this value can be used with Table 3.3 to determine the LOS for the walkway.

$$M = \frac{TS}{T_{occ}} \quad [HCM \text{ Eq. } 19\text{-}69] \quad 3.20$$

The effect of vehicles turning into the crosswalk on the LOS for pedestrians crossing the crosswalk can be estimated by assuming an 8 ft (2.4 m) swept path for the vehicle. The time the vehicle will occupy the crosswalk is assumed to be 5 sec. Equation 3.21 is used to determine the time-space of turning vehicle interference, TS_{tv}. N_{tv} is the number of turning vehicles during the green phase. The turning-vehicle time-space is subtracted from the crosswalk time-space calculated in Eq. 3.15 to determine the effective time-space, TS_E, available for pedestrians.

$$TS_{tv} = 40 N_{tv} W_E \quad 3.21$$

$$TS_E = TS - TS_{tv} \quad 3.22$$

Urban Streets

Urban street settings can occur anywhere there is a roadway with flanking sidewalks and signalized or unsignalized intersections. The analysis includes the upstream and downstream approaches to an intersection, the intersection(s), and the link between intersections. The analysis boundary is illustrated in Fig. 3.13.

Figure 3.13 Analysis Boundary for Urban Street Segment Analysis

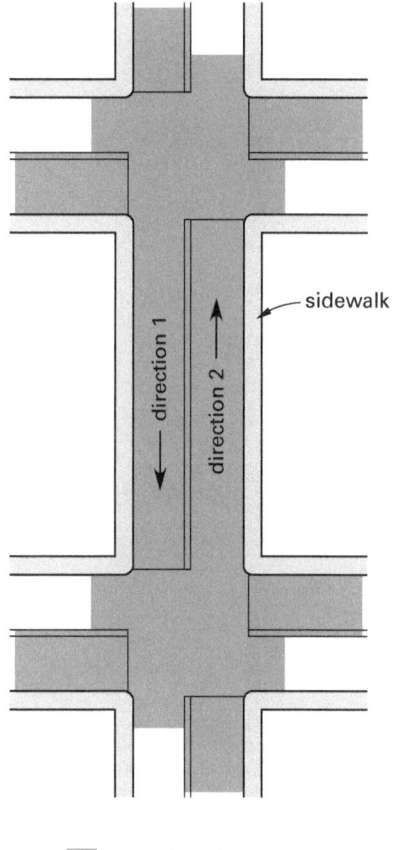

= analysis boundary

Used with permission from *Highway Capacity Manual*, 6th Edition: A Guide for Multimodal Mobility Analysis, 2016, Exhibit 18-4, by the Transportation Research Board of the National Academies of Sciences, Engineering, and Medicine, Washington DC. DOI: 10.17226/24798

The *segment* includes points and links.

A *point* is an intersection along a link.

A *link* is the section of roadway or pathway between points, where there is no influence of traffic control on pedestrian flow.

A *facility* is a length of roadway composed of contiguous urban street segments.

Segment and link conditions are illustrated in Fig. 3.14.

Figure 3.14 *Relationship of Point Performance and Link Performance*

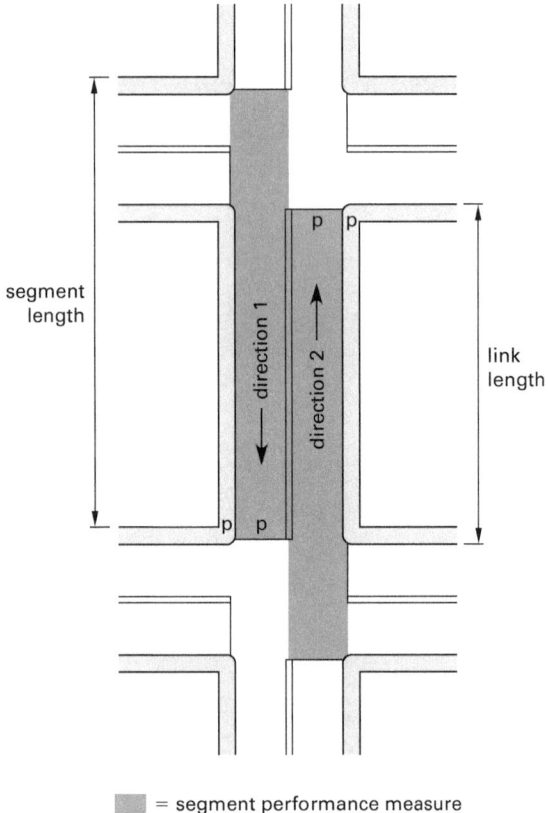

Used with permission from *Highway Capacity Manual*, 6th Edition: A Guide for Multimodal Mobility Analysis, 2016, Exhibit 18-5, by the Transportation Research Board of the National Academies of Sciences, Engineering, and Medicine, Washington DC. DOI: 10.17226/24798

Performance measures for pedestrian mode on urban street segments include
- pedestrian travel speed
- average pedestrian space
- pedestrian LOS scores for each link and segment

Normally, each side of the street is evaluated separately. The step-by-step procedures for urban street segments are illustrated in Fig. 3.15.

Figure 3.15 *Pedestrian Methodology for Urban Street Segments*

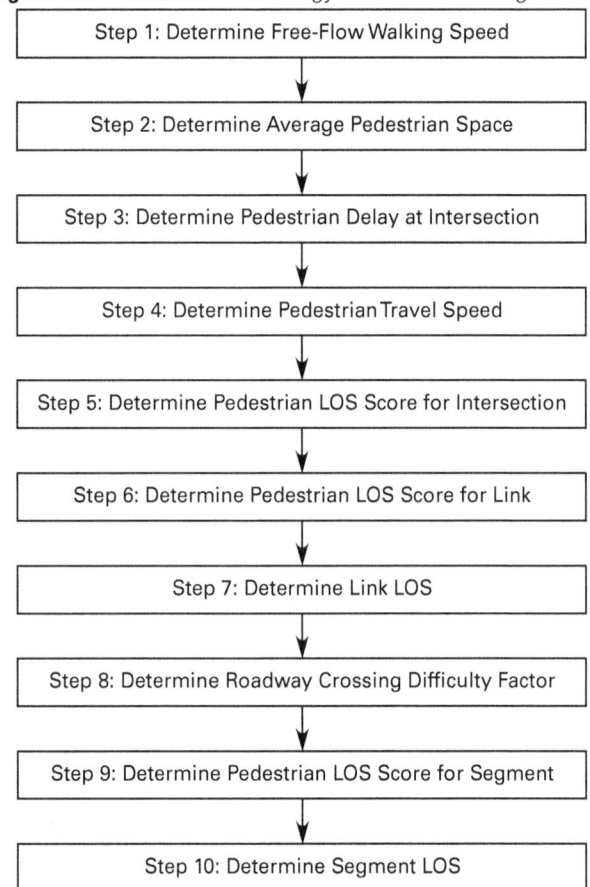

Used with permission from *Highway Capacity Manual*, 6th Edition: A Guide for Multimodal Mobility Analysis, 2016, Exhibit 18-17, by the Transportation Research Board of the National Academies of Sciences, Engineering, and Medicine, Washington DC. DOI: 10.17226/24798

For *link-based evaluation*, only steps 6 and 7 of the procedure are necessary. For all other flows, the full procedure is utilized. The following are details of the preceding methodology steps.

Step 1: Determine Free-Flow Walking Speed

The average free-flow walking speed, S_{pf}, recommended by *HCM* is

- 4.4 fps for general traffic with $\leq 20\%$ elderly persons in the flow
- 3.3 fps for flows with $> 20\%$ elderly persons in the flow

Reduce speed 0.3 fps for upgrades $\geq 10\%$.

Step 2: Determine the Average Pedestrian Space

To calculate the average pedestrian space, take into account the following factors.

Effective Sidewalk Width. Width adjustments for fixed objects are illustrated in Fig. 3.7.

The total walkway width is reduced by the effective width reductions of fixed objects to find the effective walkway width. Discontinuous objects, such as trees, poles, or benches, have an effect of about 10 ft upstream and downstream from the object itself. The effective sidewalk width is an average for the length of the link, determined by the following equations [*HCM* Eq. 18-23 to Eq. 18-27].

$$W_E = W_T - W_{0,i} - W_{0,o} - W_{s,i} - W_{s,o} \geq 0.0 \quad 3.23$$

$$W_{s,i} = \max(W_{\text{buf}}, 1.5) \quad 3.24$$

$$W_{s,o} = 3.0 p_{\text{window}} + 2.0 p_{\text{building}} + 1.5 p_{\text{fence}} \quad 3.25$$

$$W_{0,i} = w_{0,i} - W_{s,i} \geq 0.0 \quad 3.26$$

$$W_{0,o} = w_{0,o} - W_{s,o} \geq 0.0 \quad 3.27$$

Pedestrian Flow Rate per Unit Width. Use the following equation [*HCM* Eq. 18-28].

$$v_p = \frac{v_{\text{ped}}}{60 W_E} \quad 3.28$$

Average Walking Speed. Use the following equation with the free-flow walking speed [Eq. 18-29].

$$S_p = (1 - 0.00078 v_p^2) S_{\text{pf}} \geq 0.5 S_{\text{pf}} \quad 3.29$$

Pedestrian Space. Compute with the following equation [*HCM* Eq. 18-30].

$$A_p = 60 \frac{S_p}{v_p} \quad 3.30$$

Step 3: Determine Pedestrian Delay at Intersections

The following three delay considerations are considered along a link.

- d_{pp} represents the delay incurred by pedestrians who traveled through the boundary intersection along a path that is parallel to the segment centerline.

- d_{pc} represents the delay incurred by pedestrians who cross the segment at the nearest signal-controlled crossing.

- d_{pw} represents the delay incurred by pedestrians waiting for a gap to cross the segment at an uncontrolled location.

Step 4: Determine Pedestrian Travel Speed

Pedestrian travel speed is typically slower than the average walking speed, because it includes delay at the downstream boundary of the segment [*HCM* Eq. 18-31].

$$S_{Tp,\text{seg}} = \frac{L}{\frac{L}{S_p} + d_{\text{pp}}} \quad 3.31$$

Step 5: Determine Pedestrian LOS Score for Intersection

This step is covered in the upcoming "Interrupted Flow: Signalized Intersections" section of this chapter. (If a two-way intersection, score = 0.0.)

Step 6: Determine Pedestrian LOS Score for Link

Use the following equations [*HCM* Eq. 18-32 to Eq. 18-35] to find LOS for the link, $I_{p,\text{link}}$.

$$I_{p,\text{link}} = 6.0468 + F_w + F_v + F_s \quad 3.32$$

$$F_w = -1.2276 \ln\left(\begin{array}{c} W_v + 0.5 W_1 + 50 p_{\text{pk}} \\ + W_{\text{buf}} f_b + W_{aA} f_{\text{sw}} \end{array}\right) \quad 3.33$$

$$F_v = 0.0091\left(\frac{v_m}{4 N_{th}}\right) \quad 3.34$$

$$F_s = 4\left(\frac{S_R}{100}\right)^2 \quad 3.35$$

Using Table 3.8, if the condition in column 1 is satisfied, the equation in column 2 is used. If the condition is not satisfied, the equation in column 3 is used.

Table 3.8 Variables for Pedestrian LOS Score for Link

condition	variable when condition is satisfied	variable when condition is not satisfied
$p_{\text{pk}} = 0.0$	$W_t = W_{\text{ol}} + W_{\text{bl}} + W_{\text{os}}*$	$W_t = W_{\text{ol}} + W_{\text{bl}}$
$v_m > 160$ vph or street is divided	$W_v = W_t$	$W_v = W_t(2 - 0.005\, v_m)$
$p_{\text{pk}} < 0.25$ or parking is striped	$W_1 = W_{\text{bl}} + W_{\text{os}}*$	$W_1 = 10$

W_t = total width of the outside through lane, bicycle lane, and paved shoulder (ft); W_{ol} = width of the outside through lane (ft); $W_{\text{os}}*$ = adjusted width of paved outside shoulder; if curb is present $W_{\text{os}}*$ = $W_{\text{os}} - 1.5 \geq 0.0$, otherwise $W_{\text{os}}* = W_{\text{os}}$ (ft); W_{os} = width of paved outside shoulder (ft); and W_{bl} = width of the bicycle lane = 0.0 if bicycle lane not provided (ft).

Used with permission from *Highway Capacity Manual*, 6th Edition: A Guide for Multimodal Mobility Analysis, 2016, Exhibit 18-19, by the Transportation Research Board of the National Academies of Sciences, Engineering, and Medicine, Washington DC. DOI: 10.17226/24798

The LOS calculations should be determined for each sidewalk in the link. If a sidewalk is not continuous between the segment boundaries, then each subsegment should be evaluated separately. The buffer width and sidewalk width are each set to 0.0 ft for any subsegment without a sidewalk. The pedestrian LOS score for this is then computed as a weighted average of the subsegment scores, with the weight assigned to each segment corresponding to the proportion of the subsegment to the total segment length.

Step 7: Determine Link LOS

The pedestrian LOS score from step 6 and the average pedestrian space from step 2 are compared with their respective thresholds in Table 3.4 for the subject direction of travel of the link.

When a sidewalk does not exist and pedestrians must walk in the street, LOS is determined by Table 3.9, since pedestrian mode does not apply.

Table 3.9 LOS Score Criteria for Urban Link Bicycle and Transit Modes

LOS	transit LOS score
A	≤2.00
B	>2.00–2.75
C	>2.75–3.50
D	>3.50–4.25
E	>4.25–5.00
F	>5.00

Used with permission from *Highway Capacity Manual*, 6th Edition: A Guide for Multimodal Mobility Analysis, 2016, Exhibit 18-3, by the Transportation Research Board of the National Academies of Sciences, Engineering, and Medicine, Washington DC. DOI: 10.17226/24798

Step 8: Determine Roadway Crossing Difficulty Factor

For links pedestrians use to cross the roadway between boundary intersections, two conditions apply. One is that the crossing occurs at a signal-controlled location. The second is that crossing occurs between signal-controlled locations. The *diversion distance* (a pedestrian path that is not at an intersection boundary or a signalized crosswalk) is illustrated in Fig. 3.16.

For a non-signal-controlled crossing, the diversion distance, D_c, can be assumed to be one-third the distance between the nearest signal-controlled crossings; otherwise, the diversion distance is selected from observations or from best estimates by the analyst.

Diversion Delay. The diversion distance is computed as twice the distance to the nearest signalized crossing [*HCM* Eq. 18-36].

$$D_d = 2D_c \quad \quad 3.36$$

If the crossing is at location A in Eq. 3.37, the diversion distance should be increased by two increments of intersection width, W_i.

The pedestrian delay due to diversion is found using the following equation [*HCM* Eq. 18-37].

$$d_{\text{pd}} = \frac{D_d}{S_p} + d_{\text{pc}} \quad \quad 3.37$$

Figure 3.16 Diversion Distance Components at Mid-Block Crosswalks on Urban Streets

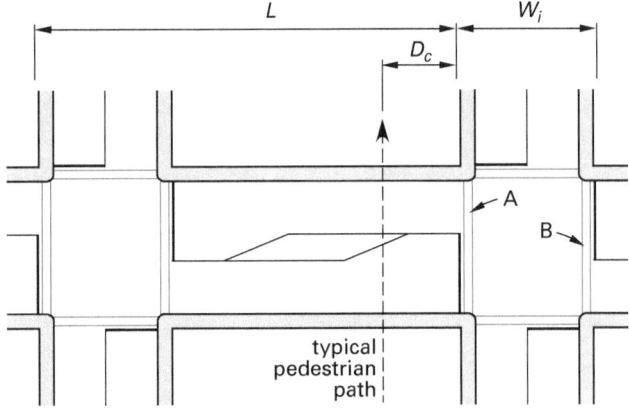

(a) divert to nearest boundary intersection

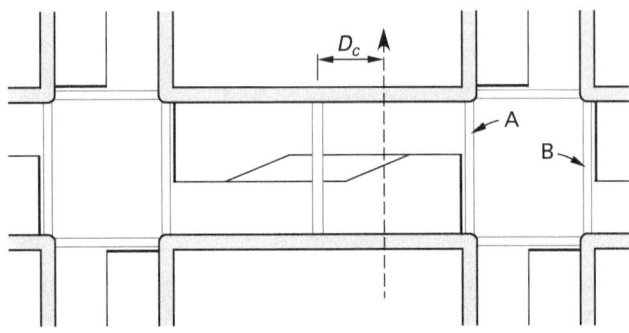

(b) divert to midsegment signalized crosswalk

Used with permission from *Highway Capacity Manual*, 6th Edition: A Guide for Multimodal Mobility Analysis, 2016, Exhibit 18-20, by the Transportation Research Board of the National Academies of Sciences, Engineering, and Medicine, Washington DC. DOI: 10.17226/24798

Roadway Crossing Difficulty Factor. The roadway crossing difficulty factor is calculated by using the following equation [*HCM* Eq. 18-38].

$$F_{\text{cd}} = 1.0 + \frac{0.10 d_{\text{px}} - (0.318 I_{p,\text{link}} + 0.220 I_{p,\text{int}} + 1.606)}{7.5} \quad 3.38$$

The factor limits are ≥ 0.80 ≤ 1.20. For illegal midsegment crossings, the waiting delay is $d_{\text{pw}} = 0.0$.

Step 9: Determine Pedestrian LOS for Segment

For the segment pedestrian LOS score, use Eq. 3.39 [*HCM* Eq. 18-39].

$$I = 0.75 \left[\frac{(F_{\text{cd}} I_{p,\text{link}} + 1)^3 \dfrac{L}{S_p} + (I_{p,\text{int}} + 1)^3 d_{\text{pp}}}{\dfrac{L}{S_p} + d_{\text{pp}}} \right] + 0.125 \quad 3.39$$

Step 10: Determine Segment LOS

Compare the LOS for the segment from step 9 and the average pedestrian space from step 2. Compare these with thresholds in Table 3.4 [*HCM* Exh. 18-2]. If a sidewalk does not exist, LOS is determined by Table 3.9 [*HCM* Exh. 18-3], because pedestrian space does not apply.

Stairways

For stairways, the horizontal speed is reduced from that of level walkways. Using a stairway capacity of 15 persons per minute per foot of width, the LOS criteria is shown in Table 3.10.

Table 3.10 LOS Criteria for Stairways

		related measures		
LOS	average space (ft^2/ped)	flow rate (ped/min/ft)[a]	v/c ratio[b]	comments
A	>20	≤5	≤ 0.33	no need to alter movements
B	>17–20	>5–6	>0.33–0.41	occasional need to adjust path to avoid conflicts
C	>12–17	>6–8	>0.41–0.53	frequent need to adjust path to avoid conflicts
D	>8–12	>8–11	>0.53–0.73	limited ability to pass slower pedestrians
E	>5–8	>11–15	>0.73–1.00	very limited ability to pass slower pedestrians
F	≤5	variable	variable	speeds severely restricted, frequent contact with other users

[a] pedestrians per minute per foot of walkway width
[b] v/c ratio = flow rate/15. LOS is based on average space per pedestrian.

Used with permission from *Highway Capacity Manual*, 6th Edition: A Guide for Multimodal Mobility Analysis, 2016, Exhibit 24-3, by the Transportation Research Board of the National Academies of Sciences, Engineering, and Medicine, Washington DC. DOI: 10.17226/24798

The design of stairs can be approached from a standpoint of the need to fit within the structural limits of a facility, or they can be designed from a standpoint of locomotion and pedestrian travel flow requirements. Usually, the *structural approach* results in stairs that fit the steepest angle, shortest tread, and highest riser permitted by codes or local ordinances, or by the architectural aspects of the facility when exceeding the minimum requirements. These codes, for example, may set the widths in multiples of 22 in (0.6 m), and a vertical angle of 32.5°, using a 7 in (17.8 cm) riser and an 11 in (27.9 cm) tread. The *pedestrian travel flow* approach views horizontal travel speed, service volume, and population agility as significant design parameters. Agility and locomotion studies have shown that a greater horizontal speed is attained in the upward direction on a vertical angle of 26.5° using a 5 in (15.2 cm) riser and a 12 in (30.5 cm) tread. Often a compromise of a 28.6° vertical angle is used, with a 5 in (15.2 cm) riser and an 11 in (27.9 cm) tread, which more closely relates to the normal average human pace length of 22 in (0.6 m). Widths determined by multiples of 30 in (0.8 m) are more comparable to the body motion of pedestrians than the minimum 22 in (0.6 m) width code requirements.

Downward speed is not greatly affected by changing from 32.5° to 28.6°, but the confidence of mobility-challenged individuals is greatly improved at less steep vertical angles, particularly in inclement weather on outdoor stairs and with significant elderly traffic.

Stair width should never be less than the effective width of approach sidewalks. Width should be increased by the equivalent width of central handrails when these are installed. When the reverse flow is a small proportion of the prevailing flow, width should be increased by 30 in (0.8 m), which is the equivalent of adding another lane. Straight runs of stairs have the highest service capacity.

Landings must be the same width as the stairs, and can be as short as 40 in (1.0 m) according to some codes. A length of 44 in (1.1 m) makes a better fit with the normal human gait, providing a greater level of comfort for mobility-challenged pedestrians. Landings should be placed at no more than 12 ft (3.7 m) of elevation change; however, this can be physically challenging for many people on long runs where the elevation change is very great. It is important to provide adequate landing space at the top and bottom of stairs in order to allow for pedestrian queues. The total elevation change from floor to floor should be divided into roughly equal sections. The placement of a few steps between relatively long approach sidewalks or interior walkways can be dangerous, even when the steps are well lighted and the tread noses are highlighted with warning colors. These conditions are best handled by short segments of walkway on an 8.33% grade.

In locations where snow and ice accumulates, vertical stair angles of 28.6° or 26.5° are safer than steeper angles. Architectural details that interfere with cleaning reduce the ability of maintenance crews to clear the steps of snow and ice, and reduce the capacity during inclement weather. In high-volume locations, additional stair width may be required to maintain fair-weather flow rates. Table 3.10 provides LOS criteria for stairways.

Exclusive Off-Street Pedestrian Walkways

Analysis of *off-street pedestrian walkways* is similar to analysis for pedestrians on an urban street segment, with the primary difference being the absence of delay due to roadway crosswalks and traffic signals. Off-street pedestrian walkways are generally located more than 35 ft (10.7 m) from an urban street, as well as streets reserved for pedestrian traffic on a full- or part-time basis. When pedestrian cross-flow traffic is considered significant, the walkway segment can also be analyzed as urban pedestrian flow. In general, off-street walkways expect to have fewer width-restricting furniture and street hardware locations, and are found in locations with conditions similar to park-like settings, large shopping malls, and walkways around stadiums. The flowchart in Fig. 3.17 shows the step-by-step analysis for off-street pedestrian facilities.

Figure 3.17 Exclusive Off-Street Pedestrian Walkway Analysis

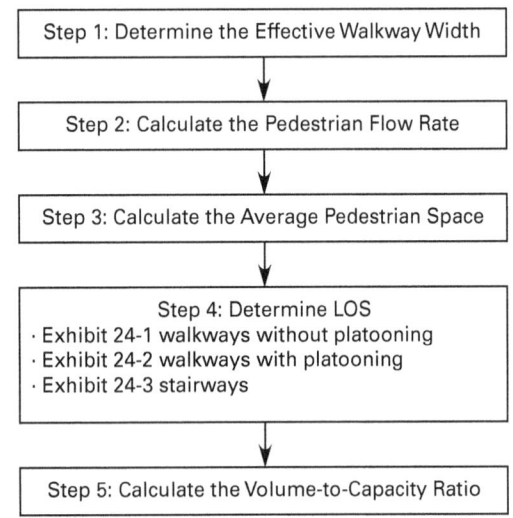

Used with permission from *Highway Capacity Manual*, 6th Edition: A Guide for Multimodal Mobility Analysis, 2016, Exhibit 24-7, by the Transportation Research Board of the National Academies of Sciences, Engineering, and Medicine, Washington DC. DOI: 10.17226/24798

Step 1: Determine Effective Walkway Width

This step is similar to step 1 in the urban street pedestrian flow analysis. Figure 3.7 and Fig. 3.16, and Eq. 3.23 through Eq. 3.27 are used to determine effective walkway width.

Step 2: Calculate the Pedestrian Flow Rate

Equation 3.28 is used to determine the pedestrian flow rate.

Step 3: Calculate the Average Pedestrian Space

Equation 3.30 is used to determine pedestrian space.

Step 4: Determine LOS

To find the LOS, select Table 3.3, Table 3.5, or Table 3.10 [*HCM* Exh. 24-1, Exh. 24-2, and Exh. 24-3], which apply to walkway conditions.

Step 5: Calculate Volume-to-Capacity Ratio

For exclusive pedestrian walkways, the following values can be used to determine volume-to-capacity (v/c) ratios.

- walkways with random flow: 23 ped/min/ft
- walkways with platoon flow, average over 5 min: 18 ped/min/ft
- cross-flow areas: 17 ped/min/ft (sum of both flows)
- stairways: 15 ped/min/ft in ascending direction

Bicycle/Pedestrian and Bicycle Facility Analysis. *Bicycle* and *bicycle/pedestrian* facility evaluations assess the effects and interactions of bicycle and pedestrian traffic with traffic signals and other events that occur along a bicycle or bicycle/pedestrian pathway. Bicycle facilities that share the pathway with pedestrians operate well below potential maximum capacity because of the divergent characteristics between bicycle and pedestrian traffic. Bicycle facilities are defined in three classes by the Federal Highway Administration (FHWA) and in five types by the *Highway Capacity Manual* as shown in Table 3.11.

Table 3.11 Design Condition Comparisons FHWA vs. HCM

design condition	FHWA	HCM
off-street, bicycle use only	Class I	bicycles only
off-street, shared use	Class I	bicycles, pedestrians, skaters
on-street, designated lane	Class II	uninterrupted flow, urban streets
on-street, designated lane	Class III	interrupted flow, urban streets
on-street, share lane with autos	Class III	assign passenger car equivalents

Off-street bicycle facilities, or paths, are called *bikeways*, whereas on-street designated bicycle facilities are called *bike lanes*. Exclusive and shared-use paths can follow the right-of-way of another facility; however, if the path is protected by a barrier or is more than 35 ft (10.7 m) from a street or highway, it is considered an off-street facility. *Shared-use* paths supplement city street circulation systems and provide recreational facilities. Shared-use paths are usually limited to bicycle, pedestrian, and other non-motorized traffic, such as skateboards and wheelchairs. Segways, golf-carts, and quads may be allowed, but are usually limited to 12.5 mph (20.1 kph) and use is regulated by local ordinances or laws. Occasionally, pedestrians are prohibited from using bike lanes, following European practice. Horse traffic is not compatible with bicycles, as bicycle gear noise can startle horses.

Bicycles have a substantially higher speed than pedestrians and have a negative effect on pedestrian flow because of their use characteristics. It is not realistic to equate the two using equivalents; therefore, each mode is evaluated from its own perspective.

The validity of input data necessary to evaluate off-street facilities follows the same hierarchy as other transportation facilities; that is, from the most accurate to the least accurate, input data sources are as follows.

- values derived from field measurements, field observations, and verified field conditions regarding the pathway environment
- locally derived values from similar or nearby facilities, either provided by the user or accumulated by the analyst and adjusted for the subject facility conditions
- default values provided by *HCM* or other creditable sources

The use of default values can provide quick estimates of output values for planning stages and preliminary engineering without the effort of extensive data collection. Difficulties with default values arise from differences between default value research local conditions. Unlike highways, there are few default values accepted as national or statewide standards for pedestrian, bicycle, and other multi-use pathways. ADA, building code, or local planning codes and regulations may have meaning for certain types of facilities. Default values used in *HCM* are shown in Table 3.12.

Uninterrupted Flow: Pedestrians on Shared-Use Paths

Pedestrian traffic on shared-use paths is evaluated using a factor of the number of bicycle passing-and-meeting events with pedestrians. A step-by-step flowchart of the procedure to evaluate pedestrian traffic on shared-use paths is shown in Fig. 3.18.

Step 1: Gather Input Data

For this analysis, data needs are very limited. For both pedestrians and bicycles, the volume demand by direction is necessary, as is the average speed of pedestrians and bicycles. The default speed for pedestrians is 4 ft/sec (1.2 m/s), reduced to 3.2 ft/sec (0.98 m/s) $\geq 20\%$ for the elderly. Average bicycle speed becomes a judgment issue, with 12.8 mph (20.6 kph) a reasonable default value on level grades (from *HCM*). Some texts have recommended using 11.2 mph (18.0 kph) as the average bicycle speed (*HCM*). Of course, actual counts and speeds taken at the facility location are preferred over default data.

Table 3.12 Suggested Default Values for Exclusive and Off-Street Multi-Use Facilities

variable	user group	default value
mode split	bicycle	55%
	pedestrian	20%
	runner	10%
	inline skater	10%
	child bicyclist	5%
mean speed by mode	bicycle	12.8 mph
	pedestrian	3.4 mph
	runner	6.5 mph
	inline skater	10.1 mph
	child bicyclist	7.9 mph
standard deviation of speed by mode	bicycle	3.4 mph
	pedestrian	0.6 mph
	runner	1.2 mph
	inline skater	2.7 mph
	child bicyclist	1.9 mph
peak hour factor	n/a	0.85

Used with permission from *Highway Capacity Manual*, 6th Edition: A Guide for Multimodal Mobility Analysis, 2016, Exhibit 24-6, by the Transportation Research Board of the National Academies of Sciences, Engineering, and Medicine, Washington DC. DOI: 10.17226/24798

Figure 3.18 Pedestrian Flow Analysis on Off-Street Shared-Use Paths

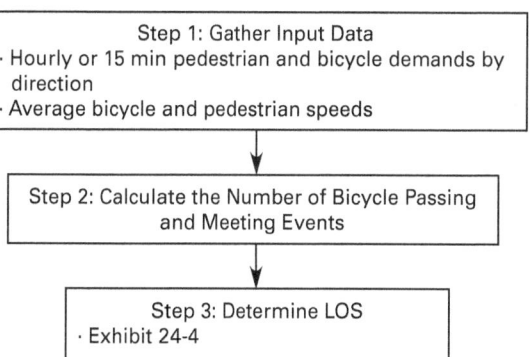

Used with permission from *Highway Capacity Manual*, 6th Edition: A Guide for Multimodal Mobility Analysis, 2016, Exhibit 24-10, by the Transportation Research Board of the National Academies of Sciences, Engineering, and Medicine, Washington DC. DOI: 10.17226/24798.

Step 2: Determine Number of Bicycle Passing and Meeting Events

Field measurements and observations should be undertaken to determine the effect of overtaking and passing behavior by path users. For one-way paths, there are no meeting events. Also, paths that are wider than 15 ft (4.6 m) operate like two one-way paths, which means meeting events have no effect on the perception of service quality for moderate higher volumes of traffic.

The passing and meeting events are calculated by Eq. 3.40 and Eq. 3.41 [*HCM* Eq. 24-5 and Eq. 24-6].

$$F_p = \frac{Q_{\text{sb}}}{\text{PHF}}\left(1 - \frac{S_p}{S_b}\right) \quad 3.40$$

$$F_m = \frac{Q_{\text{ob}}}{\text{PHF}}\left(1 + \frac{S_p}{S_b}\right) \quad 3.41$$

If the input data is available for the peak 15 min flow rate, then $\text{PHF} = 1.0$.

Meeting events tend to have less effect on pedestrians because opposing-direction traffic allows direct visual contact. A default value of 0.5 is used to adjust opposing events for this factor. Other values can be used if verified by field observations.

The total number of events per hour is found by using Eq. 3.42 [*HCM* Eq. 24-7].

$$F = (F_p + 0.5 F_m) \quad 3.42$$

Step 3: Determine LOS

The LOS is determined from Table 3.13.

Table 3.13 Pedestrian LOS for Shared-Use Paths

LOS[a]	event[b] (rate/hr)	related measure bicycle service flow rate per direction (bicycles/hr)[c]	comments
A	≤38	≤28	optimum conditions, conflicts with bicycles rare
B	>38–60	>28–44	good conditions, few conflicts with bicycles
C	>60–103	>44–75	difficult to walk two abreast
D	>103–144	>75–105	frequent conflicts with cyclists
E	>144–180	>105–131	conflicts with cyclists frequent and disruptive
F	>180	>131	significant user conflicts, diminished experience

[a]LOS is based on number of events per hour and applies to any directional split.
[b]An "event" is a bicycle meeting or passing a pedestrian.
[c]Bicycle service volumes are shown for reference and are based on a 50/50 directional split of bicycles.

Used with permission from *Highway Capacity Manual*, 6th Edition: A Guide for Multimodal Mobility Analysis, 2016, Exhibit 24-4, by the Transportation Research Board of the National Academies of Sciences, Engineering, and Medicine, Washington DC. DOI: 10.17226/24798

The LOS E/F for bicycle flow does not represent capacity, but rather a point at which the number of bicyclists meeting and passing causes a severely diminished experience for both bicyclists and pedestrians.

Example 3.1

A pedestrian crossing is located at a signalized intersection with a two-phase, 75 sec cycle. A 5 sec clearance interval is included in each phase effective green time. There are no pedestrian signals. The green time is 40 sec for the major street and 30 sec for the minor street. Data collected at the intersection during field studies are shown. Assuming a pedestrian crossing speed of 4 ft/sec and no pedestrian lost time, create a summary table showing the pedestrian LOS at the intersection based on delay and available space, and determine the location where improvements would most likely have the most impact on the intersection.

major street
   crosswalk length, L 40 ft
   effective crosswalk width, $W_{E,A}$ 14 ft
   inbound pedestrian count, $N_{p,\text{di}}$ 300 ped/15 min
   outbound pedestrian count, $N_{p,\text{do}}$ 150 ped/15 min
   phase green time, g_d 40 sec
minor street
   crosswalk length, L 20 ft
   effective crosswalk width, $W_{E,B}$ 14 ft
   inbound pedestrian count, $N_{p,\text{ci}}$ 500 ped/15 min
   outbound pedestrian count, $N_{p,\text{co}}$ 200 ped/15 min
   phase green time, g_c 30 sec
corner
   radius, R   20 ft
   sidewalk flow, $v_{A,B}$ 225 ped/15 min
   effective sidewalk width, $W_{E,A}$ 14 ft
   effective sidewalk width, $W_{E,B}$ 14 ft

Solution

Find the average delay and LOS for both the major and minor street.

For the major street crossing, from Eq. 3.6,

$$d_p = \frac{(C - g_c)^2}{2C} = \frac{(0.5)(75 \text{ sec} - 30 \text{ sec})^2}{(2)(75 \text{ sec})}$$
$$= 13.5 \text{ sec}$$

From *HCM*, the LOS is B.

For the minor street crossing, from Eq. 3.6,

$$d_p = \frac{(C-g_d)^2}{2C} = \frac{(75 \text{ sec} - 40 \text{ sec})^2}{(2)(75 \text{ sec})}$$
$$= 8.17 \text{ sec}$$

From Table 3.6, the LOS is A.

Determine the net time-space available for crossing the major and minor streets. Use Eq. 3.15(b) since there is not a pedestrian signal.

For the major street crossing, from Eq. 3.15(b),

$$\text{TS}_{cw} = L_d W_d g_{\text{walk,mi}}$$
$$= (40 \text{ ft})(14 \text{ ft})\left(30 \text{ sec} - \frac{40 \text{ ft}}{(2)\left(4 \dfrac{\text{ft}}{\text{sec}}\right)}\right)$$
$$= 14{,}000 \text{ ft}^2\text{-sec}$$

For the minor street crossing, from Eq. 3.15(b),

$$\text{TS}_{cw} = L_d W_d g_{\text{walk,mi}}$$
$$= (20 \text{ ft})(14 \text{ ft})\left(40 \text{ sec} - \frac{20 \text{ ft}}{(2)\left(4 \dfrac{\text{ft}}{\text{sec}}\right)}\right)$$
$$= 10{,}500 \text{ ft}^2\text{-sec}$$

Convert flows to pedestrians per cycle. For the major street crossing,

$$v_{p,\text{di}} = \left(\frac{N_{p,15\,\text{min}}}{60 \dfrac{\text{sec}}{\text{min}}}\right) C$$
$$= \left(\dfrac{\dfrac{300 \text{ ped}}{15 \text{ min}}}{60 \dfrac{\text{sec}}{\text{min}}}\right)\left(75 \dfrac{\text{sec}}{\text{cycle}}\right)$$
$$= 25.0 \text{ ped/cycle}$$

$$v_{p,\text{do}} = \left(\frac{N_{p,15\,\text{min}}}{60 \dfrac{\text{sec}}{\text{min}}}\right) C$$
$$= \left(\dfrac{\dfrac{150 \text{ ped}}{15 \text{ min}}}{60 \dfrac{\text{sec}}{\text{min}}}\right)\left(75 \dfrac{\text{sec}}{\text{cycle}}\right)$$
$$= 12.5 \text{ ped/cycle}$$

For the minor street crossing,

$$v_{p,\text{ci}} = \left(\frac{N_{p,15\,\text{min}}}{60 \dfrac{\text{sec}}{\text{min}}}\right) C$$
$$= \left(\dfrac{\dfrac{500 \text{ ped}}{15 \text{ min}}}{60 \dfrac{\text{sec}}{\text{min}}}\right)\left(75 \dfrac{\text{sec}}{\text{cycle}}\right)$$
$$= 41.7 \text{ ped/cycle}$$

$$v_{p,\text{co}} = \left(\frac{N_{p,15\,\text{min}}}{60 \dfrac{\text{sec}}{\text{min}}}\right) C$$
$$= \left(\dfrac{\dfrac{200 \text{ ped}}{15 \text{ min}}}{60 \dfrac{\text{sec}}{\text{min}}}\right)\left(75 \dfrac{\text{sec}}{\text{cycle}}\right)$$
$$= 16.7 \text{ ped/cycle}$$

For the corner,

$$v_{A,B} = \left(\frac{N_{p,15\,\text{min}}}{60 \dfrac{\text{sec}}{\text{min}}}\right) C$$
$$= \left(\dfrac{\dfrac{225 \text{ ped}}{15 \text{ min}}}{60 \text{ min}}\right)\left(75 \dfrac{\text{sec}}{\text{cycle}}\right)$$
$$= 18.8 \text{ ped/cycle}$$

Analyze the street corner by finding the total circulating pedestrian flow and the available time-space. From Eq. 3.7, the total pedestrian flow is

$$v_t = v_{p,\text{ci}} + v_{p,\text{co}} + v_{p,\text{di}} + v_{p,\text{do}} + v_{A,B}$$
$$= 41.7 \dfrac{\text{ped}}{\text{cycle}} + 16.7 \dfrac{\text{ped}}{\text{cycle}} + 25.0 \dfrac{\text{ped}}{\text{cycle}}$$
$$+ 12.5 \dfrac{\text{ped}}{\text{cycle}} + 18.8 \dfrac{\text{ped}}{\text{cycle}}$$
$$= 115 \text{ ped/cycle}$$

From Eq. 3.9, the available time-space on the corner is

$$\text{TS} = C(W_{E,A} W_{E,B} - 0.215 R^2)$$
$$= (75 \text{ sec})((14 \text{ ft})(14 \text{ ft}) - (0.215)(20 \text{ ft})^2)$$
$$= 8250 \text{ ft}^2\text{-sec}$$

Use Eq. 3.10 and Eq. 3.11 to find the holding-area waiting time for pedestrians waiting to cross the major and minor street. The minor-street red phase, R_{mi}, and the major-street red phase, R_{mj}, are equal to the green time plus one clearance interval.

For the major street crossing, from Eq. 3.10,

$$Q_{tdo} = \frac{N_{do} R_{mi}^2}{2C}$$

$$= \frac{\left(12.5 \ \frac{\text{ped}}{\text{cycle}}\right)(40 \text{ sec} + 5 \text{ sec})^2}{(2)\left(75 \ \frac{\text{sec}}{\text{cycle}}\right)}$$

$$= 169 \text{ ped-sec}$$

For the minor street crossing, from Eq. 3.11,

$$Q_{tco} = \frac{N_{co} R_{mj}^2}{2C}$$

$$= \frac{\left(16.7 \ \frac{\text{ped}}{\text{cycle}}\right)(30 \text{ sec} + 5 \text{ sec})^2}{(2)\left(75 \ \frac{\text{sec}}{\text{cycle}}\right)}$$

$$= 136 \text{ ped-sec}$$

Find the time-space available at the corner using Eq. 3.13.

$$TS_{c,\text{ft}^2\text{-sec}} = TS_{\text{ft}^2\text{-sec}} - 5(Q_{tdo} + Q_{tco})$$

$$= 8250 \text{ ft}^2\text{-sec} - \left(5 \ \frac{\text{ft}^2}{\text{ped}}\right)$$

$$\times \left(\begin{matrix}169 \text{ ped-sec} \\ +136 \text{ ped-sec}\end{matrix}\right)$$

$$= 6725 \text{ ft}^2\text{-sec}$$

Find the space per circulating pedestrian at the corner, then find the LOS using Table 3.3.

From Eq. 3.14,

$$M = \frac{TS_c}{4 N_{\text{tot}}} = \frac{6725 \text{ ft}^2\text{-sec}}{(4 \text{ sec})\left(115 \ \frac{\text{ped}}{\text{cycle}}\right)}$$

$$= 14.6 \text{ ft}^2/\text{ped}$$

Calculate the number of pedestrians that will cross the major and minor streets during the cycle length interval using Eq. 3.16.

For the major street crossing, from Eq. 3.16,

$$N_p = N_{do} \frac{C - g}{C}$$

$$= \left(12.5 \ \frac{\text{ped}}{\text{cycle}}\right)\left(\frac{75 \text{ sec} - 30 \text{ sec}}{75 \text{ sec}}\right)$$

$$= 7.50 \text{ ped/cycle}$$

For the minor street crossing, from Eq. 3.16,

$$N_p = N_{do} \frac{C - g}{C}$$

$$= \left(16.7 \ \frac{\text{ped}}{\text{cycle}}\right)\left(\frac{75 \text{ sec} - 40 \text{ sec}}{75 \text{ sec}}\right)$$

$$= 7.79 \text{ ped/cycle}$$

From Table 3.3, the LOS is E.

Find the crossing time needed for both the major and minor street.

Since the crosswalk width is greater than 10 ft for both the major and minor street, use Eq. 3.17. For the major street crossing,

$$t = 3.2 + \frac{L_{\text{ft}}}{S_p} + 2.7\left(\frac{N_p}{W_{\text{ft}}}\right)$$

$$= 3.2 \text{ sec} + \frac{40 \text{ ft}}{4 \ \frac{\text{ft}}{\text{sec}}} + (2.7)\left(\frac{7.50 \ \frac{\text{ped}}{\text{cycle}}}{14 \text{ ft}}\right)$$

$$= 14.6 \text{ sec}$$

For the minor street crossing,

$$t = 3.2 + \frac{L_{\text{ft}}}{S_p} + 2.7\left(\frac{N_p}{W_{\text{ft}}}\right)$$

$$= 3.2 \text{ sec} + \frac{20 \text{ ft}}{4 \ \frac{\text{ft}}{\text{sec}}} + (2.7)\left(\frac{7.79 \ \frac{\text{ped}}{\text{cycle}}}{14 \text{ ft}}\right)$$

$$= 9.70 \text{ sec}$$

Find the total crosswalk occupancy time required for crossing and the space per pedestrian at crossing for both the major and minor street crossings.

For the major street crossing, from Eq. 3.19,

$$T_{\text{occ}} = t_{\text{ps,do}}N_{\text{do}} + t_{\text{ps,di}}N_{\text{di}}$$
$$= \left(25.0\ \frac{\text{ped}}{\text{cycle}} + 12.5\ \frac{\text{ped}}{\text{cycle}}\right)(14.6\ \text{sec})$$
$$= 548\ \text{ped-sec}$$

From Eq. 3.20, the circulation space for the major street crossing is

$$M = \frac{\text{TS}}{T_{\text{occ}}} = \frac{14{,}000\ \text{ft}^2\text{-sec}}{548\ \text{ped-sec}}$$
$$= 25.5\ \text{ft}^2/\text{ped}$$

From Table 3.3, the LOS is C.

For the minor street crossing, from Eq. 3.19,

$$T_{\text{occ}} = t_{\text{ps,do}}N_{\text{do}} + t_{\text{ps,di}}N_{\text{di}}$$
$$= \left(41.7\ \frac{\text{ped}}{\text{cycle}} + 16.7\ \frac{\text{ped}}{\text{cycle}}\right)(9.70\ \text{sec})$$
$$= 566\ \text{ped-sec}$$

From Eq. 3.20, the circulation space for the minor street crossing is

$$M = \frac{\text{TS}}{T_{\text{occ}}} = \frac{10{,}500\ \text{ft}^2\text{-sec}}{566\ \text{ped-sec}}$$
$$= 18.6\ \text{ft}^2/\text{ped}$$

From Table 3.3, the LOS is D.

Complete the table. LOS using delay is based on a maximum 10 sec delay for LOS A and a maximum 30 sec delay for LOS E.

facility and activity	LOS criterion	value	LOS
corner, waiting time, crossing major street	delay	13.5 sec	B
corner, waiting time, crossing minor street	delay	8.17 sec	A
corner, circulation space	space	14.6 ft²/ped	E
crosswalk space, major street	space	25.5 ft²/ped	D
crosswalk space, minor street	space	18.6 ft²/ped	E

Comparing LOS, improving the corner for waiting pedestrians, such as enlarging the corner to provide more circulation space, would most likely provide the greatest improvement for the intersection.

2. NON-MOTORIZED FACILITIES-BICYCLE

In *HCM*, bicycle transportation is a mode that has a significant and growing influence on mobility. Bicycle mode traffic includes home shopping, commuter, business messenger, and recreational purposes. Weather conditions have a pronounced effect on home shopping and recreational purposes and a slightly less significant effect on business messenger and commuter purposes. Analysis procedures in *HCM* do not cover inclement weather conditions.

Off-Street Shared-Use Paths

The two most frequent types of bicycle facilities encountered are *on-street* and *off-street* facilities. Dirt bicycle trails are not included in these evaluations, although harder-surfaced recreational rails-to-trails facilities may operate in the same manner as paved shared-use paths. Off-street bicycle facilities, defined as paths that are 30 ft or more from a street, are evaluated by the number of passing and meeting events expected. For pedestrian shared-use paths each meeting or passing of pedestrians is considered the same as passing or meeting a bicycle.

To quantify passing and meeting events, Eq. 3.43 and Eq. 3.44 are used.

$$F_p = \frac{Q_{\text{sb}}}{\text{PHF}}\left(1 - \frac{S_p}{S_b}\right) \quad [\textit{HCM}\ \text{Eq. 24-5}] \qquad 3.43$$

$$F_m = \frac{Q_{\text{ob}}}{\text{PHF}}\left(1 + \frac{S_p}{S_b}\right) \quad [\textit{HCM}\ \text{Eq. 24-5}] \qquad 3.44$$

Where,

F_p = number of passing events (events/hr)

F_m = number of meeting events (events/hr)

Q_{sb} = bicycle demand in same direction (bicycles/hr)

Q_{ob} = bicycle demand in opposing direction (bicycles/hr)

S_p = mean pedestrian speed on path (mi/hr)

S_b = mean bicycle speed on path (mi/hr)

Additional evaluations of meeting and passing events that occur at locations with conditions other than base conditions are included in the total number of events, using procedures outlined in *HCM*.

For base conditions, the total number of events is

$$F = F_p + 0.5F_m \quad [\textit{HCM}\ \text{Eq. 24-7}] \qquad 3.45$$

F is the total events on the path in events per hour.

LOS score thresholds for levels C and D are estimated by Eq. 3.46.

$$\begin{aligned} \text{bicycle LOS score} = &\ 5.446 - 0.00809E \\ &- 15.86(\text{RW}) \\ &- 0.287(\text{CL}) - \text{DP} \end{aligned} \quad 3.46$$

[*HCM* Eq. 24-35]

Where,

E = weighted events per min, or meetings per min + 10 (active passings per min)

RW = reciprocal of path width = 1/path width (ft)

CL = 1 if trail has centerline, 0 if no centerline

DP = delayed passings per min, 1.5

The presence of bicycles and other path users on shared-use paths impinges the bicycle traffic by increasing delay, reducing bicycle traffic volumes, and reducing bicyclists' movement.

LOS E and F represent a severely diminished ability for bicyclists to use the path, but do not reflect the capacity of an off-street bicycle path.

Five or fewer weighted events are considered LOS A. Six to 10 weighted events are considered LOS B.

Bicycles on Interrupted Flow Facilities

Interrupted flow facilities are bicycle lanes that pass through signalized intersections. Input variables for bicycles are primarily approach conditions and delay. Approach conditions include dedicated bicycle lanes and mixed-use vehicle-bicycle lanes. Bicycles at intersections in mixed-use mode can be included with pedestrian criteria for very low-speed maneuvers, or evaluated similarly to single passenger car equivalents (PCE) for traffic speeds up to 15 mph. At speeds above 15 mph on wider lanes (16 ft or more) bicycles tend to separate from other vehicle traffic and operate much like bicycles in marked bicycle lanes. Bicycle-specific variables that have an effect on quality of flow include

- demand flow rate of automotive vehicles
- demand flow rate of bicycles
- right turn on red when bicycles are considered pedestrian mode
- pedestrian flow rate modified by the presence of bicycles
- number of lanes for bicycles
- pedestrian signal head reference when bicycles share pedestrian movements

In locations where bicycle traffic is very high, such as college campuses, bicycle movements can be assigned their own signal phases.

Bicycles at Unsignalized Intersections

Bicycles at unsignalized intersections may queue with motor vehicles on approaches and cross in groups rather in single file linearly. At intersections with exclusive bicycle lane approaches, the bicycles can move across the intersection concurrently with automobiles. In either case, bicycles incur fewer delays than automobile traffic. Left turns require weaving movements, either on the approach to the intersection or while other vehicles are stopped waiting for a clear interval to proceed. At very busy intersections, bicyclists, especially the less experienced, may dismount and walk alongside their bicycle to cross as a pedestrian movement. These left-turn and pedestrian-like movements increase bicycle delay. No specific procedures and anticipated bicycle LOS evaluations have been included in *HCM* for unsignalized intersections.

On-Street Bicycle Facilities

Bicycles on exclusive on-street marked bicycle lanes operate on a perceived LOS score which varies considerably depending on a multitude of factors detailed by local conditions. Perceived scores include grade, width of bicycle lane, proportion of occupied parallel parking spaces, pavement surface conditions, and, for curbside bicycle lanes, the proximity of street furniture to the curb and the condition of drain grates along the curb. Although quantifiable values of LOS thresholds have been suggested, there are considerable variations from situation to situation, such that no one value system (e.g., average speed, average delay, or number of events) have proven adequate to assign other than subjective values (i.e., rider experience) to LOS values. In general, on-street facilities may be able to be evaluated similar to off-street facilities, but only as an approximation, since off-street facilities do not tend to have as many travel interferences as on-street facilities.

3. MASS TRANSIT SYSTEM STUDIES

Mass transit is all large-scale systems of public transportation that serve a city or other metropolitan area, as well as a variety of intercity systems. *Public transportation* is any network of passenger vehicles that the general public can use. The systems operate on set routes, have set hours of operation, and charge set fares. Public transportation includes rail cars, buses, and other fixed-guideway vehicles, such as monorails. The systems operate in mixed highway traffic and on grade-separated right-of-ways that run on surface, underground, or elevated locations.

Public transit systems are generally owned and operated by government or government-sponsored entities. Government agencies include municipal, county, regional, and multistate authorities. Operating responsibility can also be contracted to private companies to manage or operate significant parts of the transit system. Public transit systems involve various forms of government financial assistance in order to cover capital, operating, and maintenance costs.

Transit systems, whether private or government operated, are heavily regulated by both state and federal programs. Service quality standards vary from region to region, depending on the primary emphasis of the prevailing government unit. Analyzing transit system service is considerably different from analyzing general automobile-mode traffic; automobile-mode traffic analysis is concerned with moving vehicles, whereas transit mode service is concerned with moving people.

Transit systems have further constraints of availability compared to pedestrian or highway vehicle facilities. A highway system, once constructed, is open for traffic and is available for use 24 hours a day, whereas transit service is available only when the transit vehicles are operating. Furthermore, fixed-route transit movements along a street or road, or through an intersection, place non-optional demands on the transit vehicle movements. Automobile drivers who find an intersection or roadway blocked or jammed with traffic may choose another route with less congestion, but a transit vehicle has no other option but to work through the congestion, either because of physical restrictions (i.e., with rail systems), or because deviating from the route would bypass transit stops and not serve the waiting transit patrons.

Transit ridership is made up of two primary groups: choice riders and captive riders. *Choice riders* are those that voluntarily use transit for their commute so as to avoid the hassles of driving and the expense of downtown parking. *Captive riders* are those that are unable to use other forms of transportation because of age, income, or physical or mental limitations.

Transit passengers become users of other forms of transportation at the ends of their transit trips. These forms include, but may not be limited to, walking, bicycles, automobiles, carpools, or motorcycles. How much these other forms play a role in the transit trip depends on the proximity of the origin and destination to the access points of the transit trip. For instance, if a desirable goal for planning is to have transit available within a 0.5 mi (0.8 km) walking distance from every resident in an urban area, to be successful there must be a network of routes that covers the residential areas at all times when transit service is desired. This coverage may not be available because of economics, topographic limits, or simply a lack of support from the local government unit. For those origins or destinations greater than 0.5 mi (0.8 km) from a transit stop, the transit patron must use another form of transportation. In order to relieve the burden of connecting services, a growing approach is to encourage the development of desirable housing close to existing transit through land-use policies that encourage densities that support transit usage.

A contrasting approach is to allow housing to develop, and then fit transit service into the land use after development occurs. The latter is mostly what happened in the United States when suburban development occurred along the interstate highway system—so much so that providing reasonable and frequent transit service in low-density housing areas resulted in insufficient ridership. Low ridership in low-density housing is considered economically unjustifiable. It is based on excess available seat-miles, or on low-passenger count per hour per transit vehicle. As a result, the form of transportation used by a great majority of riders at the home end of transit trips has become the automobile—driving to and from a park-and-ride lot at a transit terminal or near a bus stop. The transit vehicle is used for the more densely patronized portion of the trip; the automobile is used for the collection and distribution of riders to and from lower density housing areas.

Blending other transportation modes into a mass transportation system is feasible when the characteristics of other modes complement the role of the desired travel trips. An example is the development of transit villages near transit stations. The goal is to make walking feasible by building the majority of housing units within walking distance of the transit patrons. However, for the traditional park-and-ride transit station of a rail system, the housing units compete for the same limited space used for parking, which provides a great challenge to developers of the transit village concept.

System Capacity Determination

The capacity of a mass transit system is often cited for the system as a whole, or in terms of specific line segments, specific routes, or specific modes. Assuming there are sufficient vehicles available to satisfy the demand, several factors influence the capacity of a single transit line.

- line haul capacity
- station platform boarding/alighting capacity
- station access capacity to and from surrounding streets
- control delays, especially with guideway-captive vehicles (steel rail, monorail, cableway)
- junction capacity (fixed-guideway systems)
- surface traffic interference at grade crossings for fixed guideway systems, and traffic mingling for bus and light rail on city streets

Any one of the factors alone may limit capacity, or capacity can be limited by a combination of more than one. While *line haul capacity* (the uninterrupted capacity between stations) is quite often cited as the theoretical maximum of a line segment, the maximum capacity is more often restricted by station delay or signal delay. Station delay controls capacity when the station delay exceeds the *vehicle arrival spacing*, or the amount of time that passes between vehicles arriving at the station. Station delay is common on fixed-rail terminals, and in very crowded bus terminals. Signal delay controls capacity when the transit signal system or interference from surface street traffic at intersections is greater than any other delay on the line. Signal delay is most common on fixed rail systems, as minimum train spacing is necessary for the safety concerns of intersecting trains and for the stopping distance of steel rail equipment.

For surface street transit network planning, fixed-route transit movements through an intersection or line segment are non-optional demands. For instance, auto drivers finding an intersection or roadway blocked may choose another route with less congestion; whereas a transit vehicle is assigned a given route and usually has no other option but to work through the congestion. On a steel rail system, there is rarely another track to use as an alternate route should the one needed track be blocked.

For both rail and bus terminal designs, passenger egress capacity through platform and stairway exits is much greater than the maximum off-loading capacity of platforms, for obvious safety reasons. Buses on exclusive right-of-way (i.e., busways) are analyzed similar to rail on exclusive right-of-way, with the exception that busway stations can be more easily designed with bypass lanes for through buses to go around stopped buses than for rail systems. Bypass tracks for rapid rail systems are not practical because of additional delays caused by the need to traverse the special trackwork (track switches) at lower speeds, and the additional complexity of safety signal systems required for the bypass trackage. On bus and light rail systems that utilize onboard fare collection, some additional delay is encountered at the platforms. This delay has been reduced by newer-design buses and LRVs (light rail vehicles), which have lower floors and electronic fare systems onboard the vehicles themselves. Buses have the advantages of operating at a closer spacing than rail vehicles without the constraints caused by signal systems, and of serving both as the line haul and neighborhood distribution vehicle, thereby eliminating the delays caused by intermodal transfer.

Typically, buses can operate at a headway of 20 to 30 sec for extended periods, with scheduled arrivals at a rate of 90 to 150 vehicles or more during the peak hour. LRV systems are usually limited to headways of 90 to 120 sec for short peak periods, and are limited to a maximum scheduled arrival rate of 20 to 26 trains during the peak 1-hour period. LRVs have the advantage of operating in multiple-car trains to increase line haul capacity where needed. For LRVs operating in mixed-traffic streetcar mode (that is, when using on-street paved trackage without track signals), the operation is at a slower speed than open track running. LRVs in streetcar mode can be evaluated similar to on-street bus operation.

In summary, there are many factors that the analyst must take into account when determining transit system capacity. The ultimate result of capacity-planning analysis is to match the desired system or line-haul capacity to the expected travel demand.

Line Haul Direct Solutions

For flow and capacity using line haul analysis, the travel time between end points of the segment and vehicle headway must be known. While vehicle (or train) headway spacing is often controlled by schedule or by signals, travel time and travel speed are controlled by the vehicle characteristics as actually operated, which in turn determines travel time.

The time a transit vehicle spends stopped at a platform or station while passengers board and alight is called *dwell time*. *Station delay* is dwell time plus the deceleration and acceleration time required for a station stop, compared to traveling through the station without stopping.

Four elements make up a station-to-station segment: acceleration, normal travel speed between stations, deceleration, and dwell time. When curve geometry and traffic restrictions do not limit top speed, and stations are spaced at great enough distance to allow the vehicles to accelerate to top speed, the *time-velocity* diagram is plotted, as shown in Fig. 3.19. v is the normal travel speed between stations. v_0 is the initial and final speeds before leaving one station and after arriving at the next station (usually = 0).

Figure 3.19 *Time-Velocity Diagram of Vehicle Operations*

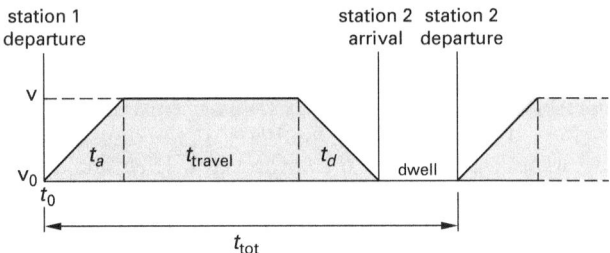

Where,

t_a = time needed for acceleration

t_d = time needed for deceleration

t_{travel} = time needed for travel

The diagram assumes that acceleration and deceleration occur at constant rates, and that normal travel speed is constant. In actual operation, acceleration and deceleration can vary with speed, and travel speed can vary depending on curve or grade restrictions. With some types of control systems on subway equipment, periodic coasting occurs after the maximum operating speed (called *balancing speed*) is attained, resulting in a saw-tooth-shaped travel speed plot. To simplify analysis, these variations are ignored. A plot of distance-velocity is shown in Fig. 3.20.

Figure 3.20 Distance-Velocity Diagram of Vehicle Operations

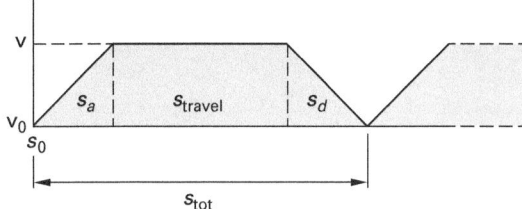

The travel time *between* stations is the sum of the time needed for acceleration, t_a, the time needed for travel, t_{travel}, and the time needed for deceleration, t_d. It is determined by applying Eq. 3.47, Eq. 3.48, and Eq. 3.49 to each element of Fig. 3.20. s_{travel} is the distance traveled at design speed.

$$t_a = \frac{v - v_0}{a} \qquad 3.47$$

$$t_{travel} = \frac{s_{travel}}{v_{initial}} \qquad 3.48$$

$$t_d = \frac{v - v_0}{d} \qquad 3.49$$

The average travel speed between stations does not include station dwell time. The *running time* is the average travel speed of a line segment including the dwell times at all stations encountered.

Applying Eq. 3.50 and Eq. 3.51 to the acceleration and deceleration rates results in the acceleration distance, s_a, and the deceleration distance, s_d. Subtracting s_a and s_d from the total distance, s_t, yields the distance traveled at full normal travel speed, s_{travel}.

$$s_a = \frac{v^2 - v_0^2}{2a} \qquad 3.50$$

$$s_d = \frac{v^2 - v_0^2}{2d} \qquad 3.51$$

If the train set or vehicle is unable to accelerate to full normal travel speed before decelerating for the next station, the distance-velocity diagram would look like a triangle, as shown in Fig. 3.21.

For this case, v, a, d, and distance, s, are related by Eq. 3.52.

$$s = \frac{v^2}{2Sa} + \frac{v^2}{2Sd} = v^2\left(\frac{1}{2Sa} + \frac{1}{2Sd}\right) \qquad 3.52$$

Figure 3.21 Distance-Velocity Diagram When the Distance Between Stations Does Not Allow Acceleration to Full Design Speed

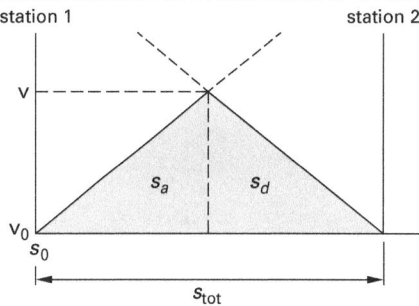

Equation 3.53 is used to solve for maximum velocity, v_{max}.

$$v_{max} = \sqrt{\frac{s}{\dfrac{1}{2a} + \dfrac{1}{2d}}} \qquad 3.53$$

Passenger comfort in vehicles follows established parameters of human comfort tolerance limits. Public transit usually has more restrictive design limits than other modes for the safety of passengers moving about and the safety of standees. As a general rule, the average acceleration rates should be limited to $\pm 0.1g$ in any direction (horizontal, vertical, and lateral) with a few exceptions. Typical *horizontal acceleration rates* can range from $0.05g$ to $0.1g$, with the higher range of acceleration being limited by horsepower and traction rates. Common *horizontal deceleration rates* vary from $0.1g$ to $0.3g$, with the higher range being limited by braking power, traction, and passenger comfort. During braking situations, passengers seem to have a somewhat higher tolerance for deceleration for short periods of time. Overhead electric-powered equipment usually has greater acceleration, whereas rubber-tired equipment has greater deceleration. During vertical acceleration, gravity tends to intensify the sense of upward acceleration and reduce the sense of downward acceleration. This sensation has led designers to increase the length of sag vertical curves over crest vertical curves for the same degree of grade change in order to reduce passenger discomfort.

Maximum Passenger Flow Rate

The maximum passenger flow rate of a given segment of a transit system can be found from Eq. 3.54. N_v is the number of transit vehicles or trains arriving per hour at the station being studied, and c_{max} is the maximum capacity of each vehicle or train serving the station.

$$v_{p,max} = N_v c_{max} \qquad 3.54$$

Once the maximum passenger flow rate is known, the number of trips per hour required to accommodate that maximum flow rate can be found from Eq. 3.55 for bus transit systems and from Eq. 3.56 for rail transit systems.

$$N_{\text{trips/hr}} = \frac{v_{p,\text{max}}}{c_{\text{max}}} \quad [\text{bus}] \qquad 3.55$$

$$N_{\text{trips/hr}} = \frac{v_{p,\text{max}}}{c_{\text{car}} N_{\text{cars}}} \quad [\text{rail}] \qquad 3.56$$

Passenger flow at a node of concentrated activity along a route or corridor may be much greater than can be accommodated at a single stop or platform location. For these locations, it is common practice to provide multiple berths (*platform locations*) so that multiple vehicles or trains can arrive and depart somewhat independently of others. The number of berths required to accommodate the service volume at a given location can be found using Eq. 3.57. The volume per passenger berth is determined by the maximum boarding and alighting flow rate at each berth position.

$$N_{\text{berths}} = \frac{v_{p,\text{max}}}{v_{\text{ps}}} \qquad 3.57$$

For multiple bus berths, the arrival and departure rate of buses is controlled by the number of available berths, the time needed for the driver to select the stopping location and maneuver the bus to the stopping location, the bus stop dwell time, and the time needed to maneuver away from the selected berth. Bus systems can operate without signal controls at multiple berth locations in all but very high traffic terminals. Rail systems using multiple platform stations with multiple tracks must have traffic sorted and controlled by signal systems in order to avoid collisions.

The capacity of a transit vehicle, transit station, or transit route can vary widely from route to route and from city to city, depending on local conditions, local regulations, and other operating requirements placed on the transit system. Capacity values should always be available from local transit information or from reliable local study data. Studies from one city should not be arbitrarily applied to another city and assumed to fit without calibration, as transit data varies much more widely between cities than other highway traffic data.

Example 3.2

A two-car light rail train has a capacity of 125 passengers per car. The scheduled arrival at a certain point on the line is at 6 min intervals. What is the maximum passenger flow rate at this location?

Solution

The number of trains per hour is found by dividing one hour by the train arrival interval.

$$N_t = \frac{60 \frac{\text{min}}{\text{hr}}}{6 \frac{\text{min}}{\text{train}}} = 10 \text{ trains/hr}$$

The maximum passenger flow rate is found from Eq. 3.56.

$$\begin{aligned} v_{p,\text{max}} &= N_t c_{\text{car}} N_{\text{cars}} \\ &= \left(10 \frac{\text{trains}}{\text{hr}}\right)\left(125 \frac{\text{passengers}}{\text{car}}\right)\left(2 \frac{\text{cars}}{\text{train}}\right) \\ &= 2500 \text{ passengers/hr} \end{aligned}$$

Example 3.3

A train with a 200 mph design speed is to run on a perfectly straight track between two cities that are 25 mi apart. The acceleration of the train is limited to $0.03g$, and the deceleration is limited to $0.02g$. (a) What is the maximum distance the train will be able to run at 200 mph? (b) If a station were to be added at 10 mi from one city, what would be the maximum speed the train would attain on the 10 mi stretch of track?

Solution

(a) Convert miles per hour to feet per second.

$$\frac{\left(200 \frac{\text{mi}}{\text{hr}}\right)\left(5280 \frac{\text{ft}}{\text{mi}}\right)}{3600 \frac{\text{sec}}{\text{hr}}} = 293.3 \text{ ft/sec}$$

Find the acceleration distance from Eq. 3.50.

$$\begin{aligned} s_a &= \frac{v_{\text{full}}^2 - v_{\text{initial}}^2}{2a} = \frac{v_{\text{full}}^2 - v_{\text{initial}}^2}{(2)(0.03g)} \\ &= \frac{\left(293.3 \frac{\text{ft}}{\text{sec}}\right)^2 - \left(0 \frac{\text{ft}}{\text{sec}}\right)^2}{(2)(0.03)\left(32.2 \frac{\text{ft}}{\text{sec}^2}\right)\left(5280 \frac{\text{ft}}{\text{mi}}\right)} \\ &= 8.43 \text{ mi} \end{aligned}$$

Find the deceleration distance from Eq. 3.51.

$$s_d = \frac{v_{full}^2 - v_{final}^2}{2d} = \frac{v_{full}^2 - v_{final}^2}{(2)(0.02g)}$$

$$= \frac{\left(293.3 \frac{ft}{sec}\right)^2 - \left(0 \frac{ft}{sec}\right)^2}{(2)(0.02)\left(32.2 \frac{ft}{sec^2}\right)\left(5280 \frac{ft}{mi}\right)}$$

$$= 12.6 \text{ mi}$$

Subtract the acceleration and deceleration distances from the total distance to find the distance traveled at full design speed, 200 mph.

$$s_{full} = s_t - s_a - s_d$$
$$= 25 \text{ mi} - 8.43 \text{ mi} - 12.6 \text{ mi}$$
$$= 3.97 \text{ mi}$$

The train will run at 200 mph for a maximum distance of 3.97 mi.

(b) For a station at the 10 mi location, the maximum speed is found using Eq. 3.53.

$$v_{max} = \sqrt{\frac{s}{\frac{1}{2a} + \frac{1}{2d}}} = \sqrt{\frac{s}{\frac{1}{(2)(0.03g)} + \frac{1}{(2)(0.02g)}}}$$

$$= \sqrt{\frac{(10 \text{ mi})\left(5280 \frac{ft}{mi}\right)}{\frac{1}{(2)(0.03)\left(32.2 \frac{ft}{sec^2}\right)} + \frac{1}{(2)(0.02)\left(32.2 \frac{ft}{sec^2}\right)}}}$$

$$\times \left(3600 \frac{sec}{hr}\right)$$

$$= 138 \text{ mph}$$

The maximum speed the train will attain is 138 mph.

Example 3.4

A busway station design uses four berths per station, allowing up to four buses to stop simultaneously at each station. Dwell times allow scheduling of three buses per minute along the busway during peak 15 min periods. One-third of the buses using the busway have an average capacity of 93 passengers. One-tenth of the buses have an average capacity of 53 passengers. The remaining buses have an average capacity of 62 passengers per bus. What is the busway's maximum passenger capacity (in passengers per hour)?

Solution

Converting the bus arrival rate per minute to an hourly rate, the hourly arrival rate is

$$N_b = \left(\frac{3 \text{ buses}}{min}\right)\left(\frac{60 \text{ min}}{hr}\right) = 180 \text{ buses/hr}$$

The maximum passenger volume is found by assigning the bus capacities in proportion to the bus populations.

$$v_{p,max} = \left(\left(\frac{1}{3}\right)\left(180 \frac{bus}{hr}\right)\right)\left(93 \frac{pass}{bus}\right)$$
$$+ \left(\left(\frac{1}{10}\right)\left(180 \frac{bus}{hr}\right)\right)\left(53 \frac{pass}{bus}\right)$$
$$+ \left(\left(1 - \frac{1}{3} - \frac{1}{10}\right)\left(180 \frac{bus}{hr}\right)\right)\left(62 \frac{pass}{bus}\right)$$
$$= 12,858 \text{ pass/hr}$$

Urban Scheduled Transit

Mass transit is also called *urban scheduled transit* or *transit mode*. From the urban planning standpoint, transit mode is a way to move large volumes of people, servicing concentrated activity centers, without increasing the burden of parking and the highway infrastructure needed to support auto travel.

Measures of transit operating effectiveness or efficiency are dependent on both the view of the observers and that of the participants. For instance, efficiency from the transit operator's viewpoint is related to a high percentage of usage for the hours that service is offered, efficient use of vehicle capacity, and the adequacy of station facilities and manpower needs. The transit rider views efficiency as having the needed frequency, reliability, and location of service, as well as a sense of overall safety and fares kept as low as possible. The overall operating authority is concerned with annual ridership, vehicle operating expense, the overall economics of passengers per revenue-mile and revenue-hour, annual operating budgets, and the justification of grant revenue streams for capital and operating expenses.

Planning transit service involves matching scheduled capacity as close as possible to the transit demand and using vehicles available in the fleet. Capacity is a function of trip frequency multiplied by vehicle capacity. Several methods are used to determine the capacity of transit vehicles. For seated loads, vehicle capacity is the number of seats provided on the vehicle. For routes or systems that accommodate standees, the capacity may be scheduled by a percentage of seated load, by an absolute number, or by the square footage of vehicle floor area. Standee limits for buses based on percentage of seats or by absolute number of standees may be in place for routes that operate across state lines. Scheduled maximum capacity determined by floor area is typical for very crowded city lines, and usually includes both seated and standing passengers. The floor area is divided by the average minimum area occupied by a transit patron. For instance, a general guide is to use 5.5 ft² (0.5 m²) per patron for combined seated and standing loads as an acceptable maximum density for planning purposes. Maximum vehicle load, or *crush load*, is a term used in transit to indicate the highest number of passengers that a vehicle can carry. Crush is not used for load scheduling purposes, as these loads only occur under extreme peak conditions for a short period of time.

Example 3.5

Determine the maximum capacity for a 40 ft long × 8.5 ft wide city transit bus with a load capacity limit of one passenger per 5.5 ft².

Solution

The maximum capacity of the bus is

$$c_{\max} = \left(\frac{1 \text{ passenger}}{5.5 \text{ ft}^2}\right)(40 \text{ ft})(8.5 \text{ ft}) = 62 \text{ passengers}$$

Transit Modes on Urban Streets

This section begins a discussion on transit modes (particularly buses) on urban streets, including traffic flow, the constraints of fixed routing, and the effects of signalization. Railroad transit, bus rail transit, and light rail transit are also covered.

Auto Equivalents

The *Highway Capacity Manual* contains a number of tables to convert bus traffic to auto equivalents for various conditions. Unlike trucks, local buses in urban mode impede traffic flow with their frequent stops, which must be taken into account when analyzing maximum lane flow rates and intersection signal timing. Bus stops are frequently placed at or near intersections. Near-side bus stops cause less intersection delay at stop-controlled intersections and at signalized intersections when buses arrive on the red signal, although near-side bus stops interfere with right-turn-on-red movements when permitted. While far-side bus stops may theoretically appear to have less effect on traffic flow from the traffic-engineering standpoint, local experience with the transit agency may find more rear-end collisions with far-side bus stops by motorists who do not expect the bus to stop after clearing the intersection on a green signal.

Mid-block pull-off stops, sometimes called *off-line stops*, delay bus service during periods of heavy traffic flow when few gaps are available for reentry. Bus drivers tend to force their way into an unyielding traffic flow, which increases motorists' aggravation and generates many complaints to municipal authorities about bus driver behavior.

The debate over near-side vs. far-side vs. mid-block stops is largely a local one, and it depends on what local drivers are used to encountering. More important to the transit patron is the location of the bus stop in relation to stop access and sufficient waiting space for bus patrons. Poor stop locations can cause increased pedestrian volumes at crosswalks, and increased danger of random pedestrian movements across the traffic flow where crosswalks are not provided. The location of bus stops (and streetcar-style LRV stops) must be carefully worked out between the responsible traffic officials and transit officials. It should be acknowledged that recognizing the needs of both motorists and transit patrons will be a source of continuing concern, requiring careful consideration to respect the needs of both parties in providing the best possible solution while minimizing compromises on the quality of the urban environment.

Bus flow along open roadways without stops is similar to truck flow. A bus can be considered the same passenger car equivalent as a single unit truck if the bus density is low and stops are infrequent.

Constraints of Fixed Routing

Urban transit buses, and for the most part intercity buses, must stay on their fixed route. This has the effect of predictability on traffic demand, but also makes for increased passenger delay when sudden congestion occurs. It is not logical to have bus drivers making their own decisions to drive around sudden congestion, moving to another street while leaving passengers stranded at the bus stops along the normal route. This constraint must be taken into account when developing a traffic management system.

Effects of Signalization

Traffic signals can help bus operations by breaking up traffic flow and allowing gaps for pedestrian movements. Signals can also help slow or stop traffic at intervals to reduce conflicts with stopped busses, and to speed up the flow of busses to reduce acceleration delays.

Timing signal phases includes allocating bus movements, particularly if the bus is making a turn through crowded pedestrian traffic. On a signal-sequenced arterial, a bus servicing a stop causes the bus to delay its progress with the sequenced platooned flow, which can delay arrival at the next signal after the green phase has timed out. A typical two-mile arterial street section with heavy retail activity can have bus-travel time delays increased 12 to 15 min when signals are improperly timed. A possible solution is to use signal preemption, activated by the arriving bus, which extends the green time to allow an approaching bus to clear the intersection. Signal preemption can improve flow capacity in certain situations.

Transit LOS Determination

HCM uses a series of analysis steps to assign LOS indicators to transit modes on urban street segments. This methodology may be viewed as an analysis of *on-street performance of transit* (usually local buses mixed with automobile traffic) from a traffic-engineering standpoint; however, it is not used for establishing *quality of service ratings* for the transit operations. LOS determination uses a series of steps assigning LOS indicators, as outlined in Fig. 3.22.

Transit operational characteristics vary from city to city, and a standard from one transit system is unlikely to apply universally. The bulk of transit variable data is derived from field observations, many of which can be provided by transit authorities.

Figure 3.22 Transit Methodology for Urban Street Segments

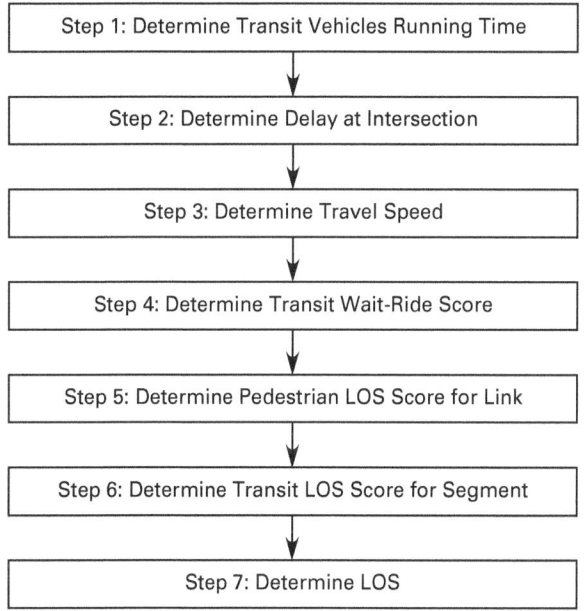

Used with permission from *Highway Capacity Manual*, 6th Edition: A Guide for Multimodal Mobility Analysis, 2016, Exhibit 18-26, by the Transportation Research Board of the National Academies of Sciences, Engineering, and Medicine, Washington DC. DOI: 10.17226/24798

Step 1: Determine Transit Vehicle Running Time. Step 1 covers segment running time, segment delays due to a stop, delays due to dwell time, and segment running computation.

Segment Running Time

Equation 3.58 [*HCM* Eq. 18-48] is independent of the proximity of stops to traffic control devices.

$$S_{Rt} = \min\left(S_R, \frac{61}{1 + e^{-1.00 + \frac{1185 N_{ts}}{L}}}\right) \quad 3.58$$

Variables

L = segment length (ft)

N_{ts} = number of transit stops on the segment for the subject route (stops)

S_R = motorized vehicle running speed = $3600L/5280t_R$ (mph)

S_{Rt} = transit vehicle running speed (mph)

t_R = segment running time (sec)

Segment Delay Due to a Stop

Segment delay due to a stop includes

- acceleration-deceleration delay
- delay due to serving passengers (stop dwell-time)
- reentry delay

Equation 3.59 [*HCM* Eq. 18-49] is applied to each stop in the segment, and summed for total stop delay for the segment.

$$d_{ad} = \left(\frac{5280}{3600}\right)\left(\frac{S_{Rt}}{2}\right)\left(\frac{1}{r_{at}} + \frac{1}{r_{dt}}\right)f_{ad} \quad 3.59$$

Variables

$f_{ad} = 1-x$ for near-side stops at roundabouts

$f_{ad} = g/c$ for near-side stops at intersections

d_{ad} = transit vehicle acceleration-deceleration delay due to a transit stop (sec)

f_{ad} = proportion of transit vehicle stop acceleration-deceleration delay not due to traffic control

x = volume-to-capacity ratio of the link's rightmost lane on a roundabout

$f_{ad} = 0.00$ for near-side stops at all all-way and major-street two-way stop-controlled intersections

$f_{ad} = 1.00$ for stops on the near side of a boundary intersection

r_{at} = transit vehicle acceleration rate (= 4.0 ft/sec² unless otherwise known)

r_{dt} = transit vehicle deceleration rate (= 4.0 ft/sec² unless otherwise known)

Acceleration-deceleration delay incurred in excess of the normal delay incurred due to a traffic control device is included in the total delay.

Delay Due to Serving Passengers (Dwell Time)

The dwell time is a function of the number of passengers served at each stop. For this analysis, the dwell time is averaged over several stops, unless a stop has significantly more dwell time than others do in the segment [*HCM* Eq. 18-51].

$$d_{ps} = t_d f_{dt} \quad 3.60$$

Variables

d_{ps} = transit vehicle delay due to serving passengers (sec)

t_d = average dwell time (sec) (usually between 2 sec and 60 sec)

f_{dt} = proportion of dwell time occurring during effective green (= g/c at near-side stops at signalized intersections and 1.00 otherwise)

Reentry Delay

Reentry delay, d_{ret}, is primarily a condition of traffic density and motorist courtesy.

- For on-line stops, reentry delay is zero.
- At off-line stops away from the influence of intersection signals, reentry delay is similar to gap-acceptance at two-way stop-controlled intersections, as if the bus were making a right turn onto the link. A critical headway of 7 sec is used to account for the acceleration of buses.
- At off-line stops near or in a signalized intersection queue, reentry delay is estimated as the queue service time.

Segment Running Computation

The segment running time is based on the segment running speed and the sum of the delays due to stops on the segment [*HCM* Eq. 18-53].

$$t_{Rt} = \frac{3600L}{5280 S_{Rt}} + \sum_{i=1}^{N_{ts}} d_{ts,i} \quad \quad 3.61$$

Variables

t_{Rt} = segment running time of transit vehicle (sec)

$d_{ts,i}$ = delay due to transit vehicle stop for passenger pickup at stop i within the segment (sec)

Step 2: Determine Delay at Intersection. For the downstream boundary intersection at the end of a segment, Eq. 3.62 [*HCM* Eq. 18-54] is used to determine through delay.

$$d_t = t_l 60\left(\frac{L}{5280}\right) \quad \quad 3.62$$

Variables

d_t = rough vehicle delay (sec/veh)

L = segment length

t_l = transit vehicle running time loss (min/mi), using Table 3.14

Step 3: Determine Transit Travel Speed. Use Eq. 3.63 [*HCM* Eq. 18-55] to determine transit travel speed, $S_{Tt,seg}$, along the segment:

$$S_{Tt,seg} = \frac{3600L}{5280(t_{Rt} + d_t)} \quad [\text{mph}] \quad \quad 3.63$$

Step 4: Determine Transit Wait-Ride Score. For segments with no transit stops, the score is score = 0.0; otherwise, the wait-ride score is made up of the headway factor and the perceived travel-time factor.

Headway Factor

The headway factor is an estimate of patronage related to the arrival rate of one transit vehicle per hour (60 min headway). The patronage is assumed to be an exponential relationship to the frequency rate as shown in Eq. 3.64 [*HCM* Eq. 18-56].

$$F_h = 4.00 e^{-1.434/v_s + 0.001} \quad \quad 3.64$$

Variables

F_h = headway factor

v_s = transit frequency for segment (vph)

Perceived Travel Time Factor

The perceived travel time factor is based on the perceived travel time rate and the expected ridership elasticity with respect to changes in the perceived travel time rate as shown in Eq. 3.65 [*HCM* Eq. 18-57] and Eq. 3.66 [*HCM* Eq. 18-58].

$$F_{tt} = \frac{(e-1)T_{btt} - (e+1)T_{ptt}}{(e-1)T_{ptt} - (e+1)T_{btt}} \quad \quad 3.65$$

$$T_{ptt} = \left(a_1 \frac{60}{S_{Tt,seg}}\right) + (2T_{ex}) - T_{at} \quad \quad 3.66$$

Equation 3.67 through Eq. 3.69 are the three forms of *HCM* Eq. 18-59. Equation 3.70 is *HCM* Eq. 18-60.

$$a_1 = 1.00 \quad F_l \leq 0.80 \quad \quad 3.67$$

$$a_1 = 1 + \frac{4(F_l - 0.80)}{4.2} \quad \text{for} \quad 0.80 < F_l \leq 1.00 \quad \quad 3.68$$

$$a_1 = 1 + \frac{4(F_l - 0.80) + (F_l - 1.00) \times \bigl(6.5 + \bigl(5(F_l - 1.00)\bigr)\bigr)}{4.2 F_l} \quad \quad 3.69$$

$$[\text{for } F_l > 1.00]$$

$$T_{at} = \frac{1.3 p_{sh} + 0.2 p_{be}}{L_{pt}} \quad \quad 3.70$$

Table 3.14 Transit Vehicle Running Time Loss

area type	transit lane allocation	traffic condition	running time loss by signal condition (min/mi)		
			typical	signals set for transit	signals more frequent than transit stops
central business district	exclusive	no right turns	1.2	0.6	1.5–2.0
		with right-turn delay	2.0	1.4	2.5–3.0
		blocked by traffic	2.5–3.0	not available	3.0–3.5
	mixed traffic	any	3.0	not available	3.0–3.5
other	exclusive	any	0.7 (0.5–1.0)	not available	not available
	mixed traffic	any	1.0 (0.7–1.5)	not available	not available

Used with permission from *Highway Capacity Manual*, 6th Edition: A Guide for Multimodal Mobility Analysis, 2016, Exhibit 18-27, by the Transportation Research Board of the National Academies of Sciences, Engineering, and Medicine, Washington DC. DOI: 10.17226/24798

Variables
a_l = passenger load weighing factor
e = ridership elasticity with respect to changes in travel time rate (= 0.40)
F_l = average passenger load factor (passengers/seat)
F_{tt} = perceived travel time factor
L_{pt} = average passenger trip length (mi) (3.7 typical)
p_{be} = proportion of stops on segment with benches (decimal)
p_{sh} = proportion of stops on segment with shelters (decimal)
$S_{Tt,seg}$ = travel speed of transit vehicles along the segment (mi/hr)
T_{at} = amenity time rate (min/mi)
T_{btt} = base travel time rate (6.0 for CBD of metro $\geq$ 5 million; otherwise 4.0) (min/mi)
t_{ex} = excess wait time due to late arrivals (min)
T_{ex} = excess wait time rate due to late arrivals (min/mi) (t_{ex}/L_{pt})
T_{ptt} = perceived travel time rate (min/mi)

Excess wait time due to late arrivals is quantified by using Eq. 3.71 [*HCM* Eq. 18-61].

$$t_{ex} = \left(t_{late}(1 - p_{ot})\right)^2 \qquad 3.71$$

Variables
p_{ot} = proportion of transit vehicles arriving within the threshold late time (default = 0.75) (min)
t_{ex} = excess wait time due to late arrivals (min)
t_{late} = threshold late time (typical = 3.80 min)

The amenity time rate reduction, T_{at}, considered earlier is computed as the equivalent time value of various transit stop amenities divided by the average passenger trip length. The average trip length is computed as the annual passenger miles divided by the annual unlinked trips. Without local data, a value of 3.7 mi (as a national average) can be used as the transit trip length.

Wait-Ride Score

Compute the wait-ride score using the following equation [*HCM* Eq. 18-62].

$$s_{w\text{-}r} = F_h F_{tt} \qquad 3.72$$

Variables
F_h = headway factor
F_{tt} = perceived travel time factor
$s_{w\text{-}r}$ = transit wait-ride score

Step 5: Determine Pedestrian LOS Score for Link. The pedestrian LOS score for the link is computed using the pedestrian methodology described in the "Urban Streets" section of this chapter.

Step 6: Determine Transit Score for Segment. The transit LOS score for the segment is found using Eq. 3.73 [*HCM* Eq. 18-63].

$$I_{t,\text{seg}} = 6.0 - 1.50 s_{w\text{-}r} + 0.15 I_{p,\text{link}} \qquad 3.73$$

Step 7: Determine LOS. Select the LOS from Table 3.9.

Limitations of LOS Methodology

The LOS methodology is limited to quantifiable traffic values. Quality of service from a transit user's standpoint includes many more features not covered by these methodologies. The demand function of transit usage is somewhat elastic, and to a great extent subject to the personal attitudes and feelings about the role of transit. The following are some features of transit not included in the *HCM* methodology.

- automobile availability to the transit user
- parking cost and convenience at destination location
- location of convenient trip ends to proximity of transit facility access
- proportion of trips involving more than one purpose
- personal comfort and feeling of security in a transit environment
- ambulatory aid requirements of the transit user
- cleanliness of transit vehicles and waiting areas
- attitude and personal behavior of transit operating personnel to the transit user
- understanding of fare system for occasional users
- overall reliability of transit adherence to schedule times, both on-peak and off-peak
- ability of transit operations to recover from adverse or severe weather disruptions
- ability of transit system to reattract riders that no longer wish to use the transit system due to a bad experience with the transit system

These and other factors can affect the usage and impacts of transit, and to a great extent, overshadow the traffic engineering statistical analysis of LOS on a particular segment. Nonetheless, the statistical LOS evaluation in *HCM* is a useful tool for on-street traffic analysis.

Railroad Transit

Railroad operations include both longer-distance suburban commuter service and intercity passenger service. For traffic planning analysis, railroad transit is commonly evaluated in terms of frequency of service, platform capacity, train capacity, corridor demand, reliability, viability of feeder service, and proximity to activity centers. Trip time and average travel speed is often compared with competing modes, such as express bus service and automobile travel on adjoining highways. Railroad transit in the United States occurs only in the largest metropolitan areas, and is virtually nonexistent elsewhere, as freight service has priority on most U.S. trackage.

Bus Rapid Transit

Bus rapid transit, or BRT, broadly includes bus-based systems of specially designed bus routes configured to handle high passenger volumes with fewer delays and higher average speeds than usually found on local city systems. BRT stops are configured more like transit stations, with adequate platform and shelter space to provide for greater numbers of riders. Buses travel on reserved lanes or roadways separated from other street traffic, permitting operation at prevailing traffic speeds. BRT movements can join existing street traffic at intersections. Intersections with BRT traffic can have some movements limited by BRT requirements.

From a systems operations standpoint, BRT on-street movements approximate LRT on-street traffic operations. In some cases, BRT is combined with HOV lane traffic.

Light Rail Transit

Light rail transit, or LRT, is a fixed-rail guided streetcar or interurban trolley system found in many dense urban centers. Constructed to vehicle standards scalable to urban traffic systems, LRT requires substantially less robust construction than needed for heavy railroad operations. LRTs operate on their own right-of-way, and can also be introduced into street situations where exclusive right-of-way is not available.

LRTs generally operate at higher speed when on their own right-of-way, and then operate at lower speeds when introduced into mixed-traffic on-street operations. Operators are specially trained to allow for the increased stopping distance required for steel-wheeled vehicles, and for the limited flexibility of streetcar movements through intersections. LRT has found a niche along corridors where patronage is higher than conventional bus systems. LRT also has a visual appeal, particularly in redeveloped downtown and near-suburban scenes, to create the ambiance of pleasant-looking modern vehicles that glide along sculptured streets and closely developed neighborhoods with stops at convenient intervals.

LRT operations primarily serve those people who live or work within walking distance of the stations. Outlying stations require park-and-ride lots in order to support capacity efficiency. The park-and-ride facilities do not need to be as large as those for heavy rail stations because ridership is generally much lower than on regional rail or subway systems. Ridership for LRT corridors typically ranges from 10,000 per day to 25,000 per day.

LRT operation requires limits to be placed on street intersection movements due to track geometry control, and often requires exclusive phases in traffic signals. LRT systems require more support personnel than for equivalent bus systems, as well as storage barns separated from bus garages. The track, signal, and power systems require daily monitoring in order to remain reliable. For these reasons, constructing an LRT system must be complete and all components kept in complete working order for it to remain operational at all times. LRT selection, like BRT selection, is a local choice, involving support from the entire urban planning and government operations units in order to be effectively integrated into the surrounding transportation culture.

4. PRACTICE PROBLEMS

1. Service disruption caused by a blockage occurs on a light rail transit line running at full capacity. The blockage requires all passengers to be transferred to buses for a portion of their trip. The LRT line operates with two-car trains in each direction, and each car has a capacity of 255 passengers/car. The schedule calls for a maximum of 28 trains/hr arriving at the station where passengers are transferred to buses. Each bus has a maximum capacity of 65 passengers. Traffic signals along the major artery traveled by the buses are timed to allow up to 10 buses in the traffic stream per 90 sec cycle. There are no bus stops at signalized intersections, all buses travel straight through the intersections, and there is no other scheduled route service along the arterial. Bus boarding times at the transfer location are as shown.

1st passenger	2.0 sec
2nd through 30th passenger	0.7 sec
31st passenger to capacity	1.4 sec
door opening	1.5 sec
door closing	2.0 sec
average interval for the next bus to move into boarding position	30 sec

(a) What is most nearly the minimum required width of the at-grade concourse walkway connecting the LRT platform and the bus platform if all passengers arriving on a two-car train are to be moved to the bus platform within 2 min of arrival?

(A) 11 ft

(B) 15 ft

(C) 23 ft

(D) 28 ft

(b) Most nearly how many bus trips will be required to carry the maximum hourly directional flow rate of passengers if each bus has a capacity of 65 passengers?

(A) 110 bus trips/hr

(B) 112 bus trips/hr

(C) 168 bus trips/hr

(D) 220 bus trips/hr

(c) Assuming the number of bus trips required by (b), what is most nearly the average number of buses traveling through the signal per cycle?

(A) 3.7 buses/cycle

(B) 5.5 buses/cycle

(C) 10 buses/cycle

(D) 40 buses/cycle

(d) What is most nearly the minimum number of bus berths required to handle the expected passenger flow rate?

(A) 2 berths

(B) 5 berths

(C) 7 berths

(D) 9 berths

2. Pedestrian intersection level of service analysis uses which of the following variables as input?

1. wait time or delay

2. available time-space

3. proportion arriving during red

4. %trucks and %buses

5. pedestrian LOS score

6. speed limit

7. base service volume

8. right-turn-on-red flow rate

(A) 1, 3, 5, 7, 8
(B) 2, 3, 4, 7, 8
(C) 2, 3, 5, 7, 8
(D) 1, 2, 3, 5, 8

3. A four-way signalized intersection analysis of time-space requirements includes which of the following inputs?

1. effective green time on the minor street approach for the pedestrian crossing of the major street
2. queue holding-area waiting time for the major street
3. v/c ratio of each approach
4. grade of the sidewalk at each crossing
5. width of the pedestrian crossings
6. sidewalk arrival flow rate
7. pavement roughness condition
8. number of outbound crossing pedestrians

(A) 1, 2, 4, 7, 8
(B) 1, 2, 5, 6, 8
(C) 1, 3, 4, 6, 7
(D) 2, 4, 5, 7, 8

SOLUTIONS

1. (a) Determine the headway of the trains.

$$h = \frac{60}{N_t} = \frac{60 \; \frac{\text{min}}{\text{hr}}}{28 \; \frac{\text{trains}}{\text{hr}}} = 2.14 \; \text{min/train}$$

Determine the maximum number of passengers traveling through the concourse in each 2 min interval, $N_{p,2\,\text{min}}$.

$$N_{p,2\,\text{min}} = \left(2 \; \frac{\text{cars}}{\text{train}}\right)\left(255 \; \frac{\text{passengers}}{\text{car}}\right)$$
$$= 510 \; \text{passengers/train}$$

The pedestrian flow rate at transit terminals is found using the platoon-adjusted flow rate shown in Table 3.5. The maximum capacity of a transit concourse is 18 pedestrians per minute per foot of width for LOS E. The minimum walkway width, $W_{\min}$, is

$$W_{\min} = \frac{510 \; \text{ped}}{\left(18 \; \frac{\text{ped}}{\text{min-ft}}\right)(2 \; \text{min})}$$
$$= 14.2 \; \text{ft} \quad (15 \; \text{ft})$$

The answer is (B).

(b) The maximum hourly directional passenger flow rate, $v_{p,\max}$, is found from Eq. 3.56.

$$v_{p,\max} = N_t c_{\text{car}} N_{\text{cars}}$$
$$= \left(28 \; \frac{\text{trains}}{\text{hr}}\right)\left(255 \; \frac{\text{passengers}}{\text{car}}\right)\left(2 \; \frac{\text{cars}}{\text{train}}\right)$$
$$= 14{,}280 \; \text{passengers/hr}$$

The hourly number of bus trips required to accommodate the maximum hourly directional flow rate is found from Eq. 3.55.

$$N_{\text{trips/hr}} = \frac{v_{p,\max}}{c_{\max}} = \frac{14{,}280 \; \frac{\text{passengers}}{\text{hr}}}{65 \; \frac{\text{passengers}}{\text{bus}}}$$
$$= 220 \; \text{bus trips/hr}$$

The answer is (D).

(c) The average number of buses traveling through the signal per cycle is

$$\left(220\ \frac{\text{buses}}{\text{hr}}\right)\left(\frac{90\ \frac{\text{sec}}{\text{cycle}}}{3600\ \frac{\text{sec}}{\text{hr}}}\right) = 5.5\ \text{buses/cycle}$$

The answer is (B).

(d) Determine the dwell time for one bus.

$$\begin{aligned}t_{\text{dw}} &= \text{first passenger time} \\ &\quad + \text{2nd through 30th passenger time} \\ &\quad + \text{31st passenger to capacity time} \\ &\quad + \text{door opening time} \\ &\quad + \text{door closing time} \\ &= (1)(2.0\ \text{sec}) + (29)(0.7\ \text{sec}) + (35)(1.4\ \text{sec}) \\ &\quad + 1.5\ \text{sec} + 2.0\ \text{sec} \\ &= 74.8\ \text{sec/bus}\end{aligned}$$

Determine the minimum service headway.

$$h_{\min} = t_{\text{dw}} + t_{\text{ave}} = 74.8\ \frac{\text{sec}}{\text{bus}} + 30\ \frac{\text{sec}}{\text{bus}} = 104.8\ \text{sec/bus}$$

Determine the passenger service volume, v_{ps}, at one berth.

$$\begin{aligned}v_{\text{ps}} &= \left(\frac{3600}{h_{\min}}\right) c_{\max} \\ &= \left(\frac{3600\ \frac{\text{sec}}{\text{hr}}}{104.8\ \frac{\text{sec}}{\text{bus}}}\right)\left(65\ \frac{\text{passengers-berth}}{\text{bus}}\right) \\ &= 2233\ \text{passengers-berth/hr}\end{aligned}$$

Given the known maximum flow rate of the system, the number of berths required is found from Eq. 3.57.

$$N_{\text{berths}} = \frac{v_{p,\max}}{v_{\text{ps}}} = \frac{14{,}280\ \frac{\text{passengers}}{\text{hr}}}{2233\ \frac{\text{passengers-berth}}{\text{hr}}}$$
$$= 6.39\ \text{berths} \quad (7\ \text{berths})$$

The answer is (C).

2. Variable 4, %trucks and %buses, is not used directly since %trucks and %buses are included in the vehicle adjustment factor to determine adjusted saturation flow rate as a direct input. Variable 6, speed limit, may vary significantly from mean vehicle headway. The mean vehicle headway, determined by mean approach speed, is used to determine the critical gap for unsignalized intersections. Variable 7, base saturation flow rate, is used as a direct input instead of base service volume.

The answer is (D).

3. Analysis does not include input 3, since the effective walk time is used to determine available pedestrian crossing gap, which is not evaluated from the v/c ratio. For input 4, *HCM* procedures assume normal walking speeds not adjusted for steeper grades. Also, eliminate input 7 because *HCM* procedures assume normal walking speeds not affected by pavement roughness.

The answer is (B).

4 Geometric Design

1. Introduction 4-2
2. Route and Corridor Alignment 4-2
3. Specifying Line Direction 4-3
4. Horizontal Curves 4-5
5. Degree of Curve 4-6
6. Stationing on a Horizontal Curve 4-7
7. Field Layout and Fitting of Horizontal Curves ... 4-9
8. Spiral Curves 4-16
9. Vertical Curves 4-18
10. Sources of Error in Route Alignment 4-22
11. Sight Distance 4-23
12. Superelevation 4-30
13. Vertical and Horizontal Clearances 4-39
14. Acceleration and Deceleration 4-42
15. Intersections and Interchanges 4-43
16. Railroad Engineering 4-46
17. Bikeway Geometric Design 4-48
18. Mass Transit Geometric Design 4-50
19. Practice Problems 4-54

Nomenclature

a	acceleration	ft/sec²	m/s²
a	given model parameter	–	–
A	absolute value of the algebraic difference in grades	%	%
A	azimuth	deg	deg
b_w	adjustment factor for number of lanes rotated	–	–
B	bearing angle	deg	deg
BC	beginning of curve	sta	sta
C	chord length	ft	m
C	maximum rate of change in lateral acceleration	ft/sec³	m/s³
C	vertical clearance	ft	m
d	distance	ft	m
D	absolute value of the algebraic difference in grades	decimal	decimal
D	degree of curve	deg	deg
D	distance	ft	m
e	superelevation rate	ft/ft	m/m
E	external distance	ft	m
E	relative elevation change	ft	m
EC	end of curve	sta	sta
f	coefficient of friction	–	–
F	force	lbf	N
g	gravitational acceleration, 32.2 (9.81)	ft/sec²	m/s²
g_c	gravitational constant, 32.2	lbm-ft/lbf-sec²	n.a.
G	grade	%	%
h_1	driver's eye height	ft	mm
h_2	object height	ft	mm
HSO	horizontal sightline offset	ft	m
I	deflection angle	deg	deg
I	intersection angle	deg	deg
k	initial tangent distance	ft	m
K	length of vertical curve per percent grade difference	ft/%	m/%
L	length of curve	ft	m
m	difference in speed between passing vehicle and vehicle being passed	mph	kph
M	middle ordinate	ft	m
n	number of deflection angles	–	–
N	number	–	–
p	lateral offset between the tangent and circular curve	ft	m
PC	point of curvature	sta	sta
PI	point of intersection, vertex	sta	sta
PT	point of tangency	sta	sta
PVC	point of beginning of vertical curve	sta	sta
PVI	point of intersection of vertical curve tangents, vertex	sta	sta
PVT	point of end of vertical curve	sta	sta
R	radius	ft	m
R	rate of change in grade	%/sta, ft/sta²	%/sta, m/sta²
S	sight distance	ft	m
S	stopping distance	ft	m
T	tangent distance	ft	m
TC	tangent to curve	sta	sta
t	time	sec	s
v	design speed	mph	kph
v	velocity	ft/sec	m/s
V	vertex	sta	sta
w	weight	lbf	N
W	width	ft	m
x	horizontal distance	ft	m

x	tangent distance	ft	m
y	tangent offset	ft	m

Symbols

α	angle	deg	deg
β	angle	deg	deg
Δ	curve deflection angle	deg	deg
Δ	maximum relative gradient	ft/ft	m/m
θ	central angle	deg	deg
θ	subtended angle of an arc	deg	deg
ϕ	angle of cross slope	deg	deg
ϕ	deflection angle	deg	deg

Subscripts

a	known
ave	average
B	backsight
c	central curve or centrifugal
D	design
F	foresight
l	central curve
L	lanes
max	maximum
min	minimum
N	north
NC	normal cross slope or normal crown
p	perception-reaction or pre-maneuver
r	roadway
R	runoff
s	side friction or spiral
S	south
t	total or tangent runout
x	desired
z	passing vehicle in opposing lane

1. INTRODUCTION

The geometric design of highways, railroads, and other surface transportation facilities involves the creation of geometric models using straight lines and curves that control the location and alignment of the facility. The alignment is controlled by precise horizontal and vertical geometry using cross-sectional templates to fit the alignment into topographic conditions. The alignment is affected by the ground-line elevations, watercourse locations, and many natural topographic land features. The available space for a new alignment or the expansion of an existing alignment is also controlled by human-made features, such as buildings, bridges, major utility lines, population centers, parks, and available property. The goal is to maximize user convenience and safety, and to minimize environmental impact. The primary references for highway facilities are *A Policy on Geometric Design of Highways and Streets* (*GDHS*) (also known as the *Green Book*), and the *Roadside Design Guide* (*RSDG*), both published by the American Association of State Highways and Transportation Officials (AASHTO). The primary references for railroad facilities are American Railway Engineering and Maintenance Association (AREMA) publications.

This chapter covers basic elements of highway geometric design, including route and corridor alignment; horizontal, spiral, and vertical curves; superelevation; intersections and interchanges; as well as the geometric design of railways, bikeways, and mass transit systems. For more information about specific elements of the geometric design of a transportation facility, state and other applicable national standards should also be consulted.

2. ROUTE AND CORRIDOR ALIGNMENT

A *route alignment* is a series of straight lines (called *tangents*) and *curves* that are laid out in a continuous path. These geometric lines comprise the horizontal and vertical alignments of a transportation facility. The details of a route's geometric layout depend on the type of facility being designed. In general, the alignments connect the terminals of two other routes or extend or revise an existing route.

A route's horizontal and vertical alignments are designed within a *corridor*, which is a collective system of transportation facilities designed for travel between two or more points. A *corridor alignment study* assesses the feasibility of placing a particular route within a corridor. There are three distinct phases in a corridor alignment study: *reconnaissance*, the *preliminary location survey*, and the *final location survey and design*.

The first phase of a corridor alignment is reconnaissance, which is the collection and review of basic information that will set the approximate location of a route. This phase includes establishing the terminal ends of the alignment and the intervening connecting points. It also includes identifying major *topographic features*, such as rivers, existing railroads and highways, buildings, mountains, historical sites, architectural features, protected lands, and anything else that will have a physical effect on the route location.

Reconnaissance also includes a review of existing mapping or other location documentation, general traffic demands, funding parameters, and general government or institutional information that can be used to determine the suitability of a facility's alignment design. For example, in highway alignment design, the number and size of trucks in the traffic flow can determine the maximum grade and the location of access points.

In the past, reconnaissance required walking the proposed corridor with a design team to establish a general location of line and grade based on visual appearances.

Measurements and notes were taken to record features affecting the alignment that may not be evident on the available mapping. Stakes and flagging were placed in the approximate location of the centerline or other horizontal reference line for future reference when returning to the site.

Reconnaissance that formerly took days, weeks, or months can now be performed in a few hours using commercially available satellite and mapping imagery. Mapping websites, such as Google maps, also have extensive files that can be referenced and reproduced on working media for a relatively small cost. Ground-level imagery of more populous areas is also available, so a preliminary record of the nature of topographic and human-made surface improvements can be prepared before ever venturing into the corridor itself. Even with these readily available resources, walking the proposed alignment remains a very important part of the route reconnaissance.

The second phase of a corridor alignment study is a *preliminary location survey*. After the reconnaissance identifies a possible location, large-scale aerial mapping suitable for laying out the preliminary location is prepared. Maps are prepared for a broad area so that several alternative alignments can be studied by the surveyors. Maps created during this phase are typically at 1:2400 (1 in = 200 ft) scale or 1:1200 (1 in = 100 ft) scale for rural locations and 1:500 (1 in = 40 ft) scale for urban locations. Aerial mapping contours are the primary means of determining ground elevations during alignment studies.

During this phase, the surveyors set control points throughout the corridor to accurately measure the stakeout during later phases. The project control traverse is surveyed and mathematically connected to the aerial map traverse in order to accurately locate objects identified on the aerial map. In some cases, it may be possible to combine the aerial map traverse with the project control traverse to reduce the probability of a measurement error between mapping images and project control location.

Alternative routes are evaluated for the best fit of line and grade to the project criteria. For each alternative, earthwork quantities are projected, bridge structures are located, soil conditions are studied, preliminary right-of-way requirements are drawn, and economic studies are performed to determine the projected cost. During this phase, community hearings are held and a preliminary environmental impact analysis is prepared.

The third phase of a corridor alignment study is the *final location survey*, which begins the *final design*. For this phase, detailed aerial mapping is prepared with narrow coverage along the selected final route at a small scale. The scale of these maps is typically 1:250 (1 in = 20 ft) for congested locations, so that engineers can design connections to existing land features and utilities. During the final location survey, the line and grade are carefully staked out in the field using the preliminary alignment as a guide. Sufficient field information is prepared to verify mapping locations, and final grades are adjusted to determine cut and fill quantities. Bridge pier locations are carefully surveyed, and subsurface investigations are performed to determine final foundation design requirements.

During the final location survey, adjustments to the alignment are made to balance earthwork, match connections to adjoining work, and allow for needed clearances for objects intersecting or adjacent to the roadway (such as overpasses or utility poles). Homes and businesses in the final right-of-way are appraised, purchased, and relocated, and environmental concerns are addressed. Community meetings are usually held during planning and prior to the final location survey to present to the community the impact of the project. These meetings often lead to minor changes in the route geometry.

During the alignment process, the roadway template is applied to the line and grade in a three-dimensional model, including allowances for cut and fill slopes, drainage ditches, and cross pipes. Allowances are also made for geometric features of intersections and interchanges and, for railroads, the geometry of connecting spurs or yard trackage, all of which can have an effect on the location of the main roadway or trackway.

3. SPECIFYING LINE DIRECTION

The direction of any line can be specified by an angle from a *principal meridian*. If the meridian is arbitrarily chosen, it is called an *assumed meridian*. If the meridian is a true north-to-south line passing through the true north pole, it is called a *true meridian*. If the meridian is parallel to the earth's magnetic field, it is known as a *magnetic meridian*. If a rectangular grid is drawn over a map with any arbitrary orientation, the vertical lines are referred to as *grid meridians*.

The difference between a true meridian and a magnetic meridian is called the *declination* (*magnetic declination* or *variation*). An *isogonic line* connects locations in a geographical region that have the same magnetic declination. If the northern end of a magnetic compass points westward of true meridian, the declination is referred to as a *west declination*, or a *minus declination*. If the northern end of a magnetic compass points eastward of true meridian, it is referred to as an *east declination*, or a *plus declination*.

The *azimuth* of a line is given as a clockwise angle from the reference direction, either "from the north" or "from the south" (for example, "NAz 320"). Azimuths never exceed 360°. Positive declinations are added to the magnetic compass azimuth, and negative declinations are subtracted from the magnetic compass azimuth. The result is called the true azimuth.

A *direction* (the variation of a line from its meridian) may be specified in several ways. (See Fig. 4.1.) Directions are specified as either bearings or azimuths, referenced to either north or south. In the United States, primary control networks use directions stated as azimuths, whereas construction projects and property surveys specify directions as bearings.

Figure 4.1 Equivalent Bearing, Angle, and Azimuth Measurements

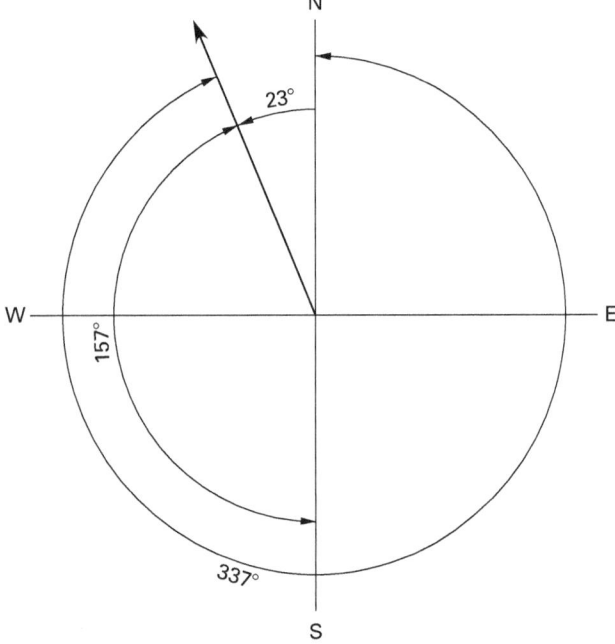

azimuth from the north: 337°
azimuth from the south: 157°
deflection angle (from the north): 23°L or −23°
bearing angle: N 23° W
angle to the right: 157°

The angle between a line and the prolongation of a preceding line is called a *deflection angle*. Such measurements are labeled as Rt for clockwise angles and L for counterclockwise angles. A clockwise deflection angle measured from the preceding line (i.e., the back line) to the following line is called the *angle to the right* or the *azimuth from the back line*. The back line may be referenced from the north, which is most common, but the choice of the back line varies between jurisdictions. In Fig. 4.1, the azimuth to the right is shown as 337° measured from due north as the back line, and 157° from due south as the back line.

The *bearing* of a line is referenced to the quadrant in which the line falls and the angle that the line makes with the meridian in that quadrant. It is necessary to specify the two cardinal directions (i.e., north or south and east or west) that define the quadrant in which the line is found. The north and south directions are specified first (for example, "N 45° E"). The angle of a bearing never exceeds 90°.

Bearings and azimuths are calculated from trigonometric and geometric relationships. From Fig. 4.2, the bearing angle, B, is given by Eq. 4.1.

$$\tan B = \frac{x_2 - x_1}{y_2 - y_1} \quad 4.1$$

Figure 4.2 Calculation of Bearing Angle

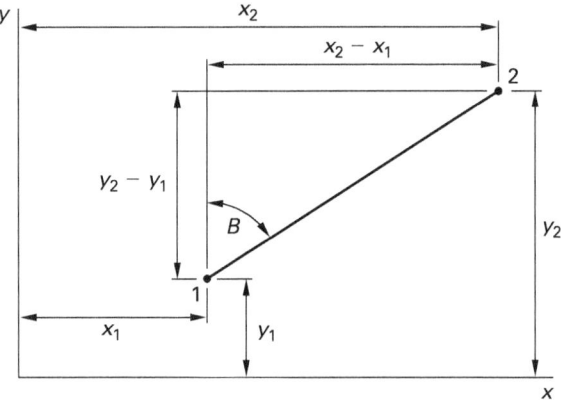

The distance, D, between two points is given by Eq. 4.2.

$$D = \sqrt{(y_2 - y_1)^2 + (x_2 - x_1)^2} \quad 4.2$$

For the denominator in Eq. 4.1, the line from point 1 to point 2 bears north if positive and south if negative. Similarly, for the numerator, the line bears east if positive and west if negative. Determining the bearing and length of a line from the coordinates of two points is called *inversing the line*.

The azimuth of a line from point 1 to 2 measured from the north meridian, A_N, is given by Eq. 4.3.

$$\tan A_N = \frac{x_2 - x_1}{y_2 - y_1} \quad 4.3$$

The azimuth of the same line from point 1 to 2 measured from the south meridian, A_S, is given by Eq. 4.4.

$$\tan A_S = \frac{x_1 - x_2}{y_1 - y_2} \quad 4.4$$

Converting from bearings to azimuths is accomplished by selecting the bearing quadrant, then adding or subtracting 0°, 180°, or 360°, depending on the azimuth meridian and the bearing quadrant. Use the following equations to find a *north meridian azimuth* from a bearing.

NE bearing: azimuth = bearing angle
SE bearing: azimuth = 180° − bearing angle
SW bearing: azimuth = 180° + bearing angle
NW bearing: azimuth = 360° − bearing angle

Use the following equations to find a *south meridian azimuth* from a bearing.

NE bearing: azimuth = 180° + bearing angle
SE bearing: azimuth = 360° − bearing angle
SW bearing: azimuth = bearing angle
NW bearing: azimuth = 180° − bearing angle

Example 4.1

Convert (a) N 35° E, (b) S 45°15′ E, (c) S 18°30′ W, and (d) N 62°45′ W to north meridian azimuths.

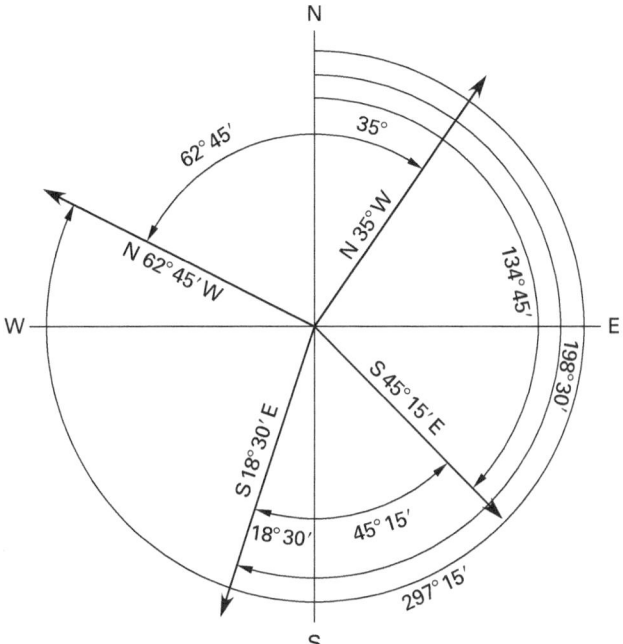

Solution

(a) N 35° E = 35° azimuth
(b) S 45°15′ E = 180° − 45°15′ = 134°45′ azimuth
(c) S 18°30′ W = 180° + 18°30′ = 198°30′ azimuth
(d) N 62°45′ W = 360° − 62°45′ = 297°15′ azimuth

4. HORIZONTAL CURVES

Horizontal curves are used to change from one heading, or direction, to another. For most horizontal work, the *circular curve*, sometimes called a *simple curve*, is used as a smooth transition to the new direction. The main feature of a circular curve is that every point on its arc is precisely the same distance from the center point. That is, the radius is constant throughout the arc length. Figure 4.3 shows the elements of a horizontal circular curve.

Figure 4.3 Elements of a Horizontal Circular Curve

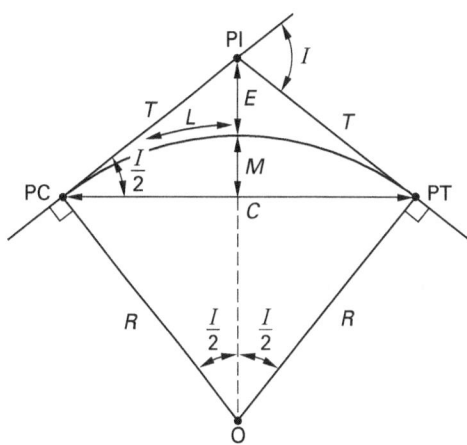

The *point of intersection*, PI, (also called the *vertex*, V) is the intersection of the two tangents to which the curve is fitted. The *deflection angle*, Δ, is the angle that the curve tangent deflects from the previous bearing to a new bearing, or heading. The deflection angle is also given the symbol I or V and called the *intersection angle*. The *point of curvature*, PC, is the point of the back tangent at which the circular curve begins. This point is also called TC, for *tangent to curve*, or BC, for *beginning of curve*. The *point of tangency*, PT, is the point at which the curve stops and the new or ahead tangent continues. Some texts refer to the PT as the TC, for tangent to curve, or EC, for *end of curve*. Lines extended at right angles to the tangents from the PC and PT intersect at the *center of the circular arc*, O. The angle subtended by the arc of the circular curve is called the *central angle*. The central angle is equal to the deflection angle. The deflection angle is called *right or positive deflection* or positive for a clockwise deflection, and *left or negative deflection* or negative for a counterclockwise deflection. The deflection can also be called the delta of the curve. The interior angle, PC-PI-PT, is 180° minus the deflection, Δ. The *tangent distance*, T, is the distance from PI to PC, or the distance from PI to PT, and is calculated from Eq. 4.5. R is the radius of the curve.

$$T = R \tan \frac{I}{2} \qquad 4.5$$

The *external distance*, E, is the distance from the PI to the midpoint of the curve and is calculated using Eq. 4.6.

$$E = R\left(\sec \frac{I}{2} - 1\right) = R \operatorname{exsec} \frac{I}{2}$$
$$= R \tan \frac{I}{2} \tan \frac{I}{4} \qquad 4.6$$

$$\operatorname{exsec} I = \sec I - 1 = \tan \frac{I}{2} \tan \frac{I}{4} \qquad 4.7$$

$$\sec I = \frac{1}{\cos I} \qquad 4.8$$

The *length of curve*, L, is the distance from PC to PT along the arc. It can be found using several relationships, as shown in Eq. 4.9.

$$L_{\text{ft}} = \frac{2\pi R_{\text{ft}} I_{\text{deg}}}{360°} = R_{\text{ft}} I_{\text{rad}} = \frac{(100 \text{ ft}) I_{\text{deg}}}{D_{\text{deg}}}$$
$$= \frac{200 \text{ ft}}{D_{\text{deg}}} \arcsin \frac{C_{\text{ft}}}{2R_{\text{ft}}} \quad 4.9$$

The *long chord*, C_{long}, sometimes represented by LC to reduce confusion with other subchords, is the distance from the PC to the PT. It is the chord length for the intersection angle, I, and is calculated from Eq. 4.10.

$$C_{\text{long}} = 2R \sin \frac{I}{2} = 2T \cos \frac{I}{2} \quad 4.10$$

The length of any chord (any straight line from one point on the curve to another), C, can be defined from the subtended intersection angle, θ, by substituting θ for I, as shown in Eq. 4.11.

$$C = 2R \sin \frac{\theta}{2} \quad 4.11$$

The *middle ordinate*, M, is the length of the ordinate from the midpoint of the long chord, C_{long}, to the midpoint of the curve arc. The middle ordinate is found from Eq. 4.12 and Eq. 4.13.

$$M = R\left(1 - \cos \frac{I}{2}\right) = R\left(\text{vers} \frac{I}{2}\right) \quad 4.12$$

$$M = \frac{C^2}{8R} = \left(\frac{C}{2}\right) \tan \frac{I}{4} \quad 4.13$$

$$\text{vers} \frac{I}{2} = 1 - \cos \frac{I}{2} \quad 4.14$$

$$\text{vers } I = 1 - \cos I \quad 4.15$$

5. DEGREE OF CURVE

Curves are specified either by the radius, R, or *degree of curve*, D. Degree of curve is most commonly used when working in U.S. units, but it can be used to determine curvature when working in SI units as well. The degree of curve is defined either by the arc or the chord of a curve, as shown in Fig. 4.4. The *arc definition* is commonly used with highways and streets, where D is the central angle subtended by a 100 ft arc. This is the same as the deflection of a 100 ft long arc, as shown in Fig. 4.4(a).

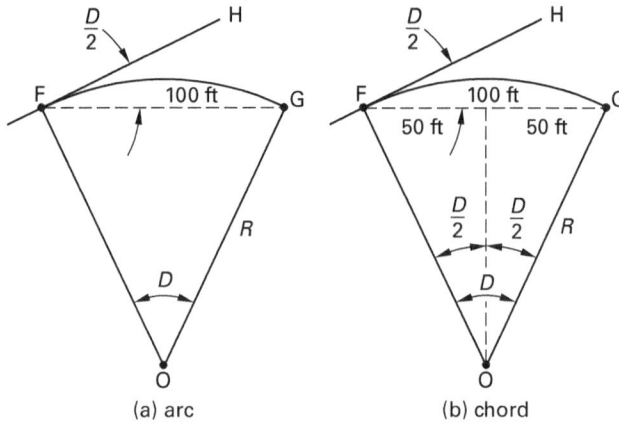

Figure 4.4 Degree of Curve Definition

The *chord definition* (see Fig. 4.4(b)) is used with railroads and was used on pre-interstate highways. For chord definition curves, the central angle is subtended by a 100 ft chord. To differentiate the chord definition from the arc definition, degree of curve for the chord definition is shown as D_c on some plans.

In practice, when layout is performed using degree of curve instead of radius, the degree is rounded to the nearest 30', or even 15' in some instances. For instance, a layout curve would be set at 2°30' instead of 2°27.8'. When layout is performed using the radius, rounded whole-numbered radii curves are used, such as 300 ft, 1000 ft, or 10,000 ft. Other curve relationships are calculated by using these rounded values as absolute, which reduces the chance for error when defining curve segments. To find the radius when the degree of curve is given, use Eq. 4.16 for arc definition curves and Eq. 4.17 for chord definition curves.

$$R_{\text{ft}} = \frac{(100 \text{ ft})\left(\dfrac{180°}{\pi}\right)}{D_{\text{deg}}} = \frac{5729.578 \text{ ft-deg}}{D_{\text{deg}}} \quad 4.16$$
$$\begin{bmatrix} \text{arc basis when } D = 1°, \\ R = 5729.58 \text{ ft} \end{bmatrix} \text{[U.S. only]}$$

$$R_{\text{ft}} = \frac{50 \text{ ft}}{\sin \dfrac{D_{\text{deg}}}{2}} \quad 4.17$$
$$\begin{bmatrix} \text{chord basis when } D = 1°, \\ R = 5729.65 \text{ ft} \end{bmatrix} \text{[U.S. only]}$$

Rearranging Eq. 4.16 and Eq. 4.17, use Eq. 4.18 and Eq. 4.19 to find D when R is known.

$$D_{\text{deg}} = \frac{(360°)(100 \text{ ft})}{2\pi R_{\text{ft}}} = \frac{(180°)(100 \text{ ft})}{\pi R_{\text{ft}}} \quad 4.18$$
[arc definition] [U.S. only]

$$\sin\frac{D_{\text{deg}}}{2} = \frac{50 \text{ ft}}{R_{\text{ft}}} \quad \begin{array}{l}\text{[chord definition]}\\ \text{[U.S. only]}\end{array} \quad 4.19$$

In jurisdictions using the metric system, distances are usually measured with a 20 m tape, though 30 m and 50 m tapes are sometimes used. Using the degree of curve, with the deflection angle for a 20 m chord, the degree of curve equations are Eq. 4.20, Eq. 4.21, and Eq. 4.22.

$$D_{\text{deg}} = \frac{(180°)(20 \text{ m})}{\pi R_{\text{m}}} \quad \begin{array}{l}\text{[arc definition]}\\ \text{[SI only]}\end{array} \quad 4.20$$

$$\sin\frac{D_{\text{deg}}}{2} = \frac{10 \text{ m}}{R_{\text{m}}} \quad \begin{array}{l}\text{[chord definition]}\\ \text{[SI only]}\end{array} \quad 4.21$$

$$L_{\text{m}} = (20 \text{ m})\frac{I}{D} \quad \text{[SI only]} \quad 4.22$$

$$R_{\text{m}} = \frac{(20 \text{ m})\left(\frac{180°}{\pi}\right)}{D} \quad \text{[SI only]} \quad 4.23$$

The length of the arc for arc or chord definition curves is slightly greater than the 100 ft chord length. The difference is greater for sharper curves and lesser for flatter curves. Arc lengths for chords on various degrees of curvature can also be found in most surveying texts.

Attempting to parallel a chord definition curve with an arc definition curve, such as fitting a highway curve adjacent to a railroad right-of-way, can be confusing. It is usually easier to convert the railroad degree of curve to a radius relationship first, then proceed with the highway curve and continue to work in radius relationships.

6. STATIONING ON A HORIZONTAL CURVE

When a route is initially laid out between PIs, the curve is undefined. The route distance is measured from PI to PI, but the route distance changes when the curve is laid out.

In route surveying, stationing is carried ahead continuously from a starting point or hub designated as station 0+00 and called *station zero plus zero zero*. The term *station* is applied to each subsequent 100 ft (or 100 m) length. Also, the term *station* is applied to any point whose position is given by its total distance from the beginning position. Thus, station 8+33.2 is a unique point 833.2 ft (or 833.2 m) from the starting point, measuring along the survey line. The partial length beyond the full station 8 is 33.2 ft (33.2 m) and is termed a *plus*. Moving or looking toward increasing stations is called *ahead stationing*. Moving or looking toward decreasing stations is called *back stationing*. Offsets from the centerline are either left or right looking ahead on stationing.

Normal curve layout follows the convention of showing ahead stationing from left to right on the plan sheet, or from the bottom to the top of the sheet. In ahead stationing, the first point of the curve is the point of curvature (PC), alternatively called the point of beginning of curve (BC). Stationing is carried ahead along the arc of the curve to the end point of the curve, called the point of tangency (PT). This end point is alternatively called the point of end of curve (EC). The PI is stationed ahead from the PC. Therefore, the difference in stationing along the back curve tangent is the *curve tangent length*. To avoid the confusion of creating two stations for the PC, the ahead tangent is rarely stationed.

From Eq. 4.24, the PT station is equal to the PC station plus the curve length, and from Eq. 4.25, the PC station is equal to the PI station minus the tangent distance.

$$\text{sta PT} = \text{sta PC} + L \quad 4.24$$

$$\text{sta PC} = \text{sta PI} - T \quad 4.25$$

Collectively, the stationing of the PI, PC, and PT, the intersection angle, the curve radius, the degree of curve, the tangent distance, the external distance, the chord length, and the length of the long chord are referred to as the complete curve specifications.

For railroad layouts, which use chord definition curves, the length of curve is defined as the distance from PC to PT along 100 ft chords. The difference between chord length and curve length for broader curves of 2° or less (radius > 3000 ft) is not great, but it is adequate for railroad work because the precise arc length of the curve is less important than chord measurements for track construction and maintenance. The approximate relationships of functions L, I, and D are shown in Eq. 4.26, Eq. 4.27, and Eq. 4.28, respectively.

$$L = 100\frac{I}{D} \quad 4.26$$

$$I = \frac{LD}{100} \quad 4.27$$

$$D = \frac{100I}{L} \quad 4.28$$

These equations should not be used for accurate calculations of other curve functions, especially for highway curves, as approximations of curve element geometry have no place in the field layout of sharper highway curves.

It is common practice to show curve geometry stationing and the lengths of curve elements to the nearest hundredth of a foot when specifying curve conditions and when locating a point on the alignment. Offsets to objects can be shown to the nearest tenth of a foot or to the nearest whole foot when the distance is not critical to layout geometry. Offsets to property corners should be shown to the nearest hundredth, if known.

Stationing for mainline railroads has been historically set up proceeding from east to west, as railroads were originally constructed. Railroad stationing starts over at each milepost; therefore, sta 52+80.00 = sta 0+00.00 of the subsequent milepost. Stationing on the interstate highway system and major freeways was set up to proceed from west to east using even numbered highways, and from south to north using odd numbers. Major interstate highways begin their stationing at the state boundary, and run continuously to the next state boundary, where the stationing starts over again. Within a state, rural road systems often start new stationing at each county line, or from the center of a major town or city, following the direction of the initial construction of the road. Some states have attempted to convert the stationing system to SI, but the practice has not caught on in the United States, and the feet-mile system remains prevalent throughout.

Example 4.2

Two horizontal tangents meet at sta 149+22.00 with a deflection angle of 65°12′. The horizontal curve between the two tangents has a radius of 2500 ft. Using arc definition, what are the complete curve specifications of the horizontal curve?

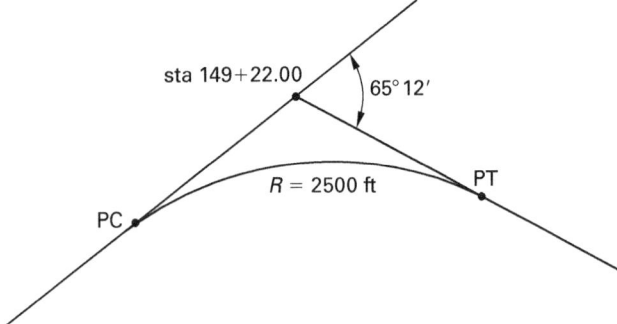

Solution

To station the PC and PT, the tangent distance and length of curve must be known. Find the nonstation characteristics first.

From Eq. 4.5, the tangent distance is

$$T = R \tan \frac{I}{2} = (2500 \text{ ft}) \tan \frac{65°12'}{2}$$
$$= 1598.82 \text{ ft}$$

From Eq. 4.10, the length of the long chord is

$$C_{\text{long}} = 2R \sin \frac{I}{2} = (2)(2500 \text{ ft}) \sin \frac{65°12'}{2}$$
$$= 2693.85 \text{ ft}$$

From Eq. 4.6, the external distance is

$$E = R \left(\sec \frac{I}{2} - 1 \right) = (2500 \text{ ft}) \left(\sec \frac{65°12'}{2} - 1 \right)$$
$$= 467.53 \text{ ft}$$

Calculate the degree of curve using Eq. 4.18.

$$D_{\text{deg}} = \frac{(360°)(100 \text{ ft})}{2\pi R_{\text{ft}}}$$
$$= \frac{(360°)(100 \text{ ft})}{(2\pi)(2500 \text{ ft})}$$
$$= 2.2918° \quad (2°17'31'')$$

From Eq. 4.9, the length of curve is

$$L_{\text{ft}} = \frac{200 \text{ ft}}{D_{\text{deg}}} \arcsin \frac{C_{\text{long,ft}}}{2R_{\text{ft}}}$$
$$= \left(\frac{200 \text{ ft}}{2.2918°} \right) \left(\arcsin \frac{2693.85 \text{ ft}}{(2)(2500 \text{ ft})} \right)$$
$$= 2844.92 \text{ ft}$$

Use Eq. 4.24 and Eq. 4.25 to station the PC and PT.

$$\begin{aligned}
\text{sta PC} &= \text{sta PI} - T \\
&= \text{sta } 149{+}22.00 \\
&\quad - 15{+}98.82 \\
& \overline{\text{sta } 133{+}23.18}
\end{aligned}$$

$$\begin{aligned}
\text{sta PT} &= \text{sta PC} + L \\
&= \text{sta } 133{+}23.18 \\
&\quad + 28{+}44.92 \\
& \overline{\text{sta } 161{+}68.10}
\end{aligned}$$

The complete curve specifications are

$$\begin{aligned}
\text{PI} &= \text{sta } 149{+}22.00 \\
\text{PC} &= \text{sta } 133{+}23.18 \\
\text{PT} &= \text{sta } 161{+}68.10 \\
I &= 65°12' \\
R &= 2500 \text{ ft} \\
D &= 2.2918° = 2°17'31'' \\
T &= 1598.82 \text{ ft} \\
E &= 467.53 \text{ ft} \\
L &= 2844.92 \text{ ft} \\
C_{\text{long}} &= 2693.85 \text{ ft}
\end{aligned}$$

Example 4.3

In laying out the centerline of a roadway, two tangents intersect, with an intersection angle of 28°18′. A curve with an external distance of approximately 40 ft must be laid out in order to clear an obstruction. Calculate the following using arc definition curves: (a) the degree of curvature to the nearest one-half degree, (b) the curve radius, (c) the actual external distance required to fit the selected curve, (d) the length of the curve, and (e) the tangential distance.

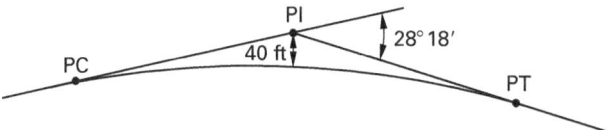

Solution

(a) The external distance is given as 40 ft.

Using Eq. 4.6, substitute for the secant and rearrange to find the radius.

$$R = \frac{E}{\frac{1}{\cos\frac{I}{2}} - 1} = \frac{40 \text{ ft}}{\frac{1}{\cos\frac{28°18'}{2}} - 1}$$
$$= 1278.35 \text{ ft}$$

Rearranging Eq. 4.16, the degree of curve is

$$D = \frac{5729.58 \text{ ft-deg}}{R_{ft}} = \frac{5729.58 \text{ ft-deg}}{1278.35 \text{ ft}}$$
$$= 4.48°$$

The nearest half-degree is 4.5°.

(b) The radius of a 4.5° curve is found using Eq. 4.16.

$$R_{ft} = \frac{(100 \text{ ft})\left(\frac{180°}{\pi}\right)}{D_{deg}} = \frac{(100 \text{ ft})\left(\frac{180°}{\pi}\right)}{4.5°}$$
$$= 1273.24 \text{ ft}$$

(c) Using the radius found in part (b), the new external distance can be found from Eq. 4.6.

$$E = R\left(\frac{1}{\cos\frac{I}{2}} - 1\right)$$
$$= (1273.24 \text{ ft})\left(\frac{1}{\cos\frac{28°18'}{2}} - 1\right)$$
$$= 39.84 \text{ ft}$$

(d) The length of the curve is found using Eq. 4.9.

$$L_{ft} = \frac{(100 \text{ ft})I_{deg}}{D_{deg}} = \frac{(100 \text{ ft})(28°18')}{4.5°}$$
$$= 628.89 \text{ ft}$$

(e) The tangent length is found using Eq. 4.5.

$$T = R\tan\frac{I}{2} = (1273.24 \text{ ft})\tan\frac{28°18'}{2}$$
$$= 321.00 \text{ ft}$$

7. FIELD LAYOUT AND FITTING OF HORIZONTAL CURVES

Horizontal curves for highways are designed based on the topography, design speed, desired sight distance around objects, and available right-of-way. All highway curves must conform with modern highway safety standards, which demand uniformity of elements throughout any given corridor.

The two most common methods of field layout are the *deflection angle method* and the *tangent offset method*. Although Global Positioning System (GPS) and Real Time Kinematic (RTK) satellite navigation systems have reduced the intensity of field work, the deflection angle method and the tangent offset method continue to be used in the field.

Topography and Fit of Alignment

Topography can be defined as level, rolling, or mountainous, as described in the *Highway Capacity Manual* (*HCM*). *Level terrain* is terrain with short grades no steeper than 1–2%. Although level terrain may be considered ideal for roadway design, large, flat regions usually require additional drainage facilities to carry rainwater away from the traveled surface.

Rolling terrain is terrain where heavy vehicles operate at reduced speeds compared to those of passenger cars but not for extended intervals. Generally, the roadway can be placed at grades of 2–4%. However, these grades do not extend more than 4000 ft (1220 m) in any one instance.

Mountainous terrain is terrain where grades occur such that heavy vehicles operate at substantially slower speeds than automobiles for extended distances. Grades steeper than 4% over a length of 4000 ft (1220 m) or more constitute mountainous terrain. The *HCM* and the *GDHS* have more information on mountainous terrain, along with recommendations for design provisions to handle traffic and safety concerns.

Design speed is a function of the terrain and the type and class of roadway to be built. All other design considerations are made in respect to the selected design speed. Interstate highways and freeways often use a minimum design speed of 70 mph (113 kph) for level, relatively straight roadways in rural areas and a minimum design speed of 50 mph (80 kph) for urban locations. Multilane highways, two-lane highways, and urban streets have cross section and design speed parameters set forth by the responsible operating and funding agency. These speeds may range from freeway-like higher speed conditions to local feeder and access road conditions with design speeds as low as 15 mph (25 kph).

Design speeds can be influenced by available sight distance around buildings, rock croppings, or other obstructions. Design speed may have to factor in available right-of-way to preserve clearances for larger vehicles.

Once a roadway cross section is selected and the design speed and sight distance parameters are set, the centerline geometry can be laid out in preliminary form on topographical maps. This centerline is tested by taking copies of the preliminary map plans to the field for the surveyor and field engineer to locate and mark the centerline. Field staking begins at a known starting point or a connection to an existing facility. The tangents and curves are fitted together one by one until the opposite end is reached, usually connecting to another facility already in place. Field measurements are taken at critical control points, and the alignment is adjusted as needed to fit the required cross section to the control point field locations. After the field alignment is staked out and recorded, the field adjustments are taken back to the design office to be applied to the plan drawings.

The preliminary alignment plan, adjusted to meet field conditions, is presented for centerline alignment approval with walls, bridges, drainage, and other structures shown in their approximate locations on the plans. A preliminary design report is prepared, and the project is organized for cost estimation and funding agency approval.

After the preliminary alignment is approved, final field surveys are performed for structures and drainage facilities. Core borings for earth samples at foundation locations and for earthwork classification require a considerable investment and are usually not taken until after the grade and alignment approval to avoid unnecessary waste of funds should the alignment not be approved. Final geometry is prepared from the preliminary alignment, incorporating minor adjustments as necessary. These changes are kept to a minimum to avoid the costs and delays of reconfiguration.

The contractor is responsible for laying out the final alignment in the field using the final geometry. The designer is no longer involved except in extreme circumstances in which the layout on the plans is in error or a field discovery requires an adjustment. In either case, a change order is required and must be carefully documented so that any additional work relying on the centerline geometry uses the correct information.

Horizontal Alignment Layout and Location Control

Alignment layout begins with horizontal geometry, which is a series of tangents connected by circular arcs. For longer distances in open country, the tangents are usually fitted first, adjusting the tangents when the curves are fitted to the topography. For more congested locations and for interchange ramps, the tangents and curves are fitted with each other; at times, the tangents are applied after the curves are located. Alignment can be thought of as a traverse with a series of tangent headings from PI to PI, with the curves smoothing the change in direction from one heading to the next. The traverse can be a closed system of headings, as in Fig. 4.5, which starts and ends at the same point. The traverse can also be an open system, such as in Fig. 4.6, which starts at a known point (A in the figure) and ends at another known point (E in the figure). Point A is a PI located on the initial heading I–I', and point E is on ending alignment, J–J', meaning the new alignment is a connection between alignments I and J. Most highway alignments are of the open traverse variety, but roadways within a network can be connected a closed survey traverse as well.

Figure 4.5 Example of a Closed Alignment Traverse

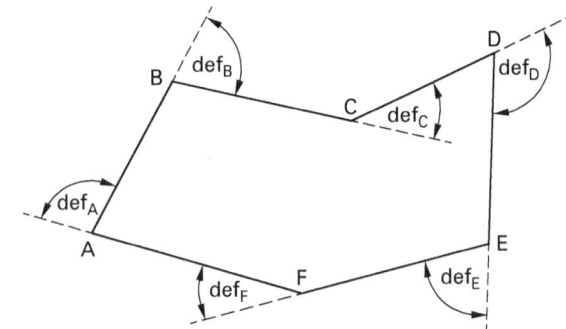

Figure 4.6 Example of Open Alignment Traverse

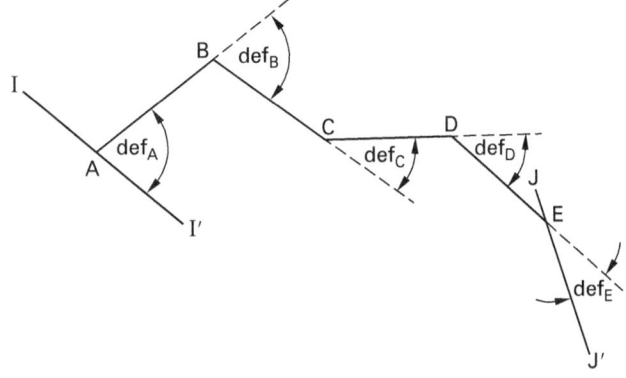

The layout engineer is most concerned with the change in direction at each PI location. The change from one heading to the next is called *deflection.* When performing traverse calculations, deflections are + for right (clockwise), and − for left (counterclockwise), as shown in Fig. 4.7. On highway and railroad plan layouts, the curve deflections, if marked for direction, are usually indicated as right or left instead of + or −. If not indicated, left or right deflections are intuitively obvious when studying the plans. PIs and curve points are tied to survey control points outside of the construction activity zone so that the alignment can be reproduced in the field for construction and ongoing maintenance during the life of the roadway. Using GPS coordinates requires information on the GPS grid being used, which is necessary to be fully noted on the alignment plan. For future reference, ties to local benchmarks should be made at periodic intervals as a check on the GPS system employed.

Figure 4.7 Illustration of Right (+) and Left (−) Deflections

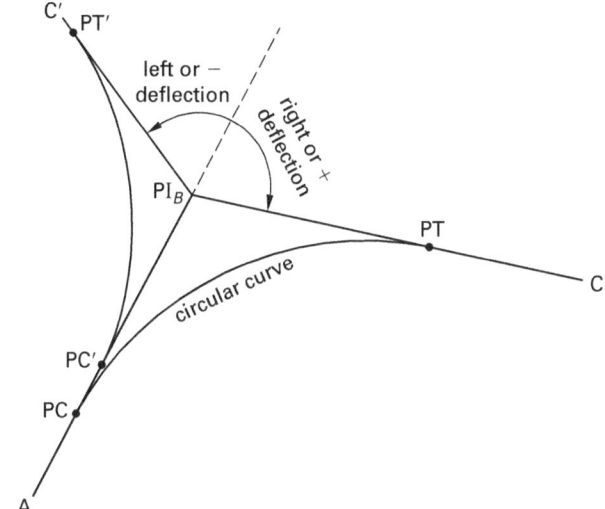

Horizontal geometry is calculated as projected to a horizontal plane, with no correction for the true length of slope distance. Adjustment for actual material quantities on slopes is made during construction estimation, but pay quantities are usually based on horizontal measurement only, such as cubic feet per station of earthwork, or square yard per station for pavement. Slope adjustments primarily affect quantities of long runs of pipe and other utilities that are ordered on a lineal foot basis.

Horizontal Curves

Figure 4.8 shows the locations of additional horizontal curve elements and introduces terminology often referred to in alignment calculations. The tangents of a curve, T, are subtangents of the complete alignment, which extends beyond the limits of the curve (i.e., beyond the PC and PT). For simplicity, these subtangents are referred to as *curve tangents*, and a specific point on either of these tangents is called a *point on subtangent*, POST. This term should not be confused with a fence post or other type of post that may be referenced in deed descriptions. A specific point on either of the tangents that is normally beyond the curve limits is called a *point on tangent*, or POT.

Figure 4.8 Additional Curve Elements

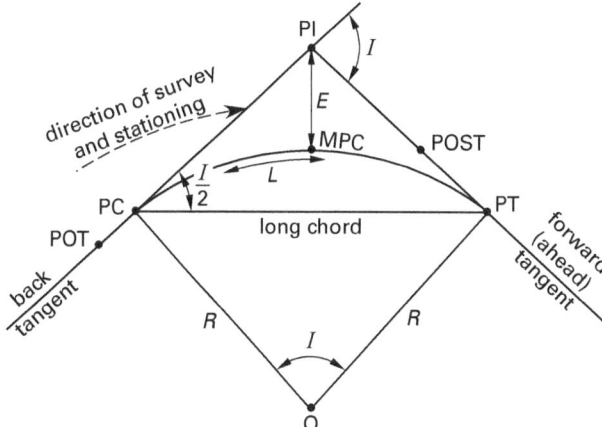

Circular curves are usually laid out on plans with the stationing proceeding from left to right, as shown in Fig. 4.8, or from bottom to top. The curve tangent from the PI to the PC is called the *back tangent*, and the curve tangent from the PI to the PT is called the *ahead tangent*. A curve is stationed continuously from the tangent entering the curve at the PC, continuing ahead along the curve to the PT, and continuing ahead to the tangent ahead of the curve. Stationing continuously along the centerline of a highway or railroad ensures an exact measurement of the distance between any two points along the corridor. The back tangent is stationed from the PC to the PI in order to locate the exact distance to the PI, but the ahead tangent from the PI to the PT is rarely stationed so as to avoid having two stations for the PC.

For horizontal curves laid out around an obstruction, the middle ordinate is sometimes referred to as the *horizontal sightline offset*, HSO, although HSO is actually taken at the center line of the inside lane of the curve. The point where the middle ordinate intersects the arc of the curve is the *midpoint of the curve* and can be labeled MPC.

Figure 4.9 shows the relationship of the various angles found in a circular curve layout. The deflection angle is positive if to the right (i.e., clockwise) and negative if to the left (i.e., counterclockwise).

Figure 4.9 Angles Used in Defining a Horizontal Circular Curve

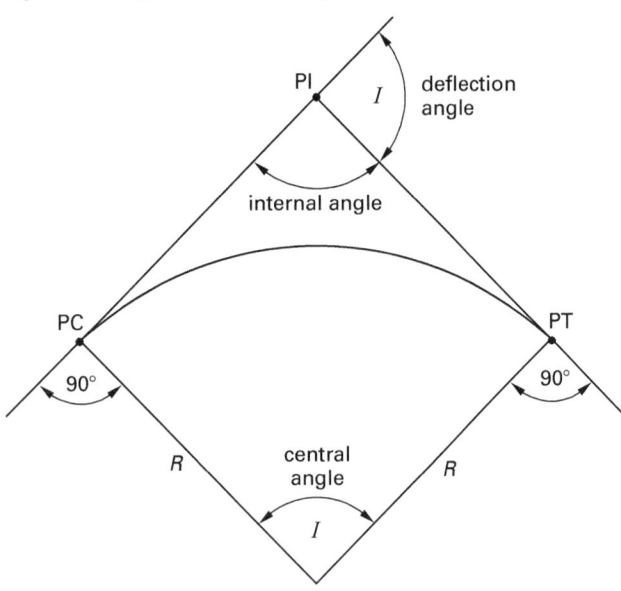

Curve Layout by Deflection

One of the most widely used methods of field layout, or stakeout, is the deflection angle method from the PC or PT. The *deflection angle* is the angle between the tangent and the chord. In Fig. 4.10, the subtended angle, θ, of arc AB is twice the angle deflecting from the tangent $\overline{AV}$ to the chord $\overline{AB}$. The deflection method is popular because it involves the fewest instrument setups.

Figure 4.10 Deflection Relationship in a Circular Curve

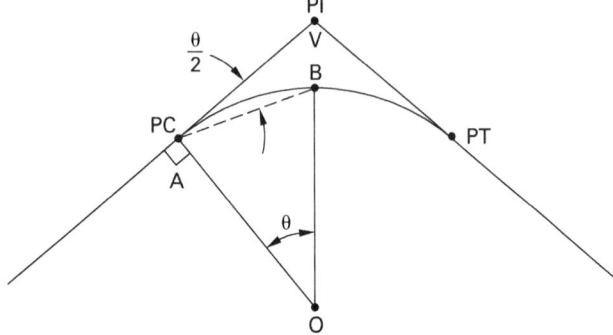

Starting from either the PC or the PT, stations are staked along the chord length. For full stations (i.e., $L = 100$ ft), $\theta = D$, as shown in Eq. 4.29. However, the PC or PT do not necessarily fall on an even station, meaning the arc length will be less than 100 ft. An arc with a chord length less than 100 ft is known as a *subarc*. The deflection angle of the subarc is determined either by multiplying the deflection angle for a full station by the proportion of the chord length to a full station (i.e., $L/100$ ft) or by using Eq. 4.30. Subsequent stations are found by adding intervals of $D/2$ to the first deflection until the last full station before the PC or PT. The final $\theta/2$ increment should make the sum of the deflections equal to half of the full deflection of the curve, $I/2$, as shown in Eq. 4.31, which also serves as a check on calculations.

$$\theta = \frac{L_{\text{ft}} D_{\text{deg}}}{100 \text{ ft}} \quad [\text{U.S. only}] \quad \quad 4.29$$

$$\frac{\theta}{2} = \frac{L_{\text{ft}} D_{\text{deg}}}{200 \text{ ft}} \quad [\text{U.S. only}] \quad \quad 4.30$$

$$\frac{I}{2} = \sum_{1}^{n} \frac{\theta_1}{2} + \frac{\theta_2}{2} + \cdots + \frac{\theta_n}{2} \quad \quad 4.31$$

When staking out a curve, a surveying instrument setup on each station is usually not necessary, as the scope on the field instrument usually provides accurate readings up to at least 300 ft (91 m), and reducing the number of setups increases productivity as well as accuracy. For a typical curve layout, once the PC is staked, the next three stations can be staked before moving the instrument—that is, the setup is moved ahead three stations, the stakeout continues for another three stations, and so on, until the PT is reached. (See Fig. 4.11.) For a setup on an intermediate station, the angle relationships are illustrated in Fig. 4.12.

Figure 4.11 Chord Deflections and Total Curve Deflection

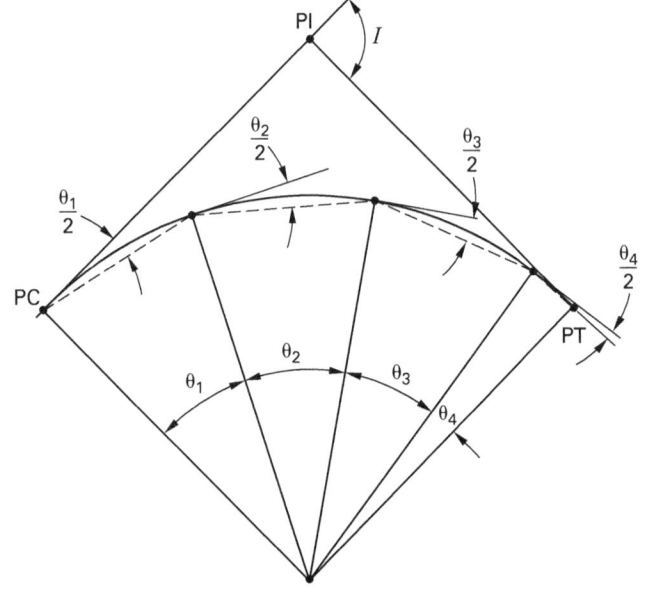

Figure 4.12 Deflection Angles from an Intermediate Point on a Circular Curve

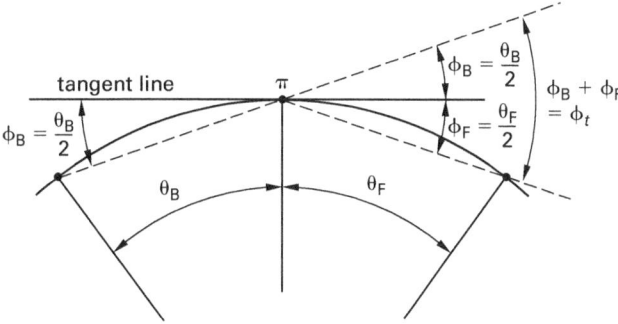

Once the PT is staked, the field instrument is moved over the PT stake. The $I/2$ angle is turned from the PC to the PI, and the scope is flopped (rotated vertically 180), allowing the forward tangent to be sighted and staked.

When the curve is long enough that an intermediate setup is required and the backsight is to the previously set point on the curve, the deflection of the back chord must be added to the deflection of the ahead chord to find the total deflection.

The points of measurement (i.e., the instrument locations) used in the field must be carefully recorded when using the deflection angle method. The instrument location is the point from which measurements are taken or about which the angle is turned. These points are referred to as stations, even if the instrument location is not recorded in units of stations (e.g., station A). Field notes are commonly recorded in tabular form, with the leftmost column heading labeled $\overline{\lambda}$ (instrument setup point) and each station listed underneath.

Example 4.4

A curve has a radius of 800 ft and a deflection angle of 32°. The PC station is 133+23.20, and the maximum sight line is 300 ft. Use sta 136+00 as the secondary instrument setup location. Using the deflection angle method, complete the tabulation of deflection angles to lay out the curve.

Solution

Calculate the necessary curve information, recording all data in tabular form as shown in *Illustration and Table for Example 4.4*. The leftmost column gives the station, $\overline{\lambda}$, of each instrument setup.

Determine the curve length using Eq. 4.9.

$$L_{\text{ft}} = \frac{2\pi R_{\text{ft}} I_{\text{deg}}}{360°} = \frac{(2\pi)(800 \text{ ft})(32°)}{360°}$$
$$= 446.80 \text{ ft}$$

From Eq. 4.18, the degree of curve for arc definition is

$$D_{\text{deg}} = \frac{(360°)(100 \text{ ft})}{2\pi R_{\text{ft}}} = \frac{(360°)(100 \text{ ft})}{(2\pi)(800 \text{ ft})}$$
$$= 7.162°$$

Use Eq. 4.24 to find the station of the PT.

$$\text{sta PT} = \text{sta PC} + L = \text{sta } 133 + 23.20 + 446.80 \text{ ft}$$
$$= \text{sta } 137 + 70.00$$

Deflections will first be made with the instrument set on the PC. To set the first full station, find the subarc length measured from the PC to sta 134+00.

$$L_{\text{subarc}} = \text{sta } 134 + 00 - \text{sta } 133 + 23.20$$
$$= 76.80 \text{ ft}$$

Calculate the deflection using Eq. 4.30.

$$\phi_{\text{sta } 134+00} = \frac{\theta}{2} = \frac{L_{\text{subarc,ft}} D_{\text{deg}}}{200 \text{ ft}} = \frac{(76.80 \text{ ft})(7.162°)}{200 \text{ ft}}$$
$$= 2.750°$$

From Eq. 4.11, the chord length for a 76.80 ft arc is

$$C_{\text{sta } 134+00} = 2R \sin \frac{\theta}{2} = (2)(800 \text{ ft}) \sin 2.750°$$
$$= 76.77 \text{ ft}$$

For sta 135+00, the deflection of a 176.80 ft arc (i.e., the subarc length plus one full station) is

$$\phi_{\text{sta } 135+00} = \frac{\theta}{2} = \frac{L_{\text{ft}} D_{\text{deg}}}{200 \text{ ft}} = \frac{(176.80 \text{ ft})(7.162°)}{200 \text{ ft}}$$
$$= 6.331°$$

The chord length for a 176.80 ft arc is

$$C_{\text{sta } 135+00} = 2R \sin \frac{\theta}{2} = (2)(800 \text{ ft}) \sin 6.331°$$
$$= 176.44 \text{ ft}$$

For sta 136+00, the deflection of a 276.80 ft arc from the tangent is

$$\phi_{\text{sta } 136+00} = \frac{\theta}{2} = \frac{L_{\text{ft}} D_{\text{deg}}}{200 \text{ ft}} = \frac{(276.80 \text{ ft})(7.162°)}{200 \text{ ft}}$$
$$= 9.912°$$

Illustration and Table for Example 4.4

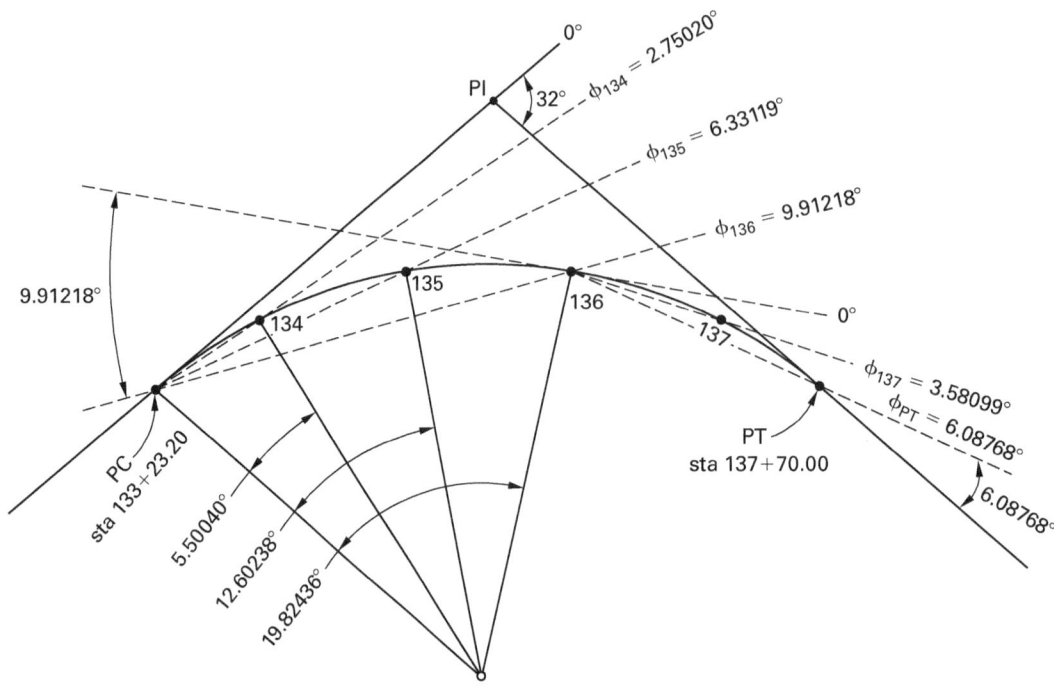

location, ⊼ (sta)	backsight (sta)	ϕ_B	foresight (sta)	chord length (ft)	ϕ_F	ϕ_t
133+23.20 PC	tan	0°	134+00	76.77	2.750°	2.750°
133+23.20 PC	tan	0°	135+00	176.44	6.331°	6.331°
133+23.20 PC	tan	0°	136+00	275.42	9.912°	9.912°
136+00	133+23.20 PC	9.912°	137+00	99.94	3.581°	13.493°
136+00	133+23.20 PC	9.912°	137+70 PT	169.69	6.088°	16.000°
137+70.00 PT	136+00	6.088°	ahead tangent	–	6.088°	12.175°
check total deflection						32.000°

The chord length for a 276.80 ft arc is

$$C_{\text{sta}\,136+00} = 2R \sin\frac{\theta}{2}$$
$$= (2)(800 \text{ ft})\sin 9.912°$$
$$= 275.42 \text{ ft}$$

Since the curve is longer than the maximum sight line of 300 ft, a second setup on the curve is necessary. The problem statement specified the second setup location as sta 136+00, but any station less than 300 ft from the PC could have been chosen. To locate sta 137+00, find the deflection angle of a 100 ft arc from a tangent located at the second setup point (i.e., sta 136+00).

$$\phi_{\text{sta}\,137+00} = \frac{\theta}{2} = \frac{L_{\text{ft}}D_{\text{deg}}}{200 \text{ ft}} = \frac{(100 \text{ ft})(7.162°)}{200 \text{ ft}}$$
$$= 3.581°$$

The chord length for a 100 ft arc is

$$C_{\text{sta}\,137+00} = 2R \sin\frac{\theta}{2}$$
$$= (2)(800 \text{ ft})\sin 3.581°$$
$$= 99.94 \text{ ft}$$

The back deflection from the tangent at sta 136+00 to the PC was found when locating sta 136+00. Add the back deflection to the ahead deflection (i.e., the deflection from the tangent at sta 136+00 to the PT) for the total deflection from the PC.

$$\phi_t = \phi_{\text{sta}\,136+00} + \phi_{\text{sta}\,137+00} = 9.912° + 3.581°$$
$$= 13.493°$$

Working from sta 136+00 to the PT,

$$\text{arc length} = \text{sta}\,137 + 70.00 - \text{sta}\,136 + 00.00$$
$$= 170.00 \text{ ft}$$

Calculate the deflection of a 170.00 ft arc from the tangent located at the setup point to the PT.

$$\phi_{\text{sta}\,137+70.00} = \frac{\theta}{2} = \frac{L_{\text{ft}} D_{\text{deg}}}{200 \text{ ft}} = \frac{(170.00 \text{ ft})(7.162°)}{200 \text{ ft}}$$
$$= 6.088°$$

The chord length for a 170.00 ft arc is

$$C_{\text{PT}} = 2R \sin \frac{\theta}{2} = (2)(800 \text{ ft}) \sin 6.088°$$
$$= 169.69 \text{ ft}$$

The table and curve are completed as shown in the *Illustration and Table for Example 4.4*.

The last line is a check of the total deflection, found by adding the angles shown in bold. The error shown is 1:114,000 deviations from 32.00000° due to the rounding in the calculations. This amounts to 1 sec of angle.

Curve Layout by Tangent Offset

The *tangent offset method* (also known as the *station offset method*) can also be used to lay out horizontal curves. This method is used to locate any point on the curve when working strictly from the tangent and PC and is typically used on short curves. The tangent offset method is well-suited to locating a single point with a high degree of accuracy or for approximating measurements using cloth, plastic, or steel tapes only. A curve's *tangent offset*, y, is the perpendicular distance from an extended tangent line to the curve. The *tangent distance*, x, is the distance along the tangent to a perpendicular point. (See Fig. 4.13.)

Figure 4.13 can be mathematically represented as Eq. 4.32.

$$y = R - \sqrt{R^2 - x^2} = R(1 - \cos 2\alpha) \qquad 4.32$$

Figure 4.13 Point on Curve Located by Tangent Offset

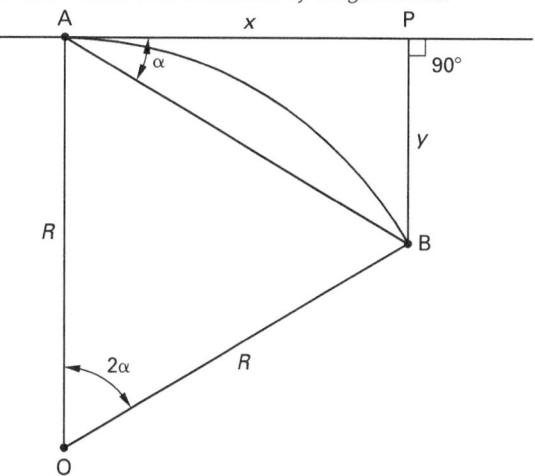

Inaccessible PI

In order to lay out a complete curve, the PI must be located and used as the instrument setup location if possible. The PI is used because it provides the most accurate calculation of the intersection angle and provides the longest sight tangent. If the PI can be occupied during curve layout, the PI is called accessible. However, the PI is at times inaccessible due to some physical obstruction or unreachable location, such as being located in a body of water, over a cliff, or in a building. The following procedure is used to compensate for an inaccessible PI.

A point on subtangent, POST, is located at any point A on the back subtangent that is visible from a point B on the ahead subtangent. The relationship between point A and point B is shown in Fig. 4.14. The sum of angles α and β must equal the intersection angle. From triangle PI-A-B, α and β can be calculated using the law of sines. The PI is found by adding the distance from A to the PI, or $\overline{A\,PI}$, to point A's station. After calculating the tangent distance, T, the PC and PT are found by measuring from points A and B to the PI. Therefore, the entire curve can be laid out accurately even though the PI cannot be occupied.

Figure 4.14 Inaccessible PI

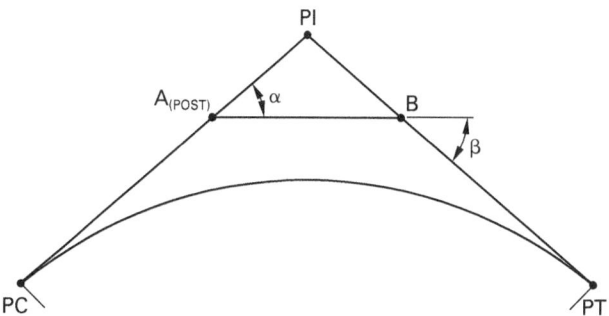

It is important to examine the location carefully before choosing the location of points A and B. Often, the subtangent $\overline{AB}$ is placed close to the curve so that it appears to be touching, or is tangent to, the curve. Making this assumption without basis will yield an inaccurate result. Also, the distances $\overline{A\,PI}$ and $\overline{B\,PI}$ do not have to be equal and rarely are. Points A and B are simply set at convenient locations along the tangents.

8. SPIRAL CURVES

Spiral curves are introduced at the ends of a circular curve (i.e., at the PC and PT) in order to provide a transition, or *easement*, between the straight tangents and the circular parts of the curve. Without an easement, the sudden change from no lateral acceleration along the tangent to the lateral acceleration necessary to travel around a curve causes an increase in lateral force on the vehicle. Spiral curves were first used on railroads to reduce the jerk caused at the ends of curves. The principle of easement is a gradual increase in lateral acceleration by decreasing the curve radius until the central curve radius is reached. *Superelevation*, which is the difference between the inner and outer elevation of a track or roadway, is also transitioned along the easement, providing a smooth and comfortable ride for passengers by changing the cross slope in proportion to the change in curve radius. (See Sec. 4.12 for more information on superelevation.)

The spiral curve adopted by AASHTO for highways is modified from the standard railroad spiral curve. The degree of curve increases linearly along the spiral transition from zero at the TS to the degree of curve of the central curve at the SC. The TS is the point of change from the tangent to the spiral, and the SC is the point of change from the spiral curve to the circular curve. The degree of the central curve at the CS or SC, D_l, can be calculated from Eq. 4.33. D_c is the degree of curve for the entire curve, L_s is the length of the spiral (i.e., the entire curve length), and L is the length of the curve from the TS or ST to the CS or SC.

$$D_l = D_c \left(\frac{L}{L_s} \right) \quad 4.33$$

The minimum length of a spiral is based on driver comfort and lateral shifts in the position of a vehicle within the curve. Spirals should be long enough that the increase in lateral acceleration as a vehicle enters the curve is comfortable to the driver. Spirals should also be long enough that the shift they cause in a vehicle's lateral position is similar to the shift produced by a vehicle's natural path. Therefore, the minimum length of a spiral should be the larger value found from the following equations.

$$L_{s,\min} = \sqrt{24 p_{\min} R} \quad [\textit{GDHS} \text{ Eq. 3-28}]$$
$$= 0.0214 \frac{v^3}{RC} \quad [\textit{GDHS} \text{ Eq. 3-29}] \quad \text{[SI]} \quad 4.34(a)$$

$$L_{s,\min} = \sqrt{24 p_{\min} R} \quad [\textit{GDHS} \text{ Eq. 3-28}]$$
$$= 3.15 \frac{v^3}{RC} \quad [\textit{GDHS} \text{ Eq. 3-29}] \quad \text{[U.S.]} \quad 4.34(b)$$

$p_{\min}$ is the minimum lateral offset between the tangent and the circular curve, v is the design speed in mph (kph), and C is the maximum rate of change in lateral acceleration. AASHTO suggests a value of 0.66 ft (0.20 m) for $p_{\min}$, as this is representative of the minimum lateral shift that occurs from the natural steering behavior of most drivers. AASHTO recommends a value of 4.0 ft/sec^3 (1.2 m/s^3) for the maximum rate of change in lateral acceleration, C.

Spiral curves must also not be too long in relation to the length of the circular curve. A spiral curve that is too long can mislead drivers about the sharpness of the upcoming curve, causing drivers to approach the curve at unsafe speeds. The maximum length of a spiral curve can be calculated from Eq. 4.35.

$$L_{s,\max} = \sqrt{24 p_{\max} R} \quad [\textit{GDHS} \text{ Eq. 3-30}] \quad 4.35$$

The maximum lateral offset between the tangent and circular curve, $p_{\max}$, is suggested by AASHTO to be 3.3 ft (1.0 m), as this value is representative of the maximum lateral shift occurring as a result of the natural steering behavior of the majority of drivers.

Table 4.1 gives roadway spiral lengths recommended in the *GDHS*. The lengths given correspond to 2.0 sec of travel time at the design speed listed, which is representative of the natural spiral path used by most drivers.

Calculations involving components of a spiral curve require an additional significant digit to the right of the decimal. A general rule is to carry a value out to three decimals for customary U.S. units and four decimals for SI units. Carrying an additional significant digit is generally sufficient to ensure accuracy of the final measurements for layout purposes.

The geometric relationships given for horizontal curves, Eq. 4.7, Eq. 4.8, Eq. 4.14, and Eq. 4.15, also apply for spiral curves. The equations for degree of curve and curve radius for horizontal curves also apply to spiral curves. Spiral curve variables shared with other curve types are distinguished using a subscript s, except for the degree of curve, which uses D_c. Figure 4.15 illustrates a fully spiraled circular curve, and Fig. 4.16 demonstrates the details of a spiral curve. Spiral curve abbreviations and variables are given in Table 4.2.

Table 4.1 Desirable Length of Spiral Curve Transition

SI units		customary U.S. units	
design speed (kph)	spiral length (m)	design speed (mph)	spiral length (ft)
20	11	15	44
30	17	20	59
40	22	25	74
50	28	30	88
60	33	35	103
70	39	40	117
80	44	45	132
90	50	50	147
100	56	55	161
110	61	60	176
120	67	65	191
130	72	70	205
		75	220
		80	235

From *A Policy on Geometric Design of Highways and Streets*, 2011, by the American Association of State Highways and Transportation Officials, Washington, D.C. Used by permission.

Figure 4.15 Fully Spiraled Circular Curve Layout

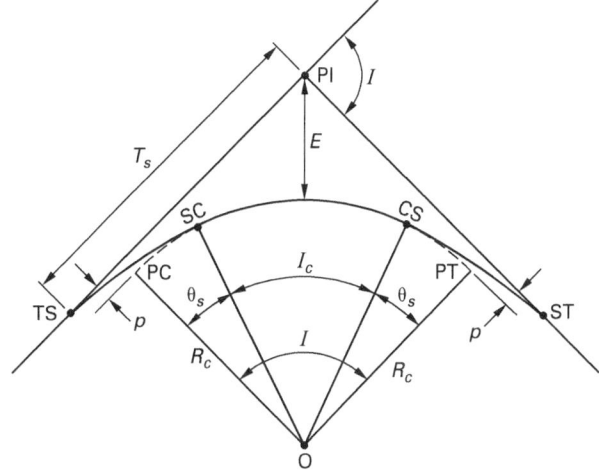

Figure 4.16 Detail Elements of a Transition Spiral

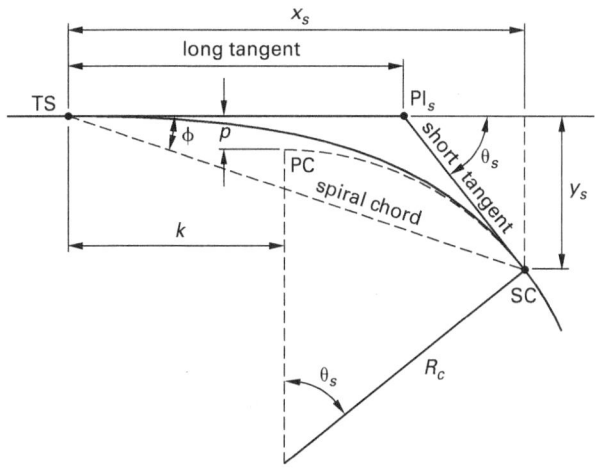

Table 4.2 Spiral Curves: Abbreviations and Terms

CS   point of change from circular curve to spiral
D_c degree of curve of spiral at SC or CS, and of the central curve
E_s external distance from the PI to the center of the circular arc (see Eq. 4.39)
I intersection angle of the tangents of the entire curve
I_c intersection angle of the included circular curve from the SC to the CS
k distance along initial tangent extended from TS to point opposite PC of shifted curve (see Eq. 4.38)
K rate of change of degree of curvature per foot (meter) of spiral
L spiral arc from the TS or ST and any point on spiral
L_s total spiral length from TS to SC, or CS to ST
LT   long tangent, distance along initial tangent extended from TS to spiral PI
p offset from initial tangent to PC of shifted circular curve (see Eq. 4.37)
PI_s spiral PI
R_c radius of central curve
SC   point of change from spiral curve to circular
SCS   point of change from one spiral to another, or an instant curve point between spirals
SS   point of change from one spiral to another, or an instant curve point between spirals
ST   short tangent, distance from spiral PI to SC
ST   point of change from spiral to tangent
T_s tangent of a spiral curve (see Eq. 4.36)
TS   point of change from tangent to spiral
x_s coordinate of SC from TS along tangent (see Eq. 4.44)
y_s coordinate of SC from TS, offset distance (see Eq. 4.45)
θ deflection angle of spiral arc of a given length, L (see Eq. 4.40)
θ_s deflection angle of total spiral arc, L_s (see Eq. 4.41)
ϕ spiral deflection angle at TS or ST from initial tangent to any point on spiral (see Eq. 4.42)
ϕ_s spiral deflection angle to SC or CS (see Eq. 4.43)

Unique to spiral curves are the points of change between spiral curves or between a spiral curve and a circular curve. SC is the point of change from a spiral curve to a circular curve, and TS is the point of change from a tangent to a spiral curve. CS is the point of change from a circular curve to a spiral curve, and ST is the point of change from a spiral curve to a tangent. SS, or SCS, is the point of change from one spiral curve to another or an instant curve point between two spiral curves.

The tangent of a spiral curve, T_s, can be calculated using Eq. 4.36. R_c is the radius of the central curve, and p is the offset from the initial tangent to the PC of the shifted circular curve calculated from Eq. 4.37. Equation 4.38 is used to calculate k, which is the distance along the initial tangent from TS to the PC of the shifted curve. I is the intersection angle of the tangent and the entire curve, and I_c is the intersection angle of the circular curve measured from the SC to the CS.

$$T_s = (R_c + p)\tan\frac{I}{2} + k \qquad 4.36$$

$$p = y_s - R_c \operatorname{vers} \theta_s \qquad 4.37$$

$$k = x_s - R_c \sin \theta_s \qquad 4.38$$

The external distance, E_s, measured from the PI to the center of the circular arc is found from Eq. 4.39.

$$E_s = (R_c + p)\sec\frac{I}{2} - R \qquad 4.39$$

The central angle of a spiral arc is calculated using Eq. 4.40 and Eq. 4.41. L is the spiral arc length from either the TS or ST to any point on the spiral. L_s is the total spiral length from TS to SC or from ST to CS. θ is the central angle of a spiral arc with a given length, L, and θ_s is the central angle of the total spiral arc length.

$$\theta = \left(\frac{L}{L_s}\right)^2 \theta_s \qquad 4.40$$

$$\theta_s = \left(\frac{L_{s,\text{ft}}}{200 \text{ ft}}\right) D_{c,\text{deg}} \quad \text{[U.S. only]} \qquad 4.41$$

Equation 4.42 is used to calculate the spiral deflection angle, ϕ, at TS or ST from the initial tangent to any point on the spiral. The spiral deflection angle to SC or CS, ϕ_s, is found from Eq. 4.43.

$$\phi = \frac{1}{3}\left(\frac{L}{L_s}\right)^2 \theta_s \qquad 4.42$$

$$\phi_s = \frac{1}{3}\theta_s \qquad 4.43$$

The length between SC and TS measured along the tangent, x_s, is found from Eq. 4.44. The tangent offset, y_s, is the vertical distance between SC and TS and is calculated from Eq. 4.45.

$$x_{s,\text{ft}} = \frac{L_{s,\text{ft}}}{100 \text{ ft}}$$
$$\times \left[100 \text{ ft} - \frac{(100 \text{ ft})\left(\frac{\pi}{180°}\right)^2 D_{c,\text{deg}}^2}{(5)(2!)} \right.$$
$$+ \frac{(100 \text{ ft})\left(\frac{\pi}{180°}\right)^4 D_{c,\text{deg}}^4}{(9)(4!)}$$
$$\left. + \frac{(100 \text{ ft})\left(\frac{\pi}{180°}\right)^6 D_{c,\text{deg}}^6}{(13)(6!)} \cdots\right] \qquad 4.44$$
[U.S. only]

$$y_{s,\text{ft}} = \frac{L_{\text{ft}}^3}{6 R_{c,\text{ft}} L_{s,\text{ft}}} \quad \text{[U.S. only]} \qquad 4.45$$

9. VERTICAL CURVES

Vertical curves are used to connect two vertical tangents in order to change the grade of a highway. Most vertical curve layouts use *parabolic curves*, which can be symmetrical or unsymmetrical. Symmetrical parabolic curves are most often used because they have a constant rate of grade change, which simplifies calculation. Calculations may also be simplified by basing control points on offsets from the vertical tangents. All calculations are based on a horizontal plane, projecting horizontal measurements up or down to the road surface, regardless of the slope of the grade. Slope distances are generally only used for quantity estimation purposes. Vertical alignment is shown as a vertical surface cut along the *horizontal control line*, which is normally the centerline of the roadway. Vertical curves are usually identified by the station of the point of vertical intersection, PVI, and the PVI elevation, such as "vertical curve 67+84 at elevation 100." The location and elevation of the PVI are the most important geometric points of a vertical curve, as the PVI location controls the grade lines.

A *sag vertical curve* is concave upward, and a *crest vertical curve* is concave downward. Grades are shown as positive (+) if the profile rises with advancing stations and negative (−) if the profile falls with advancing stations. Figure 4.17 illustrates three general cases of sag vertical curves, while Fig. 4.18 shows three general cases of crest vertical curves. The curves of Fig. 4.17(a) and Fig. 4.18(a) have *turning points*, or an instant level section, while the other curve conditions do not.

Figure 4.17 Three Conditions of Sag Vertical Curves

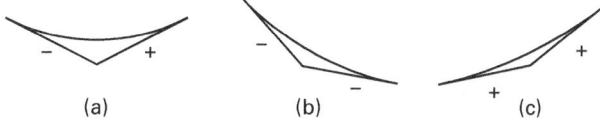

Figure 4.18 Three Conditions of Crest Vertical Curves

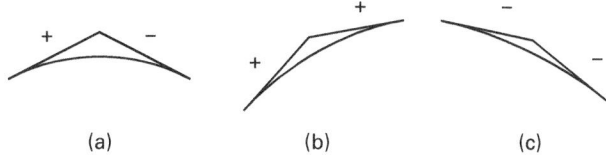

Parabolic Curves

The principle of a parabolic curve is applied directly to curve calculations. A parabolic curve consists of three basic points: the point of vertical curve, PVC, the point of vertical intersection, PVI, and the point of vertical tangency, PVT. The PVC is the beginning of the curve (also known as BVC), and the PVT is the end of the curve (also called EVC). The PVI is the intersection of the two tangents and is also referred to as the vertex, V. Offsets to a tangent, y, of a parabola are proportional to the squares of the distances from the point of tangency (i.e., the point where the curve and tangent meet), as shown in Fig. 4.19, and are calculated from Eq. 4.46. If the distance from the point of tangency is doubled, the offset from the tangent to the parabola becomes 2^2, or four times as large.

$$y = ax^2 \quad 4.46$$

Figure 4.19 Tangent Offset Relationship to Distance in a Parabolic Curve

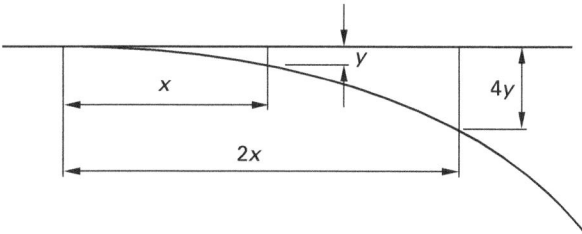

The middle ordinate and the external distance are equal for a parabolic curve, as shown in Fig. 4.20. However, the external distance is not frequently used in vertical curves. The middle ordinate, M, is found from Eq. 4.47, where A is the absolute value of the algebraic difference in grades.

$$E_{\text{ft}} = M_{\text{ft}} = \frac{|G_{2,\%} - G_{1,\%}| L_{\text{sta}}}{8 \frac{\text{sta}}{\text{ft}}} = \frac{A_{\%} L_{\text{sta}}}{8 \frac{\text{sta}}{\text{ft}}} \quad 4.47$$

[U.S. only]

Figure 4.20 Location of Middle Ordinate and External Distance on a Vertical Parabolic Curve

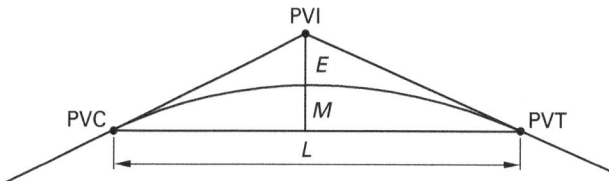

The *rate of grade change per station*, R, can be determined using Eq. 4.48. In these equations, G_2 is the grade out of the curve, and G_1 is the grade into the curve. R can be expressed in units of %/sta or ft/sta^2, since 1 sta is equivalent to 100 ft. The rate, R, will be negative for crest vertical curves and positive for sag vertical curves.

$$R = \frac{G_{2,\%} - G_{1,\%}}{L_{\text{sta}}} \quad \text{[may be negative]} \quad 4.48$$

An unknown elevation at point x, elev$_x$, on a vertical curve can be determined from a known elevation at point a, elev$_a$, and the rate of change in grade, R, as shown in Eq. 4.49.

$$\text{elev}_x = \frac{Rx^2}{2} + G_1 x + \text{elev}_a \quad 4.49$$

The x distance is the horizontal distance from the PVC, or any known elevation point on the curve, to the point of grade to be calculated. Equation 4.49 can be used to calculate the vertical curve elevations directly from a known elevation at one end. Equation 4.49 can be rewritten as Eq. 4.50, which is convenient for directly calculating the series of elevations needed to construct a curve.

$$\begin{aligned}\text{elev}_x &= \frac{Rx^2}{2} + G_{1,\text{ft/ft}} x + \text{elev}_a \\ &= \left(G_{1,\text{ft/ft}} + \frac{Rx}{2}\right) x + \text{elev}_a\end{aligned} \quad 4.50$$

When the instant slope along the curve is equal to zero, the maximum or minimum elevations will occur. This high or low point, called the *turning point*, is determined by Eq. 4.51, where x is the horizontal distance from the PVC. The elevation of the turning point can be found using either the rate of grade change or the proportional offset method detailed in the following section.

$$x_{\text{turning point,sta}} = \frac{-G_{1,\%}}{R_{\%/\text{sta}}} \quad 4.51$$

Design and Layout of Vertical Curves and Elevations

Vertical curves are considered symmetrical unless otherwise noted. The length of a symmetrical vertical curve is distributed equally on both sides of the PVI, so that the distance PVC–PVI is the same as the distance PVI–PVT. For example, a 500 ft long vertical curve would be comprised of two 250 ft half-curves located on each side of the PVI. The length of each half-curve is expressed as half the total curve length, $L/2$. The elevation of the PVC and PVT can be calculated if the PVI elevation is known by adding the product of the half-curve length, $L/2$, and the grade, G, along the curve length to the PVI elevation. Similarly, the product of the tangent length, T, and the grade can be added to or subtracted from the PVI elevation to determine the PVT and PVC elevations.

Unsymmetrical vertical curves are occasionally used to connect to existing roadways or to allow for special vertical clearance conditions, such as at an underpass (see Sec. 4.13). Unsymmetrical vertical curves are shown by designating the length of the back tangent and ahead tangent. If this information is not given, the curve is assumed to be a symmetrical vertical curve.

Both the tangent offset and the modified rate of change methods are effective ways to calculate elevations on vertical curves. However, the rate of change method is most often used when dealing with stations not at even pluses. Example 4.5 provides solutions using both methods.

Example 4.5

A symmetrical vertical curve is staked out between two grades, $\overline{AC}$ and $\overline{CB}$. The $\overline{AC}$ grade is -6.2%, and the $\overline{CB}$ grade is $+3.8\%$. The PVI is located at sta $80+73.00$ and has an elevation of 92.25 ft. The total length of the vertical curve is 2.50 sta. Write the notes for the grade stakes placed at 25 ft intervals along the curve, with the stakes not set at even pluses.

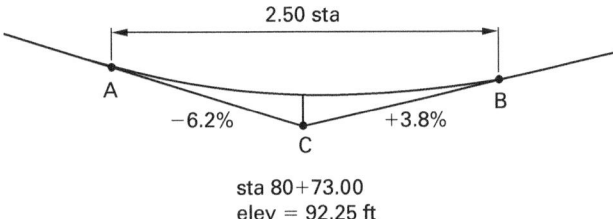

Solution

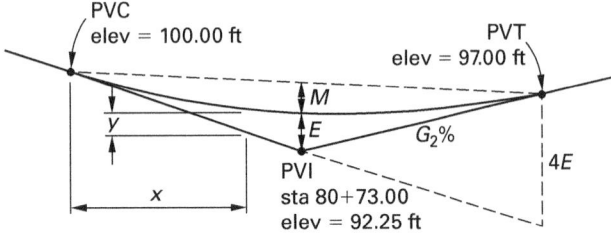

To use the method of proportional squared offsets, the solution can be worked from the PVC to the PVI and from the PVT to the PVI, or from either the PVC or PVT to the other end of the curve. This solution is worked from each end to the PVI. Tabulate all data for each station, spaced 25 ft apart, as shown in *Table for Example 4.5 Solution*.

First, the stations and elevations of the PVC and PVT are found. Since the curve is symmetrical, the PVI is at the center of the curve. Therefore, the length of the curve from the PVC to the PVI and from the PVI to the PVT is 2.50 sta/2 = 1.25 sta. The stations of the PVC and PVT are

$$\text{sta PVC} = \text{sta PVI} - \frac{L}{2}$$
$$= \text{sta } 80+73.00 - \left(\frac{2.50 \text{ sta}}{2}\right)\left(100 \frac{\text{ft}}{\text{sta}}\right)$$
$$= \text{sta } 79+48.00$$

$$\text{sta PVT} = \text{sta PVI} + \frac{L}{2}$$
$$= \text{sta } 80+73.00 + \left(\frac{2.50 \text{ sta}}{2}\right)\left(100 \frac{\text{ft}}{\text{sta}}\right)$$
$$= \text{sta } 81+98.00$$

Noting that the grades when working from the PVI to the PVC and PVT are positive, the elevations of the PVC and PVT are

$$\text{elev}_{\text{PVC}} = \text{elev}_{\text{PVI}} + G_1\left(\frac{L}{2}\right)$$
$$= 92.25 \text{ ft} + (6.2\%)\left(\frac{2.50 \text{ sta}}{2}\right)\left(1 \frac{\text{ft}}{\%\text{-sta}}\right)$$
$$= 100.00 \text{ ft}$$

$$\text{elev}_{\text{PVT}} = \text{elev}_{\text{PVI}} + G_2\left(\frac{L}{2}\right)$$
$$= 92.25 \text{ ft} + (3.8\%)\left(\frac{2.50 \text{ sta}}{2}\right)\left(1 \frac{\text{ft}}{\%\text{-sta}}\right)$$
$$= 97.00 \text{ ft}$$

The middle ordinate is

$$M_{\text{ft}} = \frac{|G_{2,\%} - G_{1,\%}| L_{\text{sta}}}{8 \frac{\text{sta}}{\text{ft}}}$$
$$= \frac{|3.8\% - (-6.2\%)|(2.50 \text{ sta})}{8 \frac{\text{sta}}{\text{ft}}}$$
$$= 3.125 \text{ ft}$$

Table for Example 4.5 Solution

station		x (sta)	elevation on tangent	$C = (2x/L)^2$	offset $= CM$	elevation on curve
PVC	79+48.00	0	100.00	0	0	100.000
	+73.00	0.25	98.45	0.04	0.125	98.575
	+98.00	0.50	96.90	0.16	0.500	97.400
	80+23.00	0.75	95.35	0.36	1.125	96.475
	+48.00	1.00	93.80	0.64	2.000	95.800
PVI	+73.00	1.25	92.25	1.00	3.125	95.375
	+98.00	1.00	93.20	0.64	2.000	95.200
	81+23.00	0.75	94.15	0.36	1.125	95.275
	+48.00	0.50	95.10	0.16	0.500	95.600
	+73.00	0.25	96.05	0.04	0.125	96.175
PVT	+98.00	0	97.00	0	0	97.000

For a parabolic curve, $M = E$. Find the elevations on tangent, working from the PVC to the PVT. The grades are negative because they descend from the PVC and PVT. The grade point calculation for sta 80+23 is shown as an example of the calculations for each desired grade point.

$$\text{elev}_{x,\text{sta } 80+23} = G_1 x + \text{elev}_{\text{PVC}}$$
$$= (-6.2\%)(0.75 \text{ sta})\left(1 \frac{\text{ft}}{\%\text{-sta}}\right)$$
$$+ 100.00 \text{ ft}$$
$$= 95.35 \text{ ft}$$

The elevation on the curve is determined by adding the vertical offset correction, CM, to the elevation on vertical tangent. The vertical correction is calculated from the proportional square of the distance from the PVC to the PVI, and the PVT to the PVI multiplied by the middle ordinate. At sta 80+23,

$$C = \left(\frac{2x}{L}\right)^2 = \left(\frac{(2)(0.75 \text{ sta})}{2.50 \text{ sta}}\right)^2 = 0.36$$

Using the tangent elevation, elev_x, at sta 80+23,

$$\text{curve elevation}_{\text{sta } 80+23} = \text{elev}_x + CM$$
$$= 95.35 \text{ ft} + (0.36)(3.125 \text{ ft})$$
$$= 96.475 \text{ ft}$$

Other grade points at 25 ft intervals are calculated and tabulated similarly.

To find the low point, first calculate R using Eq. 4.48.

$$R = \frac{G_{2,\%} - G_{1,\%}}{L_{\text{sta}}} = \frac{3.8\% - (-6.2\%)}{2.50 \text{ sta}} = 4 \text{ \%/sta}$$

Using Eq. 4.51, the low point station is

$$\text{low point station} = \text{sta}_{\text{PVC}} + \frac{-G_1}{R}$$
$$= \text{sta } 79+48.00 + \frac{-(-6.2\%)}{4 \frac{\%}{\text{sta}}}$$
$$= \text{sta } 81+03.00$$

The curve is redrawn with the essential information included, and the elevation table is completed as shown in *Table for Example 4.5 Solution*.

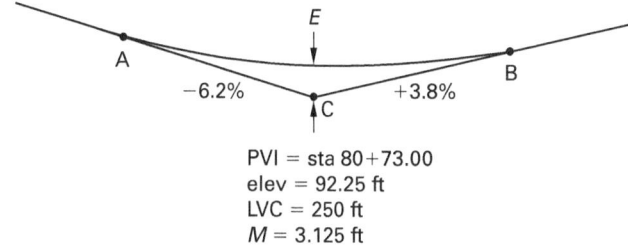

PVI = sta 80+73.00
elev = 92.25 ft
LVC = 250 ft
M = 3.125 ft

Alternate Solution

Tabulate all data for each station, spaced 25 ft apart, as shown in *Table for Example 4.5 Alternate Solution*.

Table for Example 4.5 Alternate Solution

	station	x (sta)	$G_1 + Rx/2 =$ $-6.2\% + 2x$	tangent offset $((x)(-6.2\% + 2x))$	elevation on curve
PVC	79+48.00	0	−6.2	0	100.00
	+73.00	0.25	−5.7	−1.425	98.575
	+98.00	0.50	−5.2	−2.600	97.400
	80+23.00	0.75	−4.7	−3.525	96.475
	+48.00	1.00	−4.2	−4.200	95.800
PVI	+73.00	1.25	−3.7	−4.625	95.375
	+98.00	1.50	−3.2	−4.800	95.200
	81+23.00	1.75	−2.7	−4.725	95.275
	+48.00	2.00	−2.2	−4.400	95.600
	+73.00	2.25	−1.7	−3.825	96.175
PVT	+98.00	2.50	−1.2	−3.000	97.000
LP	81+03.00	1.55	−3.1	−4.805	95.195

Find the elevations directly from the back tangent using Eq. 4.48.

$$R = \frac{G_{2,\%} - G_{1,\%}}{L_{sta}} = \frac{3.8\% - (-6.2\%)}{2.50 \text{ sta}}$$
$$= 4 \%/\text{sta}$$

Use Eq. 4.50 to find the elevation at the tangent distance. For example, to find the elevation at sta 80+23,

$$\text{elev}_x = \left(G_{1,\text{ft/ft}} + \frac{Rx}{2}\right)x + \text{elev}_a$$
$$= \left(-6.2\% + \frac{\left(4\frac{\%}{\text{sta}}\right)(0.75 \text{ sta})}{2}\right)$$
$$\times (0.75 \text{ sta})\left(1 \frac{\text{ft}}{\text{sta-\%}}\right) + 100.00 \text{ ft}$$
$$= 96.48 \text{ ft}$$

Using Eq. 4.51, the low point is

$$\text{low point station} = \text{sta}_{\text{PVC}} + \frac{-G_1}{R}$$
$$= \text{sta } 79+48.00 + \frac{-(-6.2\%)}{4 \frac{\%}{\text{sta}}}$$
$$= \text{sta } 81+03.00$$

From Eq. 4.50, the low point elevation is

$$\text{elev}_x = \left(G_1 + \frac{Rx}{2}\right)x + \text{elev}_a$$
$$= \left(-6.2\% + \frac{\left(4\frac{\%}{\text{sta}}\right)(1.55 \text{ sta})}{2}\right)$$
$$\times (1.55 \text{ sta})\left(1 \frac{\text{ft}}{\text{sta-\%}}\right) + 100.00 \text{ ft}$$
$$= 95.195 \text{ ft}$$

The tabulation can be completed as shown in *Table for Example 4.5 Alternate Solution*.

10. SOURCES OF ERROR IN ROUTE ALIGNMENT

The tangents of a route alignment between two control points (i.e., points with known coordinates) become a traverse that can be checked for accuracy. A traverse with two known control points is known as a *closed traverse*. A traverse where one point has unknown coordinates is known as an *open traverse*. Open traverses cannot be checked for closure and are associated with a lower level of accuracy than a closed traverse. Confirming the accuracy of a traverse is known as *checking for closure* and involves verifying that the error between the calculated end point and the actual end point achieves the desired level of accuracy. The error is determined as a ratio of the difference between the calculated end point and the actual coordinates of the end point, divided by the total traverse length.

The acceptable error is determined by the desired level of accuracy, which is chosen before the survey begins. The *National Geodetic Survey*, NGS, established maximum error ratios for three levels (known as *orders*) of accuracy in horizontal route alignment. A first-order survey is the most accurate, and a third-order survey is the least accurate. Second- and third-order surveys are subdivided into class I and II, with class I representing a higher level of accuracy than class II. Each order and class has a maximum error ratio that cannot be exceeded in order to obtain the chosen level of accuracy. (For example, the maximum error ratio for a second-order, class I survey is 1:50,000.)

A route alignment's accuracy is influenced by the route layout method used. The most accurate method is using the PIs of each curve as the corner points. (This method uses the least number of traverse sides, which simplifies calculation.) Figure 4.21(a) illustrates the relationship of curves to a set of alignment tangents that have been adjusted using the PIs between point A and point F. Points B, C, D, and E are PIs of the route alignment.

Figure 4.21(b) is an exaggerated view of how the curve geometry is distorted if the traverse corners are set at the PCs and PTs of the curves with the long chord of each curve as a side of the control traverse. While the curve layout may be accurate between the PC and PT of an individual curve, the tangent headings between the curves are not the same as the subtangent headings of the adjoining curves, which introduces a small angular change at each PC and PT. This angular difference introduces a new PI at each of the curve points, which results in an inaccurate layout. The location errors caused by PCs and PTs not precisely placed on the overall alignment tangent are eliminated by controlling the location of the tangents.

Another source of route alignment error is caused by rounding tangent azimuths without recalculating the curve geometry. For example, a tangent is given a preliminary heading of $47°52'25''$. The curve geometry is calculated with this heading, and coordinates are set to the resulting curve points and listed on the plan. On the plan view, the heading is rounded to $47°52'$, but the coordinates are not recalculated with the rounded heading. The error in location of a 1000 ft tangent due to heading rounding is illustrated in Fig. 4.22.

Rounding the heading 25'' shifts the location of point B by 0.12 ft. The PI intersects should be recalculated using the rounded heading as the new heading, which will accurately locate new PI coordinates. Once the curve is recalculated using the new PI coordinate, the geometry on the final plan can be recreated accurately as shown, which eliminates further layout errors and guesswork by the field engineers as to the intended final geometry.

Figure 4.21 Example of Route Alignment Layout Showing (a) Control by PI and (b) Control by PC and PT

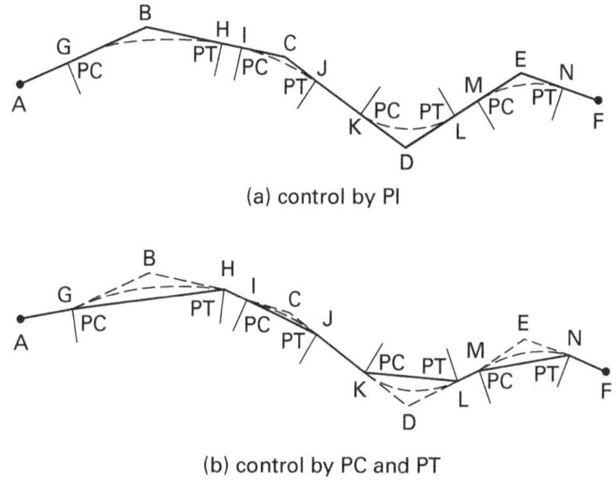

Figure 4.22 Location Error of 1000 ft Tangent Due to Heading Rounding

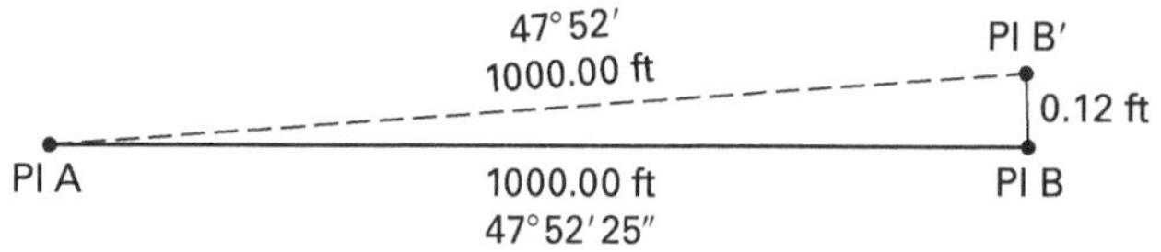

11. SIGHT DISTANCE

Sight distance is the length of roadway a driver can see ahead of the vehicle. Within this sight distance, a driver must analyze upcoming road conditions and traffic situations, select appropriate actions or maneuvers, and then complete these actions or maneuvers. The design speed of horizontal curves and vertical curves and a driver's ability to see around obstructions are based on the maximum sight distance available to a driver.

AASHTO design criteria use three types of sight distance: stopping sight distance, decision sight distance, and passing sight distance.

Stopping Sight Distance

Safe *stopping sight distance* is the total distance required for a driver traveling at design speed to stop a vehicle before reaching an object in its path. Stopping sight distance comprises two distances: the perception-reaction distance and the braking distance. The *perception-reaction distance* (also called the *PIEV distance*, an acronym for perception, identification, emotions, and volition) is the distance traveled at a constant approach speed during the perception-reaction time (known as the *PIEV time*), which is measured from the moment the driver sees an object requiring a stop to the time the brakes are applied. AASHTO refers to the PIEV distance as the *brake reaction distance* and uses an average of 2.5 sec as the *brake reaction time*, which is the time it takes a driver to apply the brakes after seeing an object. Stopping sight distance calculations are determined using the driver's eye height set at 3.5 ft (1080 mm) and the object height at 2.0 ft (600 mm), which is equivalent to the height of a passenger car's taillight. Typical stopping sight distance values for various design speeds are given in Table 4.3.

Table 4.3 Stopping Sight Distance

	SI units			
design speed (kph)	braking reaction distance (m)	braking distance on level (m)	stopping sight distance	
			calculated (m)	design (m)
20	13.9	4.6	18.5	20
30	20.9	10.3	31.2	35
40	27.8	18.4	46.2	50
50	34.8	28.7	63.5	65
60	41.7	41.3	83.0	85
70	48.7	56.2	104.9	105
80	55.6	73.4	129.0	130
90	62.6	92.9	155.5	160
100	69.5	114.7	184.2	185
110	76.5	138.8	215.3	220
120	83.4	165.2	248.6	250
130	90.4	193.8	284.2	285

	customary U.S. units			
design speed (mph)	braking reaction distance (ft)	braking distance on level (ft)	stopping sight distance	
			calculated (ft)	design (ft)
15	55.1	21.6	76.7	80
20	73.5	38.4	111.9	115
25	91.9	60.0	151.9	155
30	110.3	86.4	196.7	200
35	128.6	117.6	246.2	250
40	147.0	153.6	300.6	305
45	165.4	194.4	359.8	360
50	183.8	240.0	423.8	425
55	202.1	290.3	492.4	495
60	220.5	345.5	566.0	570
65	238.9	405.5	644.4	645
70	257.3	470.3	727.6	730
75	275.6	539.9	815.5	820
80	294.0	614.3	908.3	910

Note: Brake reaction distance predicated on a time of 2.5 s; deceleration rate of 3.4 m/s² (11.2 ft/sec²) used to determine calculated sight distance.

From *A Policy on Geometric Design of Highways and Streets*, 2011, by the American Association of State Highways and Transportation Officials, Washington, D.C. Table 3-1. Used by permission.

The *braking distance* is the distance needed to stop the vehicle once the brakes have been applied. Braking distance varies according to the approach speed and the deceleration rate of the vehicle. Tables in Chap. 3 of the *GDHS* show braking distances for a deceleration rate of 11.2 ft/sec² (3.4 m/s²), which is the maximum comfortable stopping rate. This rate takes into account wet pavement and tires with acceptable tread. Emergency stopping rates can be greater under ideal road and tire conditions but are not used for design purposes. While braking ability and stopping friction factors vary somewhat with speed, the variations are not easily quantifiable. There is sufficient variation in general traffic conditions that refinement would not warrant modifying design values. Therefore, most analysis uses a constant deceleration rate of 11.2 ft/sec² (3.4 m/s²).

The braking distance required to bring a vehicle to a complete stop is determined from the design speed, v, in miles (kilometers) per hour and the deceleration rate, a, in feet (meters) per second squared, as shown in Eq. 4.52.

$$d_{\mathrm{m}} = 0.039 \left(\frac{v_{\mathrm{kph}}^2}{a_{\mathrm{m/s^2}}} \right) \quad [\textit{GDHS}\ \mathrm{Eq.\ 3\text{-}1}] \quad [\mathrm{SI}] \quad 4.52(a)$$

$$d_{\mathrm{ft}} = 1.075 \left(\frac{v_{\mathrm{mph}}^2}{a_{\mathrm{ft/sec^2}}} \right) \quad [\textit{GDHS}\ \mathrm{Eq.\ 3\text{-}1}] \quad [\mathrm{U.S.}] \quad 4.52(b)$$

The braking distance for roadways on a grade is found from Eq. 4.53. The friction factor, f, is equal to the deceleration rate divided by the gravitational acceleration and can be substituted for a, as shown in Eq. 4.53.

$$d_{\mathrm{m}} = \frac{v_{\mathrm{kph}}^2}{254\left(\left(\frac{a_{\mathrm{m/s^2}}}{g} \right) \pm G_{\%/100\,\mathrm{m}} \right)}$$

$$= \frac{v_{\mathrm{kph}}^2}{254(f \pm G_{\%/100\,\mathrm{m}})} \quad [\mathrm{SI}] \quad 4.53(a)$$

$$[\textit{GDHS}\ \mathrm{Eq.\ 3\text{-}3}]$$

$$d_{\mathrm{ft}} = \frac{v_{\mathrm{mph}}^2}{30\left(\left(\frac{a_{\mathrm{ft/sec^2}}}{g} \right) \pm G_{\%/100\,\mathrm{ft}} \right)}$$

$$= \frac{v_{\mathrm{mph}}^2}{30(f \pm G_{\%/100\,\mathrm{ft}})} \quad [\mathrm{U.S.}] \quad 4.53(b)$$

$$[\textit{GDHS}\ \mathrm{Eq.\ 3\text{-}3}]$$

The total stopping sight distance, S, is calculated from Eq. 4.54 as the sum of the distance traveled during the perception-reaction time and the braking distance calculated from Eq. 4.52.

$$S_\mathrm{m} = \left(0.278\, \frac{\frac{\mathrm{m}}{\mathrm{s}}}{\frac{\mathrm{km}}{\mathrm{h}}}\right) \mathrm{v}_\mathrm{kph} t_{p,\mathrm{s}} + 0.039 \left(\frac{\mathrm{v}_\mathrm{kph}^2}{a_{\mathrm{m/s}^2}}\right) \quad [\text{SI}] \quad 4.54(a)$$

[*GDHS* Eq. 3-2]

$$S_\mathrm{ft} = \left(1.47\, \frac{\frac{\mathrm{ft}}{\mathrm{sec}}}{\frac{\mathrm{mi}}{\mathrm{hr}}}\right) \mathrm{v}_\mathrm{mph} t_{p,\mathrm{sec}} + 1.075 \left(\frac{\mathrm{v}_\mathrm{mph}^2}{a_{\mathrm{ft/sec}^2}}\right) \quad [\text{U.S.}] \quad 4.54(b)$$

[*GDHS* Eq. 3-2]

Decision Sight Distance

Decision sight distance is appropriate where hazards exist that require drivers to make decisions to perform maneuvers other than a stop, such as lane changes or exit ramp selections, in order for traffic to proceed in an orderly and smooth fashion. These decisions often include multiple actions to be taken simultaneously and may involve selection from several choices of action to be performed. Examples include approaches to complex intersections, multiple interchange ramps, toll booth plazas, restrictive sight distance locations, and instances in which the driver may need to be prepared for further alternative maneuvers in quick succession. Only providing sufficient sight distance for a hurried stop may increase the danger to other motorists, while not providing enough time to make an appropriate selection of an alternate path or course of action for an evasive maneuver. More decision time may also be needed where visual clutter exists, such as advertising signs and busy commercial activity found along a commercial corridor. Since decision sight distances include a margin of error in addition to the time necessary to make evasive maneuvers, decision sight distance values are typically much larger than stopping sight distances.

The response time is affected by information encountered by a driver, as illustrated in Fig. 4.23 and Fig. 4.24. The reaction time increases with the *information content*, or the amount and complexity of information, the driver must process before reacting. In order to quantify information, the unit bits is used. Expected and unexpected conditions also affect the reaction time required to process information. An expected condition, such as a signalized intersection familiar to the driver, takes less time to process and react to than an unexpected condition, such as a driver swerving into an adjacent lane.

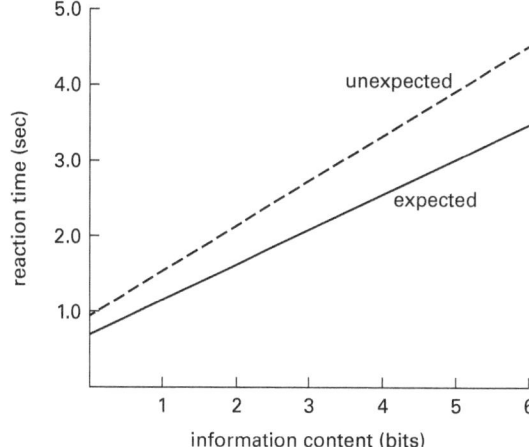

Figure 4.23 Time to React to Expected and Unexpected Information, Median Driver

From *A Policy on Geometric Design of Highways and Streets*, 2011, by the American Association of State Highway and Transportation Officials, Washington, D.C. Figure 2-26. Used by permission.

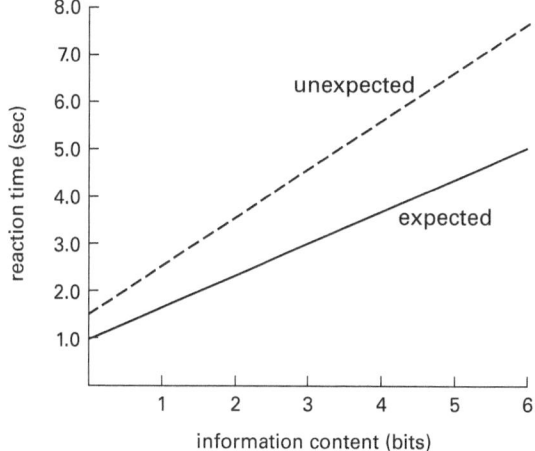

Figure 4.24 Time to React to Expected and Unexpected Information, 85th Percentile Driver

From *A Policy on Geometric Design of Highways and Streets*, 2011, by the American Association of State Highway and Transportation Officials, Washington, D.C. Figure 2-27. Used by permission.

Chapter 3 of the AASHTO *GDHS* gives approach maneuvers, which are used to calculate the decision sight distance. The decision sight distances classified by approach maneuver are shown in Table 4.4.

Avoidance maneuvers A and B are calculated using the pre-maneuver time, t_p,[1] which is the time needed for the driver to recognize the upcoming condition, identify alternative movements, and initiate a response. The pre-maneuver time, t_p, ranges from 3.0 sec to 9.1 sec, which is longer than the perception-reaction time used in stopping sight distance. Avoidance maneuvers C

---

[1]Note that stopping sight distance uses the perception-reaction time, while decision sight distance uses the pre-maneuver time, which includes the perception-reaction time. Both use the variable t_p.

through E use t_t, which is the total of the pre-maneuver and maneuver times. Typical values of t_p and t_t are given in Table 4.4.

Table 4.4 Example Decision Sight Distances

SI units

design speed (kph)	decision sight distance (m) avoidance maneuver				
	A	B	C	D	E
50	70	155	145	170	195
60	95	195	170	205	235
70	115	235	200	235	275
80	140	280	230	270	315
90	170	325	270	315	360
100	200	370	315	355	400
110	235	420	330	380	430
120	265	470	360	415	470
130	305	525	390	450	510

customary U.S. units

design speed (mph)	decision sight distance (ft) avoidance maneuver				
	A	B	C	D	E
30	220	490	450	535	620
35	275	590	525	625	720
40	330	690	600	715	825
45	395	800	675	800	930
50	465	910	750	890	1030
55	535	1030	865	980	1135
60	610	1150	990	1125	1280
65	695	1275	1050	1220	1365
70	780	1410	1105	1275	1445
75	875	1545	1180	1365	1545
80	970	1685	1260	1455	1650

Avoidance Maneuver A: Stop on rural road. $t_p = 3.0$ sec.
Avoidance Maneuver B: Stop on urban road. $t_p = 9.1$ sec.
Avoidance Maneuver C: Speed/path/direction change on rural road. t_t varies between 10.2 sec and 11.2 sec.
Avoidance Maneuver D: Speed/path/direction change on suburban road. t_t varies between 12.1 sec and 12.9 sec.
Avoidance Maneuver E: Speed/path/direction change on urban road. t_t varies between 14.0 sec and 14.5 sec.

From *A Policy on Geometric Design of Highways and Streets*, 2011, by the American Association of State Highway and Transportation Officials, Washington, D. C. Table 3-3. Used with permission.

Avoidance maneuver A is a stop condition on a rural road and uses a pre-maneuver time of 3.0 sec. *Avoidance maneuver B* is a stop condition on an urban road and uses a pre-maneuver time of 9.1 sec. The decision sight distances for maneuvers A and B are determined using Eq. 4.55, where t_p is the pre-maneuver time only, v is the design speed, and a is the driver deceleration. Most decision sight distances can be found from Table 4.4, but for cases where the given design speeds do not apply or for other special cases, Table 4.4 can be used in conjunction with Eq. 4.55.

$$d_m = \left(0.278 \frac{\frac{m}{s}}{\frac{km}{h}}\right) v_{kph} t_{p,s} + 0.039 \left(\frac{v_{kph}^2}{a_{m/s^2}}\right) \quad [SI] \quad 4.55(a)$$

[GDHS Eq. 3-4]

$$d_{ft} = \left(1.47 \frac{\frac{ft}{sec}}{\frac{mi}{hr}}\right) v_{mph} t_{p,sec} + 1.075 \left(\frac{v_{mph}^2}{a_{ft/sec^2}}\right) \quad [U.S.] \quad 4.55(b)$$

[GDHS Eq. 3-4]

Avoidance maneuver C is a speed, path, or direction change on a rural road, in which the total pre-maneuver and maneuver time, t_t, varies between 10.2 sec and 11.2 sec. *Avoidance maneuver D* is a speed, path, or direction change on a suburban road, in which t_t varies between 12.1 sec and 12.9 sec. *Avoidance maneuver E* is a speed, path, or direction change on an urban road where t_t varies between 14.0 sec and 14.5 sec. The decision sight distances for maneuvers C, D, and E are calculated using Eq. 4.56.

$$d_m = 0.278 v_{m/s} t_{t,s} \quad [GDHS \text{ Eq. 3-5}] \quad [SI] \quad 4.56(a)$$

$$d_{ft} = 1.47 v_{ft/sec} t_{t,sec} \quad [GDHS \text{ Eq. 3-5}] \quad [U.S.] \quad 4.56(b)$$

The criteria of 3.5 ft (1080 mm) eye height and 2.0 ft (600 mm) object height used for stopping sight distance are the same for decision sight distance.

Passing Sight Distance

Passing sight distance is applicable on two-lane highways when there are sufficient gaps in opposing flows to allow passing maneuvers to occur and when there are few access points, with only occasional entering traffic and sufficient sight distance to observe traffic in both lanes and both directions. Occasional access includes residential and rural driveways, minor side roads, and driveways to adjoining land uses. Passing sight distance is not applicable to multilane highways. Passing sight distances for two-lane highways are shown in Table 4.5.

Table 4.5 Passing Sight Distance for Design of Two-Lane Highways

SI units			
	assumed speeds (kph)		
design speed (kph)	passed vehicle	passing vehicle	passing sight distance (m)
30	11	30	120
40	21	40	140
50	31	50	160
60	41	60	180
70	51	70	210
80	61	80	245
90	71	90	280
100	81	100	320
110	91	110	355
120	101	120	395
130	111	130	440

customary U.S. units			
	assumed speeds (mph)		
design speed (mph)	passed vehicle	passing vehicle	passing sight distance (ft)
20	8	20	400
25	13	25	450
30	18	30	500
35	23	35	550
40	28	40	600
45	33	45	700
50	28	50	800
55	43	55	900
60	48	60	1000
65	53	65	1100
70	58	70	1200
75	63	75	1300
80	68	80	1400

From *A Policy on Geometric Design of Highways and Streets*, 2011, by the American Association of State Highway and Transportation Officials, Washington, D.C. Table 3-4. Used with permission.

Passing sight distances for passing maneuvers on two-lane highways are greater than stopping sight distances. The passing sight distances are shown in *GDHS* 2011. It should be noted that the passing sight distances currently recommended are less than what is shown in previous editions of *GDHS*, based on updated field observations of passing maneuvers. Assumptions on driver behavior are used for the recommended passing sight distances.

- The speeds of the passing and the opposing vehicles represent the design speed of the highway.
- The passed vehicle travels at a uniform speed. The passing vehicle reaches a speed of 12 mph greater than the passed vehicle.
- The passing vehicle has sufficient acceleration capability to reach the specified speed differential at the critical position, which is about 40% of the way through the passing maneuver.
- The distance between the passing vehicle and the passed vehicle is 19 ft.
- The passing driver's perception-reaction time to abort the passing maneuver is 1 sec.
- If the passing maneuver is aborted, the passing vehicle will decelerate at 11.2 ft/sec^2, the same deceleration rate used in normal stopping distance criteria.
- The headway between passed and passing vehicles is 1 sec for a completed or aborted pass.
- The minimum clearance between the passing and the opposed vehicle is 1 sec.

The recommended passing sight distances are for a single vehicle passing a single vehicle. The passing section should be as long as practical. Passing sections should be included frequently to reduce the percentage of time spent passing and to increase the average travel speed on the two-lane highway. For long sections of highway with few locations having adequate sight distance to accommodate passing zones, traffic level of service may be improved by providing four-lane segments where practical to allow slower-moving vehicles to be passed. For traffic operational analysis, the minimum passing zone lengths shown in Table 4.6 can be provided when traffic flows at the 85th percentile of posted speed.

Determining Vertical Curve Length

Vertical curve length is a function of stopping sight distance, passing sight distance, vertical acceleration comfort limits, the fit of topographic or other intersecting roadway elements, and the drainage ability of very long curves. Sight distance is the straight line distance between the driver's eye and an object or vehicle on the roadway ahead. Sag vertical curve lengths on unlit highways are often determined by headlight sight distance. AASHTO typically sets the driver's eye height at 3.5 ft (1080 mm), and the object ahead is set at 2.0 ft (600 mm) above the roadway. Design values for stopping sight distance are given in Table 4.3.

A simplified method to determine the stopping sight distance for a crest or sag vertical curve is to use the *K-value method* presented in the *GDHS*. The length of vertical curve per percent of grade difference, *K*, is the

Table 4.6 Minimum Passing Zone Lengths for Traffic Operation Analysis with Operating Speed of 85th Percentile of Posted Speed

SI units		customary U.S. units	
85th percentile speed or posted or statutory speed limit (kph)	minimum passing zone length (m)	85th percentile speed or posted or statutory speed limit (mph)	minimum passing zone length (ft)
40	140	20	400
50	180	30	550
60	210	35	650
70	240	40	750
80	240	45	800
90	240	50	800
100	240	55	800
110	240	60	800
120	240	65	800
		70	800

From *A Policy on Geometric Design of Highways and Streets*, 2011, by the American Association of State Highways and Transportation Officials, Washington, D.C. Table 3-5. Used by permission.

inverse of the rate of change, R, and is the ratio of the curve length, L, to the absolute value of the algebraic grade difference, A, as shown in Eq. 4.57.

$$K = \frac{L}{A} = \frac{L}{|G_2 - G_1|} \quad \text{[always positive]} \quad 4.57$$

A typical design situation will give two of the variables, with the third variable being the unknown. In some cases, one or more of the known conditions may have a range of available conditions, and careful selection of a value in the range will be required. This type of problem requires some logical thinking and deductive reasoning, such as determining whether the maximum or the minimum grade change is applicable to the condition at hand.

The *GDHS* presents graphs of K-values versus design speeds based on comfort criteria and stopping sight criteria. These graphs are general guidelines for transportation engineers, and data from these graphs are presented in Table 4.7 and Table 4.8 at the end of Sec. 4.18.

The minimum recommended K-values are based on sight distance. Applying K-values uniformly throughout an undulating roadway segment can improve both ride comfort and geometric appearance. While tolerance for vertical acceleration normally controls the rate of vertical change, the appearance of direction changes can be disconcerting for drivers and passengers. A graceful and smooth appearance minimizes distress, and fewer accidents are caused by startled or confused drivers.

Visually, when a curve is too short, the appearance is of no curve transition at all. It is important to also take into account the length of the approach tangent, especially on sag vertical curves. Longer approach tangents tend to draw the eye farther ahead, while a sudden increase in grade shortens the perceived sight distance. In the most extreme cases, the driver envisions the approaching grade increase with no vertical curve at all and will tend to slow considerably on the approach. This problem can be reduced by providing a vertical curve that is considerably longer than recommended. For smaller differences in grades, the length can be as much as four or five times longer than the normal recommendations. This usually does not raise the cost of construction, but may reduce costs by reducing the amount of excavation necessary to construct a sag vertical curve. Sag vertical curves with longer lengths are shallower and therefore require less excavation than a curve with a shorter length.

The upper range of K-values is controlled by the ability to drain the roadway surface. AASHTO recommends a drainage maximum of 167 ft/% (51 m/%) for crest and sag curves. Should higher K-values be necessary, other measures, such as increasing the cross slope, are necessary to adequately drain the pavement.

For lower-speed roadways where appearance is less often a factor, there are two conditions of sight distance criteria that control the minimum length of a curve: when the sight distance is shorter than the length of the vertical curve ($S < L$), and when the sight distance is longer than the length of the vertical curve ($S > L$). The equations shown in Table 4.9 are used to determine the required curve length, L, and must be initially calculated for both the $S < L$ and the $S > L$ cases. Constants are based on the driver's eye height and the object height, both in feet (millimeters).

Equation 4.58 and Eq. 4.59 may also be used to determine the curve length for sight distance on crest vertical curves.

$$L = \frac{A_\% S^2}{100(\sqrt{2h_1} + \sqrt{2h_2})^2} \quad 4.58$$
$$[S < L] \; [GDHS \text{ Eq. 3-41}]$$

$$L = 2S - \frac{200(\sqrt{2h_1} + \sqrt{2h_2})^2}{A_\%} \quad 4.59$$
$$[S > L] \; [GDHS \text{ Eq. 3-42}]$$

Stopping sight distance on sag vertical curves can be analyzed using headlight sight distances. This method assumes the maximum upward projection of the

Table 4.9 AASHTO Criteria for Minimum Vertical Curve Lengths Based on Sight Distance[a]

	stopping sight distance[b] (crest curves)	passing sight distance[c] (crest curves)	stopping sight distance (sag curves)
	SI units		
$S < L$	$L = \dfrac{AS^2}{658}$	$L = \dfrac{AS^2}{864}$	$L = \dfrac{AS^2}{120 + 3.5S}$
$S > L$	$L = 2S - \dfrac{658}{A}$	$L = 2S - \dfrac{864}{A}$	$L = 2S - \dfrac{120 + 3.5S}{A}$
	customary U.S. units		
$S < L$	$L = \dfrac{AS^2}{2158}$	$L = \dfrac{AS^2}{2800}$	$L = \dfrac{AS^2}{400 + 3.5S}$
$S > L$	$L = 2S - \dfrac{2158}{A}$	$L = 2S - \dfrac{2800}{A}$	$L = 2S - \dfrac{400 + 3.5S}{A}$

[a]$A = |G_2 - G_1|$ is the absolute value of the algebraic difference in grades, in percent.
[b]The drivers's eye is 3.5 ft (1080 mm) above the road surface, viewing an object 2.0 ft (600 mm) high.
[c]The drivers's eye is 3.5 ft (1080 mm) above the road surface, viewing an object 3.5 ft (1080 mm) high.
Compiled from *A Policy on Geometric Design of Highways and Streets*, Chap. 3, copyright © 2011 by the American Association of State Highway and Transportation Officials, Washington, D.C.

headlight high-beam will illuminate a distance of roadway ahead equal to the stopping sight distance, and oncoming traffic or roadway lighting will illuminate the roadway ahead when low-beams are used. As in crest vertical curves, the conditions of $S < L$ and $S > L$ need to be considered for headlight sight distance.

With sag curves, both gravitational and centrifugal forces act on the driver and passengers, making comfort the controlling factor in the design. Equation 4.60 can be used to calculate the length of the curve so that the added acceleration is kept below 1 ft/sec² (0.3 m/s²).

$$L_m = \frac{A_\% v_{kph}^2}{395} \quad [GDHS \text{ Eq. 3-51}] \quad [SI] \quad 4.60(a)$$

$$L_{ft} = \frac{A_\% v_{mph}^2}{46.5} \quad [GDHS \text{ Eq. 3-51}] \quad [U.S.] \quad 4.60(b)$$

Example 4.6

A downgrade of 3% intersects an upgrade of 5% at an elevation of 100.00 ft at sta 67+84.00, where a 200 ft curve will be fitted. Show complete curve specifications, including the formula used to determine each item. Include the station and elevation of the PVC and the PVT, curve length, K-value, middle ordinate, and low point station and elevation.

Solution

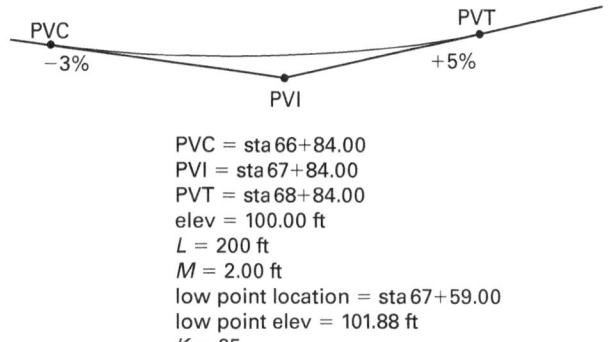

PVC = sta 66+84.00
PVI = sta 67+84.00
PVT = sta 68+84.00
elev = 100.00 ft
L = 200 ft
M = 2.00 ft
low point location = sta 67+59.00
low point elev = 101.88 ft
K = 25

The previous tabulation shows how the vertical curve information would appear on highway plans in many jurisdictions. The calculations to obtain the information are shown as follows.

From Eq. 4.48,

$$R_{\%/sta} = \frac{G_{2,\%} - G_{1,\%}}{L_{sta}}$$

$$= \frac{5\% - (-3\%)}{\dfrac{200 \text{ ft}}{100 \dfrac{\text{ft}}{\text{sta}}}}$$

$$= 4\,\%/\text{sta} \quad [\text{same as 4 ft/sta}^2]$$

As shown in Eq. 4.48 and Eq. 4.57, the K-value is the inverse of R. (Note that the units are dropped when K is shown on highway plans.)

$$K = \frac{1}{R_{\%/sta}} = \frac{(1)\left(100\,\dfrac{\text{ft}}{\text{sta}}\right)}{4\,\dfrac{\%}{\text{sta}}} = 25 \text{ ft}/\%$$

From Eq. 4.47,

$$M_{ft} = \frac{|G_{2,\%} - G_{1,\%}|\, L_{sta}}{8\,\dfrac{\text{sta}}{\text{ft}}}$$

$$= \frac{|5\% - (-3\%)|\,(200 \text{ ft})}{\left(8\,\dfrac{\text{sta}}{\text{ft}}\right)\left(100\,\dfrac{\text{ft}}{\text{sta}}\right)}$$

$$= 2.0 \text{ ft}$$

The station of the PVC is

$$\text{sta PVC} = \text{sta PVI} - \frac{L}{2}$$
$$= \text{sta } 67{+}84.00 - \frac{200 \text{ ft}}{2}$$
$$= \text{sta } 66{+}84.00$$

The elevation of the PVC is

$$\text{elev}_{\text{PVC}} = \text{elev}_{\text{PVI}} + G_1\left(\frac{L}{2}\right)$$
$$= 100.00 \text{ ft} + (0.03)\left(\frac{200 \text{ ft}}{2}\right)$$
$$= 103.00 \text{ ft}$$

The station of the PVT is

$$\text{sta PVT} = \text{sta PVI} + \frac{L}{2}$$
$$= \text{sta } 67{+}84.00 + \frac{200 \text{ ft}}{2}$$
$$= \text{sta } 68{+}84.00$$

Using Eq. 4.51, find the station of the low point of the curve.

$$\text{low point station} = \text{sta PVC} + x_{\text{turning point, sta}}$$
$$= \text{sta PVC} + \frac{-G_{1,\%}}{R_{\%/\text{sta}}}$$
$$= \text{sta } 66{+}84.00 + \frac{-(-3\%)}{4 \frac{\%}{\text{sta}}}$$
$$= \text{sta } 67{+}59.00$$

Using Eq. 4.50 and the known PVC elevation for elev_a, the elevation of the low point of the curve is

$$\text{elev}_{\text{low point}} = \frac{Rx^2}{2} + G_{1,\text{ft/ft}}x + \text{elev}_{\text{PVC}}$$
$$= \frac{\left(4 \frac{\text{ft}}{\text{sta}^2}\right)(0.75 \text{ sta})^2}{2} + (-0.03)(0.75 \text{ sta})$$
$$\times \left(100 \frac{\text{ft}}{\text{sta}}\right) + 103.00 \text{ ft}$$
$$= 101.88 \text{ ft}$$

12. SUPERELEVATION

The difference between a curve's inside and outside elevations is known as the superelevation, e. A roadway is superelevated to resist the centrifugal force acting on a vehicle as it rounds a curve, which allows the driver to comfortably maneuver the curve at a higher speed. The *angle of the slope*, ϕ, also called *curve banking* or *cross slope*, is related to the design speed, v, and the curve radius, R, as shown by Eq. 4.61.

$$e = \tan\phi = \frac{\text{v}^2}{gR} \qquad \textit{4.61}$$

When used in route design, the slope angle is usually disregarded, and the slope is described using the tangent values of the angle showing a rise to run relationship, such as 0.02 ft/ft, ¼ in/ft, or 20 mm/m.

Lateral Forces on a Moving Vehicle

A vehicle on a sloped roadway is illustrated with a free-body diagram, as shown in Fig. 4.25. Vehicles moving along a curved path experience several lateral forces. Tire friction, gravity, and the centrifugal tendency of a vehicle work in concert to provide a balance of forces within the safe and comfortable operating range of speed for each curve.

A few relationships apply to the elements shown in Fig. 4.25. Equation 4.62 illustrates the relationship between the mass and the weight of an object.

$$m = \frac{w}{g} \qquad [\text{SI}] \qquad \textit{4.62(a)}$$

$$m = \frac{wg_c}{g} \qquad [\text{U.S.}] \qquad \textit{4.62(b)}$$

The force acting on the vehicle is calculated from the acceleration and the object's mass, as shown in Eq. 4.63.

$$F = ma \qquad [\text{SI}] \qquad \textit{4.63(a)}$$

$$F = \frac{ma}{g_c} \qquad [\text{U.S.}] \qquad \textit{4.63(b)}$$

When the vehicle is steered along a curved path, the act of steering exerts a *lateral acceleration* inward, also called *radial* or *centripetal acceleration*. Lateral acceleration incorporates the friction force exerted by the vehicle's tires against the pavement and the vertical gravity acting on the mass of the vehicle. The lateral acceleration is counteracted by the *centrifugal force* exerted by the vehicle's mass.

Centripetal acceleration and centrifugal force must be properly balanced in order for the vehicle to stay in the curve and continue through it safely. In cases where steering alone is unlikely to keep the vehicle on the curve, the roadway can be sloped toward the inside of the curve to increase the vertical gravity acting on the vehicle. However, a certain amount of centrifugal force is necessary to help drivers control their vehicles while rounding the curve and to help drivers and passengers

Figure 4.25 Forces Acting on a Vehicle While Rounding a Curve

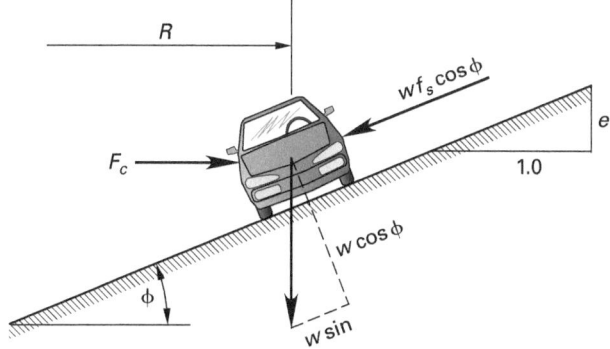

- w = weight of vehicle
- f_s = coefficient of side friction
- F_c = centrifugal force
- R = radius of curve
- φ = angle of cross slope
- e = rate of superelevation, tan φ

maintain a comfortable position in their seats. For this reason, curved roadways should never be sloped such that centrifugal force is completely negated.

The centrifugal force, F_c, is calculated from Eq. 4.64.

$$F_c = \frac{m v_t^2}{R} \quad [SI] \quad 4.64(a)$$

$$F_c = \frac{m v_t^2}{g_c R} = \frac{w v_t^2}{g R} \quad [U.S.] \quad 4.64(b)$$

The proper balance of forces acting on a vehicle while rounding a curve can be found from Eq. 4.65. R is the curve radius; v is the vehicle speed; f_s is the side friction; e is the rate of superelevation, tan φ; and g is gravitational acceleration, 32.2 ft/sec² (9.81 m/s²).

$$R \approx \frac{v_t^2}{g(e + f_s)} \quad 4.65$$

Substituting $g = 32.2$ ft/sec² (9.81 m/s²) into Eq. 4.65 and using units of miles per hour (kilometers per hour) for v gives the minimum curve radius, Eq. 4.66.

$$R_m = \frac{v_{kph}^2}{127(e + f_s)} \quad [SI] \quad 4.66(a)$$

[*GDHS* Eq. 3-10]

$$R_{ft} = \frac{v_{mph}^2}{15(e + f_s)} \quad [U.S.] \quad 4.66(b)$$

[*GDHS* Eq. 3-10]

Side Friction Factor

The *side friction factor*, f_s, is dependent on the lateral traction available where the tires meet the roadway. The side friction factor is sometimes called the *comfort factor*, as this is the unbalanced superelevation force a passenger feels as a vehicle travels around a horizontal curve. This friction factor varies with tire tread design, tire wear, roadway surface conditions, and travel speed. The general design conditions given in the AASHTO *GDHS*, Chap. 3, for the side friction factor include wet pavement (either asphalt or concrete) and average tire tread wear.

Table 4.10 extrapolates recommended design side friction factors for different types of facilities and conditions from the graphs in *GDHS* Exh. 3-11. For rural and high-speed urban highways, design friction factors range from 0.17 for a design speed of 10 mph (0.18 for a design speed of 20 kph) to 0.08 for a design speed of 80 mph (130 kph). For intersections, friction factors range from 0.38 at 10 mph (0.35 at 20 kph) to 0.15 at 45 mph (70 kph). Friction factors range from 0.30 at 20 mph (0.32 at 30 kph) to 0.16 at 45 mph (70 kph) for low-speed urban streets.

The maximum side friction factor for any design speed is based on the *point of impending skid*, the point at which a vehicle's tires would begin to skid. Although AASHTO gives studies with friction factors in the range of 0.34–0.36 for smooth tires on wet pavement at 45 mph (80 kph), recommended design values are far below this range for reasons of safety and the variability of field conditions, as well as comfort and safety. Skidding should be avoided, and a margin for error should be provided for safety reasons, so design friction factors are far below factors that approach the point of impending skid.

The friction factor reduces as speed increases. Friction factors are generally independent of vehicle weight and are instead more dependent on the roughness or smoothness of the pavement surface, the presence of surface contaminants such as oil or a film of dirt and water, tread design, and the amount of tread wear. Tire pressure above or below the tire manufacturer's recommendations changes the footprint of tire contact and also affects the friction factor.

When high friction is required to keep a vehicle on a curve, swerving becomes perceptible, the drift angle increases, and drivers experience a sensation of increased need for intense concentration that is considered undesirable. For these reasons, the AASHTO design friction factor values are conservative.

Design Speed versus Average Running Speed

Not all drivers in a traffic stream operate their vehicles at the same speed for a given design condition. Drivers seek their own comfortable speed according to many factors, such as driver capability, vehicle size and weight,

Table 4.10 Assumed Side Friction Factors, f_s, for Design Speeds

SI units

speed (kph)	rural and high-speed urban highways*	intersections*	low-speed urban streets*	new tires on wet concrete pavements
20	0.18	0.35		
30	0.17	0.28	0.32	0.50
40	0.16	0.23	0.25	0.48
50	0.16	0.19	0.22	0.46
60	0.15	0.17	0.19	0.43
70	0.14	0.15	0.16	0.41
80	0.14			0.39
90	0.13			0.37
100	0.11			0.35
110	0.10			
120	0.09			
130	0.08			

customary U.S. units

speed (mph)	rural and high-speed urban highways	intersections*	low-speed urban streets*	new tires on wet concrete pavements
10	0.17	0.38		
15	0.17	0.32		
20	0.16	0.26	0.30	0.50
25	0.16	0.23	0.25	0.48
30	0.16	0.20	0.22	0.46
35	0.15	0.18	0.20	0.44
40	0.15	0.16	0.18	0.42
45	0.14	0.15	0.16	0.40
50	0.14			0.39
60	0.12			0.35
70	0.10			
80	0.08			

*assumed for design

From *A Policy on Geometric Design of Highways and Streets*, 2011, by the American Association of State Highway and Transportation Officials, Washington, D.C. Used by permission.

familiarity with the road, and so on. All roadways have both an average running speed and a design speed. The *average running speed* is found by adding the distances traveled by all vehicles on a roadway during a period of time and dividing this number by the sum of the vehicles' running times (see Chap. 2). *Design speed* is chosen during the design of a new roadway and is the controlling factor in determining a roadway's geometric features. The average running speed of all vehicles is usually less than the design speed, with the difference between design speed and average running speed increasing as the design speed increases. The average running speed is less than the 85th percentile speed. Table 4.11 compares the distribution of average running speeds and design speeds up to 80 mph (130 kph), as described in the *GDHS*.

Table 4.11 Design Speeds and Average Running Speeds

SI units		customary U.S. units	
design speed (kph)	average running speed (kph)	design speed (mph)	average running speed (mph)
20	20	15	15
30	30	20	20
40	40	25	24
50	47	30	28
60	55	35	32
70	63	40	36
80	70	45	40
90	77	50	44
100	85	55	48
110	91	60	52
120	98	65	55
130	102	70	58
		75	61
		80	64

From *A Policy on Geometric Design of Highways and Streets*, 2011, by the American Association of State Highway and Transportation Officials, Washington, D.C. Table 3-6. Used by permission.

Distribution of Superelevation and Side Friction for Curve Design

The superelevation rate, e, and side friction factor, f_s, combine to keep a vehicle on a curved roadway. For many curves with radii greater than the recommended minimum for a given design speed, superelevation at the maximum slope is not necessary, nor is it always desirable. There are five methods of distributing e and f_s for curve design recognized by AASHTO.

Method 1: For any radius, superelevation and side friction are equal to each other. The values of f_s and e vary in proportion to the inverse of the radius.

Method 2: As the radius decreases for a given speed, superelevation is not introduced until f_s reaches maximum, then f_s remains at maximum as e increases to maximum for that speed. This method is used with urban street settings at lower speeds where drivers expect greater side friction forces, and there are more constraints involved with providing full superelevation.

Method 3: Side friction remains at zero as the radius decreases for the design speed, until e

reaches its maximum. Then, f_s increases to its maximum at the minimum radius for that speed.

Method 4: This method is the same as method 3 except it is based on average running speed instead of design speed.

Method 5: The values of f_s and e vary in inverse proportion to the radius, but in a curvilinear fashion. The distribution curve between f_s and e increases the superelevation slightly over method 1, but not nearly as much as method 3, due to the parabolic shape of the distribution curve. This method is commonly used on roadways with higher speeds, including rural highways, urban freeways, high-speed urban streets, and roadways with radii greater than the minimum for a given design speed.

Method 1 is often used because it is simplest to apply. Methods 2, 3, and 4 are less commonly used because of the tendency to choose superelevation rates that are either too high or too low for the conditions. A low superelevation rate can be taken too fast, leading to erratic driver control. A high superelevation rate can lead to negative side friction, which decreases driver and passenger comfort. Method 5 is generally favored overall for higher speed roadway conditions. Figure 4.26 shows graphical plots of the five methods. Advantages and disadvantages of all five methods are discussed in further detail in *GDHS* Chap. 3.

Superelevation Procedure Development for Method 5

The distribution of e and f_s for method 5 is a curvilinear plot. The AASHTO method used to define this curve can be studied in the *GDHS* should a more precise e and f_s distribution for various curve radii be desired.

To create a distribution curve of e and f_s for a given curve, the running speeds on the curve are first plotted in relation to design speeds using the values shown in Table 4.11. In a "balanced" condition, the superelevation rate (in ft/ft or m/m) and the side friction factor for a given curve radius, the maximum superelevation rate, and the design speed combine to exactly balance the centrifugal force exerted on a vehicle traveling over the curve. For curves with a radius greater than the minimum for the design speed and maximum superelevation rate, as shown in Table 4.12, the appropriate side friction factor from Table 4.10 is subtracted from the total superelevation. (It is assumed when plotting these distribution curves that the side friction factor for all curves is 0 until a factor from Table 4.10 is applied.) The difference between the centripetal force produced by side friction and the centripetal force needed to create a balanced condition is established by the superelevation rate, and the force generated by the superelevation is shown as the finalized distribution curve of superelevation values in ft/ft (m/m).

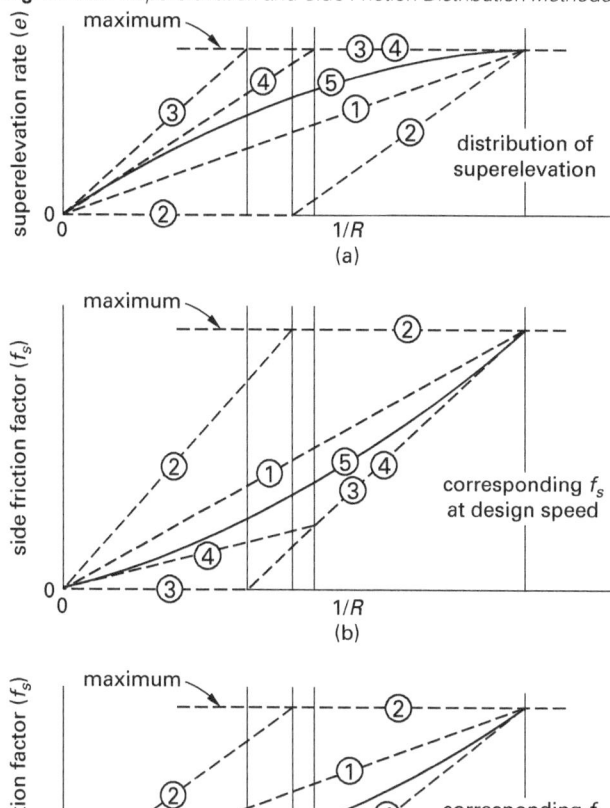

Figure 4.26 Superelevation and Side Friction Distribution Methods

From *A Policy on Geometric Design of Highways and Streets*, 2011, by the American Association of State Highway and Transportation Officials, Washington, D.C. Used by permission.

The shape of the f_s distribution curve is parabolic and can be analyzed similarly to vertical curve geometry. Three points can be established to develop the parabolic distribution curve of e and f_s. See the *GDHS* for more information on method 5.

Maximum Superelevation Rates

In general, maximum superelevation rates for highways depend on climate and terrain conditions, whether the highway is in a rural or urban area, and the frequency of slower-moving vehicles (these vehicles' operation may be impaired by high superelevation rates). Using the curve shown in Fig. 4.26 and applying superelevation method 5, AASHTO has developed a series of superelevation distribution curves

Table 4.12 Minimum Radius Using Maximum Values of e and f_s

\multicolumn{6}{c}{SI units}	\multicolumn{6}{c}{customary U.S. units}
design speed (kph)	maximum e^* (%)	maximum f_s	total $(e+f_s)$	calculated radius (m)	rounded radius (m)	design speed (mph)	maximum e^* (%)	maximum f_s	total $(e+f_s)$	calculated radius (ft)	rounded radius (ft)
15	4.0	0.40	0.44	4.0	4	10	4.0	0.38	0.42	15.9	16
20	4.0	0.35	0.39	8.1	8	15	4.0	0.32	0.36	41.7	42
30	4.0	0.28	0.32	22.1	22	20	4.0	0.27	0.31	86.0	86
40	4.0	0.23	0.27	46.7	47	25	4.0	0.23	0.27	154.3	154
50	4.0	0.19	0.23	85.6	86	30	4.0	0.20	0.24	250.0	250
60	4.0	0.17	0.21	135.0	135	35	4.0	0.18	0.22	371.2	371
70	4.0	0.15	0.19	203.1	203	40	4.0	0.16	0.20	533.3	533
80	4.0	0.14	0.18	280.0	280	45	4.0	0.15	0.19	710.5	711
90	4.0	0.13	0.17	375.2	375	50	4.0	0.14	0.18	925.9	926
100	4.0	0.12	0.16	492.1	492	55	4.0	0.13	0.17	1186.3	1190
15	6.0	0.40	0.46	3.9	4	10	6.0	0.38	0.44	15.2	15
20	6.0	0.35	0.41	7.7	8	15	6.0	0.32	0.38	39.5	39
30	6.0	0.28	0.34	20.8	21	20	6.0	0.27	0.33	80.8	81
40	6.0	0.23	0.29	43.4	43	25	6.0	0.23	0.29	143.7	144
50	6.0	0.19	0.25	78.7	79	30	6.0	0.20	0.26	230.8	231
60	6.0	0.17	0.23	123.2	123	35	6.0	0.18	0.24	340.3	340
70	6.0	0.15	0.21	183.7	184	40	6.0	0.16	0.22	484.8	485
80	6.0	0.14	0.20	252.0	252	45	6.0	0.15	0.21	642.9	643
90	6.0	0.13	0.19	335.7	336	50	6.0	0.14	0.20	833.3	833
100	6.0	0.12	0.18	437.4	437	55	6.0	0.13	0.19	1061.4	1060
110	6.0	0.11	0.17	560.4	550	60	6.0	0.12	0.18	1333.3	1330
120	6.0	0.09	0.15	755.9	756	65	6.0	0.11	0.17	1656.9	1660
130	6.0	0.08	0.14	950.5	951	70	6.0	0.10	0.16	2041.7	2040
						75	6.0	0.09	0.15	2500.0	2500
						80	6.0	0.08	0.14	3047.6	3050
15	8.0	0.40	0.48	3.7	4	10	8.0	0.38	0.46	14.5	14
20	8.0	0.35	0.43	7.3	7	15	8.0	0.32	0.40	37.5	38
30	8.0	0.28	0.36	19.7	20	20	8.0	0.27	0.35	76.2	76
40	8.0	0.23	0.31	40.6	41	25	8.0	0.23	0.31	134.4	134
50	8.0	0.19	0.27	72.9	73	30	8.0	0.20	0.28	214.3	214
60	8.0	0.17	0.25	113.4	113	35	8.0	0.18	0.26	314.1	314
70	8.0	0.15	0.23	167.8	168	40	8.0	0.16	0.24	444.4	444
80	8.0	0.14	0.22	229.1	229	45	8.0	0.15	0.23	587.0	587
90	8.0	0.13	0.21	303.7	304	50	8.0	0.14	0.22	757.6	758
100	8.0	0.12	0.20	393.7	394	55	8.0	0.13	0.21	960.3	960
110	8.0	0.11	0.19	501.5	501	60	8.0	0.12	0.20	1200.0	1200
120	8.0	0.09	0.17	667.0	667	65	8.0	0.11	0.19	1482.5	1480
130	8.0	0.08	0.16	831.7	832	70	8.0	0.10	0.18	1814.8	1810
						75	8.0	0.09	0.17	2205.9	2210
						80	8.0	0.08	0.16	2666.7	2670
15	10.0	0.40	0.50	3.5	4	10	10.0	0.38	0.48	13.9	14
20	10.0	0.35	0.45	7.0	7	15	10.0	0.32	0.42	35.7	36
30	10.0	0.28	0.38	18.6	19	20	10.0	0.27	0.37	72.1	72
40	10.0	0.23	0.33	38.2	38	25	10.0	0.23	0.33	126.3	126
50	10.0	0.19	0.29	67.9	68	30	10.0	0.20	0.30	200.0	200
60	10.0	0.17	0.27	105.0	105	35	10.0	0.18	0.28	291.7	292
70	10.0	0.15	0.25	154.3	154	40	10.0	0.16	0.26	410.3	410
80	10.0	0.14	0.24	210.0	210	45	10.0	0.15	0.25	540.0	540
90	10.0	0.13	0.23	277.3	277	50	10.0	0.14	0.24	694.4	694
100	10.0	0.12	0.22	357.9	358	55	10.0	0.13	0.23	876.8	877
110	10.0	0.11	0.21	453.7	454	60	10.0	0.12	0.22	1090.9	1090
120	10.0	0.09	0.19	596.8	597	65	10.0	0.11	0.21	1341.3	1340
130	10.0	0.08	0.18	739.3	739	70	10.0	0.10	0.20	1633.3	1630
						75	10.0	0.09	0.19	1973.7	1970
						80	10.0	0.08	0.18	2370.4	2370
15	12.0	0.40	0.52	3.4	3	10	12.0	0.38	0.50	13.3	13
20	12.0	0.35	0.47	6.7	7	15	12.0	0.32	0.44	34.1	34
30	12.0	0.28	0.40	17.7	18	20	12.0	0.27	0.39	68.4	68
40	12.0	0.23	0.35	36.0	36	25	12.0	0.23	0.35	119.0	119
50	12.0	0.19	0.31	63.5	64	30	12.0	0.20	0.32	187.5	188
60	12.0	0.17	0.29	97.7	98	35	12.0	0.18	0.30	272.2	272
70	12.0	0.15	0.27	142.9	143	40	12.0	0.16	0.28	381.0	381
80	12.0	0.14	0.26	193.8	194	45	12.0	0.15	0.27	500.0	500
90	12.0	0.13	0.25	255.1	255	50	12.0	0.14	0.26	641.0	641
100	12.0	0.12	0.24	328.1	328	55	12.0	0.13	0.25	806.7	807
110	12.0	0.11	0.23	414.2	414	60	12.0	0.12	0.24	1000.0	1000
120	12.0	0.09	0.21	539.9	540	65	12.0	0.11	0.23	1224.6	1220
130	12.0	0.08	0.20	665.4	665	70	12.0	0.10	0.22	1484.8	1480
						75	12.0	0.09	0.21	1785.7	1790
						80	12.0	0.08	0.20	2133.3	2130

*In recognition of safety considerations, use of $e_{max} = 4.0\%$ should be limited to urban conditions.

From *A Policy on Geometric Design of Highways and Streets*, 2011, by the American Association of State Highway and Transportation Officials, Washington, D.C. Used by permission.

and tables to assist in selecting a superelevation rate for a particular design curve. These plots and tables are reproduced in App. 4.A and App. 4.B.

The first step in selecting a superelevation rate is to select the maximum superelevation rate for the road conditions at hand. The highest superelevation rate commonly used for highways is 10%. Rates as high as 12% have been used in some cases. However, some truck and trailer combinations have a tendency toward overturning at these steep cross slopes. 8% is considered to be a reasonable maximum rate, particularly in locations with snow and ice.

Where traffic congestion occurs regularly or where vehicles are expected to travel at very low speeds or remain stopped for long periods of time, the maximum rate is reduced to 4% or 6% in order to reduce discomfort of passengers at steeper cross slopes. This maximum range can be applied in urban areas and along roadway segments with many driveway connections. Intersections where there are a large number of vehicles turning and crossing also have maximum superelevation rates between 4% and 6%, because the high rate of traffic movements tends to result in lower running speeds.

Superelevation is limited by how easily pavement can be drained, and areas where it is difficult to achieve proper pavement drainage cannot be superelevated. No superelevation may be applied where it is difficult to warp pavement for drainage. In some cases, the application of the normal pavement cross section results in negative superelevation, such as in a housing plan or industrial development area.

Table 4.12 shows examples of the minimum curve radius for superelevation rates of 4%, 6%, 8%, 10%, and 12% using speed-adjusted maximum friction factors and can be used to select the minimum curve radius for a given design speed.

Minimum Radius Without Superelevation

Applying Eq. 4.66 using the maximum value for f_s and the normal superelevation rate yields the minimum radius without superelevation for a given design speed. For the outside lanes on a normal sloped roadway (i.e., the lanes are sloped down from the centerline), a negative superelevation value is used. Appendix 4.D gives the minimum radii for low-speed roadways and urban streets. Negative superelevation rates of 1.5%–2.5% occur on the outside lanes of a normally crowned street without superelevation. For minimum radii of higher speed roadways, the reduced friction factors shown in App. 4.D need to be considered.

Superelevation Axis of Rotation

Before superelevation of a highway begins, the axis of rotation must be chosen. The *axis of rotation* is the point on the cross section about which the roadway will be rotated to achieve the full superelevation rate. For two-lane highways, three axes of rotation are typically used: the centerline; the lower (inside) pavement edge; or the upper (outside) pavement edge. (See Fig. 4.27.) For multilane highways separated by a median where each directional roadway is on a straight cross slope, the edges of the median are typically chosen as the axes of rotation. Therefore, each directional roadway is rotated about the median edge while the median remains level.

Figure 4.27 Typical Axes of Rotation for Highways

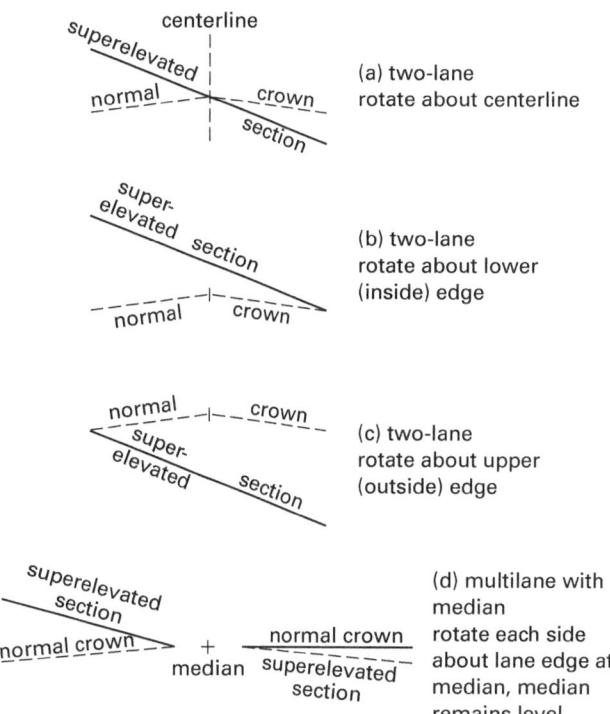

Each axis of rotation is suited to specific situations, and no one choice is applicable to all conditions. Roadway-edge appearance, sight lines, drainage, and other vertical geometry conditions usually establish which axis of rotation is selected. Using the lower edge as the axis is most common for two-lane highways because it is easier to control drainage and usually requires less excavation.

Superelevation Transitions

Superelevation is usually carried uniformly throughout the curve and is transitioned smoothly at each end from the *normal crown*, or normal cross section. The transition from the normal crown to full superelevation happens in several steps. (See Fig. 4.28.) The outer lane is first raised to a level section. This is called removing *adverse crown*, and the segment over which the removal occurs is the *tangent runout*. Next, the outer lane continues to rotate to match the slope of the inner lane. The entire roadway cross section is rotated to achieve the full superelevation. The distance required to rotate from the point of no adverse crown to full superelevation is called the *superelevation transition length*, or the *length of runoff*. This length is determined by the design speed of the roadway and the change in slope required to achieve the full superelevation rate.

When a vehicle transitions from a normal cross section to a superelevated section, the driver has to make adjustments to steering, and passengers experience a noticeable roll of the vehicle about its axis. The adjustments found to be acceptable by drivers and passengers are those carried out over a period of approximately 2 sec. One method of establishing the minimum transition length is to use the distance a vehicle travels in 2 sec at the design speed. However, this method may worsen existing drainage problems along the transition length. For this reason, AASHTO recommends using the relative difference in profile grade method.

The relative difference in profile grade between the pavement edge and the axis of rotation is used to determine the transition length to ensure the profile edges appear smooth. The *GDHS* recommends establishing a *maximum relative gradient* for the outer and inner edge, or outer edge and centerline, to avoid creating a rough transition with sudden, uncomfortable up and down movements of the vehicle. Table 4.13 shows the current AASHTO recommendations for maximum relative gradients according to design speed. Transition lengths are directly proportional to the total superelevation, which is a product of the lane width and the superelevation rate (see Eq. 4.69). The relative gradients shown are for two-lane roadways.

Another method, often used on multilane highways and other wide roads, is to run a profile along the extreme edge of the pavement and design a comfortable vertical curve into the edge transition. The transition should be checked for drainage slope, as it is possible to create level sections in the transition zone that do not drain well.

For any method used when transitioning from a negative slope to a positive slope, or vice versa, there will be a short section of pavement that is level or near level. This level section will trap water on the pavement surface. A shallow cross slope combined with a roadway profile grade of less than 0.5% can cause ponding of water, which can in turn cause drivers to lose control of their vehicles. Ponding can be minimized by keeping the transition length as short as possible, by moving the transition location ahead or back on stationing, or by changing the vertical profile of the edges of the level section.

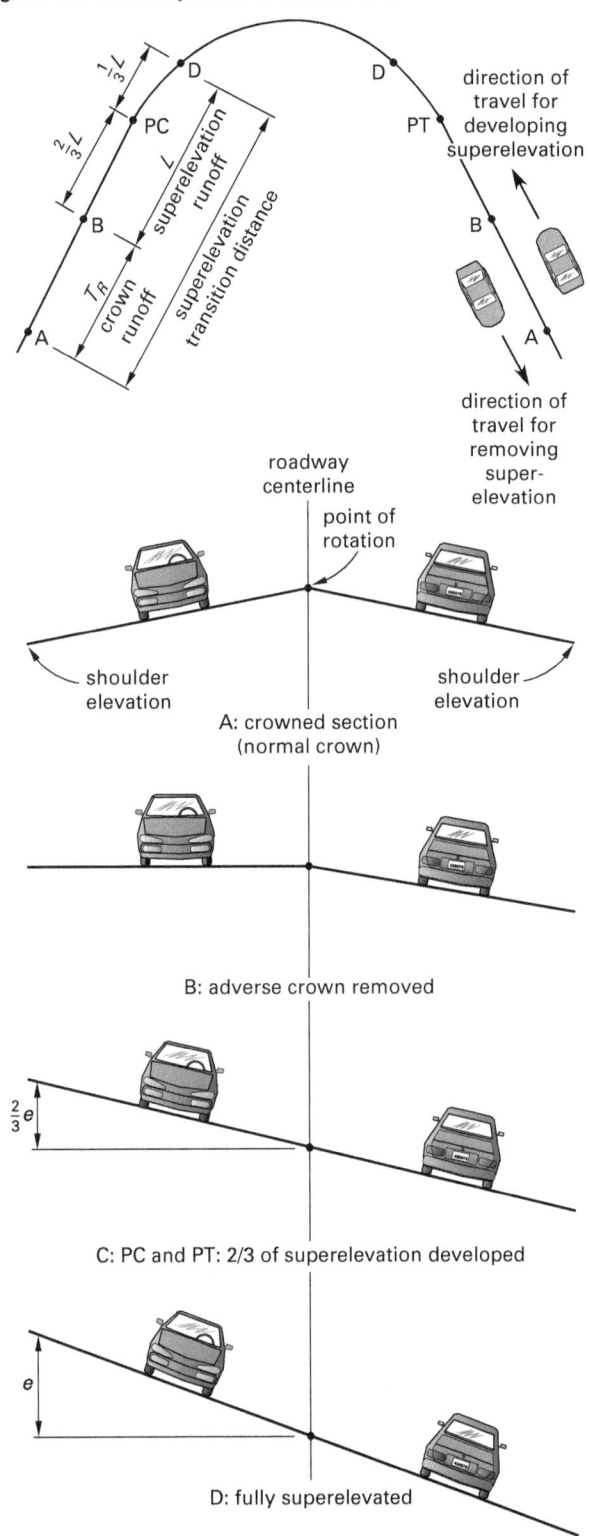

Figure 4.28 Roadway Profile Around Circular Curve

From *Civil Engineering Reference Manual*, by Michael R. Lindeburg, copyright © 2015, Professional Publications, Inc. Reproduced with permission.

Table 4.13 Maximum Relative Gradients of Pavement Edges, $\Delta_\%$, for Two-Lane Highways

SI units		
design speed (kph)	maximum relative gradient (%)	equivalent maximum relative slope
20	0.80	1:125
30	0.75	1:133
40	0.70	1:143
50	0.65	1:154
60	0.60	1:167
70	0.55	1:182
80	0.50	1:200
90	0.47	1:213
100	0.44	1:227
110	0.41	1:244
120	0.38	1:263
130	0.35	1:286

customary U.S. units		
design speed (mph)	maximum relative gradient (%)	equivalent maximum relative slope
15	0.78	1:128
20	0.74	1:135
25	0.70	1:143
30	0.66	1:152
35	0.62	1:161
40	0.58	1:172
45	0.54	1:185
50	0.50	1:200
55	0.47	1:213
60	0.45	1:222
65	0.43	1:233
70	0.40	1:250
75	0.38	1:263
80	0.35	1:286

From *A Policy on Geometric Design of Highways and Streets*, 2011, by the American Association of State Highway and Transportation Officials, Washington, D.C. Table 3-15. Used by permission.

Proportioning transition lengths according to lane widths (e.g., doubling the length for four lanes) may result in transitions far too long for practical application. To compensate for lane width, Eq. 4.67 can be used to determine the minimum runoff length, L_R. W is the width of one lane, N_L is the number of lanes rotated, and b_w is the adjustment factor for number of lanes rotated (see Fig. 4.29). $\Delta_\%$ is the maximum relative gradient (i.e., the relative difference in longitudinal grade between the axis of rotation and the pavement edge), expressed as a percentage, and e is the design superelevation rate.

$$L_R = \left(\frac{WN_L e}{\Delta_\%}\right) b_w \quad [GDHS \text{ Eq. 3-25}] \quad 4.67$$

Equation 4.68 shows a simplified method of finding the transition length using the elevation change of the outer pavement edge determined from Eq. 4.69 and the maximum relative slope. The *maximum relative slope*, Δ, is the maximum relative gradient expressed in foot per foot (meter per meter). L_R is the minimum runoff length, and E is the elevation change for a fully superelevated lane.

$$L_{R,m} = E_m \Delta_{m/m} \quad [SI] \quad 4.68(a)$$

$$L_{R,ft} = E_{ft} \Delta_{ft/ft} \quad [U.S.] \quad 4.68(b)$$

$$E = We \quad 4.69$$

Example 4.7

A freeway has three lanes, 12 ft (3.7 m) wide, in each direction. A curve with a design speed of 70 mph (110 kph) is to be superelevated to a 3.0% cross slope. Using the maximum relative gradient, determine the runoff length.

SI Solution

From Table 4.13, the maximum relative gradient for a design speed of 110 kph is 0.41%. The adjustment factor, b_w, for three rotated lanes is 0.67. Using Eq. 4.67, the runoff length is

$$L_R = \left(\frac{WN_L e}{\Delta_\%}\right) b_w = \left(\frac{(3.7 \text{ m})(3)(3.0\%)}{0.41\%}\right)(0.67)$$
$$= 54.4 \text{ m}$$

Customary U.S. Solution

From Table 4.13, the maximum relative gradient for a design speed of 70 mph is 0.40%. The adjustment factor, b_w, for three rotated lanes is 0.67. Using Eq. 4.67, the runoff length is

$$L_R = \left(\frac{WN_L e}{\Delta_\%}\right) b_w$$
$$= \left(\frac{(12 \text{ ft})(3)(3.0\%)}{0.40\%}\right)(0.67)$$
$$= 180.9 \text{ ft}$$

Example 4.8

A two-lane roadway is to be superelevated at a rate of 0.04 ft/ft (0.04 m/m). The design speed is 60 mph (100 kph), and the transition is to be designed using the equivalent maximum relative slope. The lanes are each 12 ft (3.7 m) wide. Determine the runoff length.

Figure 4.29 Superelevation Transition Adjustment Factor for Number of Lanes Rotated

number of lanes rotated (N_L)	adjustment factor* (b_w)	length increase relative to one-lane rotated ($= N_L b_w$)
1	1.00	1.0
1.5	0.83	1.25
2	0.75	1.5
2.5	0.70	1.75
3	0.67	2.0
3.5	0.64	2.25

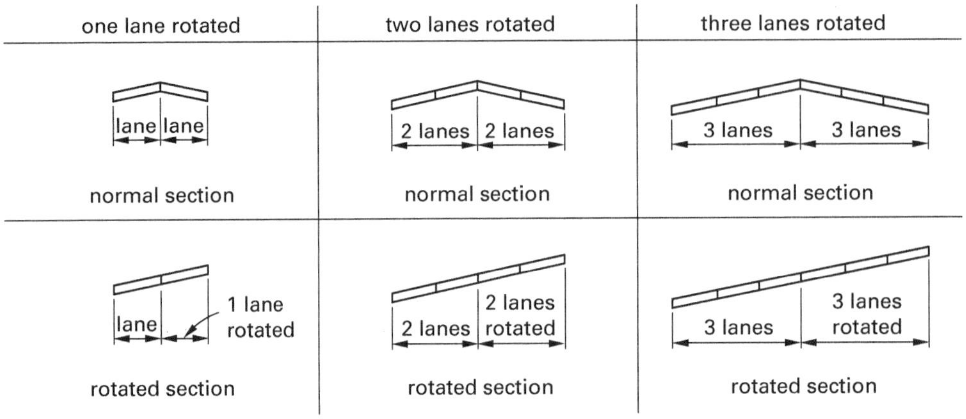

*$b_w = (1 + 0.5(N_L - 1))/N_L$

From *A Policy on Geometric Design of Highways and Streets*, 2011, by the American Association of State Highway and Transportation Officials, Washington, D.C. Table 3-16. Used by permission.

SI Solution

Determine the relative elevation change of the outer pavement edge for the fully elevated lane using Eq. 4.69.

$$E = We = (3.7 \text{ m})\left(0.04 \, \frac{\text{m}}{\text{m}}\right)$$
$$= 0.148 \text{ m}$$

From Table 4.13, the maximum relative slope for a 100 kph design speed is 1:227 or 227 m/m. Determine the runoff length using Eq. 4.68(b).

$$L_{R,\text{m}} = E_\text{m}\Delta_{\text{m/m}} = (0.148 \text{ m})\left(227 \, \frac{\text{m}}{\text{m}}\right)$$
$$= 33.6 \text{ m}$$

Customary U.S. Solution

Determine the relative elevation change of the outer pavement edge for a fully elevated lane using Eq. 4.69.

$$E = We = (12 \text{ ft})\left(0.04 \, \frac{\text{ft}}{\text{ft}}\right)$$
$$= 0.48 \text{ ft}$$

From Table 4.13, the maximum relative slope for a 60 mph design speed is 1:222 or 222 ft/ft. Determine the transition length using Eq. 4.68.

$$L_{R,\text{ft}} = E_\text{ft}\Delta_{\text{ft/ft}} = (0.48 \text{ ft})\left(222 \, \frac{\text{ft}}{\text{ft}}\right)$$
$$= 106.6 \text{ ft}$$

Minimum Tangent Runout Length

Tangent runout is the section of a roadway over which the adverse crown is removed. In order to achieve a smooth transition, the rate at which the adverse crown is removed should equal the maximum relative gradient used to determine the minimum runoff length, L_R. Therefore, the minimum length of tangent runout, L_t, can be determined by Eq. 4.70. e is the superelevation rate, and e_NC is the normal cross slope rate, both in decimals.

$$L_t = \left(\frac{e_\text{NC}}{e}\right)L_R \quad [GDHS \text{ Eq. 3-24}] \quad 4.70$$

Appendix 4.E gives a tabulation of minimum runoff lengths and tangent runout lengths for various superelevation rates and design speeds.

Example 4.9

A freeway with three 12 ft (3.7 m) wide lanes in each direction has a curve with a superelevation of 3.0%. The design speed is 70 mph (110 kph), and the minimum runoff length is 181 ft (55.1 m). If the normal section cross slope is 1.5%, what is the minimum tangent runout length?

SI Solution

Using Eq. 4.70, the minimum tangent runout length is

$$L_t = \left(\frac{e_{\text{NC}}}{e}\right)L_R = \left(\frac{0.015}{0.03}\right)(55.1 \text{ m})$$
$$= 27.6 \text{ m}$$

Customary U.S. Solution

Using Eq. 4.70, the minimum tangent runout length is

$$L_t = \left(\frac{e_{\text{NC}}}{e}\right)L_R = \left(\frac{0.015}{0.03}\right)(181 \text{ ft})$$
$$= 90.5 \text{ ft}$$

Location of Transition With Respect to the End of Curve

Transitions Without Spirals

In tangent-to-curve design, roadway designers must decide where to place the superelevation transition length in respect to the PC. Neither placing the transition entirely on the approach tangent nor placing it entirely on the circular curve is desirable. For general overall conditions, roadway engineers tend to place two-thirds (67%) of the superelevation transition length on the tangent and the remaining transition length on the curve. (See Fig. 4.28.) This practice provides a reasonable compromise between lateral acceleration and vehicle lateral motion as the driver corrects for the curve and highway cross slope for a single lane-width transition. However, this convention may not provide sufficient cross slope for wider pavements and lower design speeds. AASHTO recommends increasing the length of superelevation runoff on the tangent in some situations, as shown in Table 4.14.

Transitions with Spirals

The most effective spiral curve design for a given situation is one that closely approximates the natural spiral path drivers tend to adopt in that situation. A curve design that feels natural to drivers will result in increased driver comfort and vehicle control, both of which increase roadway safety.

The most natural spiral curves are those in which the length of the spiral and the length of the superelevation transition are equal. If a tangent runout is needed, it is most commonly placed on a length of roadway beyond the end of the spiral. Proper selection of spiral length is covered in Sec. 4.8.

Table 4.14 Portion of Runoff in Tangent to Minimize Vehicle Lateral Motion

	SI units			
	portion of runoff located prior to the curve			
design speed (kph)	no. of lanes rotated			
	1.0	1.5	2.0–2.5	3.0–3.5
20–70	0.80	0.85	0.90	0.90
80–130	0.70	0.75	0.80	0.85
	customary U.S. units			
	portion of runoff located prior to the curve			
design speed (mph)	no. of lanes rotated			
	1.0	1.5	2.0–2.5	3.0–3.5
15–45	0.80	0.85	0.90	0.90
50–80	0.70	0.75	0.80	0.85

From *A Policy on Geometric Design of Highways and Streets*, 2011, by the American Association of State Highway and Transportation Officials, Washington, D. C. Used by permission.

13. VERTICAL AND HORIZONTAL CLEARANCES

Clearances on Vertical Curves

General procedures for determining vertical clearance above or below vertical curve grade lines are covered in many design manuals, such as the *GDHS*. One complicated design condition worthy of note is the issue of sight distances for sag vertical curves at undercrossings. Figure 4.30 shows a sag vertical curve at an undercrossing. As shown in Fig. 4.30, the fascia of the undercrossing may block the line of sight at the point where the *critical clearance distance*, C, intersects the fascia, reducing available sight distance. This limited sight distance is particularly a problem when it occurs at two-lane undercrossings where passing sight distance is needed.

The available sight distance can be found graphically or by calculation. An approximation of the available sight distance can be found using a simplified geometric construction, as shown in Fig. 4.31, with the assumption that the fascia on the approach side is the location of the critical clearance distance, C, and the sight distance,

S, is greater than the length of the curve, L. A sight line can be drawn between the point where the driver's eye height, h_1, intersects the back grade point and the height of the structure, h_2, intersects the ahead grade point. The length of the sight line is the available sight distance, S. S is proportional to L, and the sum of the external distance, E, and the distance O_s is proportional to double the external distance, $2E$, as shown in Fig. 4.31 and Eq. 4.71.

Figure 4.30 Sight Distance on Sag Vertical Curves at Undercrossings

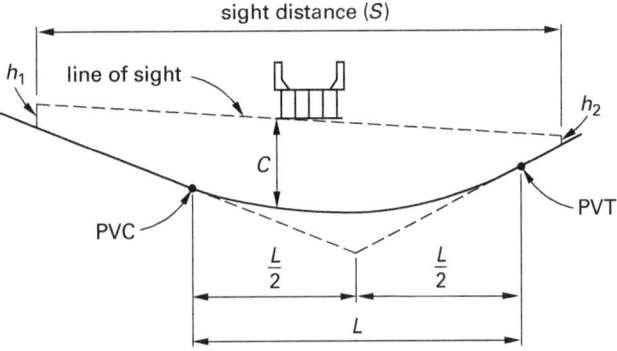

Figure 4.31 Line of Sight Below Underpass When S > L

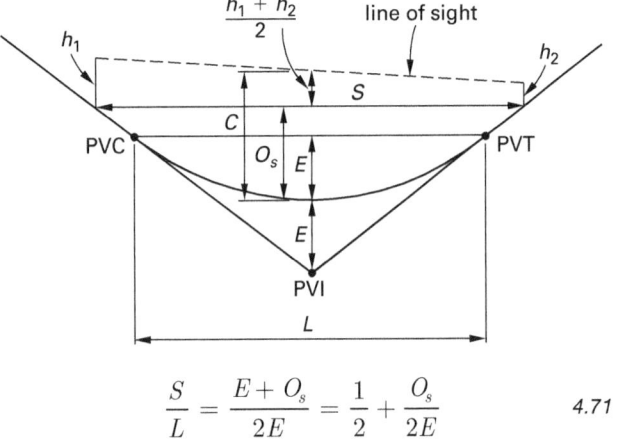

$$\frac{S}{L} = \frac{E + O_s}{2E} = \frac{1}{2} + \frac{O_s}{2E} \qquad 4.71$$

The external distance of the curve is found using Eq. 4.72. A is the absolute value of the algebraic difference in grades in percent, and L is the length of the curve in stations.

$$E_{\text{ft}} = \frac{A_\% L_{\text{sta}}}{8 \frac{\text{sta}}{\text{ft}}} \qquad \text{[U.S. only]} \qquad 4.72$$

As shown in Fig. 4.31, the O_s distance is

$$O_s = C - \frac{h_1 + h_2}{2} \qquad 4.73$$

The O_s distance can be further simplified by substituting the default values of h_1 (8.0 ft or 2.4 m) for the average truck driver eye height and h_2 (2.0 ft or 0.6 m) for the average automobile taillight height into $(h_1 + h_2)/2$, which gives an average value of 5.0 ft (1.5 m). Replacing O_s in Eq. 4.71 with $C - 5.0$ ft (1.5 m) yields Eq. 4.74, which can be rearranged into Eq. 4.75 and Eq. 4.76 to find the curve length and sight distance, respectively.

$$\frac{S_{\text{m}}}{L_{\text{m}}} = \frac{1}{2} + \frac{C_{\text{m}} - 1.5 \text{ m}}{2 E_{\text{m}}} \qquad \text{[SI]} \qquad 4.74(a)$$

$$\frac{S_{\text{ft}}}{L_{\text{ft}}} = \frac{1}{2} + \frac{C_{\text{ft}} - 5.0 \text{ ft}}{2 E_{\text{ft}}} \qquad \text{[U.S.]} \qquad 4.74(b)$$

$$L_{\text{m}} = \frac{2 S_{\text{m}}}{1 + \frac{C_{\text{m}} - 1.5 \text{ m}}{E_{\text{m}}}} \qquad \text{[SI]} \qquad 4.75(a)$$

$$L_{\text{ft}} = \frac{2 S_{\text{ft}}}{1 + \frac{C_{\text{ft}} - 5.0 \text{ ft}}{E_{\text{ft}}}} \qquad \text{[U.S.]} \qquad 4.75(b)$$

$$S_{\text{m}} = \frac{L_{\text{m}}}{2}\left(1 + \frac{C_{\text{m}} - 1.5 \text{ m}}{E_{\text{m}}}\right) \qquad \text{[SI]} \qquad 4.76(a)$$

$$S_{\text{ft}} = \frac{L_{\text{ft}}}{2}\left(1 + \frac{C_{\text{ft}} - 5.0 \text{ ft}}{E_{\text{ft}}}\right) \qquad \text{[U.S.]} \qquad 4.76(b)$$

When the sight distance below an underpass is less than the length of the curve below the underpass, the parabolic curve between PVC and PVT resembles a half circle with radius R. (See Fig. 4.32.)

Figure 4.32 Line of Sight Below Underpass When S < L

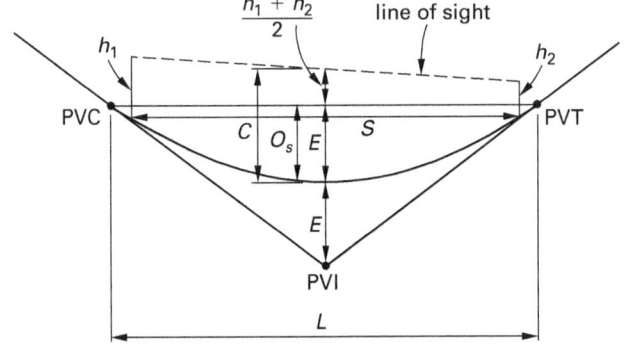

Using $h_1 = 8.0$ ft (2.4 m) for truck driver eye height and $h_2 = 2.0$ ft (0.6 m) for automobile taillight height, the equations for L and S become Eq. 4.77 and Eq. 4.78, respectively.

$$L_m = \frac{S_m^2 A_\%}{8(C_m - 1.5 \text{ m})} \quad \text{[SI]} \quad 4.77(a)$$

$$L_{ft} = \frac{S_{ft}^2 A_\%}{8(C_{ft} - 5.0 \text{ ft})} \quad \text{[U.S.]} \quad 4.77(b)$$

$$S_m = \sqrt{\frac{8 L_m (C_m - 1.5 \text{ m})}{A_\%}} \quad \text{[SI]} \quad 4.78(a)$$

$$S_{ft} = \sqrt{\frac{8 L_{ft} (C_{ft} - 5.0 \text{ ft})}{A_\%}} \quad \text{[U.S.]} \quad 4.78(b)$$

When the obstruction created by the overhead structure (known as the critical clearance distance) is not located directly above the PVI, Eq. 4.74 through Eq. 4.78 remain accurate as long as the critical clearance distance is within 200 ft (60 m) of the PVI. When the critical clearance distance is located more than 200 ft (60 m) from the PVI, calculations using these equations are more approximate than definite.

When solving directly for the length of curve needed for a given sight distance, with C being the vertical clearance, general equations [GDHS Eq. 3-52 and Eq. 3-53] are as follows.

For $S > L$,

$$L_m = 2S - \frac{800\left(C - \left(\frac{h_1 + h_2}{2}\right)\right)}{A} \quad \text{[SI]} \quad 4.79(a)$$

$$L_{ft} = 2S - \frac{800\left(C - \left(\frac{h_1 + h_2}{2}\right)\right)}{A} \quad \text{[U.S.]} \quad 4.79(b)$$

For $S < L$,

$$L_m = \frac{AS^2}{800\left(C - \left(\frac{h_1 + h_2}{2}\right)\right)} \quad \text{[SI]} \quad 4.80(a)$$

$$L_{ft} = \frac{AS^2}{800\left(C - \left(\frac{h_1 + h_2}{2}\right)\right)} \quad \text{[U.S.]} \quad 4.80(b)$$

Assuming the critical condition of a truck with an eye height of 8.0 ft (2.4 m) approaching a stopped car with a taillight height of 2.0 ft (0.6 m), the following equations [GDHS Eq. 3-54 and Eq. 3-55] result.

For $S > L$,

$$L_m = 2S - \frac{800(C - 1.5)}{A} \quad \text{[SI]} \quad 4.81(a)$$

$$L_{ft} = 2S - \frac{800(C - 5)}{A} \quad \text{[U.S.]} \quad 4.81(b)$$

For $S < L$,

$$L_m = \frac{AS^2}{800(C - 1.5)} \quad \text{[SI]} \quad 4.82(a)$$

$$L_{ft} = \frac{AS^2}{800(C - 5)} \quad \text{[U.S.]} \quad 4.82(b)$$

For underpass roadways, the drainage system becomes critical for driving safety. At an unlit underpass, a pool of quiet standing water can completely disappear from the driver's view because headlight beams are fully reflected ahead, giving the appearance of driving into a black hole. In freezing weather, water standing in the low point of an underpass often freezes, especially when protected by the shadow of the overhead bridge from warming sunshine. Trucks require additional clearance for short vertical curves, because of the bridging effect of a long trailer wheelbase.

Vertical Clearances from Overhead Obstructions

As of 2007, FHWA criteria governing the interstate highway system require that newly constructed and reconstructed mainline routes have 16 ft (4.9 m) vertical clearance. These clearances must be maintained at all times, even after a roadway has been repaved. The clearance must be across the entire roadway width, including the entire width of the shoulders. Most states allow load heights of 13.5 ft (4.1 m) without permits and 14.5 ft (4.4 m) load heights with permits. It is common practice to provide at least 1 ft (0.3 m) additional clearance above the maximum load height permitted. For locations where the clearance is less than 14.5 ft (4.4 m), advance warning signs are required, and the low-clearance point must be marked clearly with zebra or chevron safety stripes.

Local jurisdictions over local and nonmajor traffic routes often set the vertical clearance minimum as low as 12.5 ft (3.8 m), although FHWA criteria recommends 14 ft (4.3 m) be maintained throughout. Occasionally, for locations where traffic is restricted to passenger cars only, and in some parts of older cities where reconstruction costs are very high, access for taller vehicles must be provided over alternate routes or, in extreme cases, prohibited entirely.

Where a bridge or overhead obstruction is installed after an approach of many miles, or where conditions are such that drivers might perceive an overhead obstruction as too low to accommodate tall loads, there is an increased risk that truckers or others carrying tall loads will think there is insufficient clearance for their vehicle and brake suddenly, which slows traffic and increases the risk of collisions. To mitigate this problem, the vertical clearance beneath these obstructions should be increased by a minimum of 2 ft (0.6 m).

Overhead signs, utility wire crossings, and traffic signals must be designed to these requirements as well. Usually, anything that is suspended on wires and can move around in the wind is installed with a higher clearance. Many power, telephone, and cable TV companies design for at least 17 ft (5.2 m) of clearance over major arterials when installing a new crossing. Over time, cables have a tendency to sag, reducing the clearance. It is the responsibility of the utility owner to maintain the minimum clearance required by local regulations. Overhead signs rigidly mounted on frames may need to be installed with additional clearance to accommodate light fixtures and maintenance catwalks. These devices must be installed below the sign face itself to avoid blocking the view of oncoming drivers.

Clearances on Horizontal Curves

When designing horizontal curves, attention must be paid to any sight obstructions that may exist. Roadway engineers should check the conditions of each curve and make adjustments as necessary to provide adequate sight distance for the curve's design speed. The sight distance typically refers to the driver's line of sight, as shown in Fig. 4.33. Sight distances on the inside of a curve can be equated to the chord of a circular arc centered at the closest point of obstruction to the edge of the traveled lane. (See Fig. 4.33.) The radial offset distance at the point of obstruction to the centerline of the traveled lane can be considered the middle ordinate of the circular arc. The *horizontal sightline offset*, HSO, is the same as the middle ordinate, M, but is typically referred to as the horizontal sightline offset when calculating sight distance.

$$\text{HSO} = \frac{C_{\text{long}}^2}{8R} \qquad 4.83$$

Usually a radius, R, is given, and it is necessary to find either the length of the long chord, C_{long}, or the HSO. The length of the chord is often used as the line of sight. The stopping sight distance required around an obstruction is determined as the arc length of the curve and is slightly longer than the chord length, providing a margin of safety.

For analysis, the radius of the centerline of the extreme right lane must be determined. The rightmost lane is sometimes called the "inside lane," although the designation of "inside" versus "outside" varies depending on actual conditions and the jurisdiction. (For instance, on a divided highway, the leftmost lane adjacent to the median barrier is called the inside lane in some jurisdictions.) Whatever the terminology, it is important that the lane closest to any obstruction (e.g., buildings, walls, slopes, or any other sight-blocking object) has the

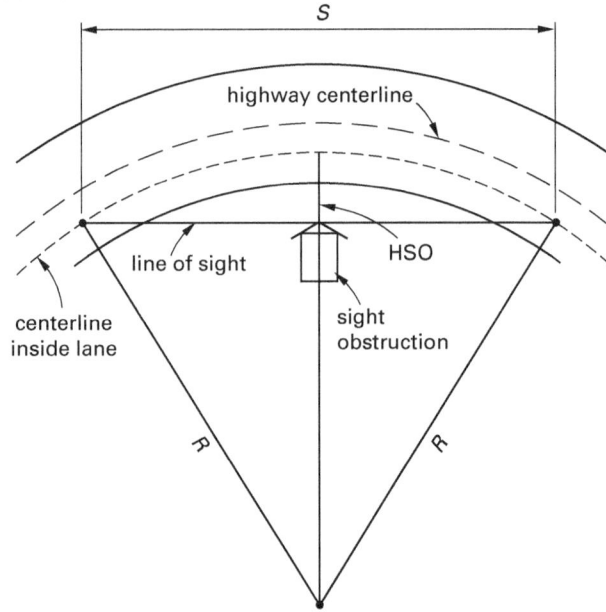

Figure 4.33 Horizontal Sight Distance Components on the Inside of a Curve

shortest sight distance. Where there is a cut slope, the horizontal sightline offset is taken at a point 2.75 ft (840 mm) above the roadway, which is the average of 3.5 ft (1080 mm) eye height and 2.0 ft (600 mm) object height. Stopping sight distance, S, and the HSO are related using Eq. 4.84.

$$\text{HSO} = R\left(1 - \cos\frac{28.65 S}{R}\right) \qquad 4.84$$

[*GDHS* Eq. 3-38]

Horizontal Clearance from Obstructions

All streets and highways are required to have a certain amount of lateral clearance. In general, all streets should have a minimum clearance of 1.5 ft (0.5 m) between the curb face and any obstructions. On streets with a continuous parking lane, a clearance of 1.5 ft (0.5 m) between the parking lane and any obstructions is desirable but not required. Rural collector streets with a design speed of 45 mph (70 kph) or less, as well as urban collector streets with shoulders but no curbs, should have *clear zones* (recovery areas free of fixed objects) of at least 10 ft (3 m). For more information on clear zones, see Chap. 6.

14. ACCELERATION AND DECELERATION

Average traffic accelerates at a rate of approximately 4.4 ft/sec^2, or 30 mph/sec^2 (1.3 m/s^2 or 4.8 kph/s^2), at speeds up to 30 mph (50 kph). The acceleration rate decreases to 2.8 ft/sec^2 (1.9 mph/sec^2) as speeds approach 70 mph (110 kph). Highway design must consider acceleration and deceleration rates and the distances necessary to safely achieve these rates. The distance required to reach a certain speed will increase for upgrade acceleration and decrease for downgrade acceleration due to driver behavior and the effects of gravity. Drivers tends to accelerate at a lower rate on upgrades and at a higher rate on downgrades. AASHTO design curves illustrating acceleration and deceleration distances are given in App. 4.F and App. 4.G. The acceleration and deceleration rates are based on low horsepower-to-weight ratio vehicles and represent the lower limit of design conditions. Both acceleration and deceleration rates will vary depending on local conditions, such as approach grade and driver familiarity, and with changing vehicle technology.

The mix of vehicles in the traffic flow can have a significant effect on acceleration rates. Modern trucks and intercity buses accelerate at nearly the same rates as automobiles, but heavily loaded large trucks and recreational vehicles tend to have much lower acceleration rates than standard automobiles. Traffic flow that contains a large number of recreational vehicles or heavily loaded large trucks will typically have lower acceleration rates than traffic flow containing only automobiles, trucks, and buses. However, a very small proportion of slower vehicles will have little effect on acceleration rates.

Deceleration rate is usually dictated by the comfort of the driver and passengers, not the braking ability of the vehicle. A comfortable deceleration rate for most drivers (approximately 90%) is 11.2 ft/sec^2 (3.4 m/s^2), which is the default deceleration value recommended by AASHTO. Rapid deceleration is discomforting to most people and tends to be used only when necessary or by select types of drivers. Deceleration rates greater than 14.8 ft/sec^2 (4.5 m/s^2) are seen only in cases where drivers are making panic stops.

When designing for deceleration, all vehicles have nearly the same capability, regardless of size or weight. For heavy vehicles such as commercial semi-trucks, the braking distance is nearly the same as for smaller vehicles. However, the total stopping distance is slightly greater because of the delay in applying air brake systems and minor variations between braking rates of multiple axle systems. Large commercial vehicles are generally driven by specially trained drivers, who know to compensate for the increased brake application time of their vehicles. Furthermore, the elevated cabs of these commercial vehicles afford the drivers an increased line of sight relative to drivers of standard automobiles. These factors are assumed to compensate for the increased brake application time, so the same stopping sight distances are used for all vehicles.

15. INTERSECTIONS AND INTERCHANGES

Intersection Geometry

An intersection of two or more roadways can create traffic flow conflicts and delays, as well as collisions. Fundamental horizontal alignment principles are explained in Chap. 1 and Chap. 9 of the *GDHS*. The primary principle governing intersection geometry is the elimination of acute angles at the intersection, which increases sight distance in one or more directions for approaching drivers and provides sufficient width for turning of the design vehicle.

Horizontal geometry involves setting the inner edge of the pavement radius to match the design speed and widening the pavement to allow for the swept path of longer vehicles. The design guides are illustrated in the appendices as follows.

- AASHTO design vehicle dimensions are shown in App. 4.H, with the minimum turning radii shown in App. 4.I.

- Selected turning path templates are shown in App. 4.J through App. 4.P.

- Turning templates and edges-of-traveled-way for various turn angles and design vehicles are given in App. 4.Q and App. 4.R.

- Curve radii for simple curves, curves with tapers, and three-centered compound curves are found in App. 4.S.

- The effect of curb radii on right-turning paths is illustrated in App. 4.U.

- Cross-street widths occupied by turning vehicles are tabulated in App. 4.V.

- The effect of crosswalk lengths and corner setbacks are illustrated in App. 4.W and App. 4.X.

Example 4.10

A cul-de-sac is being installed at the end of a suburban street to allow a conventional school bus (S-BUS36 or S-BUS11) to turn around without a backing maneuver. What is the minimum outside radius required to accommodate the school bus without allowing overhang?

Solution

From App. 4.Q, select the minimum turning radius (i.e., the path of the left front wheel), which shows a radius of 38.6 ft (11.75 m).

Example 4.11

A 120° intersection turn will have transit buses turning from the lane closest to the curb onto a cross street with 12 ft (3.6 m) wide lanes. Buses are expected to turn from the proper position in the lane and swing wide on the cross street. The curb radius is 20 ft (6 m). How wide must the cross street be for a bus to complete its turning movement?

Solution

From App. 4.V, with a transit bus turning from its proper position in the lane, case A applies. For a 120° turn and a 20 ft (6 m) curb radius, the width occupied on the cross street is 40 ft (12.2 m).

Example 4.12

A crosswalk at a 90° intersection is 3 m wide, the corner curb radius is 3 m, and the right-of-way is 3 m from the face of the curb at the traveled way. If the curb radius is increased to 8.0 m, how much length will be added to the crosswalk at each corner?

Solution

From App. 4.V, with a radius of 3 m and a crosswalk width of 3 m, the crosswalk distance added is 0.8 m for the curb return. For curb radius of 9 m, the distance added for the curb return is 8.0 m. The additional crosswalk distance necessary to increase the curb radius to 8.0 m is

$$8.0 \text{ m} - 0.8 \text{ m} = 7.2 \text{ m}$$

Interchange Geometry

Grade separation structures for crossing movements are used in interchange design to alleviate the conflict delays and safety demands caused by high volumes of traffic traveling through at-grade intersections. There are a variety of ramp configurations that can be used to connect a roadway to a crossing structure. The ramp configuration is often determined by topography or property constraints, but desired traffic operation is the most important consideration. Some configurations are proven to be more effective than others, while some configurations are actively discouraged as design standards evolve. Interchange configurations are illustrated in *GDHS* Fig. 10-1.

The layout adopted for any specific location is a function of the desired traffic movements, topographic restrictions, local expectations, cost, and available right-of-way. Project traffic conditions as forecasted by a transportation study should predominate the design layout. Figure 4.34(a) is a trumpet-shaped, three-leg interchange. Figure 4.34(b) is a three-level, three-leg directional interchange.

Figure 4.34(c) is one-quadrant, which is suitable for connections between major highways, but requires cross-traffic movements. It is not suitable for freeways and high-volume traffic, and it would not be able to handle large tuck movements well.

Figure 4.34(d) is a simple diamond interchange, suitable for freeway-to-major-highway movements of moderate volume. Figure 4.34(e) is a single-point diamond interchange using a single signalized intersection.

Figure 4.34(f) is a partial cloverleaf requiring only two cross-traffic movements. The configuration is often varied to favor larger movements with free-flow ramps, and to allow low flow movements to cross traffic lanes, thereby reducing construction costs.

Figure 4.34(g) is a full cloverleaf, one of the earliest forms of full free-flow movement interchanges. The interchange restricts weaving volume through the closely spaced entrance and exit ramp connections on the inside of the cloverleaf. The *inside loop ramps* have tighter radius curves and require considerable speed reduction, which, combined with the short weaving segments, cause this type of interchange to be unsuitable for high left-turn volumes. Some relief has been found by adding collector-distributor lanes alongside the mainline lanes.

Figure 4.34(h) is a fully directional interchange able to handle large volumes of traffic on any movement. These large interchanges result in four levels of ramp configurations, resulting in very tall bridge structure throughout the central portion of the interchange. This type of interchange completely avoids *left-hand exit ramps* from the mainline roadways, which requires slower-moving traffic to move to the left lanes that are generally used by higher-speed vehicles. Among this type of interchange's many advantages is the ability to accommodate *multilane ramps* for any or all of the movements; thereby having the capacity to move large volumes of traffic. Many other types of interchanges have been developed to suit local conditions, with the design engineer limited only by imagination to create a smoothly flowing traffic design. Diverging diamond interchanges for freeways and roundabouts for local streets are examples of relatively recent designs being analyzed for changing traffic conditions.

Advantages and disadvantages of the major interchange types are discussed in further detail in *GDHS* Chap. 10.

Intersection and Grade Separation Warrants

Interchanges and grade separations can be useful solutions to reduce traffic congestion and improve safety. However, they are also costly. To determine whether or not an interchange is justified, the *GDHS* presents six warrants to consider. Warrants for grade separations are more general than signal warrants.

1. *Design designation:* This is the universal standard for the interstate highway system. It allows no at-grade intersection to occur at any point along a designated interstate highway. Although there are many factors that influence safety, such as access control, provision of medians, and elimination of parking and pedestrian traffic, grade separation yields the largest increment of safety.

Figure 4.34 Typical Interchange Configurations

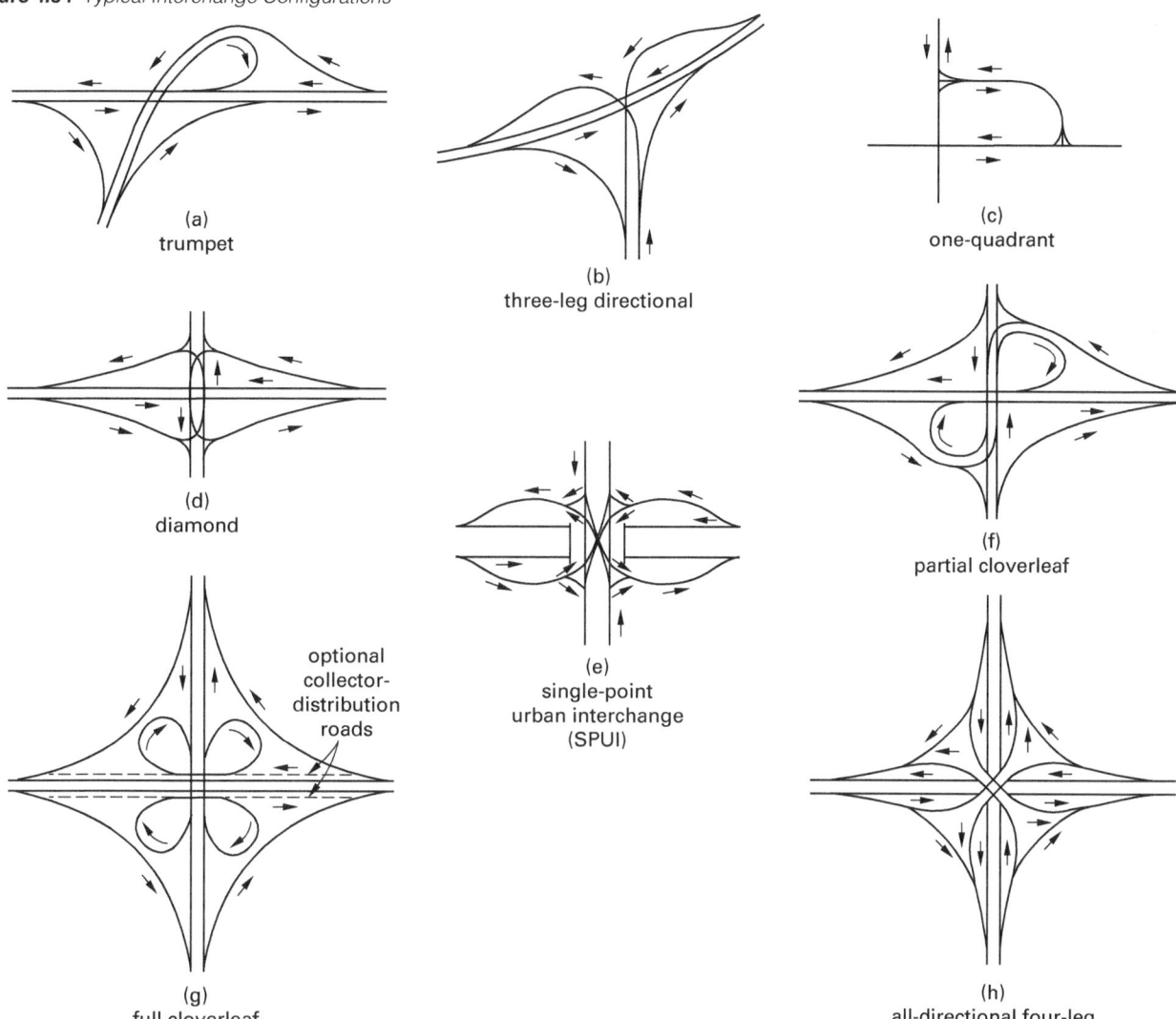

From *A Policy on Geometric Design of Highways and Streets*, Fig. 10-1, copyright © 2011, by the American Association of State Highway and Transportation Officials, Washington, D.C. Used by permission.

Therefore, any intersection that is part of a freeway should consider grade separation.

2. *Bottleneck* or *spot congestion relief:* Bottlenecks and spot congestion occur when there is insufficient capacity at intersections along heavily traveled roads. Interchanges and partial interchanges are often justified to eliminate delays caused by necessary signalization. However, it is a poor use of construction and maintenance resources to use grade separation when lesser means of access are adequate. Therefore, this warrant is usually satisfied by thorough traffic capacity and economic analysis.

3. *Safety improvement:* This warrant is often related to other conditions, such as poor sight distance or poor topographic conditions, that result in crash-prone intersection designs. A grade separation or interchange may be warranted if other less expensive methods of eliminating crashes are impractical or not likely to be effective. Safety is an especially significant factor when considering crossing grade separation of railroads and highways.

4. *Site topography:* There are locations, particularly in hilly or mountainous areas, where an at-grade intersection cannot be made with a reasonably acceptable alignment to accommodate the expected traffic. In these conditions, and when alternate access is not available, a grade separation may be warranted.

5. *Road-user benefits:* The user cost of additional travel distance caused by interchange geometry can be outweighed by the cost of delay caused by signalization at a busy intersection. Using benefit-cost analysis,

the road-user benefits are compared to the cost of construction. The greater the ratio of benefits to costs, the greater the justification for construction of the interchange.

6. *Traffic volume:* Current traffic volume or projected increased volume may be sufficient to warrant an intersection or grade separation. At-grade intersections that have volumes in excess of capacity and cross streets with high volumes of traffic benefit the most from improvements in the movement of traffic.

Underpass and Overpass Roadways

When a highway grade separation is proposed, the engineer must decide whether an overpass roadway or an underpass roadway should be built. Topography must be considered when making this decision. When an engineer looks at the topography of a site, one of three things will generally happen: (1) the influence of the topography will be great, making it necessary to closely fit the design to it; (2) the topography will favor neither an overpass nor an underpass; or (3) the alignment and gradeline of one of the highways will predominate, making it necessary to design with respect to that highway's alignment instead of the topography of the site.

When topography is not the governing factor, there are other factors that may be considered. Underpass and overpass roadways each have advantages and disadvantages. While the discussion of relative merit can be rather extensive, the basic advantages and disadvantages are as follows.

Overpass roadways

- provide greater visibility to traffic conditions ahead of the structure

- provide greater vertical clearance for large loads

- reduce the advance visibility of the upcoming interchange to approaching drivers, reducing advance notice of upcoming road conditions

- reduce the visibility of approaching and departing ramps to and from the roadway below

- increase the propensity for icing of the bridge deck (i.e., road surface) compared to the adjoining at-grade roadway

- require sturdy bridge railings to guide the movement of errant vehicles and prevent them from driving off the structure

- require careful control of drainage to avoid discharge onto the roadway below and to avoid undermining of slopes and footings from concentrated discharge points

- require more ownership and maintenance responsibility than approach roadways

Underpass roadways

- reduce the visibility of the roadway ahead of the bridge structure

- reduce the vertical clearance for large loads

- increase the advance visibility of the upcoming interchange to approaching drivers, alerting drivers to possible traffic condition changes

- increase the visibility of the departure and arrival ramps, which assist the anticipation of upcoming traffic events

- decrease the initial tendency of the roadway surface under the bridge to form ice, but also reduce the melting effect of heat from the sun

- require adequate abutment and pier setbacks, along with roadside safety protection devices for immovable piers and wall faces

- may require supplemental underbridge lighting during the daytime

- tend to collect trash and debris more than open roadways

- may require additional drainage structures, especially if the underpassing roadway dips down for additional clearance under the bridge

- may require less ownership and maintenance responsibility than an overhead structure at the same location

16. RAILROAD ENGINEERING

Railroad Design

Since railroads predate highways, railroads and highways share the same fundamentals of curves, spirals, grades, and vehicle dynamics, with the exception of the following two principles.

- Railroad tracks have a fixed *gauge*, or distance, between rails. Most measurements along the track take into account the fixed gauge and use one or the other rail as the baseline for geometry.

- Railroad (and street railway) equipment can tolerate less abrupt changes in geometry than vehicles on a roadway. To accommodate these limitations, grades are not as steep, curves are broader, and transitions are longer.

An overriding factor with all railroad design is the limited friction generated by a steel wheel on a steel rail compared to a vehicle's rubber tire on hard pavement. This factor, along with much larger vehicle masses and greater pounds per motive horsepower of trains, directly affects considerations given to grades, stopping distances, acceleration rates, and the effects of vehicle momentum. Certain rules of thumb prevail with railroad designs, such as chord definition curve layouts rather than arc definition curve layouts, and superelevation based on inches (millimeters) of outer rail elevation above the inner rail. Because axle loadings have increased greatly since the steam locomotive era, accurate rail alignment and grade control are increasingly important on modern railroads.

Typical railroad corridor design starts with chord definition curves. Grades and profiles may be evaluated using economic analysis over a several mile (kilometer) long segment to determine the most economical construction and operating cost. If the rail line is to haul passengers, an analysis of station spacing, line haul speeds, and capacity may be required.

Horizontal Curves

Chord definition curves are widely used in railroad design. The chord definition uses deflection of a 100 ft (30 m) chord instead of an arc. (See Sec. 4.5 for chord definition equations). Railroad surveyors have easy access to tables used for layout by chord deflections, precluding the need for error-prone calculations while in the field. Modern technology is also playing a role in decreasing errors, as the newest innovations in field layout employ GPS, RTK, and other remote sensing systems, which greatly improve productivity and decrease track downtime. Remote sensing is particularly applicable to railroad conditions because of the large distances involved along narrow corridors that often limit traditional line-of-sight methods.

Transitions into and out of superelevation are accomplished more gently than on highways. Additionally, when placing reverse curves (i.e., a track curving one way, then instantly curving the opposite direction) back-to-back, it is necessary to introduce a section of tangent track in order to accomplish tangent runout for each curve and also to allow the car couplers to align themselves for the next curve without causing a derailment. The minimum tangent between reverse curves is usually as long as the longest car, or the sum of the two superelevation runouts, whichever is longer. A rule of thumb is to use 100 ft (30 m) of tangent between curves, including spiral transitions.

Railroad designers must be aware of potential derailment points, such as at a *point of continuing curve*, PCC, or a PT with a sharp curve adjacent to a broad curve or tangent. When a long car (e.g., an 80 ft (24.5 m) automobile carrier) is coupled to a short car, the drawbar develops a sharp angle with the car centerline, which can pull the long car off the track at the end of the curve if no spiral easement is provided. Designs where this problem may occur must either add transitions at these points of derailment or lengthen existing transitions to reduce the risk of derailment. These design changes can be expensive, so engineers should do their best to make sure they are included in the initial design.

Grades

Grades have a greater effect on railroads than on roadways because of the higher horsepower requirements necessary to move heavy loads and the limits of friction on steel rail. With heavy freight and high-speed passenger lines, 1% is considered the maximum for the longest and heaviest trains. Mainline grades of up to 2.1% have been built in very mountainous locations, requiring the operation of shorter trains (reducing the maximum tonnage) and additional locomotives attached to the consist. Branch lines and feeder lines using short trains and with industrial access may have grades of 4% or more for short stretches. Rapid transit lines can operate on somewhat steeper grades than locomotive-hauled commuter trains, although grades are usually limited to a maximum of 4%. Grades on street railway operations are much steeper and can reach 15% for short distances. As a practical limit, street railways try to keep grades below 8%. When operation on steep or very steep grades is necessary, the vehicles must have large, powerful motors, and every axle must be powered.

When designing railroad grades, the rise in feet per mile is often cited, such as 53 ft per mi for a 1% grade. When a train that is 1 mi long is climbing a 1% grade, the locomotive is more than 50 ft above the last car, meaning the locomotive horsepower must be sufficient to lift the entire train 50 ft for each mile traveled. The grade and total train weight determine the horsepower needed to pull the load at a given speed or efficiency, and grade determinations on railroads often involve a thorough analysis to balance a variety of requirements, such as the cost of building a bridge or tunnel to change a grade versus the horsepower and operating cost requirements of moving the load.

Train operations in hilly country involve a series of carefully planned brake applications and full throttle operations so that train momentum can carry the train over short, steep segments. The locomotive may put out full power while going downhill to haul the rear of the train up the last hill and may apply brakes while going uphill in preparation for the next speed restriction several miles ahead. Improper braking technique or throttle application can derail a train through a sharp curve or break the couplings (i.e., connections) between cars.

Should a train be required to stop on a steep uphill grade, it may be necessary to uncouple cars from the rear to allow the locomotive to pull the attached cars uphill. The locomotive must then uncouple from the cars and return to pull the remainder of the train uphill. This is a time-consuming operation and is highly undesirable. Therefore, the signal location and spacing must be carefully planned in order to avoid stopping a train on an uphill climb. In contrast, for rapid transit or light rail operations, the maximum grade at any point on the line should not exceed the capacity of the vehicle or train to start from a standing stop. Backing a stalled passenger train down a hill against the prevailing traffic is an invitation for disaster that no transit operator wants to face.

17. BIKEWAY GEOMETRIC DESIGN

Bikeway Widths and Clearances

The minimum operating space for a bicycle rider should be 3.5 ft (1.1 m) wide and 8.0 ft (2.4 m) high, as suggested by the FHWA's guidelines for LOS C conditions. The minimum operating width for a single-lane, one-way bikeway should be 5.0 ft (1.5 m). For a path accommodating a bicycle lane and a pedestrian lane, the minimum width should be 6.5 ft (2.0 m), which allows a 3.0 ft (0.9 m) lane width for pedestrians.

For two or more lanes of bicycle traffic, multiples of the 3.5 ft (1.1 m) operating space width should be used. While reduction in lane width is allowed for pedestrian facilities, reductions are not recommended for multilane bikeways. Bicyclists need more space than pedestrians, as bicycling requires a certain amount of lateral movement for bicyclists to regain and maintain balance. Therefore, lateral obstructions (e.g., walls, light poles, railings, and vegetation) should be kept at least 2 ft (0.6 m) away from the edge of the pathway.

Opposing bicycle traffic designs should provide more lateral space between the opposing lanes than between same-direction, multilane paths. The minimum recommended width for a two-way, two-lane bikeway is 8.0 ft (2.4 m), increasing to 10 ft (3.0 m) in locations where there are higher volumes, higher speeds, or sharper curves. The minimum width for a three-lane, one-way bikeway should be 10 ft (3.0 m). Minimum vertical clearance of 8.0 ft (2.4 m) is necessary for seated riders. Preferred vertical clearance is 10 ft (3.0 m), which allows riders to pass under an obstruction while standing.

Design Speeds

Actual travel speeds vary greatly for bicycle riders. Recreational riders and family groups may travel at a leisurely 8–10 mph (13–16 kph) on level paths, while more serious riders may travel at speeds ranging from 14 mph to 25 mph (23 to 40 kph). Serious riding enthusiasts who are used to traveling long distances and/or performing physical workouts may regularly travel at speeds of 30 mph (48 kph) or greater.

A design speed of 20 mph (32 kph) will accommodate the majority of riders on level or nearly level ($< 1.5\%$) grades with smooth surfaces. For unpaved or rough surfaces along level paths, the design speed can be reduced to 15 mph (24 kph) because riding speeds tend to be lower along these surfaces. Usually, experienced, higher-speed bicyclists will compensate quickly for varying travel path conditions or sudden restrictions in geometry.

For grades greater than 4%, the minimum design speed should be increased to 30 mph (48 kph). Uphill riders may operate at lower speeds, while downhill riders may take advantage of the grade and operate at higher speeds.

Stopping Distance

Bicycles require the same stopping distances as automobiles, and the formulas used to determine bicycle stopping distances are the same as AASHTO's formulas for automobile stopping distances, with the exception that the coefficient of friction, f, is limited to 0.25 to account for the poor-weather braking characteristics of many bicycles.

A friction factor of 0.25 is equivalent to a deceleration rate of $0.25g$ (8 ft/sec^2; 2.4 m/s^2), which is the recommended deceleration rate for bicycles. Using a more rapid deceleration rate, such as the $0.35g$ (11.2 ft/sec^2; 3.4 m/s^2) deceleration rate used to determine stopping sight distances for highways (see GDHS Exh. 3-1), is not recommended for bicycles. Stopping at greater rates requires more skill to modulate the brakes in order to avoid skidding, and less experienced bicyclists may have difficulty maintaining balance should a skid maneuver be required. Furthermore, the small footprint of bicycle tires provides a less reliable braking surface than automobile tires.

Eq. 4.85 gives the stopping distance for bicycles. The decimal grade, G, will be positive for an ascending grade and negative for a descending grade, which will decrease or increase the braking distance, respectively. The second term of Eq. 4.85 determines the distance traveled during 2.5 sec of perception-reaction time before braking begins. Stopping distances from selected speeds, v, are shown in Table 4.15 using the default deceleration rate of $0.25g$.

$$S_{\mathrm{m}} = \frac{\mathrm{v}_{\mathrm{kph}}^2}{254(f \pm G_{\mathrm{dec}})} + \frac{\mathrm{v}_{\mathrm{kph}}}{1.4} \quad [\text{SI}] \quad 4.85(a)$$

$$S_{\mathrm{ft}} = \frac{\mathrm{v}_{\mathrm{mph}}^2}{30(f \pm G_{\mathrm{dec}})} + 3.67\mathrm{v}_{\mathrm{mph}} \quad [\text{U.S.}] \quad 4.85(b)$$

Table 4.15 Bicycle Stopping Distance for Selected Design Speeds

speed (mph)	−10%	−5%	0%	4%
10	60 ft	54 ft	50 ft	48 ft
15	105 ft	93 ft	85 ft	81 ft
20	165 ft	140 ft	127 ft	120 ft
25	230 ft	196 ft	175 ft	164 ft
30	310 ft	260 ft	230 ft	214 ft

(Multiply ft by 0.305 to obtain m.)
(Multiply mph by 1.61 to obtain kph.)

Example 4.13

Using a friction factor of 0.25, what is the design stopping distance for a bicycle traveling at a speed of 20 mph on a 4% downgrade?

Solution

From Eq. 4.85(b),

$$S_{ft} = \frac{v_{mph}^2}{30(f \pm G_{dec})} + 3.67 v_{mph}$$

$$= \frac{\left(20 \, \frac{mi}{hr}\right)^2}{(30)(0.25 - 0.04)} + (3.67)\left(20 \, \frac{mi}{hr}\right)$$

$$= 137 \text{ ft}$$

Horizontal Alignment

The minimum radius of horizontal curves depends on the design speed, v, and superelevation, e, of the bikeway. The equation relating the radius of the curve to design speed and superelevation is the same equation as is used with highways, Eq. 4.66.

Superelevation rates should be limited to a maximum of 5% (0.05 ft/ft or 0.05 m/m) for general purpose bikeways. Bicycles are harder to control on slippery surfaces than four-wheeled vehicles, so the minimum superelevation should be 2% (0.02 ft/ft or 0.02 m/m) to provide adequate drainage to eliminate standing water on the riding surface.

Side friction factors, f_s, should be no greater than 0.22–0.30 for hard paved surfaces and 0.11–0.15 for unpaved surfaces.

Sight distance around an obstruction on a horizontal curve can be determined using the sight distance equations for horizontal curves, Eq. 4.58 and Eq. 4.59. The actual stopping distance along the arc of the curve will be slightly greater than the straight-line sight distance past the obstruction.

Vertical Alignment

Vertical curves on bicycle lanes that follow street profiles will have adequate, if not generous, sight distance since streets are usually designed for speeds greater than those achieved by bicyclists. On the other hand, bikeways that don't parallel other facilities may be designed with much more abrupt vertical and horizontal alignments. Therefore, a bikeway should always be checked to ensure adequate sight distance.

Sight distance formulas for crest vertical curves (such as Eq. 4.86 and Eq. 4.87) are used with the eye height of the bicyclist set at 4.5 ft (1400 mm) and the height of the object set at 0. A is the absolute value of the algebraic difference in grade.

When the sight distance, S, is greater than the curve length, L, use Eq. 4.86 to calculate the minimum curve length.

$$L_m = 2S_m - \frac{900}{A_\%} \quad \text{[SI]} \quad 4.86(a)$$

$$L_{ft} = 2S_{ft} - \frac{280}{A_\%} \quad \text{[U.S.]} \quad 4.86(b)$$

When $S < L$, use Eq. 4.87.

$$L_m = \frac{A_\% S_m^2}{900} \quad \text{[SI]} \quad 4.87(a)$$

$$L_{ft} = \frac{A_\% S_{ft}^2}{280} \quad \text{[U.S.]} \quad 4.87(b)$$

Grades should not exceed 5% for long, unbroken distances. In hilly terrain, grades should be broken into 400–700 ft (122–213 m) segments of alternating steeper and flatter grades. Table 4.16 gives guidelines for determining maximum grades for extended segments of bikeways.

Table 4.16 Maximum Grade for Extended Bikeway Segments

length of segment (100 ft)	desirable maximum grade (%)	acceptable maximum grade (%)
0	7	10.8
1	6	9.5
2	5	8.3
3	4.3	7.4
4	3.5	6.6
5	3.2	6.2
6	2.9	5.7
7	2.8	5.4
8	2.7	5.2
9	2.6	5.0
10	2.5	4.8
12	2.4	4.6
14	2.2	4.3
16	2.0	4.0
18	1.8	3.8
20	1.6	3.5

(Multiply ft by 0.305 to obtain m.)

Example 4.14

A bicycle path has an upgrade of 2% meeting a downgrade of 3%. The design speed is 15 mph, and the friction factor is 0.25. What is the minimum vertical curve length at this location?

Solution

This is a crest vertical curve. The total grade change is

$$A = |G_2 - G_1| = |-3\% - 2\%|$$
$$= 5\%$$

Use the total decimal grade change value, -0.05, for G. From Eq. 4.85(b), the stopping distance is

$$S_{ft} = \frac{v_{mph}^2}{30(f \pm G_{dec})} + 3.67 v_{mph}$$
$$= \frac{\left(15 \, \frac{mi}{hr}\right)^2}{(30)(0.25 - 0.05)} + (3.67)\left(15 \, \frac{mi}{hr}\right)$$
$$= 92.6 \text{ ft}$$

Assume the stopping distance is greater than the length of the vertical curve. From Eq. 4.86(b),

$$L_{ft} = 2S_{ft} - \frac{280}{A_\%} = (2)(92.6 \text{ ft}) - \frac{280}{5\%}$$
$$= 129 \text{ ft}$$

Check for $S < L$ using Eq. 4.87(b).

$$L_{ft} = \frac{A_\% S_{ft}^2}{280} = \frac{(5\%)(92.6 \text{ ft})^2}{280}$$
$$= 153 \text{ ft}$$

Since 92.6 ft is less than 153 ft (i.e., $S < L$), the required length of the vertical curve is 153 ft.

18. MASS TRANSIT GEOMETRIC DESIGN

Because transit involves moving people, parameters of human comfort are the design criteria for acceleration, deceleration, and lateral forces on vehicles. Urban public transit usually has more restrictive design limits than automobiles to accommodate the comfort and safety of standing passengers. As a general rule, average accelerations should be limited to $\pm 0.1g$ in any direction, but design values should be selected carefully based on the type of service being provided.

Acceleration when starting and deceleration when stopping are also limited by passenger comfort. Starting acceleration can range from $0.05g$ to $0.1g$ for rapid transit, light rail, and buses, with the higher range of acceleration limited by horsepower and traction rates. Passengers have a somewhat higher tolerance for deceleration during braking situations. Common deceleration rates range from $0.1g$ to $0.3g$. The higher range is limited by braking power, traction (on rail vehicles), and passenger comfort. Overhead electrically powered vehicles usually have greater acceleration capability than diesel-only powered vehicles, while rubber-tired vehicles have greater deceleration capabilities than rail-mounted vehicles.

Buses operating on city streets and highways follow the existing road geometry, with operators trained to adjust for the needs of passenger comfort. Busways and bus rapid transit (BRT) roadways can be designed more like rail lines for improved comfort. Horizontal curves follow usual highway or railroad practice for superelevation and curvature. For highway curves, the design limits of comfort are covered in the AASHTO *GDHS*, and Eq. 4.66 can be used. For BRT, the side friction factor can be reduced from the values used on highways to reduce sidesway and improve the comfort of higher level seating found on buses.

The transition from an upgrade to a downgrade, and vice versa, is accomplished through a vertical curve. For train passengers on vertical curves, the vertical acceleration tolerated is less than that tolerated for highway design. AREMA has suggested Eq. 4.88 be used, where L is the length of the vertical curve, D is the absolute value of the algebraic difference in grades, expressed in decimals, v is the speed in miles per hour, and a is the vertical acceleration in feet per second squared. The recommended vertical acceleration, a, is 0.6 ft/sec^2 (0.18 m/s^2) for passenger lines, including transit.

$$L_{\text{ft}} = D\text{v}_{\text{mph}}^2 \left(\frac{2.15}{a_{\text{ft/sec}^2}} \right) \quad \text{[U.S. only]} \quad 4.88$$

Longer vertical curve lengths are often rounded to the nearest 100 ft (30 m) for ease of construction when there is negligible effect on the vertical acceleration.

Example 4.15

A bus travels at 40 mph on a curve that is superelevated 3%. The side friction factor is 0.05. What is the minimum radius for this curve?

Solution

From Eq. 4.66(b), the minimum curve radius is

$$R_{\text{ft}} = \frac{\text{v}_{\text{mph}}^2}{15(e + f_s)} = \frac{\left(40 \ \frac{\text{mi}}{\text{hr}}\right)^2}{(15)(0.03 + 0.05)}$$
$$= 1333 \text{ ft}$$

Example 4.16

A +2.0% grade meets a −1.5% grade on a transit line with a design speed of 70 mph. Using the recommended vertical acceleration of 0.6 ft/sec^2, what is the length of vertical curve required?

Solution

Using Eq. 4.88, the vertical curve length required is

$$L_{\text{ft}} = D\text{v}_{\text{mph}}^2 \left(\frac{2.15}{a_{\text{ft/sec}^2}} \right)$$
$$= |-0.015 - 0.02| \left(70 \ \frac{\text{mi}}{\text{hr}}\right)^2 \left(\frac{2.15}{0.6 \ \frac{\text{ft}}{\text{sec}^2}} \right)$$
$$= 615 \text{ ft}$$

The length of curve needed is 615 ft. However, the length can be rounded to 600 ft for ease of construction with negligible effect on the vertical acceleration.

Table 4.7 Design Controls for Crest and Sag Vertical Curves

	SI units				
	design based on stopping sight distance			design based on passing sight distance	
design speed (kph)	stopping sight distance (m)	crest vertical curve stopping sight (open road) design K-value	sag vertical curve stopping sight (open road) design K-value	passing sight distance (m)	crest vertical curve passing sight design K-value
20	20	1	3		
30	35	2	6		
40	50	4	9	140	23
50	65	7	13	160	30
60	85	11	18	180	38
70	105	17	23	210	51
80	130	26	30	245	69
90	160	39	38	280	91
100	185	52	45	320	119
110	220	74	55	355	146
120	250	95	63	395	181
130	285	124	73	440	224
	customary U.S. units				
	design based on stopping sight distance			design based on passing sight distance	
design speed (mph)	stopping sight distance (ft)	crest vertical curve stopping sight (open road) design K-value	sag vertical curve stopping sight (open road) design K-value	passing sight distance (ft)	crest vertical curve passing sight design K-value
15	80	3	10		
20	115	7	17	400	57
25	155	12	26	450	72
30	200	19	37	500	89
35	250	29	49	550	108
40	305	44	64	600	129
45	360	61	79	700	175
50	425	84	96	800	229
55	495	114	115	900	289
60	570	151	136	1000	357
65	645	193	157	1100	432
70	730	247	181	1200	514
75	820	312	206	1300	604
80	910	384	231	1400	700

From *A Policy on Geometric Design of Highways and Streets*, 2011, by the American Association of State Highway and Transportation Officials, Washington, D.C. Table 3-34, Table 3-35, and Table 3-36. Used by permission.

Table 4.8 Minimum Lengths for Vertical Curves Based on Comfort and Stopping Sight Distance

	SI units			
	crest vertical curves		sag vertical curves	
design speed (kph)	minimum length for stopping sight distance (open road) (m)	algebraic difference in grades A (%) greater than	minimum length for stopping sight distance (m)	algebraic difference in grades A (%) greater than
20	10	16	12	6.0
30	20	10	18	3.6
40	25	6.5	24	2.9
50	30	4.5	30	2.6
60	46	3.2	36	2.0
70	42	2.5	42	1.9
80	48	1.8	48	1.6
90	54	1.4	54	1.5
100	60	1.2	60	1.4
110	66	0.8	66	1.3
120	74	0.7	72	1.2
130	80	0.6	80	1.0

	customary U.S. units			
	crest vertical curves		sag vertical curves	
design speed (mph)	minimum length (ft)	algebraic difference in grades A (%) greater than	minimum length (ft)	algebraic difference in grades A (%) greater than
15	40	16	40	4.6
20	60	10	60	3.6
25	80	6.5	80	3.0
30	90	4.7	95	2.2
35	110	3.5	105	2.6
40	120	2.8	120	2.1
45	135	2.2	135	1.7
50	150	1.8	150	1.6
55	170	1.5	165	1.5
60	180	1.2	180	1.4
65	195	1.0	195	1.3
70	205	0.8	210	1.2
75	215	0.6	220	1.1
80	230	0.5	230	1.0

Interpreted from *A Policy on Geometric Design of Highways and Streets*, 2011, by the American Association of State Highway and Transportation Officials, Washington, D.C. Table 3-43 and Table 3-44. Used by permission.

19. PRACTICE PROBLEMS

1. An existing 1.5° horizontal curve will be shortened by intersecting a new back tangent at a bearing 15° north from the existing back tangent, which is at a bearing of N 80° E. The PC will relocate to PC′, and the PI will relocate to PI′. The PT will remain in its existing location at N 10,000 ft and E 50,000 ft, and the curve from the new PC location to the PT will also remain in place. The PT is at sta 38+27.00, and the ahead tangent is at bearing N 55° E.

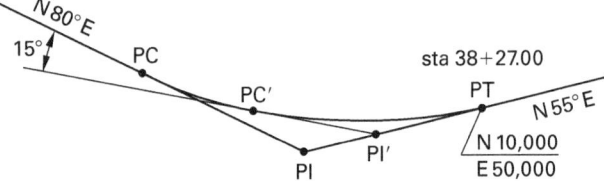

(a) The new curve length is most nearly

  (A) 15 ft

  (B) 670 ft

  (C) 1700 ft

  (D) 3800 ft

(b) The radius of the curve is most nearly

  (A) 330 ft

  (B) 670 ft

  (C) 3800 ft

  (D) 8600 ft

(c) The north coordinate of the shifted PI′ is most nearly

  (A) 330 ft

  (B) 9400 ft

  (C) 9700 ft

  (D) 9800 ft

(d) The station of the PI′ is most nearly

  (A) sta 28+26

  (B) sta 31+60

  (C) sta 34+93

  (D) sta 34+95

2. A 2100 ft radius curve is being fitted to a PI located at sta 234+34.10, with the back tangent bearing at N 80° E and the ahead tangent at S 65° E. Offset to the right of, and parallel to, the back tangent is a historic structure that has a 150 ft clearance buffer. The PI is located 100 ft back from the offset point of the corner of the structure. The proposed road construction consists of a six-lane roadway with 12 ft lanes, 10 ft shoulders, and a 16 ft wide median. The edge of the paved shoulder cannot encroach on the buffer zone.

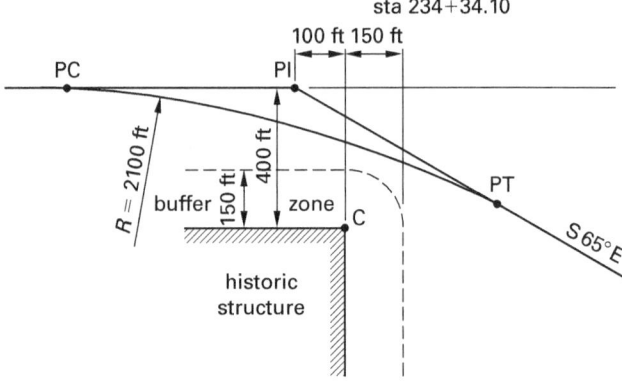

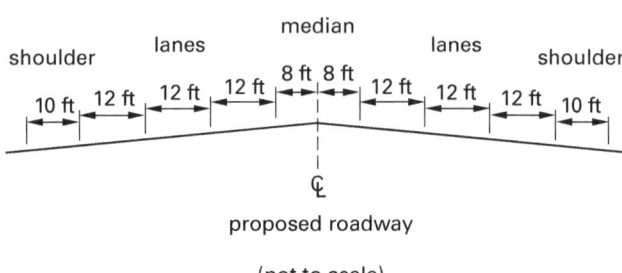

(not to scale)

(a) The actual clearance to the corner of the historical building is most nearly

  (A) 110 ft

  (B) 180 ft

  (C) 200 ft

  (D) 240 ft

(b) The station of the closest point to the building corner is most nearly

  (A) sta 233+05

  (B) sta 236+44

  (C) sta 236+57

  (D) sta 239+95

(c) The offset distance from the back tangent to the closest point to the building on the curve centerline is most nearly

(A) 140 ft
(B) 160 ft
(C) 180 ft
(D) 260 ft

3. An existing 250 ft long crest vertical curve will be connected to new roadway construction, replacing the existing part of the curve from the PVI to the PVT. The new ahead curve tangent will be 350 ft long on a -1.5% grade, replacing the existing 125 ft tangent on a -3.0% grade. The existing pavement will remain from the PVC to the PVI location, which is at elevation 105.25 ft. An existing utility pipe crosses the centerline at 70 ft ahead from the PVI location. The pipe has a cover of 2.0 ft to the existing grade line.

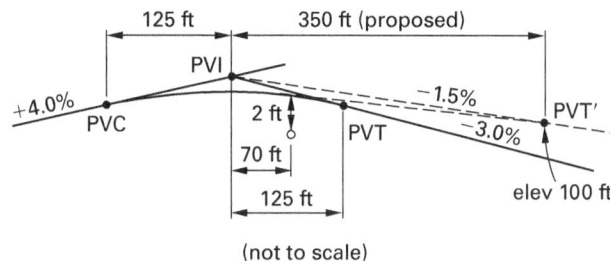

(not to scale)

(a) The middle ordinate of the existing curve is most nearly

(A) 2.2 ft
(B) 2.5 ft
(C) 2.9 ft
(D) 3.7 ft

(b) The elevation of the top of the pipe for the existing curve is most nearly

(A) 99.9 ft
(B) 100.7 ft
(C) 102.7 ft
(D) 103.2 ft

(c) The new grade line will provide a new pipe cover of most nearly

(A) 2.1 ft
(B) 5.1 ft
(C) 7.1 ft
(D) 13 ft

(d) The design speed for the new curve based on sight distance is most nearly

(A) 40 mph
(B) 50 mph
(C) 55 mph
(D) 60 mph

4. A horizontal curve with full spiral transitions has a center curve of 3° and is fitted with 150 ft spirals. The curve deflection is 31° to the right. Superelevation is to be transitioned uniformly along the spiral, with a maximum superelevation rate of 8%. The normal straight cross slope for this roadway is 2%.

(a) The curve tangent of this curve is most nearly

(A) 150 ft
(B) 530 ft
(C) 610 ft
(D) 760 ft

(b) The instant radius at the midpoint of the spiral is most nearly

(A) 960 ft
(B) 1900 ft
(C) 2500 ft
(D) 3800 ft

(c) The superelevation at a point 100 ft from the TS or ST is most nearly

(A) 0.03 ft/ft
(B) 0.06 ft/ft
(C) 0.07 ft/ft
(D) 0.08 ft/ft

5. An exit ramp from an urban freeway terminates on a 6% downgrade. The ramp has a 300 ft radius with a 75° curve to the right and a superelevation of 8.0%. There is a stop-controlled cross street intersection 200 ft downstream of the PT of the ramp curve. The intersection is not visible from the ramp until a vehicle is about two-thirds around the curve. The tangent is 250 ft long between the gore nose of the ramp (i.e., the intersection of the freeway and ramp pavements) and the PC of the curve. The ramp is also on a 200 ft vertical curve, approximately aligned with the horizontal curve. The freeway and ramp grade is +0.25% leading into the vertical curve. The ramp is elevated on a structure with 42 in high parapets on each side and a width of 18 ft between parapets. The ramp exit lane is 1500 ft long and is an extension of an upstream entrance ramp. Design speed and prevailing traffic speed on the adjoining freeway through lanes is 50 mph. Several serious collisions have occurred, including vehicles skidding into the parapet on the curve due to maintaining a speed of 50 mph and vehicles on the exit ramp broadsiding cross traffic at the stop-controlled intersection. The exit ramp has a posted speed limit of 30 mph using *MUTCD* standard exit sign positions. The ramp is being investigated for possible design deficiencies.

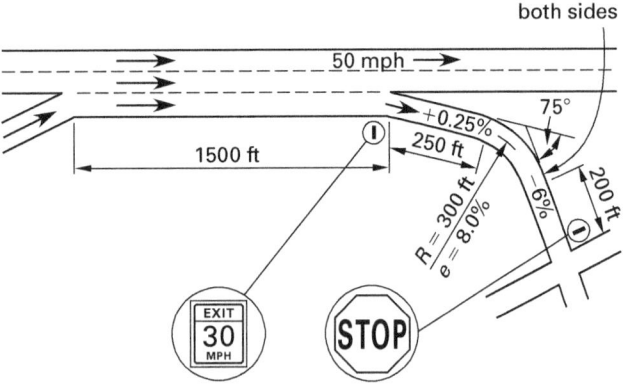

(a) The exit ramp design speed is most nearly

(A) 25 mph

(B) 30 mph

(C) 35 mph

(D) 40 mph

(b) If drivers cannot see over the ramp parapets, what is most nearly the stopping sight distance to the intersection?

(A) 70 ft

(B) 130 ft

(C) 200 ft

(D) 330 ft

(c) Using actual vehicle speed on the ramp, what is most nearly the decision sight distance for the ramp curve?

(A) 620 ft

(B) 720 ft

(C) 890 ft

(D) 1000 ft

(d) For vehicles on the exit ramp, what is most nearly the decision sight distance to the stop-controlled intersection?

(A) 220 ft

(B) 490 ft

(C) 620 ft

(D) 910 ft

(e) Based on the decision sight distance, what speed should be posted on the exit ramp?

(A) 15 mph

(B) 25 mph

(C) 30 mph

(D) 35 mph

6. An urban freeway with two lanes in each direction has two closely spaced exit ramps to local streets. The freeway also splits into separate single-lane roadways that connect to other highways. The exit ramps and roadway splits occur within a 1200 ft segment. The exit ramps are level and terminate in stop-controlled intersections, while the lane connections to other highways are free-flowing except for traffic backups. Traffic on the exit ramps backs up onto the main roadway at various times throughout the day. The backups do not necessarily occur simultaneously on both ramps, forcing approaching drivers to choose between ramp options. The approach speed on the freeway is 50 mph, the connections to other highways operate at 40 mph, and the ramps have a posted speed limit of 25 mph.

(a) What is most nearly the minimum stopping sight distance for the exit ramps to local streets?

(A) 120 ft

(B) 160 ft

(C) 200 ft

(D) 250 ft

(b) What is most nearly the decision sight distance required for the exit ramps?

(A) 430 ft

(B) 490 ft

(C) 930 ft

(D) 1000 ft

(c) What is most nearly the stopping sight distance for the ramps connecting onto other highways?

(A) 310 ft

(B) 430 ft

(C) 590 ft

(D) 690 ft

(d) What is most nearly the decision sight distance required for the lane connections to other highways?

(A) 310 ft

(B) 620 ft

(C) 830 ft

(D) 1000 ft

(e) What is most nearly the decision sight distance for the main freeway approach to the ramps and lane splits?

(A) 200 ft

(B) 620 ft

(C) 830 ft

(D) 1000 ft

SOLUTIONS

1. (a) Find the deflection angle of the curve, which is the same as the intersection angle of the curve tangents, I. First, the bearing of the new back tangent must be found. The existing back tangent lies in the NE quadrant. Therefore, shifting the back tangent 15° north results in a counterclockwise shift.

$$\text{N}\,80°\,\text{E} - 15° = \text{N}\,65°\,\text{E}$$

The deflection angle of the new curve is found by subtracting the back tangent from the ahead tangent.

$$\text{N}\,55°\,\text{E} - \text{N}\,65°\,\text{E} = -10°$$

A negative difference in the NE quadrant indicates that the deflection is to the left, or counterclockwise.

The length of the curve arc is determined using Eq. 4.9.

$$L_{\text{ft}} = \frac{(100\text{ ft})I}{D_{\text{deg}}} = \frac{(100\text{ ft})(10°)}{1.5°}$$
$$= 666.67\text{ ft} \quad (670\text{ ft})$$

The answer is (B).

(b) The radius of the curve can be found using Eq. 4.16.

$$R_{\text{ft}} = \frac{(100\text{ ft})\left(\dfrac{180°}{\pi}\right)}{D_{\text{deg}}}$$
$$= \frac{(100\text{ ft})\left(\dfrac{180°}{\pi}\right)}{1.5°}$$
$$= 3819.72\text{ ft} \quad (3800\text{ ft})$$

The answer is (C).

(c) The length of the curve tangent is found using Eq. 4.5.

$$T = R\tan\frac{I}{2} = (3819.72\text{ ft})\tan\frac{10°}{2}$$
$$= 334.18\text{ ft}$$

The difference in the north coordinate is found by multiplying the tangent distance by the cosine of the bearing of the heading. The bearing of the heading from the PT to the shifted PI is S 55° W.

$$\text{PI}'\text{ N} = \text{PT N} - T\cos(\text{heading})$$
$$= 10{,}000.00\text{ ft} - (334.18\text{ ft})\cos 55°$$
$$= 9808.32\text{ ft} \quad (9800\text{ ft})$$

The answer is (D).

(d) The station of the PI' is found by subtracting the length of the curve from the PT station to get the PC' station, then adding the tangent length to the PI' station.

$$\text{sta PT} - L + T = \text{sta } 38+27.00 - 666.67 \text{ ft} + 334.18 \text{ ft}$$
$$= \text{sta } 34+94.51 \quad (\text{sta } 34+95)$$

The answer is (D).

2. (a) It is necessary to determine the tangent of the curve, which is found from the intersection angle and the radius. The intersection angle is found by subtracting the bearing angles.

$$I = (180° - 80°) - 65° = 35°$$

The curve tangent length can be determined using Eq. 4.5.

$$T = R \tan \frac{I}{2} = (2100 \text{ ft}) \tan \frac{35°}{2} = 662.13 \text{ ft}$$

A triangle is constructed using known distances. Side $\overline{HC}$ is along the historical structure face and is parallel to the back tangent of the curve.

$$\overline{GA} = T + \overline{PI\ A} = 662.13 \text{ ft} + 100 \text{ ft}$$
$$= 762.13 \text{ ft}$$
$$\overline{OH} = R - \overline{GH} = 2100 \text{ ft} - 400 \text{ ft}$$
$$= 1700 \text{ ft}$$
$$\theta = \arctan \frac{\overline{GA}}{\overline{OH}} = \arctan \frac{762.13 \text{ ft}}{1700 \text{ ft}}$$
$$= 24.15°$$
$$\overline{OC} = \sqrt{\overline{GA}^2 + \overline{OH}^2} = \sqrt{(762.13 \text{ ft})^2 + (1700 \text{ ft})^2}$$
$$= 1863.02 \text{ ft}$$
$$\overline{CB} = R - \overline{OC} = 2100 \text{ ft} - 1863.02 \text{ ft}$$
$$= 236.98 \text{ ft}$$

The clearance length, L, is $\overline{CB}$, 236.98 ft, minus the width of the roadway, W_r, measured from the centerline to the edge of the paved shoulder. As shown in the roadway profile illustration, the median width is the distance from the centerline to the edge of the median, 8 ft.

$$W_r = 8 \text{ ft} + (3)(12 \text{ ft}) + 10 \text{ ft}$$
$$= 54 \text{ ft}$$

$$L = \overline{CB} - W_r = 236.98 \text{ ft} - 54 \text{ ft}$$
$$= 182.98 \text{ ft} \quad (180 \text{ ft})$$

The answer is (B).

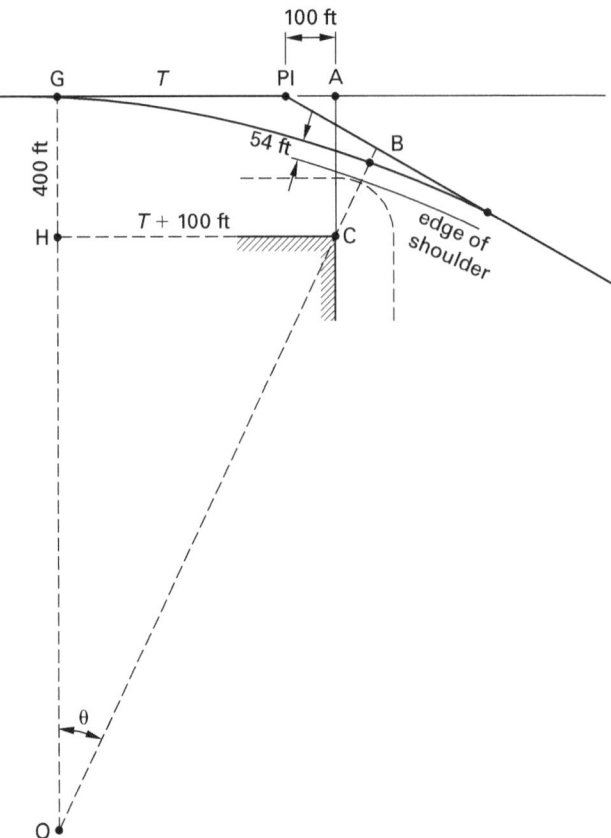

(b) The arc distance from the PC to the radial clearance point, B, is

$$\overline{PC\ B} = \frac{\theta R}{\dfrac{180°}{\pi}} = \frac{(24.15°)(2100 \text{ ft})}{\dfrac{180°}{\pi}}$$
$$= 885.14 \text{ ft}$$

The station of the closest radial point is

$$\text{sta PI} - T + \overline{PC\ B} = \text{sta } 234+34.10$$
$$- 662.13 \text{ ft} + 885.14 \text{ ft}$$
$$= \text{sta } 236+57.11 \quad (\text{sta } 236+57)$$

The answer is (C).

(c) The offset from the tangent to the radial point on the curve is shown as $\overline{EB}$ in the following illustration. The distance $\overline{AE}$ is the distance beyond the offset point to the corner of the historical structure.

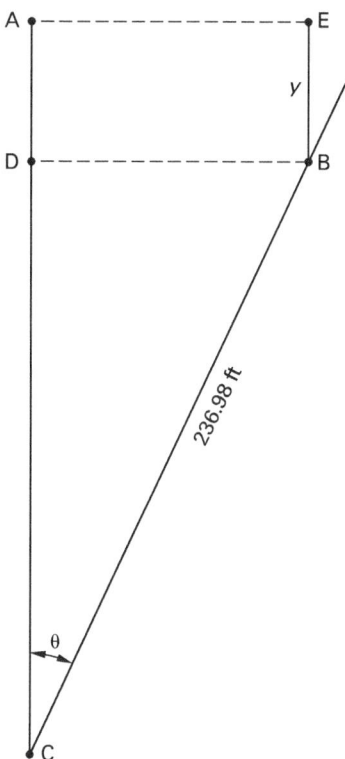

$$\overline{AE} = \overline{CB}\sin\theta = (236.98\text{ ft})\sin 24.15° = 96.95\text{ ft}$$
$$\overline{PI\ A} + \overline{AE} = 100\text{ ft} + 96.95\text{ ft} = 196.95\text{ ft}$$

The distance from the PC to point E is

$$x = T + (\overline{PI\ A} + \overline{AE}) = 662.13\text{ ft} + 196.95\text{ ft}$$
$$= 859.08\text{ ft}$$

The offset, y, from a tangent can be found using Eq. 4.32.

$$y = R - \sqrt{R^2 - x^2}$$
$$= 2100\text{ ft} - \sqrt{(2100\text{ ft})^2 - (859.08\text{ ft})^2}$$
$$= 183.76\text{ ft} \quad (180\text{ ft})$$

The answer is (C).

3. (a) The middle ordinate is determined using Eq. 4.47.

$$M_{ft} = \frac{|G_{2,\%} - G_{1,\%}|\, L_{sta}}{8\,\dfrac{sta}{ft}} = \frac{|-3.0\% - 4.0\%|\,(250\text{ ft})}{\left(8\,\dfrac{sta}{ft}\right)\left(100\,\dfrac{ft}{sta}\right)}$$
$$= 2.188\text{ ft}\quad(2.2\text{ ft})$$

The answer is (A).

(b) The elevation of the top of the pipe can be found after determining the elevation of the PVC.

$$\text{elev}_{PVC} = \text{elev}_{PVI} - G_{1,ft/ft}T_{1,ft}$$
$$= 105.25\text{ ft} - \left(0.04\,\dfrac{ft}{ft}\right)(125\text{ ft})$$
$$= 100.25\text{ ft}$$

Use Eq. 4.48 to find the rate of change in grade.

$$R = \frac{G_{2,\%} - G_{1,\%}}{L} = \frac{-3.0\% - 4.0\%}{250\text{ ft}}$$
$$= -0.028\ \%/\text{ft}$$

Use Eq. 4.50 to find the elevation of the curve above the pipe. The pipe is located $125\text{ ft} + 70\text{ ft} = 195\text{ ft}$ from the PVC.

$$\text{elev}_x = \frac{R_{\%/100\text{ ft}}x_{ft}^2}{2} + G_{1,ft/ft}x_{ft} + \text{elev}_{PVC}$$
$$= \frac{\left(\dfrac{-0.028\%}{100\text{ ft}}\right)(195\text{ ft})^2}{2} + \left(0.04\,\dfrac{ft}{ft}\right)(195\text{ ft})$$
$$+ 100.25\text{ ft}$$
$$= 102.73\text{ ft}$$

The top of the pipe elevation is

$$102.73\text{ ft} - 2.0\text{ ft} = 100.73\text{ ft}\quad(100.7\text{ ft})$$

The answer is (B).

(c) The elevation of the PVI and the grade line at the PVI are remaining the same. Therefore, the existing middle ordinate must be used to determine the grade points on the new half of the curve. Determine the elevation of the new tangent above the pipe location.

$$\text{elev}_{PVT'} = \text{elev}_{PVI} + G_2(70\text{ ft})$$
$$= 105.25\text{ ft} + (-0.015)(70\text{ ft})$$
$$= 104.20\text{ ft}$$

To determine the elevation on the new grade line, subtract the proportional square of the distance ratio multiplied by the middle ordinate from the elevation on the tangent. Subtract the elevation of the top of the pipe from the elevation on the new grade line to determine the new pipe cover.

$$\text{elev}_{x',\text{new grade}} - \text{elev}_{\text{top of pipe}}$$
$$= \text{elev}_{\text{PVT}'} - (\text{distance ratio})^2 M - \text{elev}_{\text{top of pipe}}$$
$$= 104.20 \text{ ft} - \left(\frac{350 \text{ ft} - 70 \text{ ft}}{350 \text{ ft}}\right)^2 (2.188 \text{ ft})$$
$$ - 100.73 \text{ ft}$$
$$= 2.07 \text{ ft} \quad (2.1 \text{ ft})$$

The answer is (A).

(d) To find the value of K, use Eq. 4.57.

$$K = \frac{L}{|G_2 - G_1|} = \frac{125 \text{ ft} + 350 \text{ ft}}{|-1.5\% - 4.0\%|}$$
$$= 86.36$$

Referring to Table 4.7, a K-value between 84 and 114 for a crest vertical curve indicates a 50 mph design speed.

The answer is (B).

4. (a) The components of the curve tangent equation are found first. As a matter of practice, calculations involving components of a spiral transition must carry an additional significant digit to the right of the decimal. Usually, carrying out to three decimals in the customary U.S. system and four decimals in the SI system is sufficient to control location tolerance of the final measurement for layout purposes.

Use Eq. 4.44 to find the coordinate along the tangent.

$$x_{s,\text{ft}} = \frac{L_{s,\text{ft}}}{100 \text{ ft}} \left(100 \text{ ft} - \frac{(100 \text{ ft})\left(\frac{\pi}{180°}\right)^2 D_{c,\text{deg}}^2}{(5)(2!)} + \frac{(100 \text{ ft})\left(\frac{\pi}{180°}\right)^4 D_{c,\text{deg}}^4}{(9)(4!)} + \frac{(100 \text{ ft})\left(\frac{\pi}{180°}\right)^6 D_{c,\text{deg}}^6}{(13)(6!)} \right)$$

$$= \frac{150 \text{ ft}}{100 \text{ ft}} \left(100 \text{ ft} - \frac{(100 \text{ ft})\left(\frac{\pi}{180°}\right)^2 (3°)^2}{(5)(2!)} + \frac{(100 \text{ ft})\left(\frac{\pi}{180°}\right)^4 (3°)^4}{(9)(4!)} + \frac{(100 \text{ ft})\left(\frac{\pi}{180°}\right)^6 (3°)^6}{(13)(6!)} \right)$$

$$= 149.95 \text{ ft}$$

Calculate the central angle of the total spiral arc using Eq. 4.41.

$$\theta_s = \left(\frac{L_{s,\text{ft}}}{200 \text{ ft}}\right) D_{c,\text{deg}} = \left(\frac{150 \text{ ft}}{200 \text{ ft}}\right)(3°)$$
$$= 2.25°$$

Calculate the radius of the central curve using Eq. 4.16.

$$R_{c,\text{ft}} = \frac{(100 \text{ ft})\left(\frac{180°}{\pi}\right)}{D_{c,\text{deg}}} = \frac{(100 \text{ ft})\left(\frac{180°}{\pi}\right)}{3°}$$
$$= 1909.86 \text{ ft}$$

Find the extended distance along the tangent of the shifted curve using Eq. 4.38.

$$k = x_s - R_c \sin\theta_s = 149.96 \text{ ft} - (1909.86 \text{ ft})(\sin 2.25°)$$
$$= 74.97 \text{ ft}$$

Calculate the offset distance of the full spiral length using Eq. 4.45.

$$y_{s,\text{ft}} = \frac{L_{\text{ft}}^3}{6 R_{c,\text{ft}} L_{s,\text{ft}}} = \frac{(150 \text{ ft})^3}{(6)(1909.86 \text{ ft})(150 \text{ ft})}$$
$$= 1.963 \text{ ft}$$

To find the distance of the shifted center curve, use Eq. 4.37.

$$p = y_s - R_c \text{ vers}\,\theta_s = 1.963 \text{ ft} - (1909.86 \text{ ft})(\text{vers}\,2.25°)$$
$$= 0.491 \text{ ft}$$

The spiral curve tangent can now be calculated using Eq. 4.36.

$$T_s = (R_c + p)\tan\frac{I}{2} + k$$
$$= (1909.86 \text{ ft} + 0.491 \text{ ft})\tan\frac{31°}{2} + 74.97 \text{ ft}$$
$$= 604.76 \text{ ft} \quad (610 \text{ ft})$$

The answer is (C).

(b) The degree of the central curve can be calculated from Eq. 4.33. The length of the spiral, L_s, is 150 ft, so the length of the curve, L, from the TS or ST to the midpoint of the curve is 150 ft/2 = 75 ft.

$$D_l = D_c\left(\frac{L}{L_s}\right) = (3°)\left(\frac{75 \text{ ft}}{150 \text{ ft}}\right) = 1.5°$$

Use Eq. 4.16 to calculate the curve radius.

$$R_c = \frac{(100 \text{ ft})\left(\frac{180°}{\pi}\right)}{D_{l,\text{deg}}} = \frac{(100 \text{ ft})\left(\frac{180°}{\pi}\right)}{1.5°}$$
$$= 3819.72 \text{ ft} \quad (3800 \text{ ft})$$

The answer is (D).

(c) Superelevation is transitioned uniformly along the spiral from the normal cross slope of 2% to the maximum 8%. At a point 100 ft from the TS or ST, the transition will be

$$\Delta e = \frac{L_p}{L_s}(e_{\max} - e_{\text{NC}}) + e_{\text{NC}}$$
$$= \left(\frac{100 \text{ ft}}{150 \text{ ft}}\right)\left(0.08 \frac{\text{ft}}{\text{ft}} - 0.02 \frac{\text{ft}}{\text{ft}}\right) + 0.02 \frac{\text{ft}}{\text{ft}}$$
$$= 0.06 \text{ ft/ft}$$

The answer is (B).

5. (a) The design speed is limited by the ramp geometry. The most restrictive element of the geometry is the curve radius. From Table 4.12, the minimum curve radius for a design speed of 30 mph and a superelevation of 8.0% is 214 ft, which is less than the curve radius of 300 ft. Therefore, 30 mph is not the maximum design speed. Referring again to Table 4.12, the next possible design speed is 35 mph, which has a minimum curve radius of 314 ft, so the design speed is between 30 mph and 35 mph.

However, the maximum speed for the curve is controlled by the superelevation and the side friction factor. From Table 4.10, the side friction factor is 0.20 for a design speed of 30 mph and 0.18 for a design speed of 35 mph.

Determine the design speed using Eq. 4.66(b) for a friction factor of 0.19, the average of the two possible side friction factors.

$$v_{\text{mph}} = \sqrt{R_{\text{ft}}(15)(e + f_s)}$$
$$= \sqrt{(300 \text{ ft})(15)(0.08 + 0.19)}$$
$$= 34.9 \text{ mph} \quad (35 \text{ mph})$$

The design speed is most nearly 35 mph.

The answer is (C).

(b) Find the curve length using Eq. 4.9.

$$L_{\text{ft}} = \frac{2\pi R_{\text{ft}} I_{\text{deg}}}{360°} = \frac{(2\pi)(300 \text{ ft})(75°)}{360°} = 392.7 \text{ ft}$$

The stopping sight distance is limited by the high parapets to a point two-thirds around the 300 ft radius curve. Therefore, the stopping sight distance is found by adding one-third of the curve length to the intersection approach length, 200 ft.

$$S = \left(\frac{1}{3}\right)(392.7 \text{ ft}) + 200 \text{ ft} = 330.9 \text{ ft} \quad (330 \text{ ft})$$

The answer is (D).

(c) From the footnote to Table 4.4, the avoidance maneuver for a direction change on an urban freeway is avoidance maneuver E. Using Table 4.4, for avoidance maneuver E and a 50 mph approach speed, the decision sight distance is 1030 ft (1000 ft).

The answer is (D).

(d) From Table 4.4, the stop-controlled intersection is avoidance maneuver B. The decision sight distance for the stop-controlled intersection uses the current posted speed limit of 30 mph. For avoidance maneuver B and a 30 mph design speed, the decision sight distance is 490 ft.

The answer is (B).

(e) The decision sight distance provides additional time for complex situations where drivers must recognize an obstacle, choose a course of action, and then maneuver their vehicles accordingly. However, for a ramp leading to a stop-controlled intersection, drivers only need to recognize that a stop is required. Therefore, the stopping sight distance can be used as the decision sight distance, 330.9 ft.

The footnote in Table 4.4 gives the pre-maneuver time for avoidance maneuver B as 9.1 sec. The maximum speed can be found by dividing the stopping sight distance by the pre-maneuver time.

$$v = \frac{S}{t_p} = \frac{(330.9 \text{ ft})\left(3600 \ \frac{\text{sec}}{\text{hr}}\right)}{(9.1 \text{ sec})\left(5280 \ \frac{\text{ft}}{\text{mi}}\right)}$$
$$= 24.8 \text{ mph} \quad (25 \text{ mph})$$

A speed of 25 mph should be posted on the ramp.

The answer is (B).

6. (a) The ramp is posted for 25 mph, which can be presumed to be the design speed. Based on Table 4.3, the minimum stopping sight distance for a design speed of 25 mph is 155 ft (160 ft).

The answer is (B).

(b) Based on Table 4.4, the decision sight distance would fall under avoidance maneuver B. A stop on an urban road uses a pre-maneuver time of 9.1 sec. Values for a design speed of 25 mph are not given. However, approaching traffic would be reducing speed from 50 mph when approaching the ramp and may not reach 25 mph when the decision point is reached. Therefore, using the minimum decision sight distance for a design speed of 30 mph, 490 ft, is acceptable.

The answer is (B).

(c) From Table 4.3, the design stopping sight distance for a design speed of 40 mph is 305 ft (310 ft).

The answer is (A).

(d) The decision sight distance for lane changes on ramps is avoidance maneuver E (speed or path changes on urban roads). From Table 4.4, avoidance maneuver E requires 825 ft (830 ft) of decision sight distance for a 40 mph design speed.

The answer is (C).

(e) From Table 4.4, the decision sight distance for vehicles approaching a main freeway is avoidance maneuver E (speed or path changes on urban roads), which requires 1030 ft (1000 ft) at a design speed of 50 mph.

The answer is (D).

5 Transportation Construction

1. Introduction 5-1
2. Excavation and Embankment 5-2
3. Mass Diagrams 5-7
4. Pavement Design 5-9
5. Hot Mix Asphalt 5-11
6. Marshall Mix Design 5-18
7. Hveem Mix Design 5-21
8. Superpave Mix Design Procedures 5-23
9. Value Engineering 5-27
10. Practice Problems 5-28

Nomenclature

AC	asphalt content	%	%
C	cohesiometer value	–	–
C	cost	$	$
D	deflection	in	mm
D	duration	various	various
DP	dust-to-binder ratio	–	–
EF	earliest finish	–	–
ES	earliest start	–	–
G	specific gravity	–	–
H	height	in	n.a.
L	mass	lbm	g
L_n	haul length	sta	sta
LF	latest finish	–	–
LS	latest start	–	–
m	mass	lbm	kg
N	number	–	–
p	pressure	lbf/in^2	MPa
P	percentage	%	%
Q	quantity	–	–
s	change in soil volume (shrinkage or swell)	–	–
s	temperature standard deviation	°F	°C
S	stability	–	–
T	pavement design temperature	°F	°C
T_{air}	seven-day average air temperature	°F	°C
$T_{\min}$	minimum pavement design temperature	°F	°C
V	volume	ft^3	m^3
V	volumetric ratio	–	–
VFA	voids filled with asphalt	%	%
VMA	voids in mineral aggregate	%	%
VTM	volume of air voids	%	%
w/c	water to cement ratio	–	–
W	weight	lbf	N
W	width	in	n.a.

Symbols

γ	specific weight	lbf/ft^3	n.a.
ρ	density	lbm/ft^3	kg/m^3

Subscripts

a	air voids
ave	average
b	asphalt, base, borrow, or bulk
ba	absorbed asphalt
be	effective asphalt
c	crew, cut, or concrete
ca	coarse aggregate
cap	capacity
e	effective, embankment, or excavation
f	fill
fa	fine aggregate
h	haul or horizontal
max	maximum
mb	bulk
mm	maximum (zero air voids)
p	in place
s	aggregate, specific, or supervision
sa	apparent
sb	bulk
se	effective
t	total
v	vertical
w	waste

1. INTRODUCTION

Construction of transportation facilities involves earthwork activities, drainage installation, pavement placement, erection of structures, and engineering economic analysis. Construction practice demands knowledgeable methods of construction operations and project management techniques. Successful design of rehabilitation and renewal of existing facilities often includes construction phasing to allow maintenance of traffic through the construction area in a safe manner with minimum delay.

In addition, efforts to minimize the environmental footprint of soil erosion and practices to mitigate detrimental air quality are required in the construction of new facilities, as well as the rehabilitation of existing facilities.

2. EXCAVATION AND EMBANKMENT

The alignment of a highway or railroad is established to provide connections between two or more points. When laying out highways or railroad alignments, designers attempt to follow existing topography as much as possible to minimize the construction costs and produce visually pleasing finished projects. Grade and design of horizontal and vertical curves are based primarily on the design speed, but also include adjustments for connecting roadways, drainage, right-of-way availability, and considerations for appearance, ride quality, and environmental impact.

In order to accommodate design features, the existing topography must almost always be modified. Limits of available right-of-way and clearances from topographic features, such as rivers, cliffs, and scenic views, may move the alignment to a less than desirable location, and because the natural surface of the ground is not sufficiently uniform to simply be paved, the ground surface must be altered to fit the location and shape of the new roadway. In these situations, ground that is too high must be cut down, and ground that is too low must be filled in. This shaping of the ground surface is known as *earthwork*. Excavation of existing material is known as a *cut*, and filling in a low point with an embankment is known as a *fill*.

Earthwork is often the largest and most time-consuming work element of a transportation project, and carefully fine-tuning the alignment during design to balance earthwork quantities can have a large impact on the overall cost of a job. Hauling and mass diagrams are used to determine the most economical method of performing necessary cut and fill.

Cross Sections

A *roadway template* is developed for each section of the roadway alignment. The template shows the width and slope of the roadway and the shoulder, curb and gutter, sidewalk, hinge points, slopes, ditches, the shape of the median for divided roadways, and other features depicting the roadway cross-section. Cross sections are plotted at a right angle to the centerline or base line of a roadway. On curves, the cross sections are plotted on a radial line from the arc center of the curve. Cross sections are usually drawn at intervals of 100 ft. First, the ground line is plotted, which follows the contours of mapping. Then the roadway template is drawn on each cross section using the profile grade line as a control point. Cut and fill lines are extended from the ends of the template to the ground line. The area between the cut and fill line and the ground line is calculated. A solid model is created using the cut and fill area plots and the distance between cross sections. This solid model determines the necessary earthwork quantities. Cross sections are plotted using an exaggerated vertical scale, such as 5V to 1H, in order to better visualize the roadway template features. Figure 5.1 shows a typical highway cross section with the grade line near the ground line. This cross-section depicts both cut and fill areas, which is typical for a side hill section.

Figure 5.1 Typical Two-Lane Roadway Cross Section, Side Hill Cut

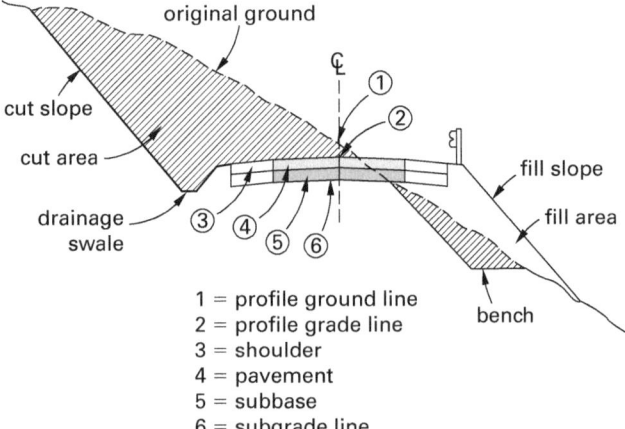

1 = profile ground line
2 = profile grade line
3 = shoulder
4 = pavement
5 = subbase
6 = subgrade line

Cut and fill slopes are determined by testing soil samples from core borings drilled along the alignment. For instance, cuts in hard rock may allow a steeper slope than cuts in soft rock or soil. Fills made from granular material with little clay can be made steeper than fills made from soil that has a large amount of clay or silt. The soil is tested in moist conditions, such as would occur in the field at the project site. The steepest slope that will be stable for the life of the project is established, and this slope is used on the cross sections to determine the minimum amount of cut or fill that must be performed.

Figure 5.2(a) shows a cross section that is entirely in cut. The cut slopes are set at 1.5:1 horizontal:vertical for the cut slope. Figure 5.2(b) is a fill section with side slopes of 2:1 horizontal:vertical. The point where the side slope intersects the original ground is the *limit of earthwork*.

The area of cut and fill at each cross section is measured using closed polygons that are bounded by the cut and fill lines and the original ground line. The cut and fill quantities are determined individually for each cross section by the *average end area method*, or *prismoidal method*, and, after adjusting for shrinkage, plotted on a mass-haul diagram to show cumulative cut and fill relationships to distance along a pavement.

Roadway Cross Sections

There are five major elements that make up a *roadway cross section*, as shown in Fig. 5.3. These elements include the subbase, subgrade, surface pavement, shoulders and curbs, and drainage systems.

Figure 5.2 Cut and Fill Sections

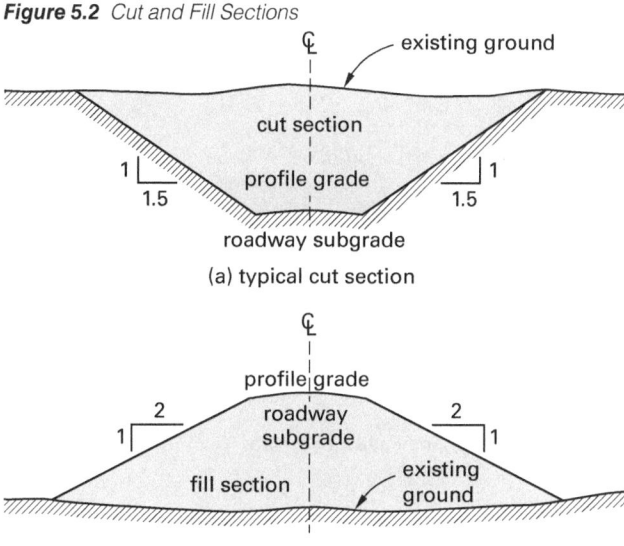

Figure 5.3 Elements of a Typical Roadway Cross Section

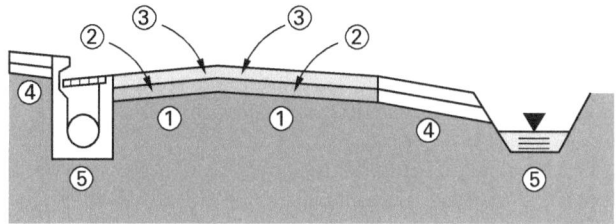

1 = subgrade
2 = subbase
3 = surface pavement
4 = shoulder and curb
5 = drainage system

The *subgrade* is the soil layer that is directly below the granular subbase. It is carefully graded and compacted, and of a drainable soil type that provides the foundation for the road. Subgrade design involves soil classification and/or the modification of soil to provide minimum bearing capabilities, stability requirements, and resistance control.

The *subbase*, or aggregate base, is composed of a free-draining, frost-free material, and is used to transmit and spread the pavement load to the subgrade. It also drains the water away from the underside of the pavement.

The *surface pavement* forms the roadway surface and also transmits wheel loads to the aggregate base. In addition to load-carrying capability, the surface pavement requires durability to withstand the fatigue of frequent axle-load passages. Surface pavement design begins with subgrade bearing capacity and applies standard loading conditions to a base and pavement combination. The loading conditions include both frequency and weight of axle passages.

Shoulders and *curbs* provide lateral support for the surface pavement edges. Shoulders carry water away from the roadway, provide a space for snow to be moved off of the roadway, and serve as a pull-off area for vehicles. Curbs control drainage flow along the pavement edge, help support sidewalks, and provide guidance for traffic flow.

Drainage systems, such as side ditches, curb gutters, inlets, and pipes, carry and keep excess water away from the roadway surface, as well as under the pavement. Procedures to determine the size, shape, and spacing of various drainage elements can be found in *PE Civil Reference Manual*.

Railroad Track Cross Sections

Railroad track cross sections have elements performing similar functions as roadway cross sections, as shown in Fig. 5.4.

As with roadways, the railroad *subgrade* is a carefully graded, compacted, and drainable type of soil that provides the foundation for the track structure. The subgrade can be sloped at 48:1 to both sides of the trackway, or entirely to one side, as the track slope does not depend on the slope of the subgrade. The subgrade can be as much as 4 ft below the top of the rail in heavy-haul mainline territory, depending on the nature of the subgrade soil. The gradation of sub-ballast stone placed on the subgrade is often similar to the top ballast, in that the sub-ballast becomes the primary drainage layer under the track. The top of the sub-ballast is kept 6 in to 9 in from the bottom of the ties so that it is not disturbed by track alignment machinery.

The *top ballast* is the base support for the ties and rail. The top ballast, along with the rail and tie system, carries the trainloads to the sub-ballast. The top ballast is manipulated by track tamping equipment to adjust the rail-riding surface in a uniform alignment according the specified grade. AREMA No. 3, No. 4, or No. 4A gradations are common for ballast.

Ballast is made from very hard, crushed stone, such as granite, to withstand the constant movement under train wheel loads. In locations where granite used in top ballast is costly, and for lower-tonnage industrial sidings, hard limestone is often substituted for granite.

Track cribs (the space between the ties) are filled with top ballast to help maintain the track location. In yard areas where crews must walk between the tracks, walkway ballast consisting of a higher percentage of smaller aggregate, such AREMA No. 5, is placed between the tracks to the level of the top of tie.

For a *railroad drainage system*, the ballast acts as a porous drain, carrying surface water down to the subgrade. Water from the subgrade must be carried away from the track structure alongside ditches and cross pipes. It is important to keep the track structure free of

Figure 5.4 Elements of a Typical Railroad Track Cross Section

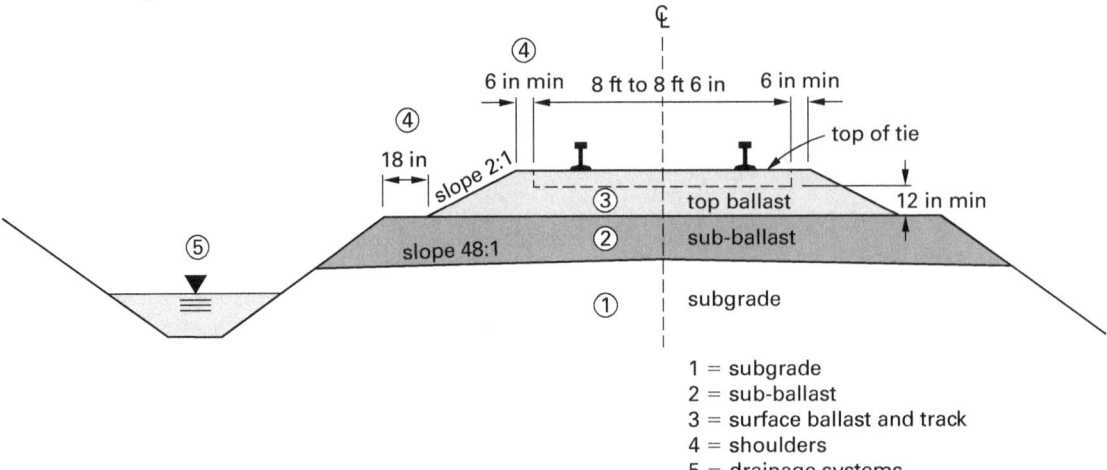

water as much as possible so that freeze-thaw cycles and the pumping action of passing rail-car axles do not shift the track alignment.

Railroad *ballast shoulders* support the ballast edges and help spread the load of succeeding layers of ballast to the subgrade. The shoulders also help restrain the ties from lateral movement under wheel loads.

Determining Cut and Fill Quantities

Generally, in determining the earthwork areas on cross sections, a horizontal line is drawn at the subgrade elevation. The subgrade line is considered the excavation or embankment limit. In a cut condition, all material above the subbase line is removed. In a fill condition, fill material is placed on the existing ground up to the subbase line. Additional excavation is needed for drainage swales and slope benches. Additional excavation may also be needed to remove unsuitable material, rock quantities that interfere with subgrade construction, detention sumps, or other proposed underground work. This unsuitable material is replaced with suitable fill.

Rock fill may be included on steep slopes, or crushed rock layers may be included for underground water drainage. Buffer zones between the roadway shoulder and pedestrian or housing areas often use landscaped berm fill sections. Berms and dikes may also be constructed to control drainage and can be a useful way to dispose of excess cut material. Unsuitable material is disposed of off-site and is not included in engineered fill sections.

The total amounts of material that must be excavated as cut or placed as fill are referred to as the *earthwork mass-haul quantities* for an earthwork project. Subbase, base, and pavement surface materials are not included in the earthwork mass-haul quantities, but are instead included in the pavement calculations. Pipe excavation is not included in the earthwork quantities, as it is normally paid for with the pipe quantity bid item.

When material is removed from its natural position, sometimes called *in situ*, its density decreases because it occupies more volume, and it is said to *swell*. When material is compacted into a fill, its density increases. Compaction reduces the volume of material from the hauled volume, and it is said to *shrink*. Generally, soil will shrink and rock will swell when compacted in a fill. Figure 5.5 illustrates these volume relationships. The in-place (*in situ* or *bank*) volume is shown as a solid line, and the dashed lines represent the change in volume from the in-place volume. Figure 5.6 further illustrates volume and density changes during earthwork activities. The percentages of change between the three material conditions shown in Fig. 5.6 are intended only as examples and are illustrated much larger than the percentages of change encountered during earthwork. When the cut and fill volumes are tabulated for each cross section, they need to be adjusted for swell and shrinkage before being plotted on a mass diagram (see Sec. 5.3.). Cut in excess of what is needed as fill, or that is unsuitable to use for fill, is called *waste*. *Borrow* refers to any additional material brought from off-site to complete a fill.

Figure 5.5 Excavation Volume Relationships

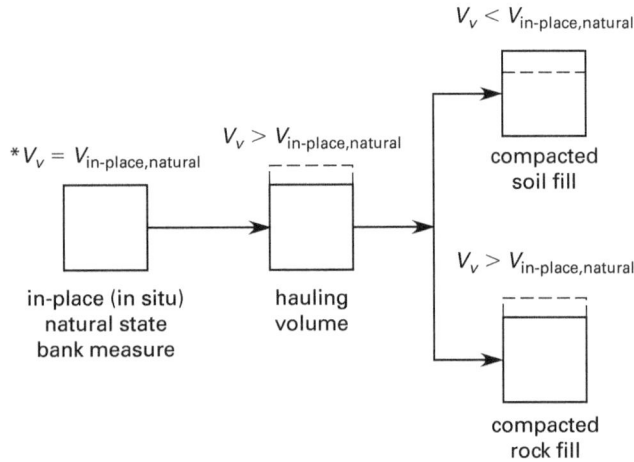

Figure 5.6 Typical Material Volume Change During Earthmoving

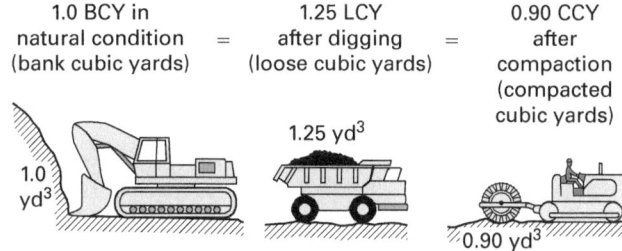

The shrinkage and swell factors for a given material are determined by soil tests on the material, and the following formulas are used to find the volume of cut or fill. s is the change in soil volume, either as shrinkage, which is negative, or swell, which is positive. The volume of fill, V_f, can be found from Eq. 5.1.

$$V_f = (1+s)V_c \qquad 5.1$$

Equation 5.1 can be rearranged into Eq. 5.2 to find the volume of cut, V_c.

$$V_c = \frac{V_f}{1+s} \qquad 5.2$$

Equation 5.1 and Eq. 5.2 determine the amount of cut or fill available at a site and represent the theoretical cut and fill volumes adjusted for shrinkage or swell. In order to determine the actual amount of material available for fill, the amount of cut produced is reduced by the volume of unsuitable excavation, which is usually wasted off-site. *Unsuitable excavation* is any material considered unsuitable for fill, such as peat or sod. The amount of fill is then adjusted for shrinkage and swell and compared to the available volume of cut to determine whether waste will be produced or borrow will be needed. If the adjusted amount of fill required is greater than the available volume of cut, the job is a borrow job (i.e., borrow is required). If the adjusted amount of fill is less than the available amount of cut, then the job is a waste job. In situations where all of the cut is waste (e.g., a site's soil is unsuitable), the volume of cut is zero, and the volume of borrow is equal to the volume of fill adjusted for shrinkage and swell.

The volume of borrow, V_b, or waste, V_w, is calculated from Eq. 5.3.

$$V_b = -V_w = V_c - \frac{V_f}{1+s} \qquad 5.3$$

Example 5.1

Earthwork for a section of highway has 7330 m³ of cut and 6960 m³ of fill. The cut material shrinks 9% when compacted into fill. How much borrow or waste occurs?

Solution

Determine the amount of borrow or waste using Eq. 5.3.

$$\begin{aligned} V_b &= V_c - \frac{V_f}{1+s} \\ &= 7330 \text{ m}^3 - \frac{6960 \text{ m}^3}{1+(-0.09)} \\ &= -318 \text{ m}^3 \end{aligned}$$

Borrow is negative. Therefore, this is a borrow job. The amount of borrow required is 318 m³.

Hauling and Material Handling

For earthwork, truck or jobsite equipment hauling involves four parts. The first part is excavating and loading. This can be accomplished by one piece of equipment, or one piece of equipment can dig up the material while another performs loading duties.

The second part is the actual *hauling*. The hauling capacity of trucks and equipment is specified as both a volume and a weight capacity. For soils of low density, including loam and organic material, the *load volume* capacity is often reached before the *maximum hauling weight*. When excavating rock or other dense material, the weight capacity most often limits the size of the load. Both capacities should be checked to determine the minimum number of load trips so that trucks and other equipment are not overloaded. The change in volume or density is compared to the original in situ density, regardless of the amount of shrinkage or swell that occurs during excavation and compaction.

Equation 5.4 can be used to find the hauling volume, V_h, and Eq. 5.5 is used to find the hauling mass, m_h. ρ_b is the in situ soil mass density. s is the shrinkage or swell of the material.

$$V_h = V_b(1+s) \qquad 5.4$$

$$m_h = \frac{\rho_b}{1+s} \qquad 5.5$$

The third part is *dumping* or unloading the trucks. Most on-road and off-road trucks dump the load by lifting and tilting the hauling end toward the rear. The load is allowed to fall out of the rear as the truck moves ahead. For large volumes, this saves time, as the trucks can keep moving through the dumping area. Some trucks unload by opening up doors in the bottom of the bed. This type of unloading is generally faster than end dumping and is safer because the truck body does not have to be elevated, which can interfere with overhead obstructions. Other off-road vehicles, such as earthmover scrapers, unload by pushing the load forward out of the scraper bowl and over the cutting edge as the scraper moves ahead. It is important to allow sufficient time for unloading when analyzing a hauling cycle to avoid congestion at the dump site.

The fourth part is the *return of haulers* to the loading point. Empty trucks can travel faster than loaded trucks, which means the return trip may not take as much time as the original haul trip.

Equipment costs are usually calculated on a per-hour basis for work within the jobsite. Costs of over-the-road travel, such as delivering materials and supplies from outside vendors, are often calculated on a *per-vehicle-mile basis*. Labor costs are usually on a *per-hour basis* and are added to the equipment cost. When comparing costs for material supplies, the delivery costs include total truck and driver time to make a round trip from the supplier's stockpile or warehouse to the jobsite.

Often, more than one material supplier is being considered or is necessary to fill the order. Typical comparisons done to choose between suppliers include the cost of material from each source and the load-haul-unload-return cycle for each truck. The most cost-effective supplier for the job is the one that can provide the lowest cost for materials delivered to the jobsite at the proper time.

Example 5.2

Earthwork requires 30,000 yd³ of borrow fill. A borrow pit is 4 mi away and has soil with an in situ density of 135 lbm/ft³. The soil in the borrow pit expands 25% when hauled and shrinks 11% when compacted in the fill.

Two contractors are being considered, each with a different type of truck to haul the material. Contractor A has a truck with a 55,000 lbm capacity, a 15 yd³ load bed, and a cost of $3.50 per mile to operate. Contractor B has a truck with an 82,000 lbm capacity, a 25 yd³ load bed, and a cost of $4.75 per mile to operate. What is the lowest cost of hauling the borrow fill?

Solution

Determine how much borrow is needed using Eq. 5.3.

$$V_b = V_c - \frac{V_f}{1+s} = 0 - \frac{30{,}000 \text{ yd}^3}{1+(-0.11)} = -33{,}700 \text{ yd}^3$$

Therefore, the amount of borrow required is 33,700 yd³. Determine the hauling volume using Eq. 5.4.

$$V_h = V_b(1+s) = (33{,}700 \text{ yd}^3)(1+0.25)$$
$$= 42{,}125 \text{ yd}^3$$

Determine the mass of the hauled material per cubic yard using Eq. 5.5.

$$m_h = \frac{\rho_b}{1+s} = \left(\frac{135 \frac{\text{lbm}}{\text{ft}^3}}{1+0.25}\right)\left(27 \frac{\text{ft}^3}{\text{yd}^3}\right)$$
$$= 2916 \text{ lbm/yd}^3$$

Determine the cost per contractor.

Contractor A

Check the capacity of the truck to determine whether mass or volume governs.

$$V_{\text{cap}} m_h = (15 \text{ yd}^3)\left(2916 \frac{\text{lbm}}{\text{yd}^3}\right)$$
$$= 43{,}740 \text{ lbm} \quad [< 55{,}000 \text{ lbm}]$$

43,740 lbm is less than 55,000 lbm. Therefore, volume governs.

The hauling cost will be the number of truck loads times the number of miles (round-trip) times the cost per mile.

$$\left(\frac{42{,}125 \text{ yd}^3}{15 \frac{\text{yd}^3}{\text{truckload}}}\right)(2)(4 \text{ mi})\left(3.50 \frac{\$}{\text{mi}}\right) = \$78{,}633$$

Contractor B

Check the capacity of the truck.

$$V_{\text{cap}} m_h = (25 \text{ yd}^3)\left(2916 \frac{\text{lbm}}{\text{yd}^3}\right)$$
$$= 72{,}900 \text{ lbm} \quad [< 82{,}000 \text{ lbm}]$$

72,900 lbm is less than 82,000 lbm. Therefore, volume governs.

The hauling cost is

$$\left(\frac{42{,}125 \text{ yd}^3}{25 \frac{\text{yd}^3}{\text{truckload}}}\right)(2)(4 \text{ mi})\left(4.75 \frac{\$}{\text{mi}}\right) = \$64{,}030$$

The lowest cost is $64,030 using contractor B's trucks.

3. MASS DIAGRAMS

Designers use a mass diagram when setting the grade line to balance the cut and fill quantities. A *mass diagram*, or *mass-haul diagram*, is a plotted curve showing the cumulative volume of excavated materials in a cut and the volume necessary for fill.

Mass diagrams are useful tools to find the most economical hauling distance of cut and fill materials. Mass diagrams use units of station-yards (station-meters), indicating one cubic yard (cubic meter) moved one station (100 ft (100 m)). Distances are plotted in stations from left to right along a horizontal axis. The adjusted cross section quantities are plotted cumulatively on the vertical axis, using positive values for excavation and negative values for fill. A rising line indicates an excess of cut over fill, and a falling line indicates an excess of fill over cut.

A mass diagram is not the same as a profile drawing, though both are used in the planning of earthwork. A *profile drawing* depicts a vertical surface cut through the centerline of the baseline of a roadway on which the original ground surface and the vertical position of the roadway are plotted. Figure 5.7 shows a typical profile of the vertical differences between a roadway grade line and a ground line with a mass diagram. The grade line is relatively straight, but the ground line rises and falls above and below the grade line. In section B, the ground must be cut down to the grade line, and in sections A and C, the ground must be filled in to meet the grade line.

Figure 5.7 Balance Line Between Two Points

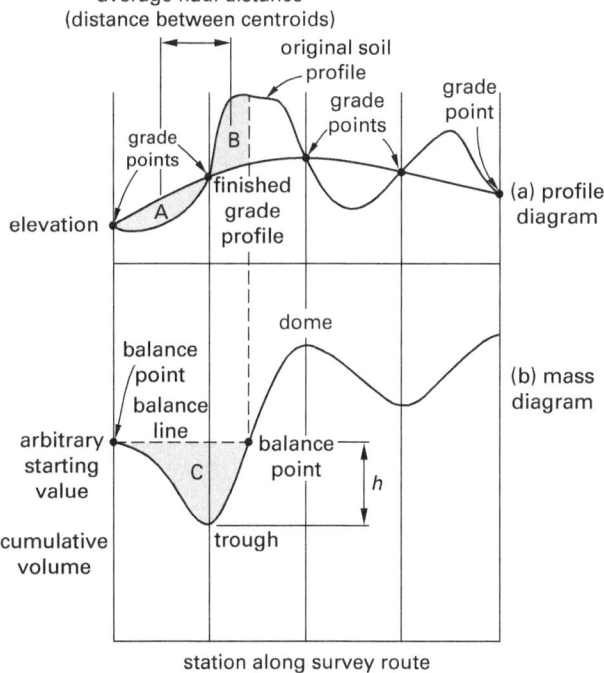

On either diagram, a *balance line* is drawn between two adjacent *balance points* in a crest or sag area. The volumes of excavation between the balance points are equal, so a contractor can use the earth volume from the excavation on one side of the grade point for the embankment on the other side of the grade point. The balance line in Fig. 5.7 intersects the mass diagram at two points, showing that the cut soil for section A can be used as fill soil for section C (assuming the soil is suitable for use as fill).

Figure 5.8 depicts a mass diagram showing the subbases for a roadway project, with distances provided in customary U.S. units. The balance point for this diagram falls between station 13 and station 14. The ordinate of station 13 is +1519, and the ordinate of station 14 is −254, so the balance line falls by 1519 yd³ + 254 yd³ = 1773 yd³ every 100 ft or 17.73 yd³/ft. The curve crosses the baseline at a distance of 1519 yd³/(17.73 yd³/ft) = 86 ft from station 13 (written as sta 13+86).

Once a job is bid, the contractor is given a *freehaul distance*, included in the bid price, that specifies the distance excavated materials will be moved by the contractor free of charge. Material moved farther than the freehaul distance is termed *paid overhaul*. Costs for overhaul will be provided in the contractor's bid. Waste and fill materials are usually balanced so that the amount of necessary overhaul is not excessive. Figure 5.9 shows a mass diagram and a profile diagram depicting the freehaul distance and overhaul volume.

The *economic overhaul limit*, also called the *limit of profitable haul*, C_{limit}, is the limit at which it is more profitable to borrow and waste rather than haul from another part of the job. This can be determined without a mass diagram, using the following equations instead. C_e is the cost of one cubic yard (cubic meter) of excavation or embankment.

$$C_{\text{limit}} = 2C_e \qquad 5.6$$

When excavation is hauled into embankment, use Eq. 5.7. C_h is the cost to haul one cubic yard (cubic meter) one station, or 100 ft (100 m), and L_n is the length of haul in 100 ft (100 m) stations.

$$C = C_e + C_h L_n \qquad 5.7$$

To find the limit of profitable haul, compare the cost calculated from Eq. 5.7 with the value of $2C_e$. If they are equal, the limit of profitable haul has been reached.

Rearranging Eq. 5.7, Eq. 5.8 can be used to find the haul length.

$$L_n = \frac{C_e}{C_h} \qquad 5.8$$

Figure 5.8 Mass Diagram Showing Subbases

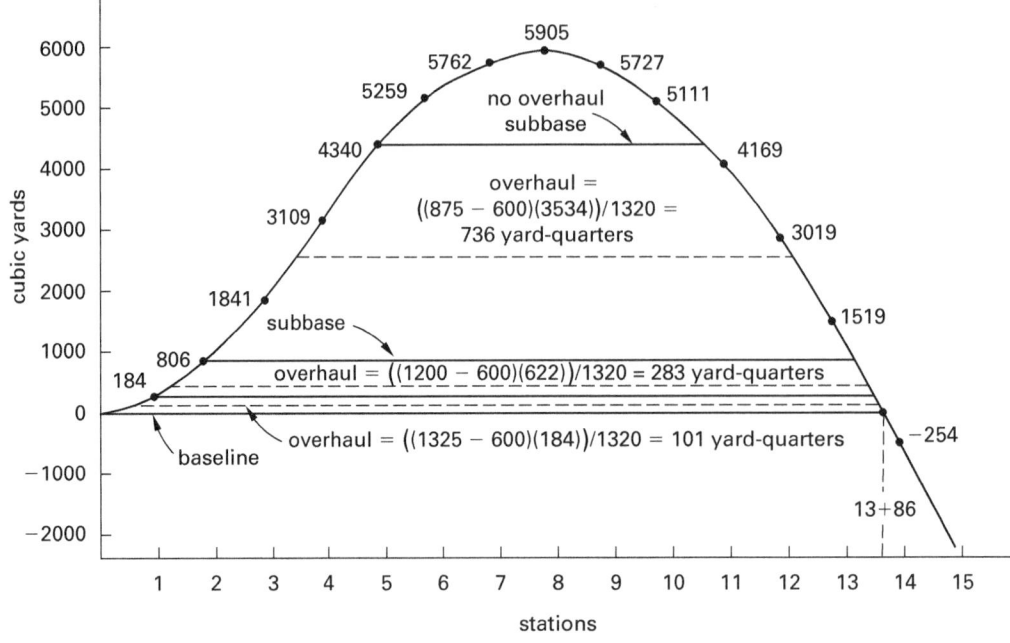

Figure 5.9 Freehaul and Overhaul

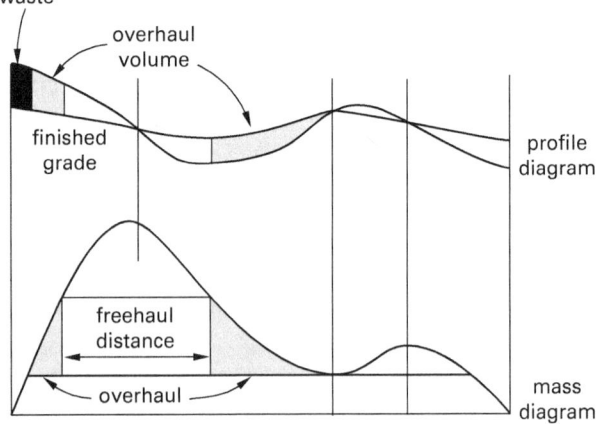

Generally, it is more common to find convenient locations for waste sites and borrow pits in rural areas than in urban areas. Freehaul distances such as 300 ft, 500 ft, or 900 ft (91.4 m, 152 m, or 274 m) were often found in specifications for railroads and interstate highways during the early to mid-20th century. However, with the completion of the interstate highway system and with fewer new railroad mainlines being constructed, a large proportion of work now involves making improvements to shorter stretches of the alignment. This must be performed frequently while maintaining traffic. Even though it has become increasingly difficult to balance cuts and fills within the limits of a single job, equipment improvements have resulted in faster handling of earthwork within the jobsite, as well as making longer hauls more manageable for on-highway equipment.

Example 5.3

The excavation or embankment cost for an earthwork project is \$10 per yd^3, and the hauling cost is \$0.65 per yd^3-sta. What is the limit of profitable haul?

Solution

Using Eq. 5.8, the haul length is

$$L_n = \frac{C_e}{C_h} = \frac{10 \; \frac{\$}{yd^3}}{0.65 \; \frac{\$}{yd^3\text{-sta}}} = 15.4 \text{ sta}$$

Check that the limit of profitable haul has been reached by comparing the cost calculated from Eq. 5.6 and Eq. 5.7.

$$C_{\text{limit}} = 2C_e = (2)\left(10 \; \frac{\$}{yd^3}\right) = \$20.0/yd^3$$

$$\begin{aligned} C &= C_e + C_h L_n \\ &= 10 \; \frac{\$}{yd^3} + \left(0.65 \; \frac{\$}{yd^3\text{-sta}}\right)(15.4 \text{ sta}) \\ &= \$20.0/yd^3 \end{aligned}$$

Therefore, the limit of profitable haul is 15.4 sta.

4. PAVEMENT DESIGN

Pavement design involves matching a project's strength and durability requirements to a given material. The base soil, or foundation, is analyzed, and the depths of various layers of pavement are designed based on the loads applied, the subgrade strength, and the frost penetration depth.

Pavement design's primary concern is vertical loads plus impacts. Designs are based on a combination of maximum expected wheel or axle loadings and the expected frequency of groups of axle loadings. *Horizontal loads*, such as those resulting from braking, acceleration, or turning movements, are not usually considered for concrete, but are included for Superpave asphalt. (See Sec. 5.5.) Pavement loadings in areas where vehicles will stand for extended periods of time, such as in parking lots, do not need to account for the frequency of wheel-load repetitions, but are instead governed by the maximum load applied for concrete and the distribution of load for asphalt.

Pavement Thickness Design

Concrete slab thickness is determined primarily by the maximum single load applied or by the repetition of loads over a long period of time, depending on which is greater. Generally, the repetition of loads over a period of time requires greater strength than a maximum single load because a highway slab with a depth of no more than 6 in (152 mm) can occasionally sustain axle loadings at low speeds considerably in excess of legal limits. A slab thickness of 5 in (127 mm) is the nominal minimum thickness for lightly traveled streets and driveways. Additionally, the relative strength of the subbase using Hveem's resistance values (*R-values*) or the *California bearing ratio* (CBR) values. influences slab thickness.

Highway pavement thickness is commonly specified in 1 in (25 mm) increments from 5 in to 10 in (130 mm to 250 mm) or greater. For heavily traveled interstates, pavement thicknesses of 12 in (300 mm) or more are used to help prevent costly traffic congestion caused by pavement rehabilitation. Where frost is a factor, highway subbase thickness should be at least 6 in and can range up to a foot or more depending on the subsoil and load applied.

In order to simplify the variety of vehicle combinations riding on highways, vehicle traffic for pavement thickness design is usually grouped according to axle load weights, spacing of axle combinations, and frequency of application. Because of the exponential relationship of axle load to pavement life, the heaviest and most frequent load combinations are used to determine the design strength. Traffic is commonly broken down into classifications of automobiles, light trucks, heavy single-unit trucks, and heavy multiple-unit trucks. Truck traffic can be further classified as commercial or industrial to indicate expected loading needs along with appropriate axle load weight classifications.

Pavement thickness design is a complex and involved process that is thoroughly covered in publications written by the American Association of State Highway and Transportation Officials (AASHTO), the American Concrete Pavement Association, the National Pavement Association, the Portland Cement Association (PCA), and the Asphalt Institute (AI).

Soils and Subgrade

Because the soil layer immediately below a pavement base has a direct effect on the performance of the pavement, the characteristics of this soil must be known in order to design a pavement that is economical and able to carry the expected loads. The best soil types have good load strengths, drain well, are not affected by frost or other types of heaving, are free of organic material, and can be shaped and compacted into a stable mass. In the AASHTO classification system, soils in the A-l and A-2 classifications are considered proper subgrade materials. Soils in the A-1 and A-2 classifications have less than 35% *fines*, or material passing the no. 200 sieve. The AASHTO classification system is given in App. 5.A, and the Unified Soil Classification System is given in App. 5.B.

Soil classification information may be obtained from a number of different systems. Generally, both pavement bearing pressure and design charts are based on *R-value* or the CBR index. When a classification is given in one system, its name in another system may be approximated using charts such as the USDA triangle or graphs relating one system to another. PCA publishes *A Soil Primer*, which also contains conversions between classification systems. The USDA triangle is given in App. 5.C.

Usually, subgrade soil is specified to be carefully excavated, compacted, and graded as a separate operation, often with its own unit price for payment. If suitable soil is not available for subgrade, cement treating or other modifications to available soil may be necessary. Once a subgrade design strength is identified, the design of the granular drainage base, subbase, and surface pavement may proceed.

Subbase Materials

Aggregate materials used for subbase construction are often locally produced to minimize hauling cost. Aggregate must be free of fines, relatively uniform in gradation, and not too large in size. A common method is to specify AASHTO no. 65 or no. 67, or some other readily available standard designation of aggregate size classification. AASHTO and AREMA size gradations are shown in App. 5.D.

In addition to size classification, subbase materials are tested for maximum organic material, acceptable chemical composition, soundness, minimum fractured faces (river gravel), frost susceptibility, shrinkage and swell, minimum crushing strength, absorption, color (architectural surfaces), alkali reaction, and other appropriate parameters.

Pavement Materials

Highway pavement materials are relatively few in number, yet their makeup and composition are directly responsible for the successful performance of a paved surface. All high-type pavements have coarse aggregate, fine aggregate, and a cementitious material to bond the aggregates.

The most common choice for high strength and durability is *concrete pavement*. Concrete is highly resistant to abrasion, chemical attack, and spilled fuel solvents. When properly placed, concrete has been known to last well over 100 years with little or no maintenance. However, concrete requires specialized knowledge to handle its unforgiving qualities (i.e., once placed, setting and curing occur within a few hours, and any mistakes made are permanently captured), and it is susceptible to the abuse of excessive deicing salts, which soften the surface and weaken the joint edges. It also can have a higher cost compared with bituminous paving mixtures with equal strength or extended life cycles. Figure 5.10 illustrates the typical concrete surfacing layers. Table 5.1 gives the commonly used maximum aggregate sizes.

Figure 5.10 Typical Concrete Surfacing Layers

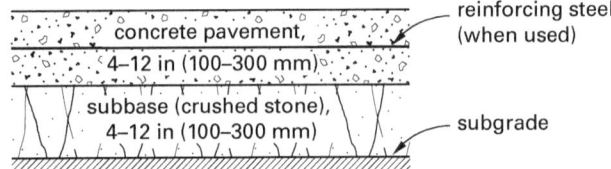

Table 5.1 Typical Aggregate Maximum Sizes for Concrete Pavements

aggregate type	size in	mm
concrete pavement	1.5–2	37.5–50
subbase	2.5	64
subgrade	4	100

Bituminous Pavement

Bituminous pavement (commonly referred to as *asphalt*, *asphalt pavement*, and *flexible pavement*) is composed of upper layers of aggregate mixed with a bituminous binder, such as asphalt. It is often preferred to concrete because it can be rapidly placed and requires less skill. It may also be installed on weaker subgrade sections, which decreases the amount of overlaying needed and defers some capital cost. However, if asphalt is not placed well, its edges can take on a ragged appearance. It also raises environmental concerns due to its hydrocarbon solvent emissions and safety concerns for workers who can become injured due to the high temperatures and irritating solvents used during the installation process. Additionally, asphalt is highly susceptible to sun damage in low traffic applications and to solvent action resulting from fuel spills. Construction costs vary regionally, making it difficult to select either asphalt or concrete pavement based on empirical data and cost alone. Figure 5.11 illustrates the layers of asphalt paving surfaces, though not all layers are present in every design. Table 5.2 gives the commonly used maximum aggregate sizes.

Figure 5.11 Typical Asphalt Surfacing Layers

*These layers may be combined according to local practice.

Table 5.2 Typical Aggregate Maximum Sizes for Asphalt Pavements

layer	size in	mm
wearing course	0.5	12.5
binder course	0.75	19
base course	1–2	25–50
stone base	3	75
subgrade	4	100

Asphalt pavement is called flexible pavement because the surface material adjusts to the load condition imposed, then elastically rebounds. Asphalt (including bitumen) is used to bond aggregate together to avoid permanent displacement under loads. Asphalt also creates an impermeable layer, protecting the base and subbase from water. The most common causes of asphalt pavement failure include

- water (frost) penetration through surface cracks
- subgrade/subbase failure due to overload
- surface cracking due to aging caused by oxidation of the asphalt cement and the effects of the sun

Materials Procedures

Pavement design usually requires calculating weights for each ingredient based on a predicted range of outcomes that are selected from known data. Bituminous mixes have a relatively narrow range of strengths and applications. Therefore, design becomes more a matter of total in-place volume. Concrete, on the other hand, can be used in a variety of ways and can have a wide range of strengths based on how its ingredients are proportioned. A commonly used reference is the AASHTO *Guide for Design of Pavement Structures*. Additional reference material is published by trade organizations that concentrate on providing technical data for their members. For concrete, PCA's *Design and Control of Concrete Mixtures* is often used as a manual of concrete mix design and should be a part of any review of mix design procedures. For asphalt pavements, the Asphalt Institute's *Mix Design Methods for Asphalt Concrete and Other Hot-Mix Types* (*MS-2*) is often referenced.

The procedure for listing data and calculating material problems is best aided by using a tabular format of the four or five main ingredients and working across the page, relating weights, volumes, densities, and correction factors.

Concrete mixtures for pavement are usually specified as requiring both a minimum compressive strength of 2500 psi to 5000 psi (17 MPa to 34 MPa), depending on the expected loads on the pavement, and a minimum entrained air content of 1.0% to 7.5%, depending on the expected weather exposure. Asphalt pavement is usually specified as requiring a percent compaction based on a maximum laboratory density, but is increasingly being specified using strength criteria based on the expected load, weather, and subgrade conditions.

For design of concrete mixes, it is necessary to know the amount of water stored in the bulk aggregate and the amount of water necessary for saturation. The water absorbed in the aggregate up to the point of saturation is not available for hydration, and water may need to be added to the total mix water to compensate for absorption, whereas water contained in a bulk pile that exceeds the saturation requirements is available for hydration and will need to be subtracted from the amount of water needed before water is added to the mix.

The amount of free water available for hydration is called the *water-cement ratio*. The water-cement ratio is used to determine strength qualities, workability, and dissolvability of a mix. Higher water-cement ratios improve workability, but reduce strength. Lower water-cement ratios increase strength, but require more power for equipment to be used in the placing and finishing of concrete slabs. Water-cement ratios of 0.4 to 0.6 by weight are common in concrete pavement design.

Water must be removed from bituminous mixtures and so is not a factor in design. Bituminous pavement design is instead performed based on the quality of the bituminous material and the quality of the aggregate available for the mix.

5. HOT MIX ASPHALT

Hot mix asphalt (HMA), also known as bituminous concrete, hot plant mix, asphalt-concrete paving, asphalt concrete, flexible pavement, blacktop, macadam, or simply "asphalt," consists of a mixture of asphalt binder and mineral aggregate (stone and sand). The asphalt cement acts as a binding agent that glues the aggregate particles into a dense mass and waterproofs the mixture. The aggregate, bound together by asphalt, becomes a stone framework with toughness able to support loads and maintain its shape. The performance of the binder and aggregate system is affected by both the properties of the individual components and their combined reaction to the system.

Asphalt is called a *viscoelastic* material because it has both viscous and elastic properties. At intermediate temperatures of asphalt paving application, the mixture displays characteristics of a low-viscosity fluid and an elastic solid.

There are several methods of designing a hot mix asphalt concrete mix. The two traditional methods are the Marshall method and the Hveem method, which use viscosity penetration-graded asphalt specifications to design asphalt concrete mixes. However, in 1987, the *Strategic Highway Research Program* (SHRP) began a five-year, $150 million research program to improve the performance and durability of asphalt roads in the United States. A major focus of the program was developing performance-graded asphalt specifications that resulted in mixes with field performances closer to the performances measured under laboratory conditions.

The result of SHRP's research was *Superpave*, short for *superior performing asphalt pavement*. Superpave is covered in more detail in Sec. 5.8. This section focuses on general information and terminology relating to HMA mix design, regardless of method.

The material presented in this section is consistent with the methods presented in *The Asphalt Handbook* (*MS-4*). Information on Superpave mix design is derived mostly from the Asphalt Institute's *Superpave Mix Design, Superpave Series No. 2* (*SP-2*). Superpave binder specifications and tests can be found in the Asphalt Institute's *Performance Graded Asphalt Binder Specification and Testing, Superpave Series No. 1* (*SP-1*).

Asphalt Pavement Behavior

Wheel loads applied to asphalt pavement produce vertical compressive stress and shear stress within the asphalt layer, and horizontal tensile stress at the bottom of the asphalt layer. The primary asphalt pavement distress conditions are permanent deformation, fatigue cracking, and low temperature cracking. HMA design should analyze these distresses in preparing a design mix.

When asphalt cement pavement hardens, either due to lower temperature or to age, it cracks more readily. When asphalt cement softens, primarily due to increased temperature of the ground surface, it tends to flow laterally under load, which causes rutting or other surface irregularities. Hardening and softening can be controlled by varying the binder mix ingredients.

Permanent deformation (*rutting*) is characterized by rutting along the wheel paths and is primarily caused by two conditions. Rutting can come from a weak subgrade where the asphalt concrete surface remains essentially intact, but the subgrade becomes overloaded and fails. It can also come from a weak asphalt layer, where repeated wheel loads cause lateral displacement of the asphalt pavement. The asphalt surface will migrate and form a hump above the original grade line adjacent to the wheel paths, making the rutting more pronounced.

Weak asphalt rutting is more pronounced during summer under higher temperatures, while *weak subgrade rutting* can occur whenever the subgrade is weakened by water intrusion. The traditional approach to reducing rutting is to increase the thickness of the asphalt pavement layers. Superpave approaches this problem by increasing control over aggregate properties to provide more friction and strength and by increasing control over binder properties to provide a wider range of temperature performance.

Fatigue cracking (also known as *alligator cracking*) is a progressive process that results from overstressing of the asphalt materials. Fatigue cracking that occurs near the end of the pavement design cycle normally indicates that the pavement has served its useful life. If it occurs earlier in the design cycle, the cracking may indicate that design loads were significantly underestimated. An early sign of fatigue cracking is longitudinal cracks that form along the wheel paths in the pavement's weakest areas. Over time, individual cracks merge together and additional cracks develop parallel and adjacent to the initial cracks, developing into cross patterns. The result is a checkerboard pattern covering the entire asphalt surface. When chunks of pavement are dislodged, potholes develop. Through these cracks and potholes, water intrudes, further weakening the subbase.

There are several ways to reduce fatigue cracking.

- Accurately predict heavy design loads during the design period.
- Use thicker pavements.
- Reduce subgrade moisture.
- Specify pavement materials that have high wet-condition strengths.
- Increase the resilience of the HMA mixture to withstand deflections.
- Use softer asphalts, which behave more like elastic materials. Hard asphalts are stiffer and have less tensile strength to resist cracking.

Low-temperature cracking is caused by adverse environmental conditions. It is characterized by intermittent transverse cracks, usually across the entire pavement width and occurring at regular intervals. Low-temperature cracks are often called *shrinkage cracks* and occur during cold weather when the tensile strength of the asphalt is less than the tensile stress. Shrinkage cracks occur more often in hard asphalts and in aged pavements. Low-temperature cracking can be reduced by specifying softer binders and controlling the air-void content of the pavement, which reduces the oxidation aging rate of the pavement.

While fatigue cracking is greatly affected by pavement structures, subgrade conditions, and traffic, permanent deformation of rutting is directly a function of shear strength and aggregate properties. Low-temperature cracking also correlates closely with binder properties.

Mix Design

The exact methods and tests used for design of hot mix asphalt for a given roadway vary depending on the method being employed for the project. All mix design methods are dependent upon the properties of the aggregates and asphalt binder used in the mix, but which properties and how they are implemented is different from method to method. The rest of this section discusses some of the general properties commonly discussed in mix design, and the following sections cover each of the three most common mix design methods and the material properties that govern the designs.

Aggregates

Mineral aggregate for asphalt pavements can come from a wide variety of sources. Two of the most common sources are natural aggregates and processed aggregates. *Natural aggregates* are mined from river or glacial deposits. They are washed and screened, and are used without further processing. *Processed aggregates* have been quarried, crushed, screened, washed, and otherwise processed to achieve specified performance criteria. *Synthetic aggregates* are made from materials not mined or quarried and are often an industrial by-product, such as slag from steel making. Other products are sometimes introduced to synthetic aggregates, such as expanded clay or shale, to increase skid resistance, or fiber reinforcement, to increase shear strength. *Reclaimed asphalt pavement* (RAP) is another source of aggregate for asphalt pavements. Using RAP as a source for aggregate is becoming common practice as costs and environmental practice make RAP more competitive.

The aggregate properties most important to mix design are shear strength and gradation density. *Shear strength*, the total load the aggregate can resist before it fails in shear, is necessary to resist repeated load applications and is a function of both the inherent strength of the aggregate and the nesting of the aggregate pieces (a "knitting action" primarily dependent on the internal friction of the compacted aggregate matrix). Rough-textured faces that are made when aggregate is crushed provide more resistance than the rounded, smooth-textured aggregate faces prevalent with river gravel. Crushed aggregate also has fewer voids than rounded aggregate, requiring less binder.

The amount of voids in aggregate is controlled by *gradation*, or the process of selecting a mixture of sizes ranging from the largest design size to the smallest to fit together in a dense, nearly solid mass. The amount of natural sand is often limited because natural sand tends to be rounded and have poor internal friction. Gradation uses a 0.45 power gradation technique to define cumulative particle size distribution. This method graphs percent passing against sieve size in millimeters raised to the 0.45 power, which results in a linear curve for most aggregates. For example, for a 19 mm maximum size, the percent passing curve would look like Fig. 5.12. The percent passing is calculated using Eq. 5.9.

$$\text{percent passing} = \left(\frac{\text{actual particle size}}{\text{maximum aggregate size}} \right)^{0.45} \quad 5.9$$

Figure 5.12 Maximum Density Gradation for 19 mm Maximum Size Using 0.45 Power for Sieve Opening Size

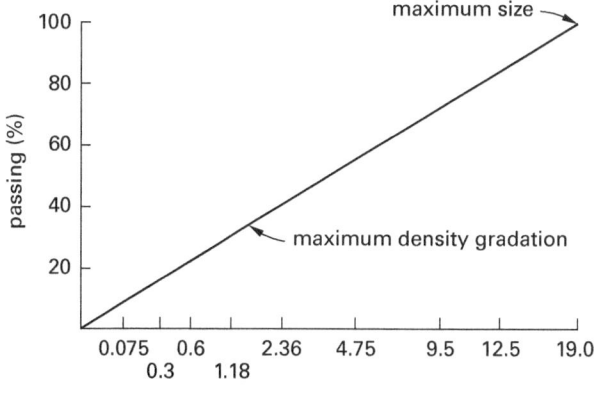

Reprinted with permission from the Asphalt Institute, *Superpave Mix Design (SP-2)*, 2001, Fig. 3.9.

The *maximum density gradation line* (also known as the *0.45 power line*) shown in Fig. 5.12 shows a gradation where the aggregates fit together in the densest configuration. Superpave size definitions include

- *maximum size:* one size larger than the nominal maximum size
- *nominal maximum size:* one sieve size larger than the first sieve to retain more than 10%

Superpave aggregates are specified by setting control points on the curve, as shown in Fig. 5.13, at the nominal maximum size, an intermediate sieve size (2.36 mm), and at the smallest sieve size (0.075 mm). The control points determine master gradation ranges. A *restricted zone* is a band set between an intermediate sieve and the 0.03 mm sieve.

Figure 5.13 Superpave Gradation Limits for 12.5 mm Superpave Mixture

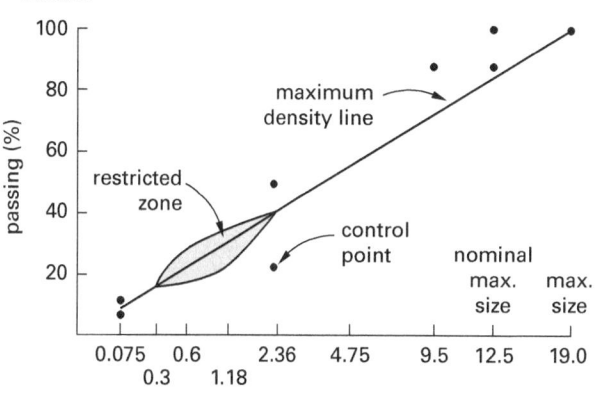

Reprinted with permission from the Asphalt Institute, *Superpave Mix Design (SP-2)*, 2001, Fig. 3.10.

Avoiding the maximum density line in the restricted zone is recommended so the mixture does not possess an excess of fine sand in relation to the total sand. An excess of fine sand creates a "sanded" mixture that can have inadequate voids in the mineral aggregate to allow enough asphalt for adequate durability.

The original Superpave specifications included the restricted zone as a guideline for HMA mixes. However, the restricted zone is not a requirement, and the value of the concept has been disputed recently. NCHRP Report 464, *The Restricted Zone in the Superpave Aggregate Gradation Specification*, suggests that the restricted zone is unnecessary if the fine aggregate angularity and all other volumetric requirements are satisfied. Studies have shown that mixtures using aggregates that fall within the restricted zone that otherwise meet all Superpave requirements can perform as well as aggregates outside the restricted zone. The restricted zone was removed from AASHTO's Superpave specification (AASHTO M323) based on the recommendations given in NCHRP Report 464.

The design aggregate structure is a distribution of aggregate sizes that meets the Superpave gradation requirements. Table 5.3 shows the five commonly used Superpave mix designations. The recommended maximum aggregate size is 0.33–0.40 times the minimum lift thickness for Superpave design and 0.40–0.50 times the minimum lift thickness for Marshall or Hveem mix designs. This maximum size is necessary to achieve the workability needed for proper compaction. Superpave aggregates require more careful stockpiling, surge-bin operations, and handling to prevent segregation, especially the larger size aggregate mixtures, in order to meet the mix performance requirements.

Table 5.3 Superpave Aggregate Mixture Gradations

Superpave designation	nominal maximum size (mm)	maximum size (mm)
37.5 mm	37.5	50.0
25.0 mm	25.0	37.5
19.0 mm	19.0	25.0
12.5 mm	12.5	19.0
9.5 mm	9.5	12.5
4.75 mm	4.75	9.5

Reprinted with permission from the Asphalt Institute, *The Asphalt Handbook, Manual Series No. 4 (MS-4)*, Table 4.3, © 2007.

Binder

Asphalt binder has three major properties: viscoelasticity, temperature susceptibility, and aging. Viscoelasticity of asphalt cement is discussed in Sec. 5.5. *Temperature susceptibility* means that the asphalt stiffness changes with variations in temperature. For this reason, almost every asphalt cement and mixture test is performed with a specified temperature component. Performance is also affected by the duration loading. That is, slow loading can be simulated by higher temperatures, and a fast loading rate can be simulated by lower temperatures.

The process of *oxidation* causes aging in asphalt binders, which are organic substances. Oxidation causes hardening, making the asphalt concrete pavement more brittle and less elastic. The aging process occurs more quickly at higher temperatures, making oxidation more of a concern in hot desert climates.

Asphalt Mixture Volumetric Properties

HMA mixture design, including Superpave, follows volumetric procedures for mix design. Volume quantities are used primarily because air voids cannot be weighed. Once the mix is designed, the volume quantities can be converted to mass to provide a job-mix formula. The definition of volumetric properties applies both to paving mixtures that have been compacted in the laboratory and to the undisturbed samples that have been cut from a pavement in the field.

Mineral aggregate used in HMA is porous and has the ability to absorb water and asphalt. The amount of absorption varies with the type of aggregate. Water can be absorbed into the surface pores, as well as internal pores. Asphalt, due to its lower viscosity, can be absorbed partially into the surface pores, but rarely permeates into the interior pores of denser aggregates. Asphalt also coats the aggregate surface. This coating seals the aggregate from moisture penetration and provides the cementitious action to bind the aggregate particles together. Figure 5.14 illustrates the coating action of aggregate in asphalt.

Figure 5.14 Asphalt Binder Coating and Aggregate Absorption

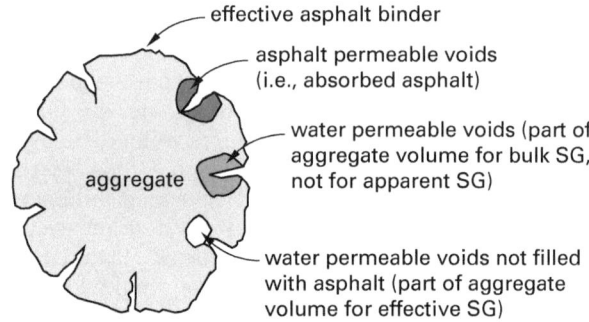

The following measurements and calculations are needed to determine the volumetric properties of a compacted paving mixture.

1. Measure the bulk specific gravity (SG) of the coarse aggregate and the fine aggregate.

2. Measure the specific gravity of the asphalt cement and the mineral filler.

3. Calculate the bulk specific gravity of the aggregate combination in the paving volume.

4. Measure the maximum specific gravity of the loose paving mixture.

5. Measure the bulk specific gravity of the compacted paving mixture.

6. Calculate the effective specific gravity of the aggregate.

7. Calculate the maximum specific gravity at other asphalt contents.

8. Calculate the asphalt absorption of the aggregate.

9. Calculate the effective asphalt content of the paving mixture.

10. Calculate the percentage of voids in mineral aggregate in the compacted paving mixture.
11. Calculate the percentage of air voids in the compacted paving mixture.
12. Calculate the percentage of voids filled with asphalt in the compacted paving mixture.

Aggregate is described using three specific gravity terms: bulk specific gravity, apparent specific gravity, and effective specific gravity. For all three, the aggregate mass and the water mass are at a specified temperature, the aggregate and the water are in air of equal density, and the water is gas-free distilled.

Bulk specific gravity, G_{sb}, is the ratio of the mass of a unit volume of permeable material to the mass of an equal volume of water. Voids in the aggregate include both those that are permeable and those that are impermeable normal to the material. The bulk specific gravity of mineral filler is difficult to determine accurately. The error is usually negligible if the apparent specific gravity of the filler is substituted for the bulk specific gravity. To calculate the bulk specific gravity of aggregate, use Eq. 5.10.

$$G_{sb} = \frac{P_1 + P_2 + \cdots + P_n}{\dfrac{P_1}{G_1} + \dfrac{P_2}{G_2} + \cdots + \dfrac{P_n}{G_n}} \quad 5.10$$

G_{sb} is the bulk specific gravity for the total aggregate; P_1, P_2, and P_n are individual percentages of mass of aggregate; and G_1, G_2, and G_n are individual bulk specific gravities of aggregate (e.g., coarse, fine).

Apparent specific gravity, G_{sa}, is the ratio of the mass of a unit volume of an impermeable material to the mass of an equal volume of water. The apparent specific gravity is found using Eq. 5.11, where m is the mass and W is weight.

$$G_{sa} = \frac{m}{V_{\text{aggregate}} \rho_{\text{water}}} \quad \text{[SI]} \quad 5.11(a)$$

$$G_{sa} = \frac{W}{V_{\text{aggregate}} \gamma_{\text{water}}} \quad \text{[U.S.]} \quad 5.11(b)$$

Effective specific gravity, G_{se}, is the ratio of the mass of a unit volume of a permeable material, excluding voids permeable to asphalt, to the mass and equal volume of water. Use Eq. 5.12 to determine the effective specific gravity of aggregate.

$$G_{se} = \frac{P_{mm} - P_b}{\dfrac{P_{mm}}{G_{mm}} - \dfrac{P_b}{G_b}} \quad 5.12$$

G_{se} is the effective specific gravity of aggregate. G_{mm} is the maximum specific gravity of paving mixture excluding air voids. P_{mm} is the percentage of mass of total loose mixture, which equals 100%. P_b is the asphalt content, the percentage of the total mass of the mixture. G_b is the specific gravity of asphalt.

The test for *maximum specific gravity*, G_{mm}, described in AASHTO T209 or ASTM D2041, should be performed on at least two, or even three, specimens that are at or very close to the estimated optimum asphalt content. Since asphalt absorption varies very little with changes in asphalt content, the effective specific gravity of the aggregate is considered constant. Averaging the G_{se} results, the maximum specific gravity for other than the optimum asphalt content can be obtained from Eq. 5.13.

$$G_{mm} = \frac{P_{mm}}{\dfrac{P_s}{G_{se}} + \dfrac{P_b}{G_b}} \quad 5.13$$

P_s is the percentage of aggregate in the total mix and is found by subtracting the percentage of asphalt, P_b, from 100%.

The volume of asphalt binder absorbed by an aggregate is almost always less than the volume of water absorbed. As a result, the value of the effective specific gravity for an aggregate should be between its bulk and apparent specific gravities. Should the effective specific gravity fall outside of these limits, its value must be assumed to be incorrect, and the composition of the mix in terms of aggregate and total asphalt content should be rechecked to find the source of the error.

The volumetric relationship of compacted asphalt is illustrated in Fig. 5.15. There are three types of voids: air voids, voids filled with asphalt, and voids filled with absorbed asphalt.

Voids in the mineral aggregate, VMA, is the volume of the intergranular void space between the aggregate particles of a compacted paving mixture, including the air voids and the effective asphalt content. VMA is expressed as a percentage of the sample's total volume.

VMA is calculated based on the bulk specific gravity of the aggregate by subtracting the volume of the aggregate, determined by its bulk specific gravity, from the bulk volume of the compacted paving mixture. The mix composition is determined as a percentage of the total volume mass, using Eq. 5.14.

$$\text{VMA} = 100\% - \frac{G_{mb} P_s}{G_{sb}} \quad 5.14$$

When the mix composition is determined as a percentage of the aggregate mass, Eq. 5.15 is used.

$$\text{VMA} = 100\% - \left(\frac{G_{mb}}{G_{sb}}\right)\left(\frac{100\%}{100\% + P_b}\right) \times 100\% \quad 5.15$$

The *effective asphalt content*, P_{be}, of a paving mixture is the asphalt coating on the outside of the aggregate. It serves as the glue that holds the matrix of aggregates

Figure 5.15 Volume Components of Compacted Asphalt

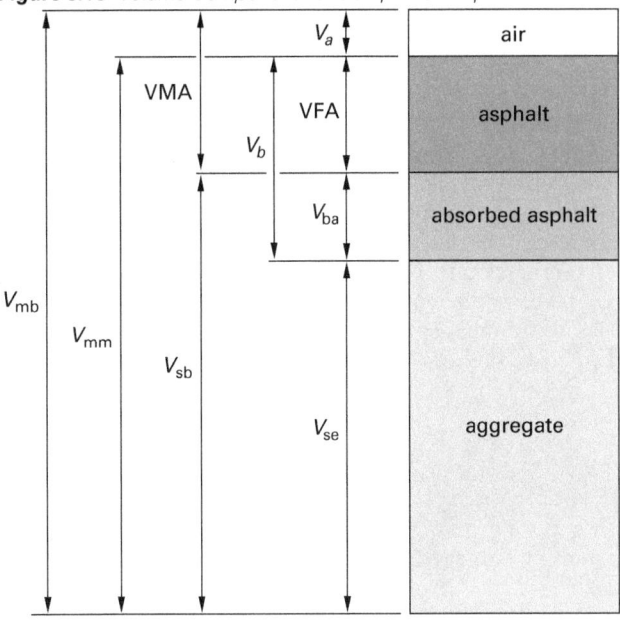

- VMA = voids in mineral aggregate
- V_{mb} = bulk volume of compacted mix
- V_{mm} = voidless volume of paving mix
- VFA = voids filled with asphalt
- V_a = air voids
- V_b = asphalt
- V_{ba} = absorbed asphalt
- V_{sb} = mineral aggregate (by bulk specific gravity)
- V_{se} = mineral aggregate (by effective specific gravity)

together and determines the performance of the mixture. The effective asphalt content is the outside coating minus the quantity lost by absorption and can be found from Eq. 5.16.

$$P_{be} = P_b - \frac{P_{ba}P_s}{100\%} \qquad 5.16$$

Air voids, V_a, is the total volume of the small pockets of air between the coated aggregate particles throughout the compacted paving mixture. V_a is expressed as a percentage of the bulk volume of the compacted paving mixture.

Asphalt absorption, P_{ba}, is expressed as a percentage of the mass of the aggregate rather than as a percentage of the total mass of the mixture. It can be found using Eq. 5.17.

$$P_{ba} = \frac{G_b(G_{se} - G_{sb})}{G_{sb}G_{se}} \times 100\% \qquad 5.17$$

The *percentage of air voids*, P_a, or the *voids in the total mixture*, VTM, is calculated using Eq. 5.18.

$$P_a = \text{VTM} = \frac{G_{mm} - G_{mb}}{G_{mm}} \times 100\% \qquad 5.18$$

Percentage of voids filled with asphalt, VFA, is the percentage of the volume of intergranular void space between the aggregate particles that is occupied by the effective asphalt. VFA does not include the absorbed asphalt and is found using Eq. 5.19.

$$\text{VFA} = \frac{\text{VMA} - \text{VTM}}{\text{VMA}} \times 100\% \qquad 5.19$$

Superpave asphalt mixtures use several AASHTO and ASTM test standards for finding the specific gravity of the mix components, as shown in Table 5.4.

Table 5.4 Specific Gravity Test Standards for Asphalt Paving Mixture Components

material	measured characteristic	symbol	test method AASHTO	test method ASTM
asphalt content	bulk specific gravity	P_b	T228	D70
mineral filler	bulk specific gravity	G_{sb}	T100	D854
coarse aggregate	bulk specific gravity	P_1	T85	C127
fine aggregate	bulk specific gravity	P_2	T84	C128
loose paving mixture	maximum specific gravity	G_{mm}	D2041	T209
compacted paving mixture	bulk specific gravity	G_{mb}	D1188-D2726	T166

Dust is calculated as the percentage of the material's mass passing through the 0.075 mm sieve (using wet-sieve analysis) divided by the effective asphalt binder content. The *dust-to-binder ratio*, also known as the *dust proportion*, DP, is calculated as

$$\text{DP} = \frac{P_{0.075\,\text{mm}}}{P_{be}} \qquad 5.20$$

$P_{0.075\,\text{mm}}$ is the aggregate content passing through the 0.075 mm sieve.

Some calculations for design of a concrete mix require the specific weight of the materials be known. The *specific weight* of a material, γ, is a customary U.S. value measured in lbf/ft³. Specific weight of a material can be found using Eq. 5.21. V is the *volumetric ratio* of the material, a dimensionless value representing the ratio of the volume of the material to the total volume of the concrete mix. G_{sb} is the bulk specific gravity of the material, and γ_{water} is the specific weight of water, 62.4 lbf/ft³.

$$\gamma = VG_{sb}\gamma_{water} \quad \text{[U.S. only]} \qquad 5.21$$

The weight of each material per sack of cement can be found from Eq. 5.22. γ is the specific weight of the material. 94 lbf/sack is the standard weight of concrete.

$$Q = \gamma \frac{94 \; \frac{\text{lbf}}{\text{sack}}}{\gamma_c} \quad \text{[U.S. only]} \quad 5.22$$

In some cases, water absorption in a concrete mix is less than the amount of moisture included with the aggregates. In these cases, there will be excess water introduced into the mix along with the aggregate, and the excess water must be subtracted from the mix. Excess water is found by simply subtracting the water absorbed by the aggregates from the total water added to the mix.

Example 5.4

Using the following table of paving mixture data, calculate the bulk specific gravity, the effective specific gravity, the maximum specific gravity, the asphalt absorption, the effective asphalt content, the percentage of air voids, the VMA, and the VFA for the aggregate mixture.

component	specific gravity	percentage of total mix (%)	percentage of total aggregate (%)
asphalt content	$G_b = 1.030$	$P_b = 5.3$	$P_b = 5.6$
coarse aggregate	$G_1 = 2.716$	$P_1 = 47.4$	$P_1 = 50.0$
fine aggregate	$G_2 = 2.689$	$P_2 = 47.3$	$P_2 = 50.0$
loose paving mixture	$G_{mm} = 2.535$	–	–
compacted paving mixture	$G_{mb} = 2.442$	–	–

Solution

Use the given table to determine values for the aggregate mixture. Using Eq. 5.10, the bulk specific gravity is

$$G_{sb} = \frac{P_1 + P_2 + \cdots + P_n}{\frac{P_1}{G_1} + \frac{P_2}{G_2} + \cdots + \frac{P_n}{G_n}}$$

$$= \frac{50.0\% + 50.0\%}{\frac{50.0\%}{2.716} + \frac{50.0\%}{2.689}}$$

$$= 2.703$$

Using Eq. 5.12, the effective specific gravity is

$$G_{se} = \frac{P_{mm} - P_b}{\frac{P_{mm}}{G_{mm}} - \frac{P_b}{G_b}}$$

$$= \frac{100\% - 5.3\%}{\frac{100\%}{2.535} - \frac{5.3\%}{1.030}}$$

$$= 2.761$$

The percentage of aggregate in the mix is

$$P_s = 100\% - P_b$$
$$= 100\% - 5.3\%$$
$$= 94.7\%$$

Using Eq. 5.13, the maximum specific gravity is

$$G_{mm} = \frac{P_{mm}}{\frac{P_s}{G_{se}} - \frac{P_b}{G_b}}$$

$$= \frac{100\%}{\frac{94.7\%}{2.761} - \frac{5.3\%}{1.030}}$$

$$= 3.430$$

Using Eq. 5.17, the asphalt absorption is

$$P_{ba} = \frac{G_b(G_{se} - G_{sb})}{G_{sb}G_{se}} \times 100\%$$

$$= \frac{(1.030)(2.761 - 2.703)}{(2.703)(2.761)} \times 100\%$$

$$= 0.8\%$$

Using Eq. 5.16, the effective asphalt content is

$$P_{be} = P_b - \frac{P_{ba}P_s}{100\%}$$

$$= 5.3\% - \frac{(0.8\%)(94.7\%)}{100\%}$$

$$= 4.54\%$$

Using Eq. 5.18, the percentage of air voids between the coated aggregate particles is

$$\text{VTM} = \frac{G_{mm} - G_{mb}}{G_{mm}} \times 100\%$$

$$= \frac{2.535 - 2.442}{2.535} \times 100\%$$

$$= 3.68\%$$

Using Eq. 5.14, the percentage of VMA is

$$\text{VMA} = 100\% - \frac{G_{\text{mb}}P_s}{G_{\text{sb}}}$$
$$= 100\% - \frac{(2.442)(94.7\%)}{2.703}$$
$$= 14.4\%$$

Using Eq. 5.19, the percentage of VFA is

$$\text{VFA} = \frac{\text{VMA} - \text{VTM}}{\text{VMA}} \times 100\%$$
$$= \frac{14.4\% - 3.68\%}{14.4\%} \times 100\%$$
$$= 74.4\%$$

6. MARSHALL MIX DESIGN

The Marshall method was developed in the late 1930s and was subsequently modified by the U.S. Army Corps of Engineers for higher airfield tire pressures and loads. It was subsequently adopted by the entire U.S. military. Although the method disregards shear strength, it considers strength, durability, and voids. It also has the advantages of being a fairly simple procedure, not requiring complex equipment, and being portable.

The primary goal of the Marshall method is to determine the optimum asphalt content. The method starts by determining routine properties of the chosen asphalt and aggregate. The specific gravities of the components are determined by standard methods, as are the mixing and blending temperatures from the asphalt viscosities. (See Fig. 5.16.) Trial blends with varying asphalt contents are formulated, and samples are heated and compacted. Density and voids are determined.

Figure 5.16 Determining Asphalt Mixing and Compaction Temperatures*

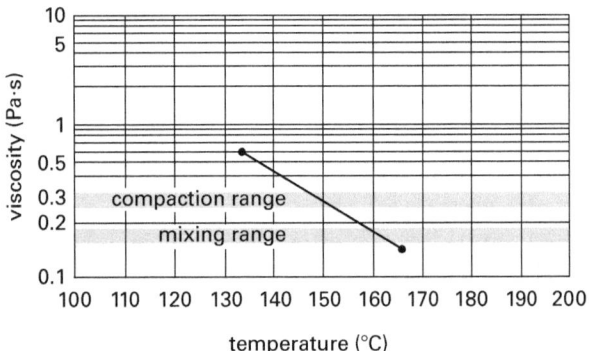

*Compaction and mixing temperature are found by connecting two known temperature-viscosity points with a straight line.

The stability portion of the Marshall procedure measures the maximum load supported by the test specimen at a loading rate of 2 in/min (50.8 mm/min). The load is increased until it reaches a maximum, then when the load just begins to decrease, the loading is stopped and the maximum load is recorded. During the loading, an attached dial gauge measures the specimen's plastic flow. The flow value is recorded in 0.01 in (0.25 mm) increments at the same time the maximum load is recorded.

Various parameters are used to determine the best mix. Three asphalt contents are averaged to determine the *target optimum asphalt content*: the 4% air voids content, the maximum stability content, and the maximum density (unit weight) content. The target optimum asphalt content is subsequently validated by checking against flow and VMA minimums.

The *Marshall test method* is a density voids analysis and a stability-flow test of the compacted test specimen. The size of a test specimen for a Marshall test is 2.5 in × 4 in (64 mm × 102 mm) diameter × length. Specimens are prepared using a specified procedure for heating, mixing, and compacting the asphalt aggregate mixtures. When preparing data for a Marshall mix design, the stability of the samples that are not 2.5 in (63.5 mm) high must be multiplied by the correlation ratios in Table 5.5. The Marshall test device and typical results are shown in Fig. 5.17.

Table 5.5 Stability Correlation Factors

specimen volume (cm³ (in³))	approximate specimen thickness (mm (in))	correlation ratio
406–420 (25.0–25.6)	50.8 (2.0)	1.47
421–431 (25.7–26.3)	52.4 (2.06)	1.39
432–443 (26.4–27.0)	54.0 (2.13)	1.32
444–456 (27.1–27.8)	55.6 (2.25)	1.25
457–470 (27.9–28.6)	57.2 (2.25)	1.19
471–482 (28.7–29.4)	58.7 (2.31)	1.14
483–495 (29.5–30.2)	60.3 (2.37)	1.09
496–508 (30.2–31.0)	61.9 (2.44)	1.04
509–522 (31.1–31.8)	63.5 (2.5)	1.00
523–535 (31.9–32.6)	64.0 (2.52)	0.96
536–546 (32.7–33.3)	65.1 (2.56)	0.93
547–559 (33.4–34.1)	66.7 (2.63)	0.89
560–573 (34.2–34.9)	68.3 (2.69)	0.86
574–585 (35.0–35.7)	71.4 (2.81)	0.83
586–598 (35.8–36.5)	73.0 (2.87)	0.81
599–610 (36.6–37.2)	74.6 (2.94)	0.78
611–625 (37.3–38.1)	76.2 (3.0)	0.76

(Multiply cm³ by 0.061 to obtain in³.)
(Multiply mm by 0.0394 to obtain in.)

Figure 5.17 Marshall Testing

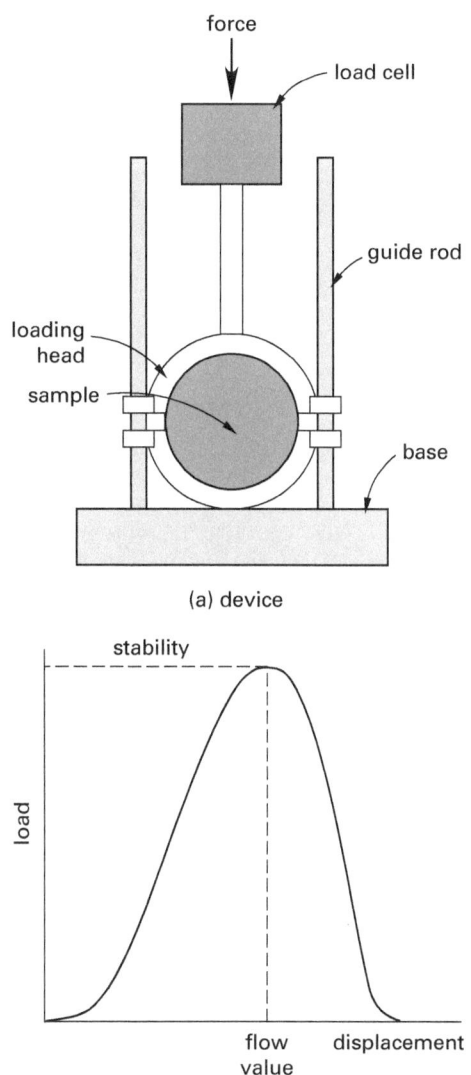

(a) device

(b) results

Figure 5.18 Typical Marshall Mix Design Test Results

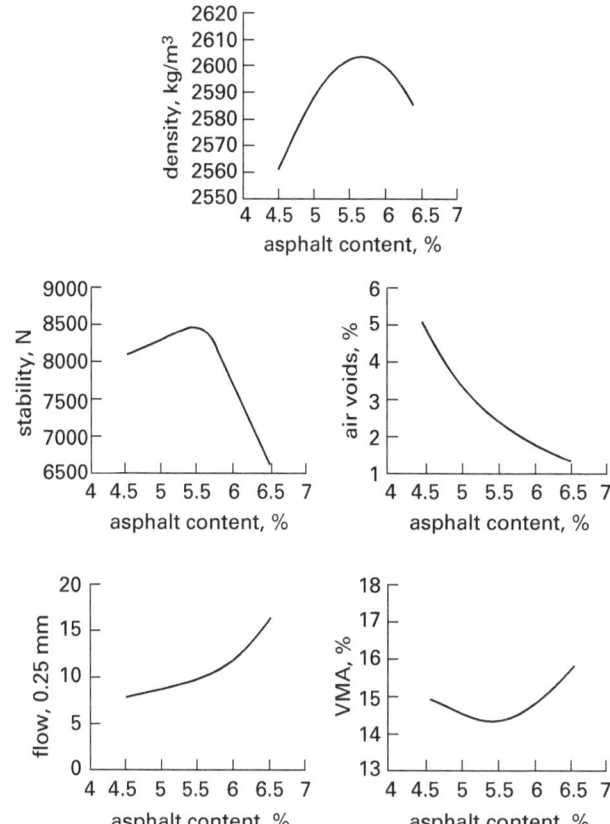

Figure 5.19 VMA Criterion for Mix Design

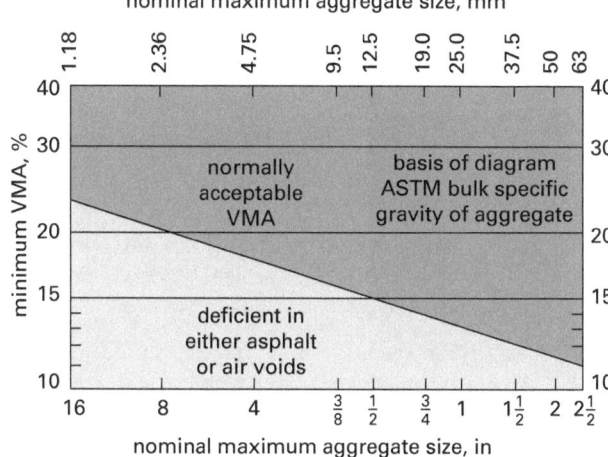

The objective of a Marshall test is to find the optimum asphalt content for the blend or gradation of aggregates. To accomplish this, a series of test specimens is prepared for a range of different asphalt contents so that when plotted, the test data will show a well-defined optimum value. Tests are performed in $\frac{1}{2}\%$ increments of asphalt content, with at least two asphalt content samples above the optimum and at least two below.

After each sample is prepared, it is placed in a mold and compacted with 35, 50, or 75 hammer blows, as specified by the design traffic category. The compaction hammer is dropped from a height of 18 in (457 mm), and after compaction, the sample is removed and subjected to the bulk specific gravity test, stability and flow test, and density and voids analysis. Results of these tests are plotted, a smooth curve giving the best fit is drawn for each set of data, and an optimum asphalt content is determined that meets the criteria of (a) maximum stability and unit weight and (b) median of limits for percent air voids from Table 5.6. (See Fig. 5.18 and Fig. 5.19.)

The *stability* of the test specimen is the maximum load resistance in pounds that the test specimen will develop at 140°F (60°C). The *flow value* is the total movement (strain) in units of $\frac{1}{100}$ in occurring in the specimen between the points of no load and maximum load during the stability test.

Table 5.6 Marshall Mix Design Criteria

mix criteria	light traffic ESALs < 10^4 surface and base min–max	medium traffic 10^4 < ESALs < 10^6 surface and base min–max	heavy traffic ESALs > 10^6 surface and base min–max
compactive effort, no. of blows/face	35	50	75
stability, N (lbf)	3336 (750)–NA	5338 (1200)–NA	8006 (1800)–NA
flow, 0.25 mm (0.01 in)	8–18	8–16	8–14
air voids, %	3–5	3–5	3–5
VMA, %	Varies with aggregate size. (See Fig. 5.19.)*		

*Multiply N by 0.225 to obtain lbf.

Reprinted with permission of the Asphalt Institute from *The Asphalt Handbook, Manual Series No. 4 (MS-4)*, 7th ed., Table 4.6, © 2007.

The *optimum asphalt content* for the mix is the numerical average of the values for the asphalt content. This value represents the most economical asphalt content that will satisfactorily meet all of the established criteria.

Use the following procedure to determine the optimum asphalt content for the mix. Examples of Marshall mix design graphs are given in Fig. 5.18.

step 1: Calculate the air voids in the mix, VTM, for each asphalt content using Eq. 5.18. Graph the VTM values, as percentages, against the asphalt content percentages.

step 2: Calculate the bulk specific gravity, G_{sb}, using Eq. 5.10, then determine the voids in mineral aggregate, VMA, for each asphalt content from Eq. 5.15. Graph the VMAs against the asphalt content percentages.

step 3: Graph the remaining data results, including the Marshall stability, theoretical density, and flow against the asphalt content percentages.

step 4: Using the graphs, determine the asphalt contents that coincide with the maximum stability, maximum unit weight, and the median air voids. The median air voids can be determined from Table 5.6, but a value of 4% is typically used. The asphalt content for VTM corresponds to the median air voids percentage.

step 5: Determine the optimum asphalt content by adding the asphalt contents determined in step 4 and dividing by three.

step 6: Verify the optimum asphalt content criteria for stability and flow against Table 5.6, then check the voids in mineral aggregate against Fig. 5.19. The optimum asphalt content is determined when all criteria are met.

Example 5.5

An asphalt mix uses an aggregate blend of 56% coarse aggregate (specific gravity of 2.72) and 44% fine aggregate (specific gravity of 2.60). The maximum aggregate size for the mixture is 3/4 in (19 mm). The resulting asphalt mixture is to have a specific gravity of 2.30. The asphalt content is to be selected on the basis of medium traffic and the following Marshall test data.

maximum content, % by weight of mix	Marshall stability (N)	flow (0.25 mm)	theoretical density (kg/m)^3	mixture specific gravity
4	4260	10.0	2300	2.42
5	4840	12.3	2330	2.44
6	5380	14.4	2340	2.44
7	5060	16.0	2314	2.40
8	3590	19.0	2250	2.30

Solution

step 1: Calculate air voids for each asphalt content and graph these values versus asphalt content. For example, for the 4% asphalt mixture, using Eq. 5.18,

$$\text{VTM}_{4\%} = \frac{G_{\text{mm}} - G_{\text{mb}}}{G_{\text{mm}}} \times 100\%$$
$$= \frac{2.42 - 2.30}{2.42} \times 100\%$$
$$= 4.96\%$$

step 2: Calculate the voids in mineral aggregate for each asphalt content. Use Eq. 5.10.

$$G_{\text{sb}} = \frac{P_{\text{ca}} + P_{\text{fa}}}{\dfrac{P_{\text{ca}}}{G_{\text{ca}}} + \dfrac{P_{\text{fa}}}{G_{\text{fa}}}} = \frac{56\% + 44\%}{\dfrac{56\%}{2.72} + \dfrac{44\%}{2.60}}$$
$$= 2.67$$

Use Eq. 5.14.

$$\text{VMA}_{4\%} = 100\% - \frac{G_{\text{mb}} P_s}{G_{\text{sb}}}$$
$$= 100\% - \frac{(2.30)(96\%)}{2.67}$$
$$= 17.3\%$$

step 3: Graph the results.

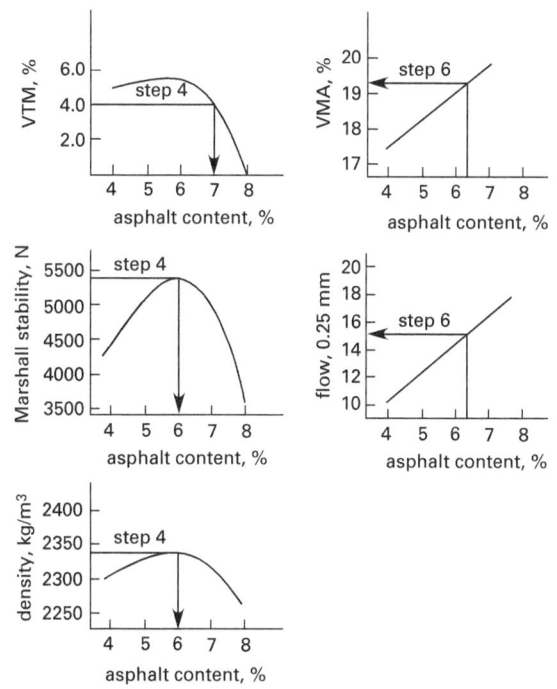

step 4: Obtain asphalt contents for maximum stability, maximum unit weight, and median air voids (4% from Table 5.6) from the graphs.

step 5: Determine the optimum AC.

$$\text{optimum AC} = \frac{\text{AC}_{\text{stability}} + \text{AC}_{\text{density}} + \text{AC}_{4\% \text{ air voids}}}{3}$$
$$= \frac{6.0\% + 6.0\% + 7.0\%}{3}$$
$$= 6.33\%$$

step 6: Check stability against the criteria for light traffic. Stability $\approx 5280 > 3336$ [OK]. Check VMA $= 19.25$ against Fig. 5.19 [OK]. Check flow $= 14.8$ against Table 5.6 [OK].

7. HVEEM MIX DESIGN

The Hveem mix design method was developed in the 1920s and has been extensively used in some western states. Like the Marshall method, the goal is to determine the optimum asphalt content. Unlike the Marshall method, it has the additional sophistications of measuring resistance to shear and considering asphalt absorption by aggregates. It has the disadvantage of requiring more specialized and nonportable equipment for mixing, compaction, and testing.

The method makes three assumptions: (1) the optimum asphalt binder content depends on the aggregate surface area and absorption, (2) stability is a function of aggregate particle friction and mix cohesion, and (3) HMA durability increases with asphalt binder content.

After selecting the materials, the *centrifuge kerosene equivalent* (CKE) of the fine aggregate and the retained surface oil content of the coarse aggregate are determined. These measures of surface absorption are used to estimate the optimum asphalt content.[1] Trial blends are formulated over the asphalt range of [CKE−1%, CKE+2%] in 0.5% increments. Then, specimens of the trial blends are prepared by heating and densifying in a California kneading compactor.[2]

A Hveem stabilometer, a closed-system triaxial test device, is used to determine the stability (i.e., the horizontal deformation under axial load). The stabilometer applies an increasing load to the top of the sample at a predetermined rate. As the load increases, the lateral pressure is read at specified intervals. The stability (*stabilometer value*), S, is calculated as

$$S = \frac{22.2}{\dfrac{p_h D}{p_v - p_h} - 0.222} \quad 5.23$$

---

[1] It is important that the trial blends include asphalt contents greater than and less than the optimum content. The purpose of determining the CKE and surface oil content is to estimate the optimum content prior to formulating trial blends. However, it is also practical to simply formulate trial blends within a range of 4–7% of asphalt, varying each blend by an increment such as 0.5%. Because of this alternative methodology of bracketing the optimum asphalt content, the CKE process has become essentially obsolete.

[2] The kneading compactor produces compactions that are more similar to pavements that have been roller-compacted with steel- and rubber-tired rollers.

In Eq. 5.23, p_v is the vertical pressure, which is typically 400 lbf/in² (2.8 MPa), and p_h is the corresponding horizontal pressure. D is the deflection in units of 0.01 in (0.25 mm). The stabilometer value ranges from 0 to 90. Zero represents a liquid, a condition where lateral pressure is equal to vertical pressure. 90 represents an incompressible solid, a condition where there is no lateral pressure regardless of the vertical pressure. Minimum stabilities vary according to a roadway's traffic volume, measured in *equivalent single-axle loads* (ESALs). Minimum stabilities of 30, 35, and 37 are typically specified for light (ESALs $< 10^4$), medium ($10^4 <$ ESALs $< 10^6$), and heavy (ESALs $> 10^6$) traffic, respectively.

HMA mixtures rarely fail from cohesion, but oil mixtures may require subsequent testing in a cohesiometer to determine cohesive strength. Basically, the sample used in the stability test is loaded as a cantilevered beam until it fails. The cohesiometer value, C, is determined from the mass of the load (shot), L, in grams, the width (diameter) of the sample, W, in inches, and the height of the specimen, H, in inches. A minimum cohesiometer value of 100 is typical for HMA, while acceptance values for oil mixtures may be 50.

$$C = \left(\frac{L}{W}\right)(0.20H + 0.044H^2) \quad \text{[U.S. only]} \quad 5.24$$

Visual observation, volumetrics (air voids, voids filled with asphalt, and voids in the mineral aggregate), and stability are used to determine the optimum asphalt content. The design asphalt content is selected as the content that produces the highest durability without dropping below a minimum allowable stability. Essentially, as much asphalt binder as possible is used while still meeting minimum stability requirements. The *pyramid method* can be used to select the optimum asphalt content.

step 1: Place all sample asphalt percentages in the base of the pyramid.

step 2: Eliminate any samples with moderate to severe surface *flushing*.[3]

step 3: Place the three highest asphalt percentages in the next pyramid level.

step 4: Eliminate any samples that do not meet the stability requirements.

step 5: Place the two highest asphalt percentages in the next pyramid level.

step 6: Choose the mixture that has the highest asphalt content while still having 4% air voids. This is the optimum mixture.

Example 5.6

An HMA sample with a minimum Hveem stability of 37 is required. The following specimens were prepared. What is the optimum asphalt content?

sample	asphalt content	stability	air voids content	comments
1	4%	37	3.9%	
2	4.5%	38	4.0%	
3	5%	39	4.1%	
4	5.5%	35	3.9%	
5	6%	38	3.8%	moderate surface flushing noted

Solution

Use the pyramid method.

step 1: Place all sample asphalt percentages in the base of the pyramid.

step 2: Eliminate any samples with more than moderate surface flushing. Sample 5 had moderate surface flushing noted, so it should be excluded.

step 3: Of the asphalt contents remaining after step 2, add the three highest percentages in the next level. This eliminates 4% from the pyramid.

step 4: Check the stability of each sample remaining after step 3 and eliminate any sample with a stability less than the stability requirements. A minimum Hveem stability of 37 is required, which eliminates sample 4, 5.5%, which has a stability of 35.

step 5: Place the highest remaining asphalt percentages on the next level. Since only 4.5% and 5% remain, move both to the next level.

step 6: The optimum asphalt content is the highest asphalt content that still has 4% air voids. Therefore, the optimum mixture has an asphalt content of 5%.

step 6			5%		
step 5		4.5%	5%		
steps 3 and 4		4.5%	5%	~~5.5%~~	
steps 1 and 2	4%	4.5%	5%	5.5%	~~6%~~

---

[3]Another name for flushing is *bleeding*. This is an indication of excessive asphalt. A specimen with light flushing will have sheen. With moderate flushing, paper will stick to the specimen. With heavy flushing, surface puddling or specimen distortion will be noted.

8. SUPERPAVE MIX DESIGN PROCEDURES

Superpave is a performance-based system for designing asphalt material with characteristics that eliminate environmental factors that decrease pavement life and increase maintenance costs, such as pavement rutting, cracking due to high or low temperatures, fatigue cracking, and heavy traffic. The Superpave process has three major components: asphalt binder specification, mixture design and analysis, and a computer software system.[4]

Superpave mix design builds upon many traditional asphalt procedures. It emphasizes a closer relationship between laboratory test procedures and in-place asphalt performance. Applicable AASHTO procedures include AASHTO T312, *Standard Method for Preparing and Determining the Density of Hot Mix Asphalt* (HMA) *Specimens by Means of the SHRP Gyratory Compactor*, and AASHTO R30, *Standard Practice for Mixture Conditioning of Hot Mix Asphalt* (HMA). In general, the procedures described by the Asphalt Institute involve

- selecting asphalt and aggregate materials that meet respective criteria
- developing several aggregate trial blends to meet Superpave gradation requirements
- blending asphalt with trial blends and short-term oven aging the mixtures
- compacting specimens and analyzing the volumetrics of the trial blends
- selecting the best trial blend as the design aggregate structure, then compacting samples of the design aggregate structure at several asphalt contents to determine the design asphalt content

An outline of Superpave mix design procedures is shown in App. 5.E.

Aggregate Properties for Superpave

The Superpave study surveyed pavement experts to determine which aggregate properties and specifications were the most effective. These properties, called *consensus aggregate properties*, were selected based on their use and specified values. Consensus aggregate properties include coarse aggregate angularity, fine aggregate angularity, flat and elongated particles in coarse aggregate, and clay content. Table 5.7 gives the required minimum values as a function of traffic level and position within the pavement. Superpave measures a roadway's traffic volume using the design ESAL, which is the estimated volume for a 20 yr design period. ESALs are used regardless of what the roadway's actual design life is and are estimated as outlined in AASHTO's *Guide for Design of Pavement Structures*. Detailed explanations of the aggregate properties can be found in *Superpave Mix Design, Superpave Series No. 2 (SP-2)*.

Coarse aggregate angularity is the percentage by weight of aggregates that are less than 0.19 in (4.75 mm) and have at least one fractured face. It has a high degree of internal friction to help prevent permanent deformation and can be measured by test procedures in ASTM D5821, *Determining the Percentage of Fractured Particles in Coarse Aggregate*.

Uncompacted void content of fine aggregate is the percentage of air voids present in loosely compacted aggregates smaller than 0.09 in (2.36 mm). Higher percentages of air voids indicate more fractured faces and can be determined using AASHTO T304, *Uncompacted Void Content of Fine Aggregate*.

Flat and elongated particles, which are characteristic of coarse aggregate, are the percentage by mass of the coarse aggregates with a length to thickness ratio greater than five. Because they tend to break during construction or under loads, it is undesirable to have large amounts in a mixture. Test methods are performed on particles greater than 4.75 mm (0.19 in) and follow procedures described in ASTM D4791, *Flat and Elongated Particles in Coarse Aggregate*.

Clay content is the percentage of clay material by weight contained in the aggregate fraction that is finer than a 4.75 mm sieve. Tests are performed according to AASHTO T176, *Plastic Fines in Graded Aggregates and Soils by Use of the Sand Equivalent Test* (ASTM D2419). The terms sand equivalent and clay content are used interchangeably when referring to the sand equivalent test.

Properties are source specific and are critical to Superpave performance. Source properties include toughness, soundness, and deleterious materials. These property values are normally set by local jurisdictions depending on the availability of aggregate material within economical reach.

Toughness is the percentage loss of material from an aggregate blend during the *Los Angeles abrasion test* (AASHTO T96, ASTM C131, or ASTM C535). For this test, coarse aggregate larger than 2.36 mm (0.09 in) is impacted and ground by steel spheres. The mass percentage of coarse material lost is then measured. An aggregate blend usually has a loss range of 35–40%.

Soundness is the percentage loss of material from weathering, simulated by immersing the test sample in saturated solutions of sodium sulfate or magnesium sulfate, followed by oven drying. The immersion and drying cycle is repeated, and the sample is measured over a required number of cycles. The solution salts penetrate into the void spaces of the aggregate, then rehydrate

---

[4]A full review of the research can be found in the *National Cooperative Highway Research Program* (NCHRP) *Report 539*, published by the Transportation Research Board, www.TRB.org.

Table 5.7 Superpave Aggregate Consensus Property Minimum Requirements

design ESALs[a] (million)	minimum coarse aggregate angularity (%)		minimum uncompacted void content of fine aggregate (%)		minimum sand equivalent (%)	maximum flat and elongated[b] (%)
	≤ 100 mm	> 100 mm	≤ 100 mm	> 100 mm		
<0.3	55/–	–/–	–	–	40	–
0.3 to <3	75/–	50/–	40	40	40	10
3 to <10	85/80[c]	60/–	45	40	45	10
10 to <30	95/90	80/75	45	40	45	10
≥ 30	100/100	100/100	45	45	50	10

(Multiply mm by 3.937×10^{-2} to obtain in.)

[a]Design ESALs (equivalent single-axle loads) are the anticipated project traffic level expected on the design lane over a 20-year period. Regardless of the actual design life of the roadway, determine the design ESALs for 20 years and choose the appropriate N_{design} level.
[b]Criterion based upon a 5:1 maximum-to-minimum ratio. (If less than 25% of a layer is within 3 in (100 mm) of the surface, the layer may be considered to be below 3 in (100 mm) for mixture design purposes.)
[c]85/80 denotes that 85% of the coarse aggregate has one fractured face and 80% has two or more fractured faces.

Reprinted with permission from the Asphalt Institute, *The Asphalt Handbook, Manual Series No. 4 (MS-4)*, 7th ed., Table 4.2, © 2007.

after drying and exert expansive forces similar to freezing water. Typical values are 10–20% loss over five cycles. The test is performed according to AASHTO T104 or ASTM C88.

Deleterious materials are contaminants such as clay lumps, wood, shale, mica, and coal. The analysis can be performed on both coarse and fine aggregate, and is covered by AASHTO T112 or ASTM C142. Using wet sieving through specified sieves, the mass percentage of material lost is reported as the percentage of clay lumps and friable particles. Allowable percentages vary from a low of 0.2% to a high of 10%, depending on the exact composition of the contaminant.

Binding Properties for Superpave

Performance testing of the binder is an important element of Superpave. AASHTO MP1 specifications (shown in App. 5.D) are selected based on the ESALs and climate conditions of the finished pavement. AASHTO asphalt binder ratings are based on the temperature requirements placed on the binder. Performance grade binders are selected in six-degree intervals (e.g., starting from a high-temperature grade of 40°C and a low-temperature grade of −10°C). For example, a binder classified as PG 58-22 means that the binder must meet physical property high-temperature requirements greater than or equal to 58°C, and low-temperature physical requirements less than or equal to –22°C. In addition to the seven common grades shown in App. 5.D, the high and low temperatures extend as far as necessary in standard six-degree increments.

There are three methods of selecting the binder grade.

- *geographic area:* An agency develops a map showing the binder grade that the designer will use based on weather and/or policy decisions.
- *pavement temperature:* The designer determines design pavement temperatures.
- *air temperature:* The designer determines design air temperatures, which are converted to design pavement temperatures.

Asphalt binder temperature grades are selected using the temperatures at 0.79 in (20 mm) below the pavement surface for the high temperature and at the pavement surface for the low temperature. For the high temperature, Eq. 5.25 is used to convert the seven-day high air temperature to the high pavement design temperature. Equation 5.25 is based on the following standard values: heat radiation to the atmosphere (0.70), radiation transmission through air (0.81), solar gain (0.90), and wind speed (14 ft/sec or 4.5 m/s).

$$T_{20\,\text{mm}} = \begin{pmatrix} T_{\text{air}} - (0.00618)(\text{lat})^2 \\ + (0.2289)(\text{lat}) \\ + 42.2°C \end{pmatrix}(0.9545) \\ -17.78°C \quad \quad 5.25$$

$T_{20\,\text{mm}}$ is the high pavement design temperature at a depth of 0.79 in (20 mm), T_{air} is the seven-day, average high air temperature in degrees Celsius, and lat is the geographical latitude of the project in degrees.

For the low design pavement temperature, the preferred method is to use the LTPPBind software developed by the Long-Term Pavement Performance (LTPP) program of the FHWA Turner-Fairbank Highway Research Center. The software deals strictly with the PG binder selection and is a Windows-based program that helps highway agencies choose the most effective and cost-effective Superpave asphalt binder performance grade for a particular site.

Temperature information is available in LTPPBind for more than 6000 weather stations in the United States and Canada and includes more than 20 years of data. The software allows input variations so that the user can quickly compare "what if" scenarios and can customize the inputs at a specific location for a specific project.

For initial estimating purposes without using the software, the pavement surface can be assumed to be equal to the low air temperature. This is a conservative approach because the pavement surface temperature is almost always warmer than the air temperature in cold weather. Another method is to use Eq. 5.26, which is used in the Asphalt Institute manuals.

$$T_{\min} = 0.859\, T_{\text{air}} + 1.7°C \qquad 5.26$$

Reliability refers to the percent probability in a single year that the actual temperature (a one-day low or a seven-day high) will not exceed the design temperature calculated from Eq. 5.25 and Eq. 5.26. The design pavement temperatures represent a 50% reliability level. Standard deviations, in degrees Celsius, can be used to achieve other reliability levels. A 95% reliability level is obtained from the design pavement temperatures plus or minus two standard deviations, $2s$, and a 98% reliability from plus or minus three standard deviations, $3s$. Agencies may select a reliability level in order to maximize specific goals. For example, funding restrictions may be such that 95% reliability is used in order to stretch available tax dollars and gain additional miles of roadway.

As described in Sec. 5.5, asphalt pavements react in a viscoelastic manner according to the duration of applied loads and the frequency of loadings. Binder grades can be adjusted to accommodate expected high traffic volumes and speeds. This practice is known as *grade bumping*, which adjusts the binder selection based exclusively on climate to ensure adequate performance. For example, the binder for slow-moving loads on a roadway with a design traffic of 3,000,000 ESALs should be one high-temperature grade higher than the binder grade selected for climate, as shown in Table 5.8. If the original binder grade were PG 52, PG 58 would be used instead. These adjustments can reduce, for example, the washboard effect caused by a large volume of trucks waiting at a signalized intersection or buses waiting with their engines running to allow passengers at a heavily patronized bus stop to board. Table 5.8 summarizes AASHTO's grade-bumping policy as presented in the Asphalt Institute's *SP-2*.

Table 5.8 *Binder High Temperature Grade Selection Increase Based on Traffic Speed and Traffic Load*[a]

design ESALs[b] (millions)	adjustment to binder PG grade[c] traffic load rate		
	standing[d]	slow[e]	standard[f]
<0.3	−[g]		
0.3 to <3	2	1	
3 to <10	2	1	
10 to <30	2	1	−[g]
≥30	2	1	1

[a]Practically, performance graded binders stiffer than a PG 82-XX should be avoided. In cases where the required adjustment to the high-temperature binder grade would result in a grade higher than a PG 82, consideration should be given to specifying a PG 82-XX and increasing the design ESALs by one level (e.g., 10 to <30 million increased to ≥30 million).
[b]Design ESALs are the anticipated project traffic levels expected on the design lane over a 20-year period. Regardless of the actual design life of the roadway, determine the design ESALs for 20 years and choose the appropriate N_{design} level.
[c]Increase the high-temperature grade by the number of grade equivalents indicated (one grade equivalent to 42°F or 6°C). Do not adjust the low-temperature grade.
[d]Standing traffic—where the average traffic speed is less than 11 mph (20 kph).
[e]Slow traffic—where the average traffic speed ranges from 11 mph to 38 mph (20 kph to 70 kph).
[f]Standard traffic—where the average traffic speed is greater than 38 mph (70 kph).
[g]Consideration should be given to increasing the high-temperature grade by one grade equivalent.

Reprinted with permission from the Asphalt Institute, *Superpave Mix Design (SP-2)*, 2001, Table 3.1.

Other factors may influence pavement performance, which the engineer should take into consideration when selecting a binder. Unlike fatigue cracking, which is typically affected by pavement structure, subgrade conditions, and traffic loadings, permanent deformation (i.e., rutting) is a function of shear strength and aggregate properties. Low-temperature cracking also correlates closely with binder properties. Careful selection of binder properties can reduce pavement damage and extend the life of the asphalt.

Software such as LTPPBind can be used to help determine the appropriate grade, as can the AASHTOWare, which assists in the entire Superpave mix design process, including quality control and quality assurance guidelines.

Example 5.7

A midwestern location with a latitude of 41.42° has a mean seven-day maximum air temperature of 31°C and a standard deviation of 2°C. In an average year, there is a 50% chance that the seven-day maximum air temperature will exceed 31°C. The low design temperature in the same location has a one-day minimum temperature of −20°C and a standard deviation of 4°C. What are the design temperatures for 50% and 98% reliabilities?

Solution

For 50% reliability, calculate the design high temperature using Eq. 5.25.

$$\begin{aligned}T_{20\,\text{mm}} &= \begin{pmatrix} T_{\text{air}} - (0.00618)(\text{lat})^2 \\ + (0.2289)(\text{lat}) + 42.2°\text{C}\end{pmatrix}(0.9545) \\ &\quad - 17.78°\text{C} \\ &= \begin{pmatrix} 31°\text{C} - (0.00618)(41.42°)^2 \\ + (0.2289)(41.42°) + 42.2°\text{C}\end{pmatrix}(0.9545) \\ &\quad - 17.78°\text{C} \\ &= 51°\text{C}\end{aligned}$$

For the design low temperature, use Eq. 5.26.

$$\begin{aligned}T_{\min} &= 0.859\, T_{\text{air}} + 1.7°\text{C} \\ &= (0.859)(-20°\text{C}) + 1.7°\text{C} \\ &= -15°\text{C}\end{aligned}$$

For the 50% reliability level, the design pavement temperatures are 51°C and −15°C. A 98% reliability level is obtained from the mean plus or minus three standard deviations.

$$\begin{aligned}51°\text{C} + 3s &= 51°\text{C} + (3)(2°\text{C}) \\ &= 57°\text{C} \\ -15°\text{C} - 3s &= -15°\text{C} - (3)(4°\text{C}) \\ &= -27°\text{C}\end{aligned}$$

Therefore, for the 98% reliability level, the design pavement temperatures are 57°C and −27°C.

Superpave Gyratory Compactor

Most of the test equipment required for the preparation of a Superpave asphalt-mixture design is carried over from traditional asphalt test equipment, such as ovens, mixers, scales, and so forth. One of the major elements introduced for the Superpave mix design process is laboratory testing with a *Superpave gyratory compactor* (SGC). The SGC was developed to emulate field conditions of wheel pass repetitions. It involves placing a 4.6 in (150 mm) diameter asphalt mix sample in a gyrating cylinder and applying an 87 psi (600 kPa) load during rotational gyrations. The sample is angled at 1.25° from the axis of the compactor and is rotated at 30 rpm. The test specimen's height is measured throughout the test, and a compaction characteristic is developed for the subject sample.

The SGC is designed to impart a certain degree of lateral loading on the test sample to emulate rutting caused by tire movement, in addition to the vertical load carrying characteristics. The SGC can also provide data during compaction to help determine reactions of a particular mix while compaction is occurring.

Superpave asphalt mixtures are designed at a specific level of compactive effort, which is a function of the design number of gyrations, N_{design}. N_{design} is a function of climate and traffic level. Climate condition is a function of the average high air temperature determined using the average seven-day maximum air temperature for the project conditions. This is the equivalent temperature at 50% reliability. The traffic level is the design ESALs, taken from actual traffic data or from other nearby data and adapted to the project location. N_{initial} is an estimation of the mixture's ability for compaction. $N_{\max}$ is a laboratory density determined by using additional SGC specimens of the selected design as a check to help guard against plastic failure caused by traffic levels that exceed the design level. Table 5.9 shows the range of values for determining N_{initial}, N_{design}, and $N_{\max}$. N_{initial} and $N_{\max}$ are determined by Eq. 5.27 and Eq. 5.28.

$$\log_{10} N_{\text{initial}} = 0.45 \log_{10} N_{\text{design}} \qquad 5.27$$

$$\log_{10} N_{\max} = 1.10 \log_{10} N_{\text{design}} \qquad 5.28$$

Design Selection

Individual asphalt and aggregate materials are selected from locally available and approved sources. Using local selections, trial blends are made according to Superpave gradation requirements. Gradation blends for the five Superpave aggregates are shown in App. 5.D.

While there is no set number of trial blends, trials usually begin with three blends as their starting point. Once selected, the initial blend's volumetric analysis is performed on each of the individual three blends. In order to save time, the Asphalt Institute suggests that the asphalt binder content for aggregates have a bulk specific gravity of 2.65, as shown in Table 5.10. Aggregate with significantly higher bulk specific gravity may need less asphalt, while aggregate with a lower bulk specific gravity may need more asphalt.

Once the design aggregate blend is selected from the trial samples, specimens are compacted at varying asphalt binder contents. The mixture properties are evaluated to determine the design asphalt binder content.

To obtain the design asphalt binder content for Superpave design, a minimum of four asphalt contents must be evaluated. A minimum of two specimens are compacted at the trial blend's estimated asphalt content. A minimum of one specimen each is prepared at +0.5%, +1.0%, and −0.5% of the estimated asphalt content. The process is repeated until a satisfactory level of performance is achieved on the aged samples. Table 5.11 shows Superpave design requirements recommended by the Asphalt Institute.

Table 5.9 Superpave Design Gyratory Compaction Effort

design ESALs* (millions)	compaction parameters $N_{initial}$	N_{design}	N_{max}	typical roadway applications
< 0.3	6	50	75	Applications include roadways with very light traffic volumes, such as local roads, county roads, and city streets where truck traffic is prohibited or at a very minimal level. Traffic on these roadways would be considered local in nature, not regional, intrastate, or interstate. Special-purpose roadways serving recreational sites or areas may also be applicable to this level.
0.3 to < 3	7	75	115	Applications include collector roads or access streets. Medium-trafficked city streets and the majority of country roadways may be applicable to this level.
3 to < 30	8	100	160	Applications include many two-lane, multilane, divided, and partially or completely controlled access highways. Among these are medium to heavily trafficked city streets, many state routes, U.S. highways, and some rural intersections.
≥ 30	9	125	205	Applications include the vast majority of the U.S. Interstate System, both rural and urban in nature. Special applications such as truck-weighing stations or truck-climbing lanes on two-lane highways may also be applicable to this level.

*The significant project traffic level expected on the design lane over a 20-year period. Regardless of the design life of the roadway, determine the design ESALs for 20 years.

Reprinted with permission from the Asphalt Institute, *The Asphalt Handbook, Manual Series No. 4 (MS-4)*, 7th ed., Table 4.4, © 2007.

Table 5.10 Typical Asphalt Binder Content for Aggregate with G_{sb} of 2.65

nominal maximum aggregate size (mm)	trial asphalt binder content (%)
37.5	3.5
25.0	4.0
19.0	4.5
12.5	5.0
9.5	5.5

(Multiply mm by 3.937×10^{-2} to obtain in.)

9. VALUE ENGINEERING

Value engineering (VE), as outlined by the SAVE International (formerly the Society of Value Engineers), is a process of quickly evaluating a project as a solution to a set of primary objectives. VE is a system of *value methodology*, which is often referred to by a variety of terms, including value engineering, value analysis, and value management. The process takes place in a workshop setting using a team of persons with expert experience in several disciplines, such as structures, drainage, building construction, electrical, or other applicable disciplines. The team is given a time schedule to go through a prescribed valuation process and make a recommendation about the relative value of the project elements compared to objectives, which are set forth in the first hours of the workshop session. VE can be performed at any time during a project, but is usually most effective when performed about two-thirds through completion of the project design. There needs to be enough of a definition of what the final project will look like. Also, there must be a commitment to embrace the value engineering recommendations, even if it means making changes necessary to increase the project's value. The premise of the VE process is based on Pareto's 80/20 rule, stating that 80% of the project cost is determined by 20% of the project elements. Therefore, the VE team needs to look at only the largest 20% of work items to optimize the overall value of the project.

Since the 1980s, projects receiving government funds are required to have VE performed. This was based on the expectation of as much as 30% reduction in project costs. The cost of the VE study was to be paid for with the project savings. In practice, this goal was met or even exceeded at times. Quite often, project savings come from the design process—not considering viable alternative solutions, or from misapplying a solution from another project that is not optimal for the project at hand. There can be other experiences, though. Projects can actually have an increase in cost after value engineering is applied, based on the ability to execute the project within constraints, such as site topography. Third-party review systems, such as VE, can indeed

Table 5.11 Superpave Volumetric Design Requirements

design ESALs[a] (million)	required density (% of theoretical maximum specific gravity)			minimum voids in the mineral aggregate (%)					voids filled with asphalt (%)	dust-to-binder ratio[b]
	$N_{initial}$	N_{design}	N_{max}	nominal maximum aggregate size (mm)						
				37.5[c]	25.0[d]	19.0	12.5	9.5[e]		
<0.3	≤ 91.5	–	–	–	–	–	–	–	70–80	0.6–1.2
0.3 to < 3	≤ 90.5	96.0	≤ 98.0	11.0	12.0	13.0	14.0	15.0	65–78	0.6–1.2
3 to < 10	≤ 89.0	96.0	≤ 98.0	11.0	12.0	13.0	14.0	15.0	65–75	0.6–1.2
10 to < 30	≤ 89.0	96.0	≤ 98.0	11.0	12.0	13.0	14.0	15.0	65–75	0.6–1.2
≥ 30	≤ 89.0	96.0	≤ 98.0	11.0	12.0	13.0	14.0	15.0	65–75	0.6–1.2

(Multiply mm by 3.937×10^{-2} to obtain in.)

[a]Design ESALs are the anticipated project traffic level expected on the design lane over a 20-year period. Regardless of the actual design life of the roadway, determine the design ESALs for 20 years, and choose the appropriate N_{design} level.
[b]If the aggregate gradation passes beneath the boundaries of the aggregate restricted zone, consideration should be given to increasing the dust-to-binder ratio criteria from 0.6–1.2 to 0.8–1.6.
[c]For 37.5 mm (1.5 in) nominal maximum size mixtures, the specified lower limit of the VFA shall be 64% for all design traffic levels.
[d]For 25.0 mm (1 in) nominal maximum size mixtures, the specified lower limit of the VFA shall be 67% for design traffic levels < 0.3 million ESALs.
[e]For 9.5 mm (0.375 in) nominal maximum size mixtures, the specified VFA range shall be 73% to 76% for design traffic levels ≥ 3 million ESALs.

Reprinted with permission from the Asphalt Institute, *The Asphalt Handbook, Manual Series No. 4*, (MS-4), 7th ed., Table 4.5, © 2007.

reduce project costs, or on occasion, increase costs, with the goal of making the project more viable. The increased use of design-build procedures has improved the opportunity for applying project designs that are in close context with local constructability, capability, and efficiency, thereby reducing much of the unnecessary effort and expense caused by inefficient use of local resources to execute unfamiliar design conditions. In effect, design-build can be considered a method of direct application of VE principles.

Experienced designers who use optimization techniques throughout their design have nothing to fear from VE, as most often the case is that the value engineer finds little to optimize within a well-thought-out and well-designed project. Drawbacks with a process, such as VE being performed as a one-time activity, include the danger that projects can change greatly after the value engineering is performed, with the changes possibly bypassing VE review. Another drawback is when project designers anticipate VE will discover design errors before the project is delivered to the client, there can be a trend to compress the design schedule and eliminate quality checks and accuracy reviews normally performed during the design process. Another opinion is that VE is not engineering at all, but rather second-guessing under pressure of a single goal of cost reduction. Thus, the confusion of value is too often related to cost alone. In spite of misgivings, VE can be an effective quality control tool, and continues to be used as a peer review process that leads to project cost optimization based on project objectives.

Value engineering is often performed by contractors and construction managers after the project is bid. In this type of value engineering, the contractor or project manager proposes a design variation that is expected to reduce cost or reduce completion time. This type of value engineering on an as-bid project should result in a change order that does not violate the original goals and objectives of the project. An example of a conflict in objectives is when the owner and designer specify a contract item or procedure because of longevity or maintenance costs; whereas the construction contractor sees only the project itself and feels no responsibility for longevity or maintenance procedures.

Value engineering is also used in design-build projects in which there is one contract combining the detailed final design with the construction process. Value engineering, in these cases, is a part of the project management, and it is easier to control compliance with the project's original objectives.

10. PRACTICE PROBLEMS

1. A project's schedule is to be adjusted in order to minimize cost. Currently, the project will complete on schedule, assuming no delays arise that affect the critical path, but the contract contains a bonus/penalty clause that the contractor would like to take advantage of to achieve maximum benefit-cost. The activity-on-arc network for the project has been simplified by grouping component activities into logical sequence groups using three labor crews.

To expedite the work, crew 1 can increase daily output using overtime at 1.5 times the normal daily rate (for simplicity use 1.5 times the total daily rate). Crew 2 and crew 3 must be doubled in size in order to increase daily output, but will not accrue further cost increase from overtime. Crew costs are as follows.

crew	normal daily rate ($)	expedited daily rate ($)
1	800	1200
2	1000	2000
3	650	1300

Supervision and on-site support cost $500 for each day of on-site work activity. The activity schedule is

activity	duration (days)	predecessors	successors	crew
A	3	–	B, F	3
B	6	A	C, E	1
C	21	B	J, D	1
D	12	C, H	L	1
E	0	B	G	–
F	4	A	G	2
G	14	E, F	H, I	2
H	20	G	D, J	2
I	5	G	K	3
J	3	C, H	K	2
K	5	I, J	L	3
L	5	D, K	–	3

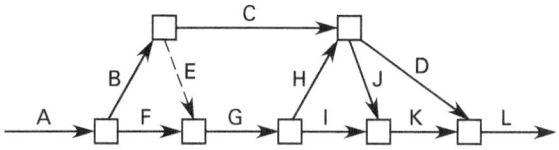

The expedited schedule reduces certain activities to the following durations. These reduced-duration activities use the expedited daily rates to determine the cost of the activities.

activity	normal days	expedited days
B	6	5
C	21	18
D	12	8
G	14	9
H	20	12

The bonus/penalty clause in the contract states that the contractor will receive a bonus if the project is completed more than three days early and will be penalized if the project is completed more than three days late. The bonus is $1000 per day early the project is completed, not including the first three days, and the penalty is $2500 per day late the project is completed, not including the first three days.

(a) What is the critical path for the normal schedule?

(A) A-B-E-G-H-D-L

(B) A-F-G-H-D-L

(C) A-B-C-D-L

(D) A-F-G-I-K-L

(b) What is the normal, nonexpedited duration of the project?

(A) 45 days

(B) 60 days

(C) 62 days

(D) 77 days

(c) The original labor crew cost, including cost of supervision, is most nearly

(A) $30,000

(B) $84,000

(C) $94,000

(D) $114,000

(d) Comparing total costs, how much more or how much less does the expedited schedule cost than the normal schedule?

(A) $200 less

(B) $200 more

(C) $2150 less

(D) $2150 more

(e) Including the project bonus or penalty, what is most nearly the cost of the expedited schedule compared to the cost of the normal schedule?

(A) $14,800 less

(B) $14,800 more

(C) $12,800 less

(D) $12,800 more

(f) Which of the five activities that can be expedited will NOT help decrease the duration of the project if expedited?

(A) B

(B) C

(C) H

(D) none of the above

2. Superpave binder is being selected for a location at 42.2° latitude with a mean seven-day high temperature of 31.9°C and a low air temperature of −21.1°C. The high temperature standard deviation is 1.7°C, and the low temperature standard deviation is 3.9°C. Traffic load rate is projected at 3,000,000 ESALs, with fast traffic operating at greater than 70 kph.

(a) What is the expected pavement high temperature at 20 mm below the pavement surface at the 95% reliability confidence level?

(A) 32°C

(B) 35°C

(C) 56°C

(D) 58°C

(b) What is the binder selection for this location at the 95% reliability level?

(A) PG 52-20

(B) PG 56-20

(C) PG 58-22

(D) PG 58-28

3. Aggregate blends are being prepared for asphalt paving. Coarse aggregate is stored in three cold bins with the following characteristics.

bin	bulk specific gravity, G_{sb}	apparent specific gravity, G_{sa}
1	2.720	2.787
2	2.689	2.755
3	2.450	2.510

A trial blend is prepared with a mix of 25% from bin 1, 35% from bin 2, and 40% from bin 3. The blend has 5% asphalt binder by total mass of mixture. Use 1.030 as the specific gravity of asphalt, G_b. The maximum specific gravity, G_{mm}, of the trial blend is 2.434.

(a) What is most nearly the bulk specific gravity of the trial blend?

(A) 2.385

(B) 2.595

(C) 2.689

(D) 2.720

(b) What is most nearly the effective specific gravity of the trial blend?

(A) 2.510

(B) 2.595

(C) 2.622

(D) 2.992

(c) What is most nearly the percentage of asphalt absorption?

(A) 0.4%

(B) 0.8%

(C) 4%

(D) 5%

(d) What is most nearly the effective asphalt content of the mixture?

(A) 0.046%

(B) 4.4%

(C) 4.6%

(D) 5.0%

4. A concrete mix is being prepared with the following material properties.

material	specific weight	oven dry bulk specific gravity, G_{sb}	moisture by weight (%)	absorption by weight (%)
portland cement	1.00	3.14	–	–
fine aggregate	2.27	2.66	3.25	0.41
coarse aggregate	3.65	2.72	2.08	0.25

The effective water content is 5.83 gal/sack of cement. There are 3% air voids in the final mix.

(a) What is most nearly the water-cement ratio of the mix?

(A) 0.016

(B) 0.420

(C) 0.520

(D) 32.0

(b) Most nearly how much water per sack of cement must be added to obtain the effective water content?

(A) 37.7 lbf/sack

(B) 39.9 lbf/sack

(C) 47.7 lbf/sack

(D) 48.5 lbf/sack

(c) What is most nearly the effective yield per sack of cement?

(A) 0.256 yd^3/sack

(B) 0.264 yd^3/sack

(C) 7.13 yd^3/sack

(D) 16.5 yd^3/sack

5. An asphalt mixture has the following properties.

bulk specific gravity of the aggregate mix, $G_{sb} = 2.687$

maximum specific gravity of asphalt mix,
$G_{mm} = 2.526$

bulk specific gravity of the compacted mixture,
$G_{mb} = 2.438$

asphalt content, $P_b = 5.25\%$

specific gravity of the asphalt, $G_b = 1.035$

volume of water absorbed into the bulk aggregate = 2.76%

volume of water absorbed by the aggregate = 2.72%

asphalt absorption factor, $P_{ba} = 0.7$

(a) What is most nearly the effective asphalt content?

(A) 4.54%

(B) 4.60%

(C) 5.25%

(D) 5.95%

(b) What is most nearly the volume of air voids (VTM) in the mixture?

(A) 2.76%

(B) 3.48%

(C) 3.61%

(D) 5.90%

(c) What is most nearly the volume of voids in the mineral aggregate (VMA)?

(A) 0.91%

(B) 2.76%

(C) 14.0%

(D) 35.3%

(d) What is most nearly the volume of voids filled with asphalt (VFA)?

(A) 10.5%

(B) 14.0%

(C) 28.8%

(D) 75.2%

SOLUTIONS

1. (a) To find the critical path for the project, find the total duration of each path through the CPM network and select the path that has the longest total duration. Possible paths include path A-B-C-D-L, path A-B-C-J-K-L, path A-B-E-G-H-D-L, path A-B-E-G-H-J-K-L, path A-B-E-G-I-K-L, path A-F-G-H-D-L, path A-F-G-H-J-K-L, and path A-F-G-I-K-L.

Path A-B-C-D-L has a total duration of

$$D = 3 \text{ days} + 6 \text{ days} + 21 \text{ days} + 12 \text{ days} + 5 \text{ days}$$
$$= 47 \text{ days}$$

Path A-B-C-J-K-L has a total duration of

$$D = 3 \text{ days} + 6 \text{ days} + 21 \text{ days} + 3 \text{ days}$$
$$+ 5 \text{ days} + 5 \text{ days}$$
$$= 43 \text{ days}$$

Path A-B-E-G-H-D-L has a total duration of

$$D = 3 \text{ days} + 6 \text{ days} + 0 \text{ days} + 14 \text{ days}$$
$$+ 20 \text{ days} + 12 \text{ days} + 5 \text{ days}$$
$$= 60 \text{ days}$$

Path A-B-E-G-H-J-K-L has a total duration of

$$D = 3 \text{ days} + 6 \text{ days} + 0 \text{ days} + 14 \text{ days} + 20 \text{ days}$$
$$+ 3 \text{ days} + 5 \text{ days} + 5 \text{ days}$$
$$= 56 \text{ days}$$

Path A-B-E-G-I-K-L has a total duration of

$$D = 3 \text{ days} + 6 \text{ days} + 0 \text{ days} + 14 \text{ days} + 5 \text{ days}$$
$$+ 5 \text{ days} + 5 \text{ days}$$
$$= 38 \text{ days}$$

Path A-F-G-H-D-L has a total duration of

$$D = 3 \text{ days} + 4 \text{ days} + 14 \text{ days} + 20 \text{ days}$$
$$+ 12 \text{ days} + 5 \text{ days}$$
$$= 58 \text{ days}$$

Path A-F-G-H-J-K-L has a total duration of

$$D = 3 \text{ days} + 4 \text{ days} + 14 \text{ days} + 20 \text{ days} + 3 \text{ days}$$
$$+ 5 \text{ days} + 5 \text{ days}$$
$$= 54 \text{ days}$$

Path A-F-G-I-K-L has a total duration of

$$D = 3 \text{ days} + 4 \text{ days} + 14 \text{ days} + 5 \text{ days}$$
$$+ 5 \text{ days} + 5 \text{ days}$$
$$= 36 \text{ days}$$

The critical path is path A-B-E-G-H-D-L.

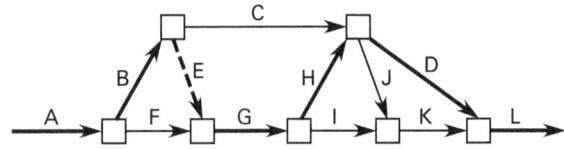

The answer is (A).

(b) The duration of a project is equal to the duration of its critical path, so the duration of this project is 60 days.

The answer is (B).

(c) To find the original labor crew cost, multiply the number of days required for each activity by the normal daily rate of the crew required for that activity, and total up all the costs to find the total project crew cost.

The crew cost for activity A is

$$C_{A,c} = \left(650 \ \frac{\$}{\text{day}}\right)(3 \text{ days}) = \$1950$$

The crew cost for activity B is

$$C_{B,c} = \left(800 \ \frac{\$}{\text{day}}\right)(6 \text{ days}) = \$4800$$

The crew cost for activity C is

$$C_{C,c} = \left(800 \ \frac{\$}{\text{day}}\right)(21 \text{ days}) = \$16,800$$

The crew cost for activity D is

$$C_{D,c} = \left(800 \ \frac{\$}{\text{day}}\right)(12 \text{ days}) = \$9600$$

The crew cost for activity E is 0.

The crew cost for activity F is

$$C_{F,c} = \left(1000 \ \frac{\$}{\text{day}}\right)(4 \text{ days}) = \$4000$$

The crew cost for activity G is

$$C_{G,c} = \left(1000 \ \frac{\$}{\text{day}}\right)(14 \text{ days}) = \$14,000$$

The crew cost for activity H is

$$C_{H,c} = \left(1000 \ \frac{\$}{\text{day}}\right)(20 \text{ days}) = \$20{,}000$$

The crew cost for activity I is

$$C_{I,c} = \left(650 \ \frac{\$}{\text{day}}\right)(5 \text{ days}) = \$3250$$

The crew cost for activity J is

$$C_{J,c} = \left(1000 \ \frac{\$}{\text{day}}\right)(3 \text{ days}) = \$3000$$

The crew cost for activity K is

$$C_{K,c} = \left(650 \ \frac{\$}{\text{day}}\right)(5 \text{ days}) = \$3250$$

The crew cost for activity L is

$$C_{L,c} = \left(650 \ \frac{\$}{\text{day}}\right)(5 \text{ days}) = \$3250$$

The total crew cost for the original project schedule is

$$\begin{aligned}C_{t,c} =\ & \$1950 + \$4800 + \$16{,}800 + \$9600 + \$4000 \\ & + \$14{,}000 + \$20{,}000 + \$3250 + \$3000 \\ & + \$3250 + \$3250 \\ =\ & \$83{,}900\end{aligned}$$

The cost of supervision for the project is

$$C_s = \left(500 \ \frac{\$}{\text{day}}\right)(60 \text{ days}) = \$30{,}000$$

The total cost of the project is

$$C_t = \$83{,}900 + \$30{,}000 = \$113{,}900 \quad (\$114{,}000)$$

The answer is (D).

(d) Find the total crew cost of the expedited project. Only activities B, D, G, and H will be expedited, so new costs only need to be found for these four activities.

The expedited crew cost for activity B is

$$C_{B,c} = \left(1200 \ \frac{\$}{\text{day}}\right)(5 \text{ days}) = \$6000$$

The expedited crew cost for activity D is

$$C_{D,c} = \left(1200 \ \frac{\$}{\text{day}}\right)(8 \text{ days}) = \$9600$$

The expedited crew cost for activity G is

$$C_{G,c} = \left(2000 \ \frac{\$}{\text{day}}\right)(9 \text{ days}) = \$18{,}000$$

The expedited crew cost for activity H is

$$C_{H,c} = \left(2000 \ \frac{\$}{\text{day}}\right)(12 \text{ days}) = \$24{,}000$$

The total crew cost for the expedited project is

$$\begin{aligned}C_{t,c} =\ & \$1950 + \$6000 + \$16{,}800 + \$9600 + \$4000 \\ & + \$18{,}000 + \$24{,}000 + \$3250 + \$3000 \\ & + \$3250 + \$3250 \\ =\ & \$93{,}100\end{aligned}$$

Find the cost of supervision for the expedited project. The total cost of supervision will decrease, depending on the number of days shorter the expedited schedule is than the normal schedule.

Find the total duration of the expedited schedule. Changes in duration for the expedited activities will not cause a change in critical path, so the duration can be found by recalculating the total duration of path A-B-E-G-H-D-L using the expedited durations for activities B, G, H, and D.

The total duration of the expedited schedule is

$$\begin{aligned}D_t =\ & 3 \text{ days} + 5 \text{ days} + 0 \text{ days} + 9 \text{ days} + 12 \text{ days} \\ & + 8 \text{ days} + 5 \text{ days} \\ =\ & 42 \text{ days}\end{aligned}$$

The total cost of supervision for the expedited project is

$$\begin{aligned}C_{t,s} &= \left(500 \ \frac{\$}{\text{day}}\right)(42 \text{ days}) \\ &= \$21{,}000\end{aligned}$$

The total cost of the expedited schedule, before bonuses or penalties, is

$$\begin{aligned}C_t &= \$93{,}100 + \$21{,}000 \\ &= \$114{,}100\end{aligned}$$

This is a cost increase over the normal project. Subtract the total cost of the normal project from the total cost of the expedited project to find the exact amount of the cost increase.

$$\$114{,}100 - \$113{,}900 = \$200$$

Expediting the project will cost $200 more than adhering to the normal schedule.

The answer is (B).

(e) A duration of 42 days means that the project will finish 18 days ahead of schedule. According to the bonus/penalty clause, this means the contractor will receive a bonus of

$$\left(1000 \ \frac{\$}{\text{day}}\right)(18 \text{ days} - 3 \text{ days}) = \$15{,}000$$

The total cost of the expedited schedule is

$$C_t = \$114{,}100 - \$15{,}000 = \$99{,}100$$

This is a savings over the normal schedule. Subtract the total cost of the normal schedule from the total cost of the expedited schedule to find the exact amount of the savings.

$$\$113{,}900 - \$99{,}100 = \$14{,}800$$

The answer is (A).

(f) For the expediting of an activity to decrease the duration of the project, it must decrease the total duration of the critical path. Therefore, only activities that are on the critical path for the expedited schedule should be expedited; expediting other activities will not affect the total duration and will result in an unnecessary increase in cost.

Activity C is not on the critical path for the normal schedule. For activity C to become part of the critical path for the expedited schedule, its expedited duration must be less than the total expedited duration of activities E, G, and H. The expedited duration of activity C is 18 days. The total expedited duration of activities E, G, and H is 21 days. Activity C is not on the critical path for the expedited schedule, meaning activity H is, and expediting activity C will not affect the critical path and would result in an unnecessary increase in cost.

Both activity B and activity H are on the critical path for the normal schedule. For activity B to not be on the critical path for the expedited schedule, its expedited duration must be less than the duration of activity F, the only alternative path from the completion of activity A. For activity H to not be on the critical path for the expedited schedule, either activity C must become a part of the critical path rather than activities E, G, and H, or the expedited duration of activity H must be less than the duration of activity I, the only alternative route from the completion of activity G.

The expedited duration of activity B is 5 days, which is greater than the duration of activity F, so activity B remains on the critical path when expedited, and expediting activity B should help to decrease the project duration.

The expedited duration of activity H is 12 days, which is greater than the duration of activity I, so activity G remains on the critical path unless activity C supplants path E-G-H, and expediting activity G should help to decrease the project duration.

The answer is (B).

2. (a) For 95% reliability, first calculate the high pavement design temperature at a 50% reliability level using Eq. 5.25.

$$T_{20\,\text{mm}} = \begin{pmatrix} T_{\text{air}} - (0.00618)(\text{lat})^2 \\ + (0.2289)(\text{lat}) + 42.2°\text{C} \end{pmatrix}(0.9545)$$
$$\qquad - 17.78°\text{C}$$
$$= \begin{pmatrix} 31.9° - (0.00618)(42.2°)^2 \\ + (0.2289)(42.2°) + 42.2°\text{C} \end{pmatrix}(0.9545)$$
$$\qquad - 17.78°\text{C}$$
$$= 51.7°\text{C}$$

The 95% reliability level is found from the high pavement design temperature plus two standard deviations.

$$51.7°\text{C} + 2s = 51.7°\text{C} + (2)(1.7°\text{C}) = 55.1°\text{C} \quad (56°\text{C})$$

The answer is (C).

(b) In order to select a binder grade, calculate the low air design temperature at a 95% reliability level. From Eq. 5.26, the low pavement design temperature is

$$T_{\min} = 0.859\,T_{\text{air}} + 1.7°\text{C} = (0.859)(-21.1°\text{C}) + 1.7°\text{C}$$
$$= -16.4°\text{C}$$

The low pavement design temperature at 95% reliability is

$$-16.4°\text{C} - 2s = -16.4°\text{C} - (2)(3.9°\text{C})$$
$$= -24.2°\text{C} \quad (-24°\text{C})$$

Using the high pavement design temperature of 55.1°C calculated in part (a) and the low pavement design temperature of −24.2°C, select the asphalt binder grade from App. 5.D. For 95% reliability, the grade is PG 58-22. (There is no PG 56-22, and 56°C is higher than 52°C; therefore, use PG 58-22.)

The answer is (C).

3. (a) To determine the bulk specific gravity of a combination of aggregates, use Eq. 5.10.

$$G_{sb} = \frac{P_1 + P_2 + \cdots + P_n}{\dfrac{P_1}{G_1} + \dfrac{P_2}{G_2} + \cdots + \dfrac{P_n}{G_n}}$$

$$= \frac{0.25 + 0.35 + 0.40}{\dfrac{0.25}{2.720} + \dfrac{0.35}{2.689} + \dfrac{0.40}{2.450}}$$

$$= 2.595$$

The answer is (B).

(b) The effective specific gravity is found using Eq. 5.12.

$$G_{se} = \frac{P_{mm} - P_b}{\dfrac{P_{mm}}{G_{mm}} - \dfrac{P_b}{G_b}} = \frac{100\% - 5.0\%}{\dfrac{100\%}{2.434} - \dfrac{5.0\%}{1.030}}$$

$$= 2.622$$

The answer is (C).

(c) The percentage of asphalt absorption is found using Eq. 5.17.

$$P_{ba} = \frac{G_b(G_{se} - G_{sb})}{G_{sb}G_{se}} \times 100\%$$

$$= \frac{(1.030)(2.622 - 2.595)}{(2.595)(2.622)} \times 100\%$$

$$= 0.409\% \quad (0.4\%)$$

The answer is (A).

(d) The percentage of aggregate in the mix is

$$P_s = 100\% - P_b = 100\% - 5\%$$

$$= 95.0\%$$

The effective asphalt content of the mixture is found using Eq. 5.16.

$$P_{be} = P_b - \frac{P_{ba}P_s}{100\%}$$

$$= 5.0\% - \frac{(0.409\%)(95.0\%)}{100\%}$$

$$= 4.61\% \quad (4.6\%)$$

The answer is (C).

4. (a) The water-cement ratio is found by comparing the unit weights of each. One sack of cement weighs 94 lbf, and water weighs 8.33 lbf/gal.

$$w/c = \frac{\text{weight of water per unit}}{\text{weight of cement per unit}}$$

$$= \frac{\left(5.83 \,\dfrac{\text{gal}}{\text{sack}}\right)\left(8.33 \,\dfrac{\text{lbf}}{\text{gal}}\right)}{94 \,\dfrac{\text{lbf}}{\text{sack}}}$$

$$= 0.520$$

The answer is (C).

(b) Determine the weight ratio of the components using Eq. 5.21.

The specific weight of cement in the mix is

$$\gamma_c = VG_{sb,c}\gamma_{water}$$

$$= (1.00)(3.14)\left(62.4 \,\dfrac{\text{lbf}}{\text{ft}^3}\right)$$

$$= 196 \text{ lbf/ft}^3 \text{ of concrete}$$

The specific weight of fine aggregate in the mix is

$$\gamma_{fa} = VG_{sb,fa}\gamma_{water}$$

$$= (2.27)(2.66)\left(62.4 \,\dfrac{\text{lbf}}{\text{ft}^3}\right)$$

$$= 377 \text{ lbf/ft}^3 \text{ of concrete}$$

The specific weight of coarse aggregate in the mix is

$$\gamma_{ca} = VG_{sb,ca}\gamma_{water}$$

$$= (3.65)(2.72)\left(62.4 \,\dfrac{\text{lbf}}{\text{ft}^3}\right)$$

$$= 620 \text{ lbf/ft}^3 \text{ of concrete}$$

Adjust the specific weights to ratios per 94 lbf sack of cement using Eq. 5.22.

The adjusted net quantity of the cement is

$$Q_c = \gamma_c \left(\dfrac{94 \,\dfrac{\text{lbf}}{\text{sack}}}{\gamma_c}\right) = \left(196 \,\dfrac{\text{lbf}}{\text{ft}^3}\right)\left(\dfrac{94 \,\dfrac{\text{lbf}}{\text{sack}}}{196 \,\dfrac{\text{lbf}}{\text{ft}^3}}\right)$$

$$= 94 \text{ lbf/sack}$$

The adjusted net quantity of the fine aggregate is

$$Q_{\text{fa}} = \gamma_{\text{fa}}\left(\dfrac{94\,\dfrac{\text{lbf}}{\text{sack}}}{\gamma_c}\right) = \left(377\,\dfrac{\text{lbf}}{\text{ft}^3}\right)\dfrac{94\,\dfrac{\text{lbf}}{\text{sack}}}{196\,\dfrac{\text{lbf}}{\text{ft}^3}}$$

$$= 181\ \text{lbf/sack}$$

The adjusted net quantity of the coarse aggregate is

$$Q_{\text{ca}} = \gamma_{\text{ca}}\left(\dfrac{94\,\dfrac{\text{lbf}}{\text{sack}}}{\gamma_c}\right) = \left(620\,\dfrac{\text{lbf}}{\text{ft}^3}\right)\dfrac{94\,\dfrac{\text{lbf}}{\text{sack}}}{196\,\dfrac{\text{lbf}}{\text{ft}^3}}$$

$$= 297\ \text{lbf/sack}$$

Adjust the aggregate weight to account for water included in the stockpiled aggregate.

The adjusted weight of the fine aggregate is

$$W = \left(181\,\dfrac{\text{lbf}}{\text{sack}}\right)(1 + 0.0325) = 187\ \text{lbf/sack}$$

The adjusted weight of the coarse aggregate is

$$W = \left(297\,\dfrac{\text{lbf}}{\text{sack}}\right)(1 + 0.0208) = 303\ \text{lbf/sack}$$

Water absorption is less than the amount of moisture included with each aggregate. Therefore, there will be excess water introduced into the mix along with the aggregate.

$$\text{total water} - \text{absorbed water} = \text{excess water}$$

The excess water included with the fine aggregate is

$$3.25\% - 0.41\% = 2.84\%\ \text{excess water}$$

The excess water included with the coarse aggregate is

$$2.08\% - 0.25\% = 1.83\%\ \text{excess water}$$

Determine the total weight of excess water.

$$W_t = (0.0284)\left(187\,\dfrac{\text{lbf}}{\text{sack}}\right) + (0.0183)\left(303\,\dfrac{\text{lbf}}{\text{sack}}\right)$$

$$= 10.9\ \text{lbf/sack excess water}$$

The final adjusted water to be added is

$$\left(5.83\,\dfrac{\text{gal}}{\text{sack}}\right)\left(8.33\,\dfrac{\text{lbf}}{\text{gal}}\right) - 10.9\,\dfrac{\text{lbf}}{\text{sack}} = 37.7\ \text{lbf/sack}$$

The answer is (A).

(c) Using the material quantities adjusted for water content, convert the yield to cubic yard per sack.

$$\dfrac{\left(\dfrac{\gamma_c}{G_{\text{sb},c}\gamma_w} + \dfrac{\gamma_{\text{fa}}}{G_{\text{sb,fa}}\gamma_w} + \dfrac{\gamma_{\text{ca}}}{G_{\text{sb,ca}}\gamma_w}\right)(1 + \text{fraction air})}{27\,\dfrac{\text{ft}^3}{\text{yd}^3}}$$

$$= \dfrac{\left(\dfrac{196\ \text{lbf}}{(3.14)\left(62.4\,\dfrac{\text{lbf}}{\text{ft}^3}\right)} + \dfrac{377\ \text{lbf}}{(2.66)\left(62.4\,\dfrac{\text{lbf}}{\text{ft}^3}\right)} + \dfrac{620\ \text{lbf}}{(2.72)\left(62.4\,\dfrac{\text{lbf}}{\text{ft}^3}\right)}\right)\left(1 + \dfrac{3\%}{100\%}\right)}{27\,\dfrac{\text{ft}^3}{\text{yd}^3}}$$

$$= 0.264\ \text{yd}^3/\text{sack}$$

The answer is (B).

5. (a) Using water absorption to indicate the air voids within the aggregate particles, the apparent specific gravity, G_{sa}, and effective specific gravity, G_{se}, can be estimated.

To find the effective asphalt content, the percentage of aggregates in the mix must be known. The asphalt content is already known, so the percentage of aggregates is

$$P_s = 100\% - P_b = 100\% - 5.25\%$$
$$= 94.75\%$$

Use Eq. 5.16 to find the effective asphalt content.

$$P_{\text{be}} = P_b - \dfrac{P_{\text{ba}}P_s}{100\%} = 5.25\% - \dfrac{(0.7)(94.75\%)}{100\%}$$
$$= 4.58\%\quad (4.60\%)$$

The answer is (B).

(b) The volume of air voids can be found using Eq. 5.18.

$$\text{VTM} = \frac{G_{mm} - G_{mb}}{G_{mm}} \times 100\%$$
$$= \frac{2.526 - 2.438}{2.526} \times 100\%$$
$$= 3.484\% \quad (3.48\%)$$

The answer is (B).

(c) The volume of voids in mineral aggregate is found using Eq. 5.14.

$$\text{VMA} = 100\% - \frac{G_{mb} P_s}{G_{sb}}$$
$$= 100\% - \frac{(2.438)(94.75\%)}{2.687}$$
$$= 14.03\% \quad (14.0\%)$$

The answer is (C).

(d) The volume of voids filled with asphalt is found using Eq. 5.19.

$$\text{VFA} = \frac{\text{VMA} - \text{VTM}}{\text{VMA}} \times 100\%$$
$$= \frac{14.03\% - 3.484\%}{14.03\%} \times 100\%$$
$$= 75.17\% \quad (75.2\%)$$

The answer is (D).

6 Traffic Safety

1. Traffic Safety and Roadway Design Analysis...... 6-2
2. Traffic Data Used for Crash Analysis 6-2
3. Roadway Elements for Safe Design 6-2
4. Crash Analysis 6-4
5. Encroachment Crash Analysis......................... 6-5
6. Roadside Clearance Analysis........................... 6-8
7. Economic Analysis 6-19
8. Driver Behavior and Performance................. 6-21
9. Work Zone Safety....................................... 6-24
10. Temporary Traffic Control Zones 6-26
11. Conflict Analysis 6-27
12. Practice Problems...................................... 6-28

Nomenclature

A	annual amount	–	–
AADT	average annual daily traffic	veh/day	veh/d
ACC	annual crash cost	$	$
B	benefit	–	–
C	calibration factor	–	–
C	cost	$	$
CEI	cost effectiveness index	–	–
CZ_c	clear zone on outside curvature	ft	m
d	deceleration	ft/sec²	m/s²
EUAC	equivalent uniform annual cost	$	$
F	force	lbf	N
g	annual growth rate	%	%
g	gravitational acceleration, 32.2 (9.81)	ft/sec²	m/s²
g_c	gravitational constant, 32.2	lbm-ft/lbf-sec²	n.a.
i	effective interest rate	decimal	decimal
$i\%$	effective interest rate	%	%
K	constant or factor	–	–
K_{cz}	curve correction factor	–	–
L	length, distance, or taper length	ft	m
L_1	obstruction distance	ft	m
L_2	specified flare rate	–	–
L_3	lateral distance from edge of roadway to near side of object	ft	m
L_A	distance from traveled edge to far side of object	ft	m
L_C	lateral extent of clear zone extending beyond fixed object	ft	m
L_R	runout length	ft	m
L_S	shy distance from edge of roadway	ft	m
m	mass	lbm	kg
n	number	–	–
N	number	–	–
P	present worth	–	–
P	probability	–	–
R	crash rate	various	various
s	sample standard deviation	various	various
S	speed limit	–	–
SF	safety factor	–	–
SPF	safety performance function	–	–
v	velocity or design speed	mph	kph
V	traffic volume	vph	vph
VP_E	expected encroachment frequency	–	–
X, Y	coordinates of end of barrier need	–	–
Y	lateral offset	–	–

Symbols

α	vehicle orientation angle	deg	deg
θ	encroachment angle	deg	deg

Subscripts

a	after
ave	average
A	annualized crash or societal alternative
b	before
cr	crash
cz	clear-zone
d	distance
D	annualized direct
exp	expected
E	encroachment
i	level i
int	intersection
M	mean
n	year n
PRT	perception-reaction time
s	stopping
seg	segment
SI	injury severity
th	threshold

1. TRAFFIC SAFETY AND ROADWAY DESIGN ANALYSIS

Traffic safety is a fundamental part of transportation design and affects multiple aspects of planning, including capacity analysis, design standards, construction, operations, and maintenance. Safe design practices include providing adequate roadside clearances, analyzing and planning for traffic management in conflict zones, and providing work zone safety during construction and maintenance activity. Once a roadway network is in place, the emphasis on safety continues in order to mitigate crashes, which often result in property damage, injury, and/or fatalities. Local governments implement *countermeasures* to reduce the severity of crashes or prevent them altogether. When crashes do occur, crash causation analysis should be undertaken to identify factors and mechanisms that increase crash risk and to prioritize changes and improvements that will prevent crashes in the future.

Traffic safety begins with applying highway safety principles during design. The comfort, convenience, and safety of a highway can be greatly improved by designs that work to eliminate surprises, such as sudden sharp curves or rapid lane shifts. One of the greatest past improvements to highway safety was the elimination of at-grade intersection crossings so that all traffic enters and leaves the main roadways via high-speed ramps, and all traffic crossing the roadway is carried on grade-separated bridges. This allows the mainline flow to continue unimpeded (i.e., in "free flow") in all but high-density traffic conditions, hence the term "freeway."

Observations of consistent driver behavior patterns are quantified by AASHTO and in design guides. In particular, AASHTO's *Green Book* provides charts, tables, and graphs for acceleration rates, braking rates, comfortable travel speeds, perception-reaction times, signage visibility, and blending of horizontal and vertical curve geometry. This information is beneficial to designing roadways with safety in mind. AASHTO has adopted the design philosophy that accommodating 85% of the usual driving behavior is a cost-effective goal. Although accommodating the upper or lower 5–10% of behavior beyond the normal range is not necessarily ignored, unusual or erratic behavior and extreme driving patterns are generally not accounted for due to cost or other considerations.

2. TRAFFIC DATA USED FOR CRASH ANALYSIS

The methodologies and time periods used in data collection can have a marked effect on crash analysis. In particular, engineers should consider the various definitions of "average" traffic when selecting and applying traffic data to a location. While the terms *average daily traffic* (ADT) and *annual average daily traffic* (AADT) are frequently used synonymously, the way that average traffic is defined in a study can have a significant impact on the resulting crash statistics.

Average annual daily traffic is the total traffic for one year divided by 365 days. Average daily traffic is a traffic volume that occurs on a given average day, and can be determined by counting traffic over a period of days or hours and converting to average daily volumes. ADT is normally used for design analysis, and must be determined for the actual design time period. For instance, an ADT value may be a weekday daylight average that takes into account the a.m. and p.m. peak flows, and this could vary considerably from a seven-day, 24-hour count. Another possible source of variation is the average daily directional flow and its effect on the average traffic overall. The suitability of the count period and how representative the data are of the prevailing traffic conditions must be considered before proceeding with the analysis using ADT data.

Traffic data that are seasonally adjusted are called *average seasonal daily traffic* (ASDT) data. ASDT data are useful in resort areas with heavy tourist traffic, locations with extreme weather conditions such as large amounts of winter snowfall, and other areas where there are large variations in traffic flow from season to season. For example, when ASDT recorded for tourist areas varies substantially from the non-tourist season, some jurisdictions use this data to adjust signal timing, parking regulations, lane configuration, or other roadway features for the peak tourist season.

For most general conditions, AADT is used instead of ADT. The total annual count can be found using many methods, but the most common is using a continuous counter set up in a semi-permanent or permanent location. The total annual traffic count on a roadway segment (i.e., a full 12-month count) can also be extrapolated from a network of counting stations using a variety of count adjustment methods. Counts obtained for partial-year periods can be expanded to represent a full year. These numbers become the AADT figures that highway offices use for traffic management, funding applications, crime and crash statistics, and a variety of other purposes. AADT is frequently the most reliable of the various count statistics and is perhaps the most universally accepted as a comparative statistic for describing the traffic on a roadway segment. However, ADT is used for specific elements of design since ADT most closely represents the actual traffic condition.

3. ROADWAY ELEMENTS FOR SAFE DESIGN

On-Road Elements of Safe Design

The basis of safe roadway design should be no surprises for the driver so that sudden and erratic moves are unnecessary for maintaining the prevailing speed at a reasonable level of comfort and safety. Lack of consistency and uniformity of design elements can distract drivers and lead to collisions. All roadway features are selected based on the *design speed*, which is typically set for the 85th percentile driver. The following are the main concepts for designing safe roadways.

Curve radius and *superelevation* must be set according to the minimum design speed or greater.

Curve widening, especially for two-lane roads, is necessary to account for off-tracking of larger vehicles, to minimize unnecessary slowing of traffic around curves, and to avoid collisions.

Sight distance must be maintained for the design speed, whether the minimum for stopping or a longer distance for passing and making complex decisions.

Cross section elements must be matched to the volume, design speed, and vehicle mix. The number of lanes, provisions for curbs and shoulders, and the width of the roadway must match at least the minimum requirements for the desired service level.

Intersections and *ramps* must provide adequate sight distance for the design speed, provide sufficient room for the vehicles to maneuver, and allow the expected volume of traffic to flow in a normal fashion without requiring the drivers to make extraordinary moves.

Weaving and maneuvering sections, or areas where one-way traffic streams cross at contiguous access points, must be long enough to accommodate the expected flow volumes, especially on freeways and expressways.

Off-Road Elements for Crash Avoidance or Reduction

Once a vehicle leaves the normal roadway, there are a number of design features that can either help the driver avoid a crash or reduce damage or injury should a crash occur.

Clearance to obstructions is important because where more clearance is provided for a driver to recover and regain vehicular control, the likelihood of a severe crash is reduced (or a crash may be prevented altogether).

Objects in the clear zone, such as light poles and signs, should be designed with frangible (i.e., breakaway) fittings. These fittings minimize deceleration forces and help prevent hardware from penetrating the vehicle and increasing crash severity.

Ditches, *drain grates*, and *endwalls* should incorporate designs that allow a vehicle to pass over them without snagging or causing the vehicle to vault or overturn.

Earthwork side slopes within the clear zone should be designed to allow a vehicle to traverse the slope with as little potential for overturning as possible.

Guiderails and other protective devices should incorporate impact attenuators and should employ designs known to deflect and redirect vehicles instead of trapping and stopping with blunt force action.

Delineation—which includes pavement markings, signage, reflective surfaces, and appropriate warning of approaching hazards—provides improved guidance for the driver, especially under low light and adverse weather conditions.

Surrounding clutter and *distractions* can play a large role in traffic safety, as they tend to draw drivers' attention away from the road. Common examples of distractions are animated advertising signs near eye level or within the cone of vision of the driver, lighted signs that distract attention from traffic signals, driveways and store entrances along the roadway in commercial districts, nighttime lighting that glares into oncoming motorists' windshields, and bars and restaurants with busy patronage and night entertainment spilling onto the sidewalks and roadway.

The *clear zone* includes shoulders, bike lanes, and auxiliary lanes, except those auxiliary lanes that function like through lanes. After many of the obstacles in the clear-zone distance were removed, relocated, redesigned, or shielded by barriers or crash cushions, it became apparent that in some limited locations, embankments sloping downward significantly caused vehicles to be drawn farther away from the roadway, and the 30-ft clear zone might not be adequate. On the other hand, for many low-speed and urban facilities, the 30-ft clear zone becomes excessive when considering engineering, economic, and environmental justifications. Specific information on slope conditions will be explained later in this chapter.

For roadside hardware, AASHTO's *Manual for Assessing Safety Hardware* (MASH) provides recommendations for testing and evaluating highway features and hardware safety performance. This includes longitudinal barriers, terminals and end sections, crash cushions, work zone elements, and breakaway structures. *MASH* recognizes changes in vehicle fleet, and updates guidelines published in *NCHRP Report 350*. FHWA policy requires all roadside appurtenances such as traffic barriers, barrier terminals and crash cushions, bridge railing, sign and light pole supports, and work zone hardware used on the National Highway System (NHS) to meet the performance criteria described in *NCHRP Report 350* or *MASH*. It is recommended that hardware that has not met *NCHRP Report 350* or *MASH* criteria be upgraded during resurfacing, rehabilitation, or restoration (3R) projects, or when the hardware is damaged beyond repair.

The selection of best approach among several alternative designs or hardware is a function of careful judgment among the design engineer, the owning agency, and the local policy affecting field conditions at the subject location. AASHTO's *Roadside Design Guide*

(*RSDG*) is a resource for current information in the area of roadside design, and one reference on which to build the roadside design criteria best suited to the particular location and project.

4. CRASH ANALYSIS

Crashes (also known as *accidents*) involve vehicles colliding with other vehicles, stationary objects, or nonstationary objects. There are three categories of crashes: *property damage*, *injury*, or *fatality*. Crashes often involve more than one category and vary in severity. To eliminate small claims, most jurisdictions have a minimum reportable crash threshold for crashes involving property damage or injury. For property damage, this minimum threshold may be as low as $500. Reportable injuries vary by jurisdiction based on nearby emergency medical care, insurance coverage, or other regional differences. Reportable injuries commonly include severe bleeding, head trauma, loss of extremities, broken bones, breathing difficulties, heart attack, stroke, loss of mobility, loss of consciousness, exposure to extreme heat or cold, burns, eye damage, hearing damage, or some other debilitating injury. Because of these reporting thresholds, minor crashes and "near miss" situations are rarely reported.

Crashes are classified according to type and occurrence and are ranked according to exposure. *Exposure* is the number of vehicles that travel over a roadway in a period of time and/or that travel a specific distance. The *crash rate*, R, is the ratio of the number of crashes to the exposure. The general equation for crash rate is Eq. 6.1.

$$R = \frac{N_{cr}}{\text{exposure}} \quad \text{6.1}$$

Crash data are commonly grouped into either intersection data or segment data. For intersections, the crash rate is the rate of crashes per number of entering vehicles per year and is found from Eq. 6.2. To make the numbers more manageable, the rate is reported as the rate per million entering vehicles (RMEV) from all directions.

$$R_{\text{int}} = \frac{N_{cr}(10^6)}{(\text{AADT})N_{yr}\left(365 \, \frac{\text{days}}{\text{yr}}\right)} \quad \text{6.2}$$

For a *highway segment*, the number of crashes is reported as the ratio of the number of crashes per year to the AADT per mile of length, L, and is found from Eq. 6.3. To make numbers more manageable, the rate is calculated per 100 million vehicle-miles (HMVM).

$$R_{\text{seg}} = \frac{N_{cr}(10^8)}{(\text{AADT})N_{yr}\left(365 \, \frac{\text{days}}{\text{yr}}\right)L_{mi}} \quad \text{6.3}$$

Locations with the highest numbers of crashes when compared with other locations of similar design and function are given a high crash rate ranking. Screening processes using classic statistical methods are often performed to select critical locations for study and improvements based on crash rates and/or the severity of the crashes that occur. Once locations are ranked, there may be breaks in the ranking list separated by larger differences in frequency rates. These breaks may indicate a natural grouping of locations by crash type or severity.

Analysis is performed with the assumption that the crash rate follows a standard normal probability distribution as described in the Institute of Transportation Engineers' (ITE's) *Manual of Transportation Engineering Studies*. Logically, the locations with the highest crash rates are likely to benefit the most from a detailed study to find improvements. However, in order to make the best use of available funds for remedial action, there must be a consistent criterion upon which locations are judged and selected. The criterion commonly used is to select locations that appear to have a crash frequency significantly higher than the mean frequency of the rate of occurrence. Locations are ranked according to the rate of occurrence, which is calculated by comparing the crash rate, R, to the threshold rate, R_{th}. If the crash frequency rate is greater than the threshold rate, then the location likely qualifies for improvement funding. If the crash frequency rate is less than the threshold rate, the location does not meet the criterion. Locations where the threshold rate falls below the crash frequency rate are not considered for funding unless special circumstances exist, such as community pressure to make an improvement.

Use Eq. 6.4 to determine how the crash frequency rate compares to the threshold rate. R_M is the mean crash rate, and s is the sample standard deviation. Table 6.1 gives values for the constant, K, at common levels of confidence. The units of the standard deviation will be the same as the units of the crash rate.

$$R_{th} = R_M + Ks \quad \text{6.4}$$

Table 6.1 Values of K at Selected Levels of Confidence Using Two-Tailed Limits

level of confidence (%)	K
90	1.645
95	1.960
99	2.58

When selecting locations for data analysis, attention must be paid to how broad of an area is being selected, as well as how broad the criteria for selection are. For example, statewide data may exist only in relation to intersection type, but if criteria are narrowed and closer proximity limits to comparison locations are developed, it may become easier to find comparable locations by roadway type or traffic condition, as well. Doing this, however, will also decrease the sample size, which will make statistical analysis more difficult.

The following steps are used to analyze crash data.

step 1: Obtain all crash data for a time period of two years or more.

step 2: Prepare a summary of the crash data, including the time and date of each occurrence, weather conditions, road conditions, crash type, type of vehicles involved, driver actions, and other useful information from the report forms.

step 3: Prepare collision diagrams to illustrate the crash patterns using scale layouts of the intersection or road segment.

step 4: Prepare a diagram of the study location showing the location of relevant physical features such as traffic control devices, utility poles, fire hydrants, building lines, and street furniture. Pavement conditions that could have affected the crash should also be noted.

step 5: Obtain data showing traffic volumes, parking, speeds, driveway access, vehicle classifications, and signal timing.

step 6: Visit the site to become familiar with specific characteristics and conditions of traffic, or other information not readily available from reports or apparent from sketches. Take photographs to supplement and illustrate the report data.

The crash analysis should cover the direct effects of pertinent site features, such as sight distance limits, signal timing, or traffic patterns, to establish a cause and effect relationship. If necessary, statistical analysis can be weighted for the severity of crashes and to account for observed near-miss situations. The report should focus on the effects of possible engineering improvements and the prediction of crash reduction due to these improvements. The effectiveness of the improvements is presented in a before-and-after evaluation using the best approximations of the improvements' effects, obtained from Eq. 6.5.

$$\text{effectiveness} = \frac{N_{cr,b} - N_{cr,a}}{N_{cr,b}} \times 100\% \quad 6.5$$

Incidents are related to the location and can include crashes with property damage only, crashes with bodily injury, and crashes involving fatalities. The significance of the change in effectiveness should be based on statistical test procedures. This is especially true if there is a small sample of crashes to establish the effectiveness base.

Example 6.1

A highway segment has a crash rate of 257 HMVM. The mean crash rate for all highway sections in the region with similar roadway characteristics is 139 HMVM. The standard deviation of this mean is 68 HMVM. The segment will qualify for funding consideration if the location rate exceeds the threshold criteria using a 90% confidence level. Does this location qualify?

Solution

Use Table 6.1 to determine K, and solve for threshold rate using Eq. 6.4.

$$R_{th} = R_M + Ks = 139 \text{ HMVM} + (1.645)(68 \text{ HMVM})$$
$$= 251 \text{ HMVM}$$

The location crash rate (257 HMVM) is in excess of the threshold rate (251 HMVM). Therefore, the location qualifies for funding.

5. ENCROACHMENT CRASH ANALYSIS

Extensive investigation during the early days of interstate highway operations revealed that most off-road vehicle incursions did not result in a crash when the vehicle traveled 30 ft (9 m) or less from the edge of the shoulder. When the landscape adjoining the shoulder had a gentle slope and could support a vehicle, the vehicle often returned to the roadway unharmed. Further analysis of crash data has resulted in clear-zone distance design curves that take into account the design speed, the AADT volume, and the steepness of the slope beyond the paved shoulder. These findings, along with design procedure guidelines, are described in the *RSDG*. The *RSDG* covers detailed procedures to evaluate road design conditions for potential off-pavement crashes and contains guidelines for mitigation procedures to reduce the potential for these crashes.

The National Highway Traffic Safety Association (NHTSA) and the Insurance Institute for Highway Safety (IIHS) collect comprehensive data on crashes so that engineers and safety officials can better predict crash occurrence and severity. Figure 6.1, Table 6.2, and Table 6.3 give examples of available data. Approximately 20% of motor vehicle deaths result from a vehicle leaving the roadway and hitting a fixed object (e.g., a tree or utility pole) alongside the road. Generally, fixed-object deaths occur in single-vehicle crashes, as opposed to multiple-vehicle crashes.

Figure 6.1 shows the percentage distribution of fixed-object crashes in 2015 by object struck. Trees are by far the most common fixed object struck, followed by utility poles and traffic barriers. Alcohol is frequently a contributing factor in these crashes, but excessive speed, falling asleep, and inattention also contribute to vehicles leaving the roadway. Table 6.2 can be used with Table 6.3 to predict fatality rates for various off-pavement features based on past occurrences and traffic volume.

Figure 6.1 Percentage Distribution of Fixed-Object Crash Deaths by Object Struck, 2015

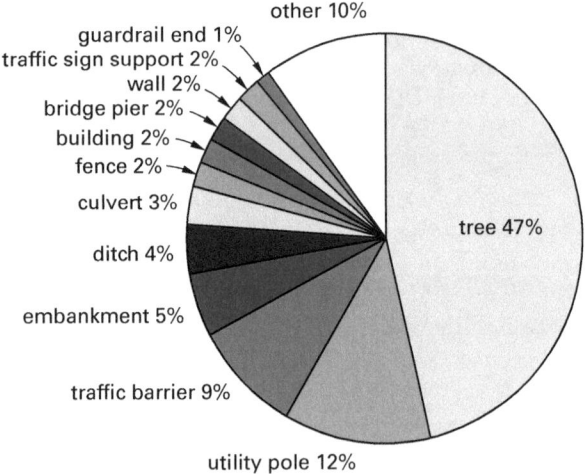

Adapted from Fatality Analysis Reporting System (FARS), U.S. Department of Transportation. iihs.org/iihs/topics/t/roadway-and-environment/fatalityfacts/fixed-object-crashes/2015

Table 6.2 Deaths in Fixed-Object Crashes as a Percent of All Motor Vehicle Deaths, 1979–2015

year	fixed object crash deaths no.	%	other deaths no.	%	all motor vehicle deaths no.
1979	10,550	21	40,543	79	51,093
1980	10,968	21	40,123	79	51,091
1981	9464	19	39,837	81	49,301
1982	8417	19	35,528	81	43,945
1983	8333	20	34,256	80	42,589
1984	8796	20	35,461	80	44,257
1985	9319	21	34,506	79	43,825
1986	9851	21	36,236	79	46,087
1987	9662	21	36,728	79	46,390
1988	9865	21	37,222	79	47,087
1989	9537	21	36,045	79	45,582
1990	9355	21	35,244	79	44,599
1991	8894	21	32,614	79	41,508
1992	8221	21	31,029	79	39,250
1993	8256	21	31,894	79	40,150
1994	8142	20	32,574	80	40,716
1995	8524	20	33,293	80	41,817
1996	8442	20	33,623	80	42,065
1997	8381	20	33,632	80	42,013
1998	8401	20	33,100	80	41,501
1999	8431	20	33,286	80	41,717
2000	8899	21	33,046	79	41,945
2001	9011	21	33,185	79	42,196
2002	9580	22	33,425	78	43,005
2003	9433	22	33,451	78	42,884
2004	9242	22	33,594	78	42,836
2005	9176	21	34,334	79	43,510
2006	9303	22	33,405	78	42,708
2007	9289	23	31,970	77	41,259
2008	8792	23	28,631	77	37,423
2009	7928	23	25,955	77	33,883
2010	7529	23	25,470	77	32,999
2011	7378	23	25,101	77	32,479
2012	7697	23	26,085	77	33,782
2013	7553	23	25,341	77	32,894
2014	7518	23	25,226	77	32,744
2015	7627	22	27,465	78	35,092

From Insurance Institute for Highway Safety (IIHS), Highway Loss Data Institute, Highway Safety Research & Communications, www.iihs.org/iihs/topics/t/roadway-and-environment/fatalityfacts/fixed-object-crashes/2015.

Table 6.3 Deaths, Crashes, and Total Number of Vehicles Involved, 1975–2015

year	deaths	crashes	motor vehicles
1975	44,525	39,161	55,534
1976	45,523	39,747	56,084
1977	47,878	42,211	60,516
1978	50,331	44,433	64,144
1979	51,093	45,223	64,762
1980	51,091	45,284	63,484
1981	49,301	44,000	62,698
1982	43,945	39,092	56,449
1983	42,589	37,976	55,103
1984	44,257	39,631	57,970
1985	43,825	39,196	58,271
1986	46,087	41,090	60,792
1987	46,390	41,438	61,836
1988	47,087	42,130	62,702
1989	45,582	40,741	60,870
1990	44,599	39,836	59,292
1991	41,508	36,937	54,794
1992	39,250	34,942	52,227
1993	40,150	35,780	53,777
1994	40,716	36,254	54,906
1995	41,817	37,241	56,524
1996	42,065	37,494	57,347
1997	42,013	37,324	57,037
1998	41,501	37,107	56,920
1999	41,717	37,140	56,820
2000	41,945	37,526	57,594
2001	42,196	37,862	57,918
2002	43,005	38,491	58,426
2003	42,884	38,477	58,877
2004	42,836	38,444	58,729
2005	43,510	39,252	59,495
2006	42,708	38,648	58,094
2007	41,059	37,435	55,926
2008	37,423	34,172	49,151
2009	33,883	30,862	45,540
2010	32,999	30,296	44,862
2011	32,479	29,867	44,119
2012	33,782	31,006	45,960
2013	32,894	30,203	45,102
2014	32,744	30,056	44,950
2015	35,092	32,166	48,923

Adapted from Insurance Institute for Highway Safety (IIHS), Highway Loss Data Institute, Highway Safety Research & Communications, www.iihs.org/iihs/topics/t/general-statistics/fatalityfacts/overview-of-fatality-facts/2015.

Encroachment Crash Prediction

Crash prediction estimates how many off-road encroachments result in crashes. *Encroachments* refer to the off-pavement travel of vehicles onto the roadside beyond the paved shoulder. Random occurrence and simulations, such as the *Monte Carlo method*,[1] are used based on distributions-derived average data, such as those given in the *RSDG*.

In order to predict crashes, traffic growth over the life of the project must be established. The traffic growth adjustment factor averages the traffic volume over the life of the project and is found from Eq. 6.6. g is the annual percentage growth rate and n is the number of years in the project life.

$$\text{average traffic growth adjustment factor} = \sum_{i=1}^{n} \frac{\left(1 + \dfrac{g}{100\%}\right)^n}{n} \quad 6.6$$

The traffic volume for a given year, allowing for increases in traffic, is adjusted from the base year AADT, using Eq. 6.7.

$$\text{AADT}_n = \text{AADT}_1 \left(1 + \frac{g}{100\%}\right)^n \quad 6.7$$

Figure 6.2 shows typical encroachment frequency curves used by the AASHTO Roadside Safety Analysis Program (RSAP), which can also be used in manual calculations. The encroachments are expressed as the number of encroachments per mile per year per AADT for undivided and divided highways. The encroachment frequency curves were developed by observing tire tracks in medians and roadsides. The encroachment frequency curves were adjusted upward by a ratio of 2.466 for two-lane, undivided highways and 1.878 for multilane, divided highways to account for under-reporting of encroachments due to paved shoulders. The percentage of uncontrolled encroachments is often assumed to be 60% based on a study of reported versus unreported crashes involving longitudinal barriers. Therefore, the encroachment frequency is multiplied by a factor of 0.6 to account for the lack of ability to detect the difference between controlled and uncontrolled encroachments.

Vehicle speed, the angle of the encroachment path, and vehicle orientation are determined by distributions estimated from crash data. In the absence of reliable local data, statewide or national data can be used and adjusted for local conditions if they vary from the national averages.

---

[1]The Monte Carlo method is a class of computational algorithms that make repeated random sampling to calculate results.

Figure 6.2 Encroachment Frequency Curves Used by the RSAP

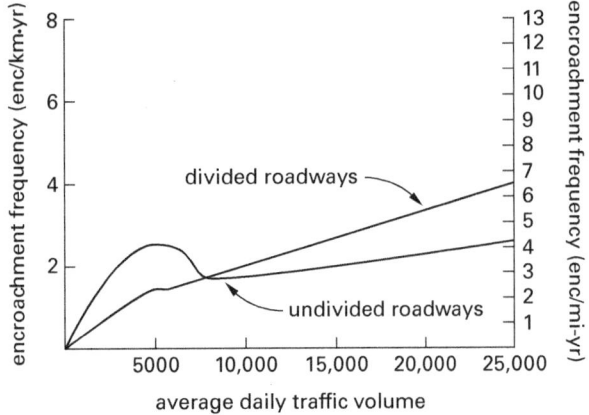

Adapted from *Roadside Design Guide*, 2011, by the American Association of State Highway and Transportation Officials, Washington, D.C. Used by permission.

Figure 6.3 Lateral Extent of Encroachment Distribution

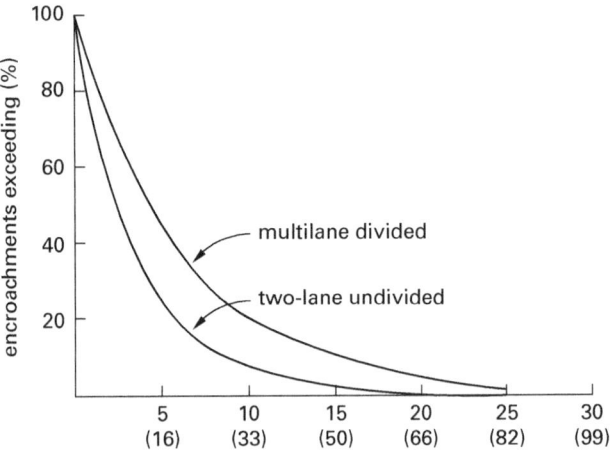

From *Roadside Design Guide*, 2011, by the American Association of State Highway and Transportation Officials, Washington, D.C. Used by permission.

The *impact envelope*, also known as the *encroachment path*, is the path of the vehicle once it leaves the roadway and is assumed to be a straight line. The *encroachment angle* is the angle between the edge of traveled way and the outer edge of the impact envelope, which is measured from the corner of the vehicle. The *vehicle orientation angle* describes the orientation of the vehicle in relation to the impact envelope and varies based on the impact envelope and the vehicle dimensions.

The probability of a crash is related to the encroachment path coverage and the lateral extent of the encroachment. If an encroachment occurs and the vehicle does not impact an object, the encroachment will not result in a crash. The path area that results in an object at a certain lateral distance being impacted, compared with the total roadside area at that same lateral distance in the study zone, can be used to approximate the proportion of encroachments that will result in a crash.

It is significant to note that virtually all of the available crash data indicate that nearly 100% of encroachments penetrate no farther than 82 ft (25 m), and only 30–45% of encroachments penetrate farther than 16 ft (5 m). Using the rule of thumb 80% of encroachments penetrate no farther than 30 ft (10 m), as shown in Fig. 6.3.

To produce results that show reliability significance, the data should be based on many encroachments over several years. Once the database expands to 10,000 encroachments, the distributions are compared with pre-established distributions to check if they are within the specified level of convergence. Outputs that do not meet the convergence criteria are ignored in the final output. The process can be performed manually using a few iterations with general measures of reliability to obtain approximate crash predictions for planning purposes.

6. ROADSIDE CLEARANCE ANALYSIS

Clear-Zone Distance

The *RSDG* defines a *clear zone* as the total roadside area, beginning at the edge of the roadway, that allows drivers to stop safely or regain control of a vehicle that leaves the roadway. This area may consist of a shoulder, a recoverable slope, a non-recoverable slope, and/or a clear runout area.

Recoverable slopes are slopes that generally allow a motorist to regain control of a vehicle by slowing or stopping. Slopes flatter than 1V:4H are considered recoverable. *Non-recoverable slopes* are considered traversable, but the vehicle will continue to the bottom of the slope. Embankment slopes between 1V:3H and 1V:4H are considered traversable (as long as they are smooth and free of obstructing objects) but non-recoverable. A *clear runout area* is the area at the toe of a non-recoverable slope that is available for the vehicle's safe use. Slopes steeper than 1V:3H are not considered traversable and are not considered part of the clear zone.

Using design speed and design AADT, the recommended clear-zone distance can be selected from Table 6.4. AASHTO's *Green Book* recommends a clear-zone distance between 7 ft and 10 ft (2 m and 3 m) on local roads and streets without curbs. For collectors without curbs, a 10 ft (3 m) minimum is recommended.

The term *backslope* is used to describe an upward slope away from the shoulder and the term *foreslope* for a downward slope away from the shoulder. These are the terms used by the *RSDG*, although the terms *cut slope* and *fill slope*, respectively, are more commonly used by

Table 6.4 Clear-Zone Distance from the Edge of Through Traveled Lane

SI units

design speed	design AADT	foreslopes[a] 1V:6H or flatter	foreslopes[a] 1V:5H to 1V:4H	foreslopes[a] 1V:3H	backslopes 1V:5H to 1V:4H	backslopes 1V:6H or flatter
60 kph or less	under 750[b]	2.0–3.0	2.0–3.0	2.0–3.0	2.0–3.0	2.0–3.0
	750–1500	3.0–3.5	3.5–4.5	3.0–3.5	3.0–3.5	3.0–3.5
	1500–6000	3.5–4.5	4.5–5.0	3.5–4.5	3.5–4.5	3.5–4.5
	over 6000	4.5–5.0	5.0–5.5	4.5–5.0	4.5–5.0	4.5–5.0
70–80 kph	under 750[b]	3.0–3.5	3.5–4.5	2.5–3.0	2.5–3.0	3.0–3.5
	750–1500	4.5–5.0	5.0–6.0	3.0–3.5	3.5–4.5	4.5–5.0
	1500–6000	5.0–5.5	6.0–8.0	3.5–4.5	4.5–5.0	5.0–5.5
	over 6000	6.0–6.5	7.5–8.5	4.5–5.0	5.5–6.0	6.0–6.5
90 kph	under 750[b]	3.5–4.5	4.5–5.5	2.5–3.0	3.0–3.5	3.0–3.5
	750–1500	5.0–5.5	6.0–7.5	3.0–3.5	4.5–5.0	5.0–5.5
	1500-6000	6.0–6.5	7.5–9.0	4.5–5.0	5.0–5.5	6.0–6.5
	over 6000	6.5–7.5	8.0–10.0[c]	5.0–5.5	6.0–6.5	6.5–7.5
100 kph	under 750[b]	5.0–5.5	6.0–7.5	3.0–3.5	3.5–4.5	4.5–5.0
	750–1500	6.0–7.5	8.0–10.0[c]	3.5–4.5	5.0–5.5	6.0–6.5
	1500–6000	8.0–9.0	10.0–12.0[c]	4.5–5.5	5.5–6.5	7.5–8.0
	over 6000	9.0–10.0[c]	11.0–13.5[c]	6.0–6.5	7.5–8.0	8.0–8.5
110 kph[d]	under 750[b]	5.5–6.0	6.0–8.0	3.0–3.5	4.5–5.0	4.5–5.0
	750–1500	7.5–8.0	8.5–11.0[c]	3.5–5.0	5.5–6.0	6.0–6.5
	1500–6000	8.5–10.0[c]	10.5–13.0[c]	5.0–6.0	6.5–7.5	8.0–8.5
	over 6000	9.0–10.5[c]	11.5–14.0[c]	6.5–7.5	8.0–9.0	8.5–9.0

customary U.S. units

design speed	design AADT	foreslopes[a] 1V:6H or flatter	foreslopes[a] 1V:5H to 1V:4H	foreslopes[a] 1V:3H	backslopes 1V:5H to 1V:4H	backslopes 1V:6H or flatter
40 mph or less	under 750[b]	7–10	7–10	7–10	7–10	7–10
	750–1500	10–12	12–14	10–12	10–12	10–12
	1500–6000	12–14	14–16	12–14	12–14	12–14
	over 6000	14–16	16–18	14–16	14–16	14–16
45–50 mph	under 750[b]	10–12	12–14	8–10	8–10	10–12
	750–1500	14–16	16–20	10–12	12–14	14–16
	1500–6000	16–18	20–26	12–14	14–16	16–18
	over 6000	20–22	24–28	14–16	18–20	20–22
55 mph	under 750[b]	12–14	14–18	8–10	10–12	10–12
	750–1500	16–18	20–24	10–12	14–16	16–18
	1500–6000	20–22	24–30	14–16	16–18	20–22
	over 6000	22–24	26–32[c]	16–18	20–22	22–24
60 mph	under 750[b]	16–18	20–24	10–12	12–14	14–16
	750–1500	20–24	26–32[c]	12–14	16–18	20–22
	1500–6000	26–30	32–40[c]	14–18	18–22	24–26
	over 6000	30–32[c]	36–44[c]	20–22	24–26	26–28
65–70 mph[d]	under 750[b]	18–20	20–26	10–12	14–16	14–16
	750–1500	24–26	28–36[c]	12–16	18–20	20–22
	1500–6000	28–32[c]	34–42[c]	16–20	22–24	26–28
	over 6000	30–34[c]	38–46[c]	22–24	26–30	28–30

[a]Because recovery is less likely on the unshielded, traversable 1V:3H foreslope on a fill section, fixed objects should not be present in the vicinity of these slopes. Recovery of high-speed vehicles that encroach beyond the edge of the shoulder may be expected to occur beyond the toe of slope. Determination of the width of the recovery area at the toe of slope should consider right-of-way availability, environmental concerns, economic factors, safety needs, and crash histories. Also, the distance between the edge of the through traveled lane and the beginning of the 1V:3H slope should influence the recovery area provided at the toe of slope. While the application may be limited by several factors, the foreslope parameters that may enter into determining a maximum desirable recovery area are illustrated in Fig. 3.2. A 3 m recovery area at the toe of slope should be provided for all traversable, non-recoverable fill slopes.

[b]For roadways with low volumes, it may not be practical to apply even the minimum values found in *RSDG 2011* Table 3-1. Refer to *RSDG 2011* Chap. 12 for additional considerations for low-volume roadways and Chap. 10 for additional guidance for urban applications.

[c]When a site-specific investigation indicates a high probability of continuing crashes or when such occurrences are indicated by crash history, the designer may provide clear-zone distances greater than the clear zone shown in *RSDG 2011* Table 3-1. Clear zones may be limited to 9 m for practicality and to provide a consistent roadway template if previous experience with similar projects or designs indicates satisfactory performance.

[d]When design speeds are greater than the values provided, the designer may provide clear-zone distances greater than those shown in *RSDG 2011* Table 3-1.

Reprinted with permission from the American Association of State Highway and Transportation Officials, *Roadside Design Guide*, Table 3-1, copyright © 2011.

design engineers. The *RSDG* corrections for curves, or *clear-zone correction factor* values, K_{cz}, are shown in Table 6.5 and can be derived from Eq. 6.8.

$$CZ_c = L_{cz}K_{cz} \qquad 6.8$$

L_{cz} is the clear-zone distance found from Table 6.4, and CZ_c is the clear zone on the outside curvature. The clear-zone correction factor is applied to outside curves only. Curves that are flatter than 2860 ft (900 m) do not need to be adjusted.

Table 6.5 Horizontal Curve Adjustments*

	K_{cz}, SI units [U.S. units]					
	design speed (kph/[mph])					
radius (m/[ft])	60 [40]	70 [45]	80 [50]	90 [55]	100 [65]	110 [70]
900 [2950]	1.1	1.1	1.1	1.2	1.2	1.2
700 [2300]	1.1	1.1	1.2	1.2	1.2	1.3
600 [1970]	1.1	1.2	1.2	1.2	1.3	1.4
500 [1640]	1.1	1.2	1.2	1.3	1.3	1.4
450 [1475]	1.2	1.2	1.3	1.3	1.4	1.5
400 [1315]	1.2	1.2	1.3	1.3	1.4	–
350 [1150]	1.2	1.2	1.3	1.4	1.5	–
300 [985]	1.2	1.3	1.4	1.5	1.5	–
250 [820]	1.3	1.3	1.4	1.5	–	–
200 [660]	1.3	1.4	1.5	–	–	–
150 [495]	1.4	1.5	–	–	–	–
100 [330]	1.5	–	–	–	–	–

*Clear zone on outside of curvature, CZ_c, clear zone distance, L_c, and curve correction factor, K_{cz}, are shown in the following equation, $CZ_c = L_cK_{cz}$

The clear-zone correction factor is applied to the outside of curves only. Corrections are typically made only to curves with a less than 900 m (2950 ft) radius.

Reprinted with permission from the American Association of State Highway and Transportation Officials, *Roadside Design Guide*, Table 3-2, copyright © 2011.

For every object or feature impacted by a crashing vehicle, the condition of the impact determines the outcome and the severity of the crash. For instance, if a vehicle traveling at high speed impacts a guiderail at a low angle and is redirected, the probability of injury will be less than if the impact were at a high angle and the vehicle were redirected at the same speed. The performance of the roadside feature must also be taken into account (e.g., whether or not the impact object was designed to absorb impact). Details of specific types of guiderails and protective barriers, including crash data for effectiveness, are found in *RSDG*.

Objects in the Clear Zone

There are several options for how to treat objects in the clear zone. Whenever possible, the best option is to remove the object. However, if the object serves a necessary function, such as a utility pole or sign fixture, the object must be designed so that the lowest degree of personal injury occurs upon impact. For instance, breakaway design or shielding by a longitudinal barrier or crash cushion may be employed to mitigate injury.

Placement of large, fixed objects within the clear zone, such as piers, abutments, or large sign structures, must be protected by impact attenuators, guiderails, or other protective measures. Delineation can also be employed to improve the visibility of the object for oncoming drivers.

Objects should also have adequate shy distance. *Shy distance* is measured beginning at the edge of the traveled way. Objects with adequate shy distance should be far enough from the edge of the traveled way that they will not be perceived as an immediate hazard and cause drivers to slow or change the placement of their vehicle.

Fill slopes, or foreslopes, should be traversable whenever possible. When a traversable slope is not possible, the slope can be steepened with a guiderail placed at the top of the slope. The shoulder can also be widened by 2 ft to 4 ft (0.6 m to 1.2 m) to allow for additional vehicle-run-off recovery. Generally, slopes flatter than 1V:6H do not require guiderails. Slopes as steep as 1V:4H or 1V:3H are recoverable if a suitable runout is provided at the bottom of the slope, whether or not the driver can regain control of the vehicle within the space provided without a collision. Guiderail locations should be analyzed to determine if the potential damage caused by the guiderail would be greater or less than the potential damage if the guiderail were not present. If the guiderail has an equal or greater potential for causing damage, then the guiderail should not be installed.

Cut slopes, or backslopes, follow the same rules of recoverable versus non-recoverable as fill slopes. However, cut slopes may redirect the vehicle back onto the roadway, which can result in a secondary crash. Both cut and fill slopes are considered non-recoverable if they are planted with large trees or if vegetation has become so dense that vehicles would not penetrate it upon impact but instead be brought to an abrupt stop.

To reduce construction and right-of-way costs, it may be appropriate to design the cut or fill slope with a *variable-slope design*. This design places a recoverable slope adjacent to the paved shoulder for a certain distance, then allows for a non-recoverable slope with a recovery zone at the bottom of the steeper slope, as shown in Fig. 6.4. A recoverable slope followed by a non-recoverable slope is also called a *barn roof section*. This type of design is more cost-effective than a continuous flat foreslope and safer than a continuous steep foreslope, so it provides a good compromise between the two extremes.

Figure 6.4 Clear Zone for Non-recoverable Parallel Foreslope with Runout

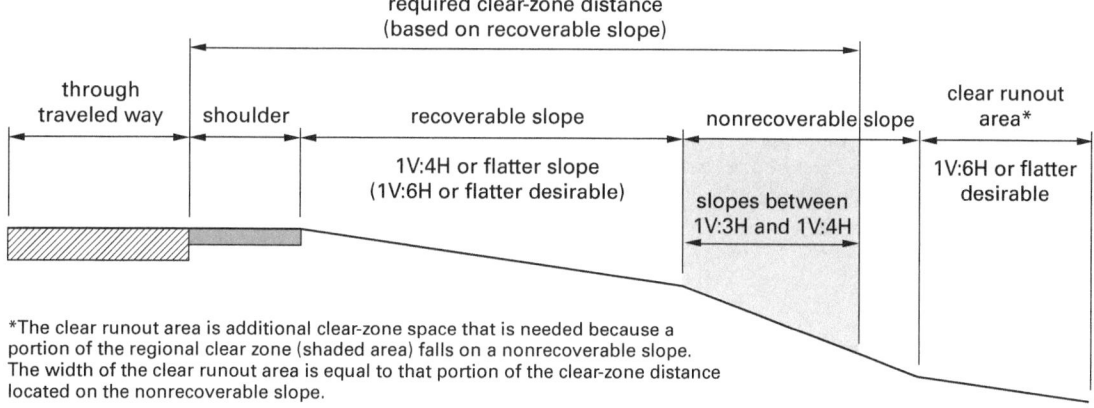

Reprinted with permission from the American Association of State Highway and Transportation Officials, *Roadside Design Guide*, Fig. 3.2, copyright © 2011.

Example 6.2

A freeway is designed for a speed of 60 mph (100 kph) and an AADT of 4500. There is an 8 ft (2.4 m) wide paved shoulder with a 0.04 ft/ft (0.04 m/m) cross slope. The distance from the edge of the traveled lane to the right-of-way line is 37 ft (11 m). The ground line at the right-of-way line is 5.5 ft (1.7 m) below the edge of the traveled lane. The design calls for variable slopes, with the section adjacent to the paved shoulder being 1V:8H and as wide as possible. The final slope against the right-of-way line is to be a maximum of 1V:5H. What will the width of the 1V:8H slope be, and what is the equivalent average slope from the edge of the running lane?

SI Solution

Draw the cross section.

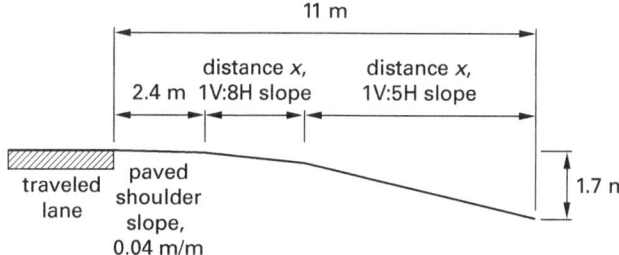

From Table 6.4, for a 100 kph design speed, an AADT of 4500, and a 1V:6H or flatter slope, the clear zone is 8–9 m. For slopes 1V:5H to 1V:4H, the clear zone is 10–12 m. Selecting a 1V:5H slope 6 m wide will gain 1.2 m in elevation. The remaining 0.50 m will be taken in the 2.4 m shoulder and the 1V:8H section. The average slope from the edge of the traveled lane to the right-of-way line is 1V:6.7H, which is within the criteria given in Table 6.4 for a slope between 1V:6HV and 1V:5H. Rounding the slope ratios to the nearest two significant digits is adequate for the analysis.

For the shoulder, the elevation change is

$$(2.4 \text{ m})\left(0.04 \ \frac{\text{m}}{\text{m}}\right) = 0.096 \text{ m}$$

For the 1V:5H slope, the elevation change is

$$(6 \text{ m})\left(0.2 \ \frac{\text{m}}{\text{m}}\right) = 1.20 \text{ m}$$

For the 1V:8H slope, the width is

$$11 \text{ m} - 6 \text{ m} - 2.4 \text{ m} = 2.6 \text{ m}$$

The elevation change of the 1V:8H slope is

$$(2.6 \text{ m})\left(0.125 \ \frac{\text{m}}{\text{m}}\right) = 0.325 \text{ m}$$

The elevation gain calculated by the slope ratio is

$$0.096 \text{ m} + 1.20 \text{ m} + 0.325 \text{ m} = 1.621 \text{ m}$$

Check the elevation difference.

$$1.7 \text{ m} - 1.621 \text{ m} = 0.079 \text{ m}$$

The elevation difference of 0.079 m is made by slightly adjusting either of the foreslopes, as the earthwork elevation in the clear zone will be within 0.03 m, which is a reasonable tolerance for this location.

The 11 m width has an elevation change of 1.7 m. The average slope is

$$\frac{11 \text{ m}}{1.7 \text{ m}} = 6.47 \quad (6.5)$$

Therefore, the slope ratio is 1V:6.5H.

Customary U.S. Solution

Draw the cross section.

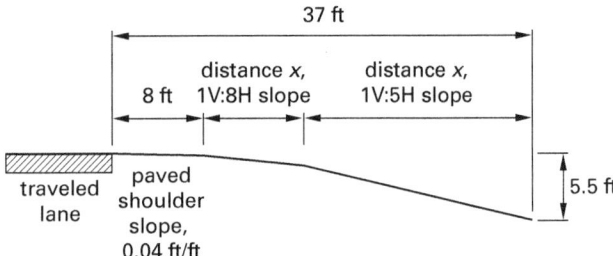

From Table 6.4, for a design speed of 60 mph, an AADT of 4500, and a 1V:6H or flatter slope, the clear zone is 26–30 ft. For slopes 1V:5H to 1V:4H, the clear zone is 32–40 ft. Selecting a 1V:5H slope 20 ft wide will gain 4.0 ft in elevation. The remaining 1.5 ft will be taken in the 8 ft shoulder and the 1V:8H section. The average slope from the edge of the traveled lane to the right-of-way line is 1V:6.7, which is within the criteria shown in Table 6.4 for a slope between 1V:6H and 1V:5H. Rounding the slope ratios to the nearest two significant digits is adequate for the analysis.

For the shoulder, the elevation change is

$$(8 \text{ ft})\left(0.04 \, \frac{\text{ft}}{\text{ft}}\right) = 0.32 \text{ ft}$$

For the 1V:5H slope, the elevation change is

$$(20 \text{ ft})\left(0.20 \, \frac{\text{ft}}{\text{ft}}\right) = 4.0 \text{ ft}$$

For the 1V:8H slope, the width is

$$37 \text{ ft} - 20 \text{ ft} - 8 \text{ ft} = 9 \text{ ft}$$

The elevation change of the 1V:8H slope is

$$(9 \text{ ft})\left(0.125 \, \frac{\text{ft}}{\text{ft}}\right) = 1.125 \text{ ft}$$

The elevation gain calculated by the slope ratio is

$$0.32 \text{ ft} + 4.0 \text{ ft} + 1.125 \text{ ft} = 5.445 \text{ ft}$$

Check the elevation difference.

$$5.5 \text{ ft} - 5.445 \text{ ft} = 0.055 \text{ ft}$$

The elevation difference of 0.055 ft is made by slightly adjusting either of the foreslopes, as the earthwork elevation in the clear zone will be within 0.1 ft, which is a reasonable tolerance for this location.

The 37 ft width has an elevation change of 5.5 ft. The average slope is

$$\frac{37 \text{ ft}}{5.5 \text{ ft}} = 6.73 \quad (6.7)$$

Therefore, the slope ratio is 1V:6.7H.

Geometry of Roadside Features

The roadside area, which covers the area beyond the actual traveled lanes, may consist of everything from an entirely flat, featureless surface to complex combinations of foreslopes, backslopes, drainage structures, walls or barriers, vegetation, and other manmade or natural features. Each of these features has an effect on the trajectory of an encroaching vehicle. Foreslopes (i.e., fill slopes) may be considered *recoverable*, *non-recoverable*, or *critical*. A recoverable foreslope is smooth and traversable so that an encroaching vehicle may be stopped, or slowed enough to return to the roadway in a safe manner.

Recoverable foreslopes do not have protruding culvert endwalls or other devices that project above the finished surface to snag the vehicle. Recoverable foreslopes also have sufficient surface load-bearing strength to avoid vehicle wheels becoming bogged, causing the vehicle to undergo unsafe rapid deceleration or to overturn from its wheels sinking into the surface.

Non-recoverable foreslopes are traversable but the vehicle will not be able to stop or return to the roadway easily. On foreslopes in the range of 1V:4H to 1V:3H, the vehicle can be expected to reach the toe of the foreslope.

Critical foreslopes are those on which an encroaching vehicle is expected to overturn. Foreslopes steeper than 1V:3H fall into this category. Barriers are recommended for the top of these types of slopes. A so-called "barn roof" configuration is often employed, providing a relatively flat recovery area near the traveled lane, followed by a 1V:3H slope with a recovery area at the bottom of the steeper slope. Figure 6.5 illustrates this concept.

Recoverable backslopes (i.e., cut slopes) are smooth enough to be traversable (1V:4H or flatter) without rocks or other protrusions, allowing vehicles to return to the traveled lane undamaged. Most cut slopes are littered with loose or embedded fallen rocks or other debris, and are not considered recoverable.

Transverse slopes are slopes caused by median barriers, driveways, median crossovers, or intersecting side roads. Figure 6.5 illustrates various roadside geometry features, including transverse slope locations. Transverse slopes of less than 1V:10H are preferred, with 1V:6H the recommended maximum in order to accommodate drainage fixtures. An errant vehicle encountering a transverse slope can be stopped abruptly on steeper transverse slopes, or can be catapulted into oncoming or cross traffic with poor design of the transverse slope. *RSDG* explains suggestions to reduce the severity and to improve the recoverability of transverse slope designs.

Figure 6.5 Roadside Feature Geometry

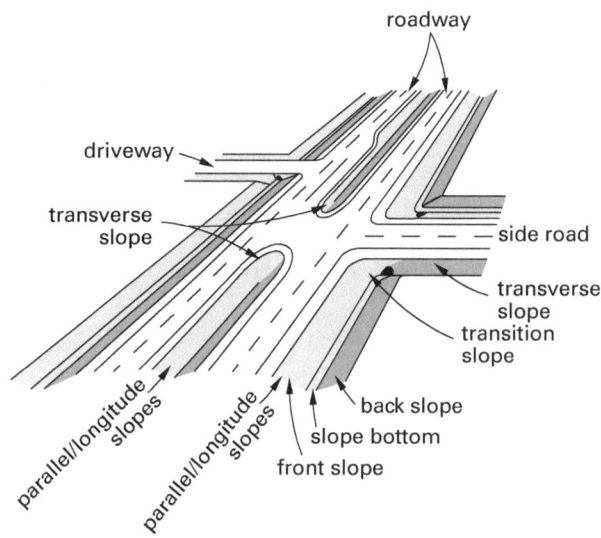

Drainage Channels

Transverse drainage ditches can create non-recoverable backslopes if not considered in setting up the roadway template. In some cases, where there are frequent interruptions of the drainage channel with intersecting driveways and street connections, changing the design from an open channel to a closed-pipe system may eliminate the non-recoverable condition.

Drainage channels parallel to the roadway lanes collect surface water from the roadway and from the adjoining right-of-way, and then convey the runoff to acceptable outlet points. Figure 6.6 and Fig. 6.7 show preferred cross sections for V-bottom and trapezoidal or rounded channels. Trapezoidal (flat-bottom) or rounded-bottom channels are preferred over V-bottom channels if a suitable right-of-way is available. For V-bottom channels, there is an abrupt slope change from foreslope to backslope at the channel bottom, which increases the probability of stopping an errant vehicle abruptly. The traversability of this type of channel can be improved by designing the channel slopes to be flatter than 1V:3H, as shown in Fig. 6.6. For trapezoidal and rounded channels, the foreslope and backslope are separated, which allows an errant vehicle an increased ability to traverse diagonally across the channel. *RSDG* suggests that rounded channel bottoms should be greater than 8 ft in width, and trapezoidal channels should have a bottom width of at least 4 ft. For trapezoidal channels, the backslope (that is, the channel side farthest from the traveled roadway) can be at a steeper slope than 1H:3V as long as the foreslope remains 1H:3V or flatter.

Emphasis on reducing the runoff flow rate by installing ditch flow checks, silt catchment ponds, and other flow-restricting features, can change the effectiveness of even the most carefully applied clear-zone concepts. Runoff control and other environmental concerns need to be integrated with the clear-zone design process so that there remains a cooperative and coordinated effort to reach mutually compatible objectives among these and other project functions.

Figure 6.6 Preferred Cross Sections for V-bottom Channels with Abrupt Slope Changes at Ditch Line

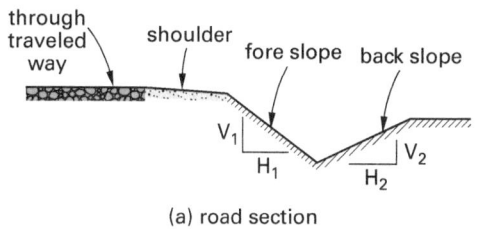

(a) road section

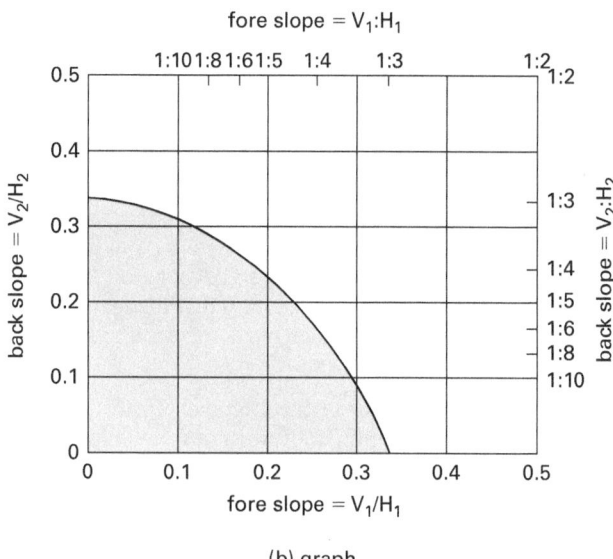

(b) graph

Figure 6.7 Preferred Cross Sections for Circular and Trapezoidal Channels with Gradual Slope Changes at Ditch Line

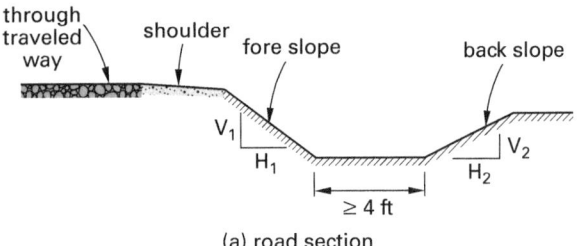

(a) road section

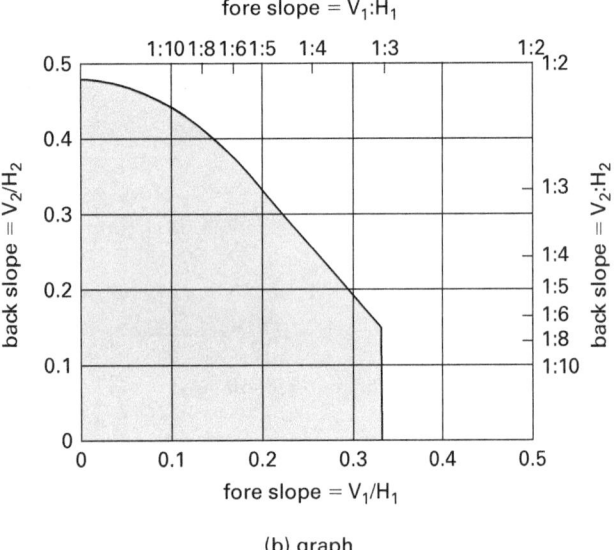

(b) graph

Barrier Protection for Fixed Objects

For conditions in which a fixed object (such as a bridge pier or a large overhead sign support) must remain in the clear zone, a barrier is installed to guide errant vehicles away from the fixed object. A general layout of barrier protection is shown in Fig. 6.8. This layout is for cases where there is no other need for a barrier on the approach to the fixed object.

The runout length, L_R, is the distance from the fixed object to the closest point that an errant vehicle is likely to leave the roadway and strike the fixed object. This distance has been found to vary with the design speed and the average daily traffic (ADT). Table 6.6 shows L_R distances suggested by *RSDG*. Local practices and conditions may suggest other values applicable to the condition at hand.

The barrier length sufficient to intersect the trajectory of an errant vehicle is adjusted for the flare rate of the barrier, shown as distance x in the diagram. The length of a crashworthy terminal is added to this needed barrier length. The *shy line* is the distance from the edge of the roadway that a roadside object will not be perceived to be an obstacle for which the driver will change speed or change position in the roadway. This distance varies with speed. Suggested offsets are shown in Table 6.7.

Barrier flare within the shy distance should be at less of an angle than outside of the shy distance. Suggested flare rates are shown in Table 6.8.

The area of concern extends upstream of the obstruction distance, L_1, which is where the taper ends and the parallel barrier placement begins. This distance is set by local jurisdictions, and may be restricted by local conditions. AASHTO research suggests driver reaction time to expected bits of information requires 1 to 1.5 sec for 1 bit, and 2 to 2.5 sec for 2 bits, simultaneously. Using this information, the change in clear distance should end about 1 to 1.5 sec before passing the obstruction itself. Adjusting for speed, this approach suggests L_1 should be, for instance, at least 88 ft for a running speed of 60 mph when the guiderail offset, L_2, lies within the shy distance. Slopes between the edge of the pavement and the face of guiderail should be no greater than 1H:10V to avoid the possibility of a vehicle vaulting the barrier or wedging under the barrier. The most effective height of the barrier is the location where the vehicle contacts the barrier at or near the height of the vehicle's normal center of gravity.

Figure 6.8 Fixed-Object Approach Barrier Layout

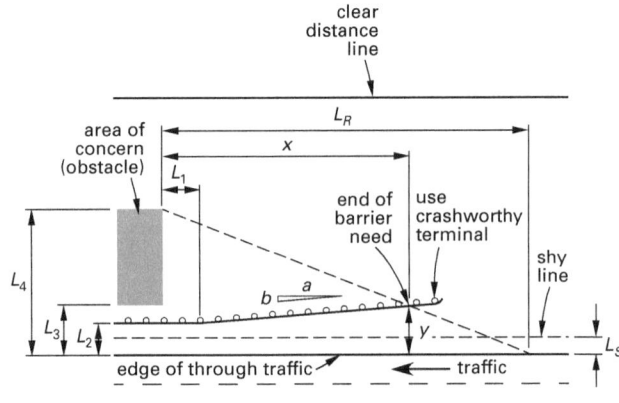

key
L_1 = tangent length of barrier upstream of the obstacle
L_2 = specified flare rate
L_3 = lateral distance from edge of roadway to near side of object
L_A = distance from traveled edge to far side of object
L_C = lateral extent of clear zone that extends beyond the fixed object
L_R = lateral extent of runout length
L_S = shy distance from edge of roadway
X, Y = coordinates of end of barrier need

Adapted from AASHTO: *Roadside Design Guide*, Fourth ed., 2011, Figure 5-39. American Association of State Highway and Transportation Officials, Washington, DC.

Table 6.6 Suggested Runout Lengths, L_R for Barrier Design

SI units

design speed (kph)	runout length, L_R, given traffic volume, ADT, (m)			
	over 10,000 veh/day	5000 to 10,000 veh/day	1000 to 5000 veh/day	under 1000 veh/day
130	143	131	116	101
110	110	101	88	76
100	91	76	64	61
80	70	58	49	46
60	49	40	34	30
50	34	27	24	21

customary U.S. units

design speed (mph)	runout length, L_R, given traffic volume, ADT, (ft)			
	over 10,000 veh/day	5000 to 10,000 veh/day	1000 to 5000 veh/day	under 1000 veh/day
80	470	430	380	330
70	360	330	290	250
60	300	250	210	200
50	230	190	160	150
40	160	130	110	100
30	110	90	80	70

Reprinted with permission from the American Association of State Highway and Transportation Officials, *Roadside Design Guide*, Tables 5-10a and 5-10b, copyright © 2011.

Table 6.7 Suggested Shy Distances from RSDG

design speed		shy-line offset, L_s	
kph	[mph]	m	[ft]
130	[80]	3.7	[12]
120	[75]	3.2	[10]
110	[70]	2.8	[9]
100	[60]	2.4	[8]
90	[55]	2.2	[7]
80	[50]	2.0	[6.5]
70	[45]	1.7	[6]
60	[40]	1.4	[5]
50	[30]	1.1	[4]

Reprinted with permission from the American Association of State Highway and Transportation Officials, *Roadside Design Guide*, Table 5-7, copyright © 2011.

Table 6.8 Suggested Barrier Flare Rates

design speed		flare rate for barrier inside shy line	flare rate for barrier at or beyond shy line	
kph	[mph]		A	B
110	[70]	30:1	20:1	15:1
100	[60]	26:1	18:1	14:1
90	[55]	24:1	16:1	12:1
80	[50]	21:1	14:1	11:1
70	[45]	18:1	12:1	10:1
60	[40]	16:1	10:1	8:1
50	[30]	13:1	8:1	7:1

A = suggested maximum flare rate for rigid barrier system

B = suggested maximum flare rate for semi-rigid barrier system

The MGS has been tested in accordance with the NCHRP Report 350 TL-3 at 5:1 flare.

Flatter flare rates for the MGS installations also are acceptable. The MGS should be installed using the flare rates shown or flatter for semi-rigid barriers beyond the shy line when installed in rock formations.

Reprinted with permission from the American Association of State Highway and Transportation Officials, *Roadside Design Guide*, Table 5-9, copyright © 2011.

The barrier should be installed as far away from the roadway as possible, with enough space between the fixed object and the barrier to allow barrier deflection on impact. With barriers on the right side of the roadway (the "blind side"), many drivers have more of a problem with depth perception than with the left side, and they tend to move their vehicle farther from the barrier than is necessary. For barriers on the left side of the roadway, the driver has a better perception of the distance between the barrier and the vehicle, allowing the shy distance to be reduced slightly.

The required length of need, X, is determined with Eq. 6.9 [*RSDG* Eq. 5-1].

$$X = \frac{L_A + \frac{b}{a}L_1 - L_2}{\frac{b}{a} + \frac{L_A}{L_R}} \quad 6.9$$

For parallel installations with no flare, the equation becomes Eq. 6.10 [*RSDG* Eq. 5-2]

$$X = \frac{L_A - L_2}{\frac{L_A}{L_R}} \quad 6.10$$

The lateral offset, Y, from the edge of the traveled lane to the beginning of the length-of-need is calculated using Eq. 6.11 [*RSDG* Eq. 5-3]

$$Y = L_A - \frac{L_A}{L_R} X \qquad 6.11$$

HSM Terminology

In the AASHTO *Highway Safety Manual* (*HSM*), a transportation *project* is a series of steps taken to evaluate and improve safety. The term *crash* is preferred over *accident* and *collision*, although common usage remains inconsistent. Crashes include nonimpact incidents.[2]

The *crash type* refers to the nature of crash (e.g., rear-end, sideswipe). *Network* (e.g., *transportation network*) refers to all of the roadways and elements within the project. *Highway* and *roadway* are used interchangeably to refer to any road, arterial, or connector. *Element* (e.g., *roadway network element*) describes such features as intersections, roadway segments, facilities, ramps, ramp terminal intersections, and at-grade rail crossings. *Segment* refers to a defined stretch of roadway or ramp. A *node* is an intersection in a network. A *facility* is a combination of segments and elements. A *site* is a specific crash or element location. *Site conditions* (*field conditions*) refer to characteristics that can be observed. A *treatment* is a change made to an element. A *countermeasure* is a treatment specifically intended to improve safety. *Base condition* represents the condition before a treatment is applied.

HSM Incident Descriptors

A specific incident (i.e., crash) will usually be described by its crash data, facility description, and traffic volume. *Crash data* (*accident data*) includes the location; time and date; severity; collision type; and a description of the roadway, vehicles, and people involved. *Facility data* describes the crash site by specifying the roadway classification, number of lanes, length, presence of medians, and shoulder width. The *traffic volume data* usually specifies annual average daily traffic (AADT) volume at a crash site.

For the purpose of uniformly defining terms among jurisdictions, the *HSM* standardizes *crash severity* using the *KABCO severity scale* keyed to the level of injury and/or property damage: K–fatal injury (FI); A–incapacitating injury; B–noncapacitating injury; C–possible injury; O–no injury, or property damage only (PDO).

HSM Crash Estimation Methods

Crash estimation is the calculation of a past or future crash frequency for a new or existing roadway under existing or alternate conditions. Historically, both the crash frequency and crash rate have been used as primary indicators of roadway safety.[3] *Crash frequency* is calculated from Eq. 6.12.

$$N = \frac{\text{no. of crashes}}{\text{period in years}} \quad [\textit{HSM Eq. 3-1}] \qquad 6.12$$

The crash rate is calculated from Eq. 6.13. The most common highway safety *exposure* is the highway segment length in miles (kilometers).

$$R = \frac{\text{average crash frequency in a period}}{\text{exposure in same period}} \qquad 6.13$$
$$[\textit{HSM Eq. 3-2}]$$

Statistical methods can also be used to estimate future crashes at a site from the current crash rate and other frequency data. The potential for future crashes at a site can also be determined indirectly and qualitatively by observing the conflicts and other factors that exist at the site.

These three techniques (historical crash rate analysis, statistical extrapolation, and qualitative observations) are commonly affected by limited data, dissimilar sites and elements, and the nonlinear effect of traffic volume on frequency. To address these limitations, standardized, quantitative, and predictive methods should be used. These predictive methods use statistical base models with data from similar sites and adjustments for local site conditions to predict an estimate of the average expected crash frequencies.[4]

In the *HSM*, the statistical models are referred to as *safety performance functions* (SPFs). SPFs are correlations used to arrive at frequencies and rates, N_{SPF}. The *HSM* has developed SPFs for three types of facilities (rural two-lane roads, rural multiline highways, and urban/suburban arterials) and four main site types (undivided roadway segments, divided roadway segments, signalized intersections, and unsignalized intersections.) Jurisdictional (locally developed) SPFs may also be used. Equation 6.14 is an example of an SPF for a rural two-lane highway segment.[5]

$$N_{\text{SPF,rural,crashes/yr}} = \text{AADT}_{\text{veh/day}} \times L_{\text{mi}}$$
$$\times 365 \times 10^{-6} \times e^{-4.865} \qquad 6.14$$
$$[\textit{HSM Eq. 3-4}]$$

---

[2] In this chapter, "crash" can be substituted for "accident" in most cases without any significant loss in meaning.
[3] Although the terms are often used interchangeably, a crash *frequency* should be used to describe the number of crashes per unit time (e.g., crashes/yr). A crash *rate* is used to describe the number of crashes per unit time per unit distance (e.g., crashes/mi-yr).
[4] Statisticians will recognize the redundancy and overlap of the terms "predict," "estimate," "average," and "expected." These terms are loosely used interchangeably, singularly, and together in the literature.
[5] Equation 6.14 illustrates an SPF. Insufficient background is presented in this chapter for its use.

The adjustments are referred to as *crash modification factors* (CMFs). A CMF is a ratio of the effectiveness at one condition to the effectiveness at another condition, with all other factors being held equal. A multiplicative *calibration factor*, C, is included to account for specific site conditions. The *HSM* provides Eq. 6.15 for calculation of the calibration factor. (See the *HSM* Part C Appendix.) The summations are taken over all sites.

$$C = \frac{\sum N_{\text{observed}}}{\sum N_{\text{predicted}}} \qquad 6.15$$

The *HSM* predictive method formula takes on the form of Eq. 6.16.

$$\begin{aligned} N_{\text{predicted}} = N_{\text{SPF}} &\times C \\ &\times (\text{CMF}_1 \times \text{CMF}_2 \times \cdots \times \text{CMF}_n) \end{aligned} \qquad 6.16$$
[*HSM* Eq. 3-3]

HSM Crash Modification Factors

Site modifications that are intended to reduce crash frequency are known as *countermeasures*. Countermeasures that would require CMF adjustments include adding more street lights, increasing roadway width, signalizing an intersection, and modifying signal cycle lengths.

CMFs adjust crash frequency for changes in one condition at a site, holding all other conditions the same. Therefore, a CMF is the ratio of crash frequencies at a site for two different conditions. A CMF with a value less than 1.00 represents an effective safety treatment, since crash frequency is reduced. From Eq. 6.17, a CMF is calculated as

$$\text{CMF} = \frac{\text{expected average crash frequency with modified site}}{\text{expected average crash frequency with base site}} \qquad 6.17$$
[*HSM* Eq. 3-5]

CMFs are multiplicative. For example, if the CMF for adding street lights along a unit increment of highway is known, the CMF for n increments is given by Eq. 6.18 as

$$\text{CMF}_n = (\text{CMF}_1)^n \quad [\textit{HSM} \text{ Eq. 3-7}] \qquad 6.18$$

The percentage reduction in crash frequency due to a particular countermeasure is given by Eq. 6.19. For example, changing the traffic control of an urban intersection from a two-way, stop-controlled intersection to a roundabout has a CMF of 0.61 for all collision types and crash severities. This value indicates that the expected average crash frequency will decrease by 39% after converting the intersection control.

$$\begin{aligned} &\text{percent reduction in crash frequency} \\ &\quad = (1.00 - \text{CMF}) \times 100\% \quad [\textit{HSM} \text{ Eq. 3-6}] \end{aligned} \qquad 6.19$$

HSM Network Screening

In the *HSM* quantitative predictive method, *network screening* is the process of reviewing a transportation network in order to identify the sites most likely to benefit from countermeasures. Network screening consists of five steps.

step 1: *Establish focus* by defining the purpose of the analysis project. Typically, the purpose is either to identify and rank sites where improvements have potential to reduce crashes, or to identify sites with a particular crash type.

step 2: *Identify the network* by specifying the elements (e.g., intersections) to be screened. Then categorize, separate, and group them by type (e.g., unsignalized and signalized intersections).

step 3: *Select network screening performance measures* by identifying one or more performance measures that will be used to quantify safety. Common measures include crash frequency, crash rate, crash severity, property damage, and fatality rate, among others. When estimating past performance, the choice of measures is driven primarily by the data available.

step 4: *Select screening methods* by identifying which segments, nodes, or elements can most benefit from countermeasures. There are three standardized methods. All three methods can be used with segments and facilities; however, simple ranking must be used with nodes.

- The *sliding window method* uses a "window" of specified length, which is moved along the roadway segment.
- The *peak searching method* divides a roadway segment into separate, nonoverlapping windows.
- The *simple ranking method* calculates and orders the performance measures for all sites.

step 5: *Screen and evaluate results* by ordering the sites from most- to least- likely to benefit from countermeasures.

One screening method used to evaluate a network of facilities for sites likely to respond to safety improvements is the *excess expected average crash frequency with empirical Bayes* (EB) *adjustment*. This performance measure combines predictive model crash estimates with historical crash data to obtain a more reliable estimate of crash frequency. An advantage of this method is that it can be used to create a single ranking for a mix of facility types and traffic volumes.

With this method, each site is evaluated to determine how much the predicted average crash frequency for the site differs from the long-term EB adjusted expected average crash frequency for the same site. This difference is referred to as the *excess value*. Sites with a high excess value are the most likely to respond to safety improvements because they theoretically experience more crashes than other similar sites. The basic procedure contains three steps. (See the *HSM* Part C Appendix for details on these steps.)

step 1: For each site, calculate the *predicted average crash frequency* from the appropriate SPF. Take into consideration property damage only (PDO) crashes and fatal injury (FI) crashes.

step 2: For each site, calculate the *expected average crash frequency* using the EB method.

step 3: Calculate the excess value from Eq. 6.20.

$$\text{excess} = \left(N_{\text{expected } n(\text{PDO})} - N_{\text{predicted } n(\text{PDO})}\right) + \left(N_{\text{expected } n(\text{FI})} - N_{\text{predicted } n(\text{FI})}\right) \quad 6.20$$

[*HSM* Eq. 4-45]

HSM Network Diagnosis

In the *HSM diagnosis procedure*, causes of the crashes are identified by analyzing crash patterns, past studies, and physical site characteristics. The diagnosis procedure involves the following three steps.

step 1: *Perform safety data review* by collecting data (often from police reports) and looking for patterns in the crash data (e.g., time of day, direction of travel), crash severity, crash type (e.g., rear-end, sideswipe), type of vehicle, driver information, and roadway conditions (e.g., pavement, traffic control, lighting). *Collision diagrams*, as shown in Fig. 6.9, may be used. The crash frequency data may be summarized in a graph, as in Fig. 6.10.

step 2: *Assess supporting documentation* by reviewing current traffic counts, as-built construction plans, inventories of field conditions, historic weather patterns, land use maps, and other records that may provide additional insight. For instance, documentation may show that the frequency of rear-end collisions increased after a stop sign was added to an intersection.

step 3: *Assess field conditions* by conducting a first-hand field investigation of site conditions. Direct observations may reveal important aspects of pedestrian and vehicle travel at the site. For instance, a field observation may reveal a blind spot for drivers. Checklists are used to ensure a consistent and comprehensive field assessment of the various *site conditions*, which are observable characteristics. Some of these site conditions include

- *roadway and roadside characteristics:* signing and striping, posted speed limits, overhead lighting, pavement conditions, sight distances, lane and shoulder widths, roadside furniture

- *traffic conditions:* types of users (e.g., cars, trucks, bicyclists, pedestrians, student drivers), travel conditions (e.g., free-flow, congested), queue storage, speed, traffic control, signal clearance

- *travel behavior:* aggressive driving, speeding, ignoring traffic control devices, bicyclists using the sidewalk instead of the bike lane, pedestrians jaywalking

- *roadway consistency:* roadway cross section is consistent with intended use

- *land uses:* adjacent land use is consistent with road travel conditions, driveway placement, types of users associated with land use (e.g., schools, parks, residential areas, senior facilities)

- *weather conditions:* seasonal and historical weather patterns that may affect roadway conditions

- *evidence of problems:* broken glass, skid marks, damaged signs, damaged guardrails, damaged road furniture

Figure 6.9 Collision Diagram

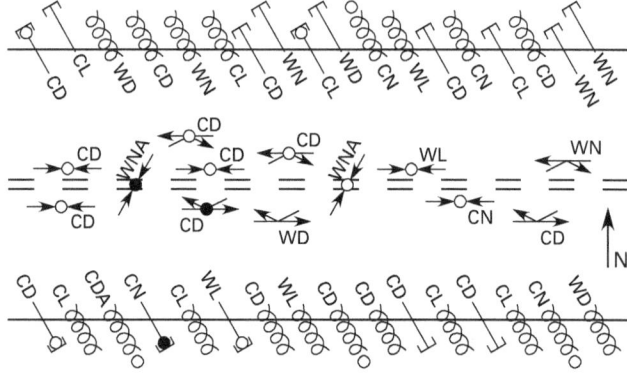

key			
A	alcohol/drug related		sideswipe, opposite direction
C	clear, dry pavement		head-on
D	day		roll-over
L	dawn/dusk		fixed object
N	night	○	injury
W	wet pavement	●	fatal

From *Highway Safety Manual*, 2010, by the American Association of State Highway and Transportation Officials, Washington, D.C. Used by permission.

Figure 6.10 Graphical Summary of Crash Frequency Data

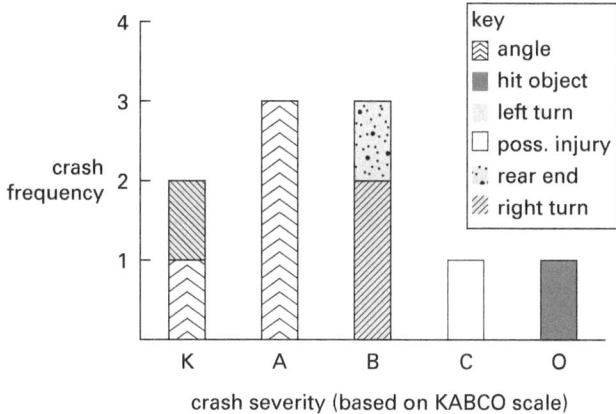

From *Highway Safety Manual*, 2010, by the American Association of State Highway and Transportation Officials, Washington, D.C. Used by permission.

HSM Economic Appraisal

In the *HSM analysis process*, a traditional *cost-benefit analysis* may be used to select the preferred treatment from all of the countermeasures with CMFs greater than 1.0. The present worth of all costs and benefits is determined first. Costs are associated with right-of-way acquisition, construction materials, labor, grading, utility relocation, environmental impacts, engineering and design, and maintenance. Benefits depend on reductions in crash frequencies and values of the various crash severities.[6] Table 6.9 presents representative values from the *HSM* that are categorized by the KABCO scale.

Table 6.9 Societal Crash Cost Estimates Versus KABCO Crash Severity

collision type	comprehensive societal crash costs
fatal (K)	$4,008,900
disabling injury (A)	$216,000
evident injury (B)	$79,000
fatal/injury (K/A/B)	$158,200
possible injury (C)	$44,900
PDO (O)	$7400

From *Highway Safety Manual*, 2010, by the American Association of State Highway and Transportation Officials, Washington, D.C. Used by permission.

The cost-benefit (or benefit-cost ratio) analysis is calculated for each countermeasure evaluated.[7] The traditional benefit-cost ratio, as calculated from Eq. 6.21 using present worth (present values), P, should have a value greater than 1.0.

$$\frac{B}{C} = \frac{P_{\text{benefits}} - P_{\text{disbenefits}}}{P_{\text{cost}}} \quad [\textit{HSM Eq. 7-4}] \quad 6.21$$

Another approach, known as the *cost-effectiveness method*, avoids the difficulties of establishing values for crashes of different severities and for values of human life. This method combines the crash frequency reduction directly with project costs.[8] The *cost-effectiveness index* (CEI) is used in comparing various alternatives. According to this method, the most cost-effective countermeasure will have the smallest CEI. Equation 6.22 is used to calculate CEI.

$$\text{CEI} = \frac{P_{\text{cost}}}{N_{\text{predicted}} - N_{\text{observed}}} \quad [\textit{HSM Eq. 7-5}] \quad 6.22$$

7. ECONOMIC ANALYSIS

Crash studies are performed to estimate the number of crashes expected at a given location. Translating this data into cost analysis involves establishing a Severity Index (SI) for types of crashes. Using the severity index, the costs associated with the crash can be calculated. Benefit-cost and engineering economic analysis can be applied to roadside safety-design features to determine a project's priority ranking. Economic analysis is most often used to compare different strategies to find the most economical alternative to implement on the jobsite. Economic evaluation of roadside safety using benefit-cost analysis requires data related to

- encroachments
- roadside geometry
- crash costs

Encroachments are primarily affected by traffic volume, roadway alignment, and lane widths. The number of encroachments that result in crashes is estimated from actual crash data (*crash statistics*) from the study location or from data derived from similar locations. The crash rates may be adjusted using the angle of departure from the roadway and the types of vehicles involved, as well as engineering judgment.

---

[6]Property damage is easier to value than loss of life. Table 6.9 is an example of possible sets of crash values that might be used. Statutory and legislated values are rare. Values are greatly affected by locale. In reality, the value of an individual life is often a matter of litigation.
[7]A list of benefit-cost ratios cannot and should not be used to rank countermeasures simplistically. Incremental benefit-cost analyses must be used to compare competing alternatives.
[8]The cost-effectiveness method effectively assumes that all crashes included have the same crash cost.

Roadside geometry encompasses the physical characteristics of the surface and appurtenances that a stray vehicle is likely to traverse. It is important to take roadside geometry into account when performing an economic analysis of a roadway. Objects in the clear zone deserve special consideration. There are several options for how to treat objects in the clear zone, including relocating them or eliminating them entirely, mounting them on breakaway devices, shielding them, modifying them, or giving them a greater degree of visibility. The decision for how to treat an object in the clear zone should be based not only on the safety and feasibility of any option, but also on economic considerations.

Crash costs are a function of the number of crashes and the severity of crashes. Dollar values are applied to the severity indices using actual crash cost data from sources such as National Safety Council (NSC), NHTSA, and FHWA and IIHS databases. Using the data from these sources, Table 6.10 shows a general range of costs for various SI index classifications. This information must be applied with careful engineering judgment to be sure of a good fit with the study location characteristics.

Table 6.10 Crash Cost Figures

crash severity	cost range
fatal crash (K)	$1,000,000–$2,600,000
severe injury crash (A)	$200,000–$180,000
moderate injury crash (B)	$12,500–$36,000
slight injury crash (C)	$3750–$19,000
PDO2	$3125–$2000
PDO1	$625–$2000

To be sure of a good fit, this information must be applied with careful engineering judgment of the study location's characteristics. For instance, a fatal crash involving multiple vehicles, multiple injuries, and property damage may involve a variety of medical costs for the injured, funeral costs for the deceased, vehicle damage costs, and property damage costs. Crash investigators must sort out costs and values assigned to each fatality and each injured person. This assignment of costs will determine the outcome of the analysis model.

Measures of Cost-Effectiveness

Benefit-cost analysis can be performed using available data and a prediction model developed for the study location. The AASHTO RSAP can help reduce the tedium of repetitive evaluations of locations and alternatives. The program is adaptable to local data and site-specific applications for the selection of alternative treatments during the planning and design process of a project. Predictions can also be performed manually using readily available traffic crash data from the Bureau of Transportation Statistics or from individual state transportation department sources.

Crash cost is estimated using an encroachment probability model, which is unique to roadside safety cost-effectiveness procedures. Crash-cost predictions are based on the concept that the "run-off-the-road" crash frequency can be directly related to the encroachment frequency.

Encroachment Cost

The expected crash cost, $C_{\exp}$, can be calculated using Eq. 6.23.

$$C_{\exp} = \sum_{i=1}^{N_{\text{SI}}} VP_E P_{\text{cr}/E} P_{\text{SI},i/\text{cr}} C_{\text{SI},i} \qquad 6.23$$

V is the traffic volume, P_E is the probability of an encroachment, $P_{\text{cr}/E}$ is the probability of a crash given an encroachment, $P_{\text{SI},i/\text{cr}}$ is the probability of injury severity level i given a crash, $C_{\text{SI},i}$ is the cost associated with injury severity level i, and N_{SI} is the number of injury severity levels. The term VP_E is the expected encroachment frequency.

Benefit-Cost Determination

Using the predicted severity of a crash, the annualized crash costs, C_A, are calculated from Eq. 6.24 by multiplying the probability of each level of injury by the cost associated with that level of injury.

$$C_A = \sum_{i=1}^{N_{\text{SI}}} P_{\text{SI},i} C_{\text{SI},i} \qquad 6.24$$

$P_{\text{SI},i}$ is the probability of injury severity level i, $C_{\text{SI},i}$ is the cost associated with injury severity i, and N_{SI} is the total number of injury severity levels. Examples of crash cost figures are shown in Table 6.10.

The concept of benefit-cost analysis is that public funds should be invested in projects in proportion to the benefits exceeding the direct costs of the project. Benefits are considered the reduction in crash costs or societal costs due to the decrease in number and severity of crashes. Direct project costs are the costs of installation, annual maintenance, and crash repairs. Comparing the benefit-cost ratio of one alternative to another is generally used for the primary ranking of mitigation measures and determining whether or not a public investment in the improvement is appropriate. Incrementing benefit-cost (B/C) is determined by comparing alternatives as shown in Eq. 6.25.

$$B/C_{2\text{-}1} = \frac{C_{A,1} - C_{A,2}}{C_{D,1} - C_{D,1}} \qquad 6.25$$

$C_{A,1}$ and $C_{A,2}$ are the annualized crash or societal costs of alternatives 1 and 2, and $C_{D,1}$ and $C_{D,2}$ are the annualized direct costs of alternatives 1 and 2.

8. DRIVER BEHAVIOR AND PERFORMANCE

Driving is an acquired skill that relies on auditory and visual awareness, motor skill coordination, and the ability to judge situations and react accordingly. Therefore, driver education plays an important role in expected on-road driver behavior. Skills are taught and demonstrated, then studied and practiced. Although the quality of instruction is important in driver training, the driver's attitude toward the rules of the road has a greater effect on how well the driver's actions will conform to general traffic behavior. Attempts to change nonconforming behavior can range from passive measures, such as road designs and signs, to active measures, such as law enforcement.

Transportation systems should be designed to accommodate the *design driver*, a term that encompasses a range of drivers whose differences and limitations are accounted for during road, vehicle, and traffic control designs. Historically, designers have used the 85th percentile of driver behavior as the cutoff to determine transportation system designs and controls. The 85th percentile is used to represent the average driver, or the *reasonable worst case driver*, though this percentile may need to be adjusted depending on regional population characteristics. For instance, regions with large elderly populations may require more careful consideration of sign placement and sign lettering, as well as an increase in standard parking stall size and signal timing design.

Driver behavior is also influenced by vehicle design. Drivers of large commercial vehicles tend to drive conservatively due to the size of the vehicle, limited maneuverability, and the nature and value of the carried cargo. Similarly, bus drivers generally make driving decisions based on the comfort and safety of the passengers. Of standard passenger cars, the faster acceleration and maneuverability of performance cars allow and even encourage their drivers to engage in higher-risk driving behaviors than drivers of sedan-like cars.

Of the three major elements making up a highway system—the driver, the vehicle, and the road—the driver is perhaps the most variable and least predictable. Therefore, it is important to understand how the following internal and external factors influence driver performance so that they can be accounted for in a transportation system's design.

Perception-Reaction Responses

Perception-reaction refers to how quickly a driver can react to a situation. Most studies divide the perception-reaction response into four subprocesses—perception, identification, emotion, and volition—also known as PIEV. *Perception* first occurs when the driver sees a control device, sign, object, or another vehicle on the road. *Identification* follows perception when the driver classifies the nature of the perceived object. *Emotion* encompasses a driver's decision-making process to determine what action needs to be taken. Finally, *volition*, or *reaction*, occurs at the point at which the driver executes the action determined to be necessary.

These subprocesses take small but measurable amounts of time, together known as the *perception-reaction time* (PRT). The average necessary time to react and execute a response to a roadway condition or object message must be accounted for in the design process, such as placement of warning signs at the proper distance ahead of the condition being warned of.

PRT is important in determining the necessary sight distance on curves and for setting the amber phase of traffic signals. It is also a necessary factor to include when determining the safe following distance between vehicles. PRT increases with age, but is not uniform across the driver population at any age group. Experienced and alert drivers can react in 1.5 sec or less, while more passive drivers may take as long as 3 sec to respond to the same situation. AASHTO design guides generally use 2.5 sec as an average reaction time for determining stopping distances. However, PRT can vary significantly depending on roadway types and conditions. 2.5 sec may not be adequate under adverse or very complex conditions. For instance, an intersection approach may be complicated enough that the amber phase must be increased by 1–2 sec to eliminate running the red phase. Other instances where a longer PRT may exist include confusing or incorrect information on warning or directional signs. Drivers have generally learned to expect that adequate warning time will be given in order for them to react to a situation and make adjustments in vehicle operation. When drivers come across a situation that deviates from this norm (i.e., detour signs or unanticipated construction), they will likely need more time to react than usual, as the situation will come as a surprise. A *surprise* is when the time allowed for adjustment to a change in road condition, traffic condition, or visibility is less than the required normal perception-reaction time. The goal of good highway design is to eliminate surprises whenever possible.

Visual and Recognition Acuity

Visual acuity refers to the sharpness with which a person can see an object, while *recognition acuity* requires the viewer to not only see an object, but recognize the meaning of the object as well. For a normal eye, visual acuity is greatest near the visual center of the retina, which is along the visual axis of the eye. The cone of greatest visual acuity is about 3°, or about 1.5° in each direction from the center axis of vision. *Peripheral vision* can extend as far as 180° or more in some people, but 160° is the accepted normal limit of peripheral vision. In the *peripheral zone*, objects are visible and important for orientation and awareness, but may be blurred compared to the central cone of vision. Traffic signs and critical road markings should be placed so that they are within the normal central cone of vision in order to be clearly read by drivers.

The standard *Snellen chart* is used universally for checking visual acuity during an eye examination. A person reads letters of different heights on the chart from a specified distance. Charts used to determine visual acuity are designed so that a person with normal vision can read lettering $1/3$ in (0.85 cm) in height from a distance of 20 ft (6.1 m). This ratio is called *20/20 vision*. When a person needs an object twice as large to be visible at 20 ft (6.1 m), the ratio is given as 20/40. Visual and recognition acuity can deteriorate because of age, fatigue, eye disease, medication side effects, and/or alcohol and drug use.

The criterion for letter height on signs is often cited as 1 in (25 mm) for each 50 ft (15 m) of viewing distance. However, this criterion is incomplete without including letter spacing and letter shape. The letter shapes used on interstate highway signs were developed after extensive visibility research during the development of the early interstate highway system. The FHWA has developed the *federal alphabet* with respect to within letter spaces and between letter spaces. *Within letter spaces* are the openings in loop letters such as a, p, and so on. *Between letter spaces* are the spaces that exist between letters when they are put together to form words. Each letter in the federal alphabet has been created with an optimum amount of white space around and, if necessary, within it. By this design, when letters are put together, the white space between them optically balances with the white space within them, producing optimum readability and legibility. These guidelines allow traffic signs to be viewed from a distance of 100 ft (30 m) for each 1 in (25 mm) of letter height, or twice the viewing distance when using other lettering styles. This standard has not been replicated in most current electronic digital lettering signs, meaning the viewing distance of many of these changeable lettering signs is much less than the fixed lettering version. Standards for a sign's height, color, letter spacing, message content, panels, layout, and location are given in further detail in the *Manual on Uniform Traffic Control Devices* (*MUTCD*).

Color Vision

The colors chosen for traffic control devices and informational signs were selected to minimize the effects of color blindness on their readability. *Color blindness* refers to a difficulty in perceiving differences between colors. Driver color blindness exists in various types and degrees that range from a difficulty in distinguishing between minor hue differences to a complete lack of color definition, in which the driver sees everything as varying intensities of gray. Background and letter coloring have been researched to provide the greatest amount of contrast so that lettering does not disappear when viewed by someone with color-impaired vision. For example, the green background with white lettering on roadway direction signs has become a standard on U.S. federal highways, eliminating the effect of color versus contrast sensitivities. Similarly, traffic signals place the red, amber, and green lenses in standardized positions so that a driver with total or red/green color blindness will still be able to determine signal phases.

Glare Vision

Glare can be caused by a direct shaft of light in the field of vision or by reflected light. When too much light enters the eye, the pupil closes to reduce the amount of incoming light. *Glare recovery* can take 6 sec or more for a normal eye, though factors such as age, congenital disabilities, or drug and alcohol use can increase the eye's recovery time.

Glare also affects the eye's visual contrast ratio between light and dark objects. The human eye cannot discern a light-intensity-contrast ratio greater than 3 to 1. *Glare blindness* occurs when the presence of an intense light source at one part of the visual plane causes images in the rest of the visual plane to disappear. For example, glare blindness can occur when driving at night along a dark road. Objects are illuminated by the headlight's radius, but objects appear invisible outside that radius.

When designing transportation systems, designers must account for changes between lit and unlit sections of the roadway and allow sufficient time for drivers' eyes to acclimate to new conditions before introducing critical movement, such as exit ramps or abrupt geometry conditions. Sources of light off of, but near to, roadways, such as advertising lighting and lighting in shopping center parking lots, can also create glare and affect the safety of traffic.

Depth Perception

Depth perception allows a person to estimate the distance to an object by using either stereo vision for objects closer to the eye or by estimating the relative size, position, and movement of objects farther away. The human eye has difficulty measuring the speed and the relative distances of objects that are more than a few feet away. This is especially true of approaching objects that are moving in a straight line, such as two oncoming vehicles occupying the same lane. Traffic signs and marking devices have standardized locations to account for the driver's viewing distance, which helps maintain uniformity and predictability of approaching roadway conditions. One example is the placement of a stop sign on the corner of a minor street intersecting a major street.

Hearing

Sounds received by the human ear can help a driver adjust speed, vehicle location, and other factors in relation to certain circumstances, such as an oncoming vehicle or the sound of a car horn. They can also warn drivers of unusual occurrences, such as an approaching emergency vehicle. Drivers with limited or no hearing are generally able to drive quite well, as long as they take extra precautions by using more visual feedback to compensate for the lack of hearing.

Driver Error and Unavoidable Situations

When a driver encounters traffic or road conditions that are outside of the range of normal conditions, collisions, off-road excursions, or other out-of-the-ordinary travel may occur. A pattern of similar crashes, particularly if involving drivers unfamiliar with the local conditions, may indicate a specific need for countermeasure development.

For example, if drivers frequently miss a turn at an intersection and drive off the side of the road, there may be a sight-limiting condition that is blocking the view at the driver's eye level. If the condition is primarily weather-related, such as icing, there may be a need to redirect drainage or recontour the intersection.

When a specific countermeasure cannot be developed to prevent a crash, or when crashes are random, countermeasures may include providing a safer environment for the most damaging or injury-prone type of occurrence.

Crash Attenuators

The chance that a vehicle will strike an object, such as a bridge pier or light pole, increases when the object is placed nearer to the edge of a high-speed roadway. Objects can be designed to yield under impact (e.g., light poles), or they can be designed to withstand considerable impact (e.g., bridge piers). In order to reduce the number of injuries or deaths of vehicle occupants, protective devices can be installed ahead of fixed objects to attenuate, or weaken, the decelerating force on the vehicle. Protective devices include barrels of sand, water-filled tubes, plastic foam cartridges, lightweight concrete, and a variety of sliding metal shapes.

The objective for all protective systems is to provide a stopping or deflecting action to the vehicle at a predetermined *maximum g-force* (usually 8.0 g). The *g*-force associated with an object is its acceleration in standard earth gravities, 32.2 ft/sec² (9.81 m/s²). The maximum survivable *g*-force is also dependent on the duration of the force being applied. Higher *g*-forces can be tolerated for a shorter duration, while lower *g*-forces can be endured longer. Personal injury, vehicle control, rebound, damage to other vehicles, reusability, reparability, installation cost, maintenance cost, and liability are some of the factors considered in the design and installation of protective devices. To determine the minimum length of deceleration, an engineer selects the maximum *g*-force and the probable impact speed along with other predetermined criteria, such as the design vehicle weight, and calculates the required working length of the selected device. The loss of kinetic energy is equated to the work in deforming the attenuator, or crash cushion. Some energy is also expended in the deformation of the vehicle. However, this can vary from vehicle to vehicle and is not always included in the calculations. Equation 6.26 can be used for deceleration of a vehicle during a crash barrier impact. v is the design speed, and L is the unit length of the attenuator.

$$d = \frac{v^2}{2L(0.75)} \quad\quad 6.26$$

The 0.75 factor in the denominator is the attenuation efficiency factor, which takes into account the portion of the length of the attenuator that is occupied by the energy absorption mechanism. Efficiency can be set at any level, but 75% is commonly used.

Once the engineer determines the required deceleration distance needed, a unit is selected from the vendor's catalog based on the specific requirements of the application. The engineer must then determine the design force on the backer wall or anchor system for the attenuator. A safety factor, SF, is also commonly applied to the anchor system or backer wall. Therefore, the stopping force, F_s, is calculated from Eq. 6.27 as

$$F_s = (\text{SF})md \quad\quad [\text{SI}] \quad 6.27(a)$$

$$F_s = \frac{(\text{SF})md}{g_c} \quad\quad [\text{U.S.}] \quad 6.27(b)$$

Example 6.3

An energy cushion unit is being considered for a design speed of 50 mph (80 kph) and a maximum deceleration of 8.0 g. The design vehicle weighs 4500 lbm (2040 kg). The unit is 14.5 ft (4.4 m) long, and 75% efficiency is to be used. (a) Determine whether the cushion unit length is sufficient, and (b) using a factor of safety of 1.5, determine the force requirement for the backer wall supporting the unit.

SI Solution

(a) From Eq. 6.26, the deceleration in m/s² is

$$d = \frac{v^2}{2L(0.75)}$$

$$= \frac{\left(\left(80\ \frac{\text{km}}{\text{h}}\right)\left(\frac{1000\ \frac{\text{m}}{\text{km}}}{3600\ \frac{\text{s}}{\text{h}}}\right)\right)^2}{(2)(4.4\ \text{m})(0.75)}$$

$$= 74.6\ \text{m/s}^2$$

Determine the g-force.

$$d_g = \frac{d}{g} = \frac{74.6 \frac{\text{m}}{\text{s}^2}}{9.81 \frac{\text{m}}{\text{s}^2 \cdot g}}$$

$$= 7.6 \text{ g} \quad \begin{bmatrix} <8.0 \text{ g. The cushion unit} \\ \text{length is sufficient.} \end{bmatrix}$$

(b) Determine the stopping force of a 2040 kg vehicle on the backup wall from Eq. 6.27(a), using a factor of safety of 1.5.

$$F_s = (\text{SF})md = (1.5)(2040 \text{ kg})\left(74.6 \frac{\text{m}}{\text{s}^2}\right)$$

$$= 228\,000 \text{ N}$$

Customary U.S. Solution

(a) From Eq. 6.26, the deceleration in ft/sec^2 is

$$d = \frac{\text{v}^2}{2L(0.75)} = \frac{\left((50 \frac{\text{mi}}{\text{hr}})\left(\frac{5280 \frac{\text{ft}}{\text{mi}}}{3600 \frac{\text{sec}}{\text{hr}}}\right)\right)^2}{(2)(14.5 \text{ ft})(0.75)}$$

$$= 247 \text{ ft/sec}^2$$

Determine the g-force.

$$d_g = \frac{d}{g} = \frac{247 \frac{\text{ft}}{\text{sec}^2}}{32.2 \frac{\text{ft}}{\text{sec}^2 \cdot g}}$$

$$= 7.7 \text{ g} \quad \begin{bmatrix} <8.0 \text{ g. The cushion unit} \\ \text{length is sufficient.} \end{bmatrix}$$

(b) Determine the stopping force of a 4500 lbm vehicle on the backup wall from Eq. 6.27(b), using a factor of safety of 1.5.

$$F_s = \frac{(\text{SF})md}{g_c}$$

$$= \frac{(1.5)(4500 \text{ lbm})\left(247 \frac{\text{ft}}{\text{sec}^2}\right)}{32.2 \frac{\text{lbm-ft}}{\text{lbf-sec}^2}}$$

$$= 52{,}000 \text{ lbf}$$

Example 6.4

A directional sign is needed before a highway turnoff. The highway has an 85th percentile speed of 50 mph. Vehicles normally make the turn at the design speed of 25 mph, and cars travel at a constant speed of 50 mph during the perception-reaction time. Using a perception-reaction time of 1.5 sec to read and understand the sign and a deceleration (or braking) rate of $0.3g$, what is the minimum distance the directional sign should be placed in advance of the turnoff?

Solution

The total distance traveled from the moment when the sign is observed includes the perception-reaction distance plus the braking distance needed to decelerate to the turnoff speed of 25 mph. Add the perception-reaction time to the deceleration distance. Since the car travels at a constant speed of 50 mph during the perception-reaction time, t_{PRT}, the total distance is

$$s_d = \frac{\text{v}_{\text{full}}^2 - \text{v}_{\text{final}}^2}{2d} + \text{v}_{\text{full}} t_{\text{PRT}}$$

$$= \frac{\left(50 \frac{\text{mi}}{\text{hr}}\right)^2 - \left(25 \frac{\text{mi}}{\text{hr}}\right)^2}{(2)(0.3)\left(32.2 \frac{\text{ft}}{\text{sec}^2}\right)}\left(\frac{5280 \frac{\text{ft}}{\text{mi}}}{3600 \frac{\text{sec}}{\text{hr}}}\right)^2$$

$$+ \left(50 \frac{\text{mi}}{\text{hr}}\right)(1.5 \text{ sec})\left(\frac{5280 \frac{\text{ft}}{\text{mi}}}{3600 \frac{\text{sec}}{\text{hr}}}\right)$$

$$= 319 \text{ ft}$$

The car will travel 110 ft before braking begins, and the car requires 209 ft to brake from 50 mph to 25 mph. Therefore, the sign should be placed at least 209 ft before the turnoff.

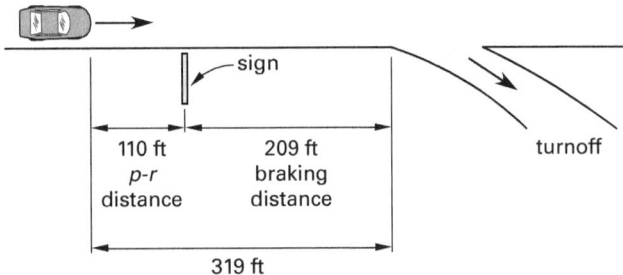

9. WORK ZONE SAFETY

Work zones increase crashes and congestion, and account for approximately 2% of all roadway fatalities each year. With the majority of the fatalities being motorists, the FHWA created the National Highway Work Zone Safety Program, which focuses on standardization and evaluation to improve work zone safety.

Work zone standardization is set by the FHWA in both traffic control and work zone safety devices. All safety standards are contained in the *MUTCD*.

The National Work Zone Safety Information Clearinghouse is a public outreach organization that provides information and resources about work zone safety. Managed by the FHWA Office of Safety, it contains the largest online database of accident and crash rate data and statistics, training programs and materials, research services, latest technologies and equipment, best practices, safety engineer contact information, laws and regulations, and public education.

The *Manual for Assessing Safety Hardware* (MASH) contains the federal standards and guidelines for all work zone safety devices, including traffic barriers, end treatments, crash cushions, and breakaway devices. *MASH* provides a wide range of test procedures to evaluate these devices, with different test levels defined for various classes of roadside safety features. It provides enhanced measurement techniques related to occupant risk, as well as guidelines for device installation and test instrumentation. Evaluation criteria to assess the performance of work zone safety devices include occupant risk, occupant compartment integrity, test article debris, and vehicle stability.

A significant part of work zone safety includes temporary traffic control devices. In order to be effective, *MUTCD* describes the basis for these devices.

- Fulfill a need.
- Command attention.
- Convey a clear, simple meaning.
- Command respect from road users.
- Give adequate time for proper response.

Construction, maintenance, utility work, survey crews, and incident zones inevitably disrupt the normal flow of traffic and create unexpected or unusual conditions for road users. Plans must be prepared in advance for each type of work requiring *temporary traffic control* (TTC). These plans must be coordinated with all affected parties, including law enforcement, transit, railroads, and other highway users. In some instances, it may be necessary to reroute commercial traffic, or even to reroute all traffic using detours. Relocating transit activity and providing for pedestrian and bicycle traffic may be included as well. For TTC zones, *MUTCD* offers guidance for applying the principles of TTC.

- Provide safety for motorists, bicyclists, pedestrians, workers, and emergency equipment.
- Traffic movement should be disrupted as little as possible.
- Road users should be guided in a clear and positive manner while approaching and within construction, maintenance, and utility work areas.
- Routine inspection and maintenance of traffic control elements should be performed both day and night.
- At least one person from both the contractor and the contracting agency should have the day-to-day responsibility for assuring that the traffic control elements are operating effectively, and that any needed operational changes are brought to the attention of their supervisors. These persons must be qualified and familiar with *MUTCD* principles and should receive safety training in TTC operations, including incident management.
- All persons whose actions affect TTC zone safety should receive training appropriate to the job decisions each individual is required to make.
- Good public relations should be maintained through cooperation of various news media resources, adequate advance notice of upcoming work activity, accommodating emergency services through the TTC zone, and recognizing ongoing community events that may affect access to or work within the TTC zone.

TTC that extends over a long distance will benefit from provisions for temporary pull-off areas for rescue and recovery. TTC in heavy urban traffic locations may require disabled-vehicle rescue services stationed periodically throughout the TTC site to reduce the time required to respond to incident management.

MUTCD suggests five categories of work duration.

- long-term stationary work that occupies a location more than three days
- intermediate-term stationary work that occupies a location more than one daylight period up to three days, or nighttime work lasting more than one hour
- short-term stationary work that occupies a location for more than one hour in a single daylight period
- short duration work that occupies any location for less than one hour
- mobile work that moves intermittently or continuously

Long-term and intermediate-term work that requires operations to extend into nighttime must have reflective and/or illuminated devices. Permanent markings or devices for any category of work that interfere with TTC markings or devices are to be covered or removed during the work period in order to avoid confusion for the motorist. For short-term work, the time required to set up devices may be greater than the time required to perform the work. A reduction in the number of devices for short-term work may be offset by the use of more dominant devices, such as high-intensity rotating, flashing, or strobe lights on work vehicles.

Mobile operations are to have high-intensity rotating, flashing, oscillating, or strobe lights on the equipment, or may use a separate vehicle with appropriate warning devices. A shadow vehicle with an arrow board or a sign should follow the work vehicle, especially when vehicular speeds or volumes are high. For high-volume conditions, work should be scheduled as much as possible during off-peak hours.

In general, the closer the work zone is to road users, the greater the number of TTC devices are needed. Work locations are described as

- outside the shoulder
- on the shoulder with no lane encroachment
- within the median
- within the travel lanes

10. TEMPORARY TRAFFIC CONTROL ZONES

For work at fixed locations, properly implemented temporary traffic control zones are composed of four sections (*areas*): (1) advance warning area, (2) transition area, (3) activity area, and (4) termination area. These areas are illustrated in Fig. 6.11.

Except in areas where the advance warning area may be eliminated because the activity area does not interfere with normal traffic, road users will learn of the traffic control zone in the *advance warning area*. The warning can be accomplished by signage, flashing light trailers, or rotating lights and strobes on parked vehicles. Three consecutive points of warning are required, as illustrated in Fig. 6.11. Table 6.11 summarizes the *MUTCD*'s suggested distances A, B, and C, between the three consecutive warning signs.

Table 6.11 *Meaning of Letter Codes on Typical Application Diagrams*

road type	distance between signs[a]		
	A	B	C
urban (low speed)[b]	100 ft	100 ft	100 ft
urban (high speed)[b]	350 ft	350 ft	350 ftt
rural	500 ft	500 ft	500 ft
expressway/freeway	1000 ft	1500 ft	2640 ft

[a]The column headings A, B, and C are the dimensions shown in Fig. 6H-1 through Fig. 6H-46. The A dimension is the distance from the transition or point of restriction to the first sign. The B dimension is the distance between the first and second signs. The C dimension is the distance between the second and third signs. (The "first sign" is the sign in a three-sign series that is closest to the TTC zone. The "third sign" is the sign that is farthest upstream from the TTC zone.)
[b]Speed category to be determined by highway agency.

In the transition area, vehicles are redirected out of their normal paths by the use of *tapers*. A taper is created by using channelization devices and/or other pavement markings. Channelization design is specified by

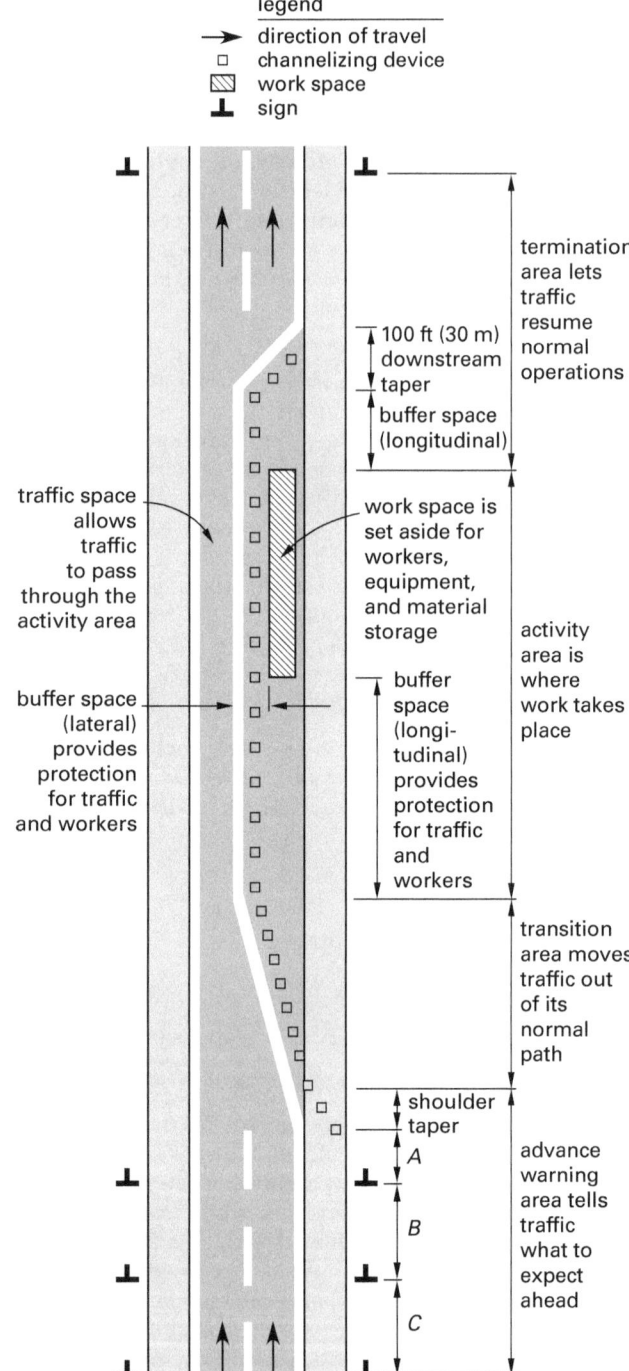

Figure 6.11 *Sections of a Temporary Traffic Control Zone*

Reprinted with permission from Fig. 6C-1 of the *Manual on Uniform Traffic Control Devices*, 2009 ed., U.S. Department of Transportation, Federal Highway Administration, 2009.

the distance between devices and the overall length of the transition area. The *MUTCD* specifies that, except for downstream and one-lane, two-way tapers, the maximum distance between channelization devices is 1 ft for every mph of speed (0.2 m for every kph). For downstream and one-lane, two-way tapers, the maximum distance between channelization devices is approximately 20 ft (6.1 m).

The minimum taper length depends on the type of transition. Table 6.12 summarizes the *MUTCD*'s suggested *taper lengths*, *L*. In Eq. 6.28 and Eq. 6.29, *S* is the posted speed limit, the off-peak 85th percentile speed prior to work starting, or the anticipated operating speed.

$$L = \frac{WS^2}{60} \quad\quad 6.28$$

$$L = WS \quad\quad 6.29$$

Table 6.12 Taper Length Criteria for Temporary Traffic Control Zones

type of taper	taper length, L (as calculated from Eq. 6.28)
merging taper	L minimum
shifting taper	$0.5\,L$ minimum
shoulder taper	$0.33\,L$ minimum
one-lane, two-way traffic taper	50 ft (15 m) minimum, 100 ft (30 m) maximum
downstream taper	50 ft (15 m) minimum, 100 ft (30 m) maximum per lane

(Multiply ft by 0.3048 to obtain m.)

From *Manual on Uniform Traffic Control Devices*, Table 6C-3, 2009 Ed., U.S. Department of Transportation, Federal Highway Administration.

Merging tapers (where two lanes merge) are longer than *nonmerging tapers* and require the full taper length since vehicles are required to merge into a common road space. When more than one lane must be closed, merged traffic should be permitted to travel approximately $2L$ before the second lane is merged. *Shifting tapers* (where traffic moves to an entirely different travel path) require a minimum length of $L/2$ but can benefit from longer lengths. *Shoulder tapers* preceding shoulder construction on an untraveled improved shoulder with a minimum width of 8 ft (2.4 m) that might be mistaken as a driving lane require a minimum length of $L/3$. When two-directional traffic on a two-way road is slowed, stopped, and alternately channeled into a single lane by a *flagger*, *two-way traffic tapers* with a maximum length of 100 ft (30 m) are used, one at each end. *Downstream tapers* are useful in providing a visual cue to drivers but are optional. When used, they have lengths of 100 ft per lane.

As shown in Fig. 6.11, the *activity area* consists of the actual work space, traffic space, and buffer space. The *work space* is the part of the highway that is closed to road users and is set aside for workers, equipment, and materials. It also includes a *shadow vehicle* if one is used

upstream. Work spaces are separated from passing vehicles by channelization devices or temporary barriers.

The *traffic space* is the part of the highway in which road users are routed through the activity area. The *buffer space* is a lateral and/or longitudinal (with respect to the work space) area that separates road user flow from the work space or an unsafe area. It might provide some recovery space for an errant vehicle. The buffer space should not be used for storage of equipment, vehicles, or supplies. If a shadow vehicle is used, the buffer space ends at the bumper first encountered. If a lateral buffer area is provided, its width is chosen using engineering judgment.

11. CONFLICT ANALYSIS

Conflicts in traffic streams occur at any location where two traffic flows cross, merge, or diverge. Conflicts can be within the same type of traffic, such as automobile traffic, or between different types of traffic, such as between auto traffic and pedestrian traffic. Conflicts involve gaps in the traffic flow and the drivers' judgment on whether the gap is acceptable or not. Diverging moves, such as for turning vehicles at intersections, are the least disruptive conflicts as long as the diverging paths operate at about the same speed. If one path is considerably slower than the other, a conflict occurs with the faster flow. Figure 6.12 shows vehicle streams at a three-way intersection.

Crossing and merging flows require larger gaps and have the greatest potential for collision. At a typical intersection of a two-lane street crossing another two-lane street (see Fig. 6.13), there are 32 conflict points. 16 points are crossing conflicts, 8 points are merging conflicts, and 8 points are diverging conflicts. Pedestrian crossings add another 20 conflict points to the intersection.

Figure 6.12 Vehicle Streams at a Three-Way Intersection

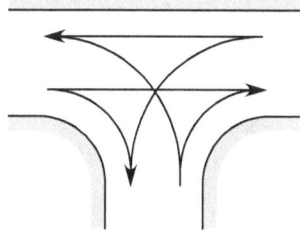

Khisty, C. Jotin; Lall, B. Kent; *Transportation Engineering: An Introduction*, 3rd edition, © 2003, p. 292. Reprinted by permission of Pearson Education, Inc., Upper Saddle River, NJ.

Figure 6.13 illustrates where vehicle conflict points occur for a four-way intersection and two three-way intersections. Replacing a four-way intersection with two three-way intersections may reduce some conflict points for certain moves. However, this improvement creates two intersections. The delay caused by eliminating one of the through-movements is more than offset by the extra maneuvering required to perform an offset through-movement. For a three-way intersection, right turns and through-movements along the lanes opposite the entering traffic are the least impeded by other vehicle movements.

Figure 6.13 Vehicle Intersection Conflict Points

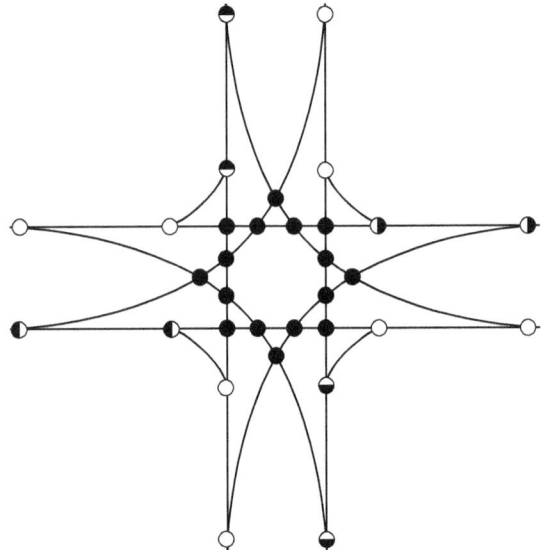

four-way intersection
● 16 crossing conflicts
○ 8 merging conflicts
◐ 8 diverging conflicts

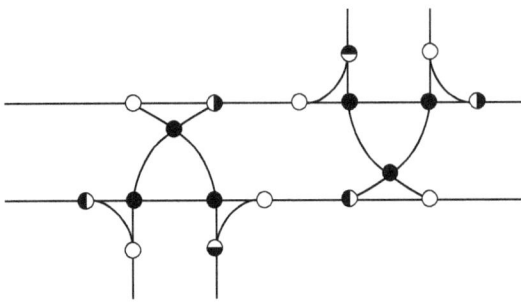

two three-way intersections
● 6 crossing conflicts
○ 6 merging conflicts
◐ 6 diverging conflicts

Khisty, C. Jotin; Lall, B. Kent; *Transportation Engineering: An Introduction*, 3rd edition, © 2003, p. 292. Reprinted by permission of Pearson Education, Inc., Upper Saddle River, NJ.

Conflict analysis involves reviewing data on conflicting traffic stream movements that show the flow density of each movement, acceptable gaps in following distance, the capacity limits of conflicting movements, and the delays caused by conflicting movements. Conflict analysis can be used to develop strategies to reduce delay, improve average speed, increase capacity, and improve safety.

12. PRACTICE PROBLEMS

1. A 5 mi segment of highway is analyzed for inclusion in a funding program. Included in the analysis is a statistical test to determine if the crash rate is significantly greater than the statewide average. The crash statistics over the previous four years are as follows.

year	fatal crashes	injury crashes	total crashes	AADT (veh/day)
1	4	162	320	38,800
2	3	176	340	40,000
3	2	190	365	41,200
4	5	185	352	42,400

The statewide average total property damage crash rate for similar roadway types is 106 crashes/HMVM for the 4 yr period. The standard deviation corresponding to this mean rate is 56 crashes/HMVM.

(a) What is most nearly the average crash rate per HMVM involving property damage only for the 4 yr period?

(A) 8.0 HMVM

(B) 44 HMVM

(C) 220 HMVM

(D) 650 HMVM

(b) At what confidence level do the statistics exceed the statewide average crash rate for the 4 yr period?

(A) 90%

(B) 95%

(C) 99%

(D) more than 99%

2. A four-lane urban freeway with a 9 m wide median and a 95 kph design speed is being studied for construction of additional lanes for capacity. The alternative proposed is to construct an additional 3.6 m lane on the left side in each direction with a concrete barrier, leaving a left shoulder of 1.3 m between the edge of the left lane and the median barrier. The freeway is 16 km long, lies in flat terrain, and has an AADT of 10,000. Construction cost of the alternative will be $30,000,000, with a life expectancy of 15 years. The annualized maintenance cost will be $150,000/km·yr for the project life. The discount rate is 4% per year, and the traffic growth factor is projected at 2.5% per year. Crash statistics are projected as shown in the table.

crash severity	current roadway crash rate per 10^6 veh·km	alternate crash rate per 10^6 veh·km	crash cost
K—fatal crash	0.0172	0.0258	$1,000,000
A—disabling injury crash	0.0487	0.0688	$200,000
B—evident injury crash	0.0602	0.0860	$12,500
C—possible injury crash	0.119	0.150	$3750
PDO2	0.150	0.173	$625
PDO1	0.0989	0.120	$3125

(a) The number of additional crashes per year projected for the proposed alternative, averaged over the life of the project, is most nearly

(A) 9.0 crashes/yr

(B) 34 crashes/yr

(C) 44 crashes/yr

(D) 78 crashes/yr

(b) The average annualized cost of construction and maintenance for the life of the project is most nearly

(A) $2,900,000

(B) $5,100,000

(C) $220,000,000

(D) $240,000,000

(c) What is most nearly the average projected crash cost per year for the proposed alternative?

(A) $57,000

(B) $69,000

(C) $89,000

(D) $110,000

(d) Using benefit-cost analysis, compare the annualized cost of implementing the proposed alternative with the additional crash cost.

(A) 0.24:1

(B) 0.43:1

(C) 2.3:1

(D) 2.8:1

(e) What information would best complete the benefit-cost analysis for this study?

(A) data on right shoulder crashes

(B) effects of changing design speed

(C) number of trucks and buses

(D) benefits of added lane for increased capacity

3. A 5 mi long, nearly straight highway segment through flat lowlands has experienced a high number of off-road crashes due to very heavy traffic, frequent icing conditions, and poor visibility caused by fog. The AADT of 55,000 operates at LOS C during the midday, and a peak period flow of about 8000 vph is at LOS E for the prevailing direction in the morning and evening periods. The cross section consists of a four-lane divided freeway with a wide median and a 10 ft paved right shoulder. The clear zone measures 20 ft wide from the traveled lane edge to the top slope of a drainage ditch running parallel to the roadway. The clear zone beyond the shoulder has a foreslope of 12H:1V, and the drainage ditch has 3H:1V side slopes along its 2 ft average depth. Typical crashes involve a vehicle being forced off the roadway and into the clear zone, traversing the drainage ditch and landing in an adjoining pasture. About 35% of the ditch crossings involve vehicle upset, and about 2% of heavy vehicle crashes involve at least one fatality, primarily due to vehicle upset while crossing the drainage ditch. The posted speed limit is 65 mph for passenger cars and 55 mph for trucks. Several alternative countermeasures have been proposed.

Alternative 1 is to lower the speed limit to 50 mph for all vehicles, along with strict enforcement. No other changes to the roadway cross section are proposed.

Alternative 2 is to place a guiderail behind the shoulder 10 ft from the edge of the traveled lanes.

Alternative 3 is to place a guiderail near the top of the slope of the drainage ditch, about 18 ft from the edge of the traveled lanes.

Alternative 4 is to change the slope of the drainage ditch sides to 6H:1V, narrowing the ditch bottom to 3 ft. This alternative also involves narrowing the clear zone adjacent to the shoulder from 20 ft wide to 16 ft wide and increasing its slope to 8H:1V. In certain limited areas, the entire 16 ft wide clear zone may be as steep as 6H:1V in order to maintain the needed ditch invert. The ditch bottom would be 31 ft from the traveled lanes, versus 30 ft for the current design. Right-of-way is available to perform the ditch and slope changes.

Which alternative would be the preferred countermeasure based on AASHTO guidelines?

(A)
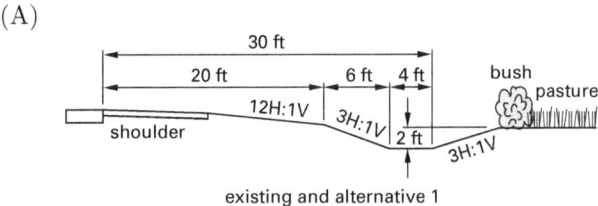
existing and alternative 1

(B)
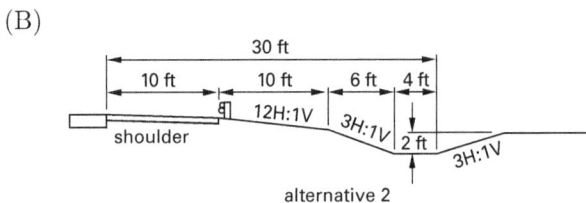
alternative 2

(C)
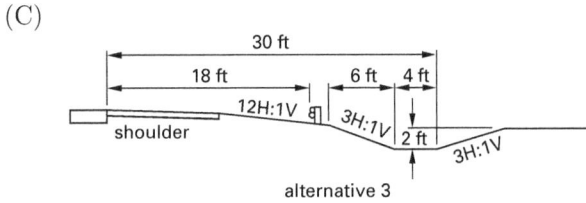
alternative 3

(D)
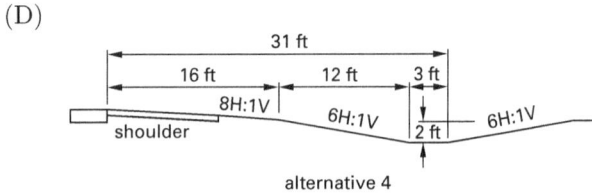
alternative 4

4. An intersection near a school with several conflict points has an increasing number of crashes. The intersection has stop sign control, and a school guard is stationed at the major school crossing for 3 hr each school day. An additional problem is the intersection flooding at least four times a year due to clogged and inadequately sized drainage pipes, which requires cleanup after every flood. Community groups are demanding improvements, including a traffic signal and increased school guard presence. Crash costs are given in the following table.

crash type	average cost per incident
K	$1,000,000
A	$200,000
PDO1	$3000

Crash type K involves a fatality, crash type A involves a disabling injury, and crash type PDO1 involves property damage only, including the result of swerving to avoid pedestrians.

Several alternatives have been proposed to reduce the conflict between vehicles, pedestrians, and/or other vehicles.

Alternative 1 is to do nothing with the intersection geometry. The drain pipes would be cleaned, but not repaired or replaced. The intersection would be given a maintenance overlay and re-striped in the existing traffic lane configuration. The historical crash rate is expected to continue for the next 2 yr. This alternative is used for comparison purposes.

Alternative 2 provides drainage improvements and a grade separation for the major cross traffic flows, with left-turning bays and slip ramps to accommodate the traffic movements. A pedestrian bridge is provided to eliminate the conflict with young children crossing the main traffic flow. However, a few older children cross at other places. The school guard would be eliminated with this alternative.

Alternative 3 is to provide drainage improvements, perform intersection geometric improvements, and install a traffic signal. A school guard would be stationed at the main school crossing for a total of 3 hr per school day.

Alternative 4 is to perform drainage improvements, make modest geometric improvements, and install a blinking traffic beacon, which would flash amber for the main movement and red for all other movements. A school guard would be stationed at the intersection for a total of 5 hr per school day.

Funding is available from the FHWA, the state transportation department, and local sources, sharing costs as shown in the following table. Crash occurrence and annual maintenance costs are also shown. School guard costs are entirely local. The projected crash statistics are for a 2 yr period following construction.

	alternative 1	alternative 2	alternative 3	alternative 4
crashes				
K	5	0	0	0
A	45	4	12	16
PDO1	160	14	28	40
crash years (total)	5 yr	2 yr	2 yr	2 yr
total construction cost	$100,000	$2,500,000	$650,000	$450,000
construction cost share				
federal	0%	50%	0%	0%
state	0%	30%	80%	80%
local	100%	20%	20%	20%
total annual maintenance	$50,000	$50,000	$45,000	$30,000
maintenance cost share				
state	0%	60%	20%	20%
local	100%	40%	80%	80%
school guard annual cost (local)	$30,000	$0	$30,000	$45,000
design life	3 yr	20 yr	12 yr	7 yr

Cost escalation is projected to be 5%/yr for construction and maintenance activities and 3%/yr for school guard wages.

(a) Using equivalent uniform annual cost (EUAC), which alternative will have the lowest local annual cost?

(A) alternative 1

(B) alternative 2

(C) alternative 3

(D) alternative 4

(b) Using a benefit-cost analysis of annual crash reduction cost compared with EUAC construction cost, what is most nearly the highest benefit-cost ratio?

(A) 17:1 for alternative 2

(B) 23:1 for alternative 3

(C) 27:1 for alternative 4

(D) 51:1 for alternative 3

SOLUTIONS

1. (a) The number of crashes that cause property damage for each year is found by subtracting the number of fatal crashes and the number of injury crashes from the total number of crashes.

$$N_{property} = N_t - N_{fatal} - N_{injury}$$

year 1: $N_{property,1} = 320 - 4 - 162 = 154$

year 2: $N_{property,2} = 340 - 3 - 176 = 161$

year 3: $N_{property,3} = 365 - 2 - 190 = 173$

year 4: $N_{property,4} = 352 - 5 - 185 = 162$

$N_{cr} = 1$ since there is only one period covering four years of data. The crash rate is determined by Eq. 6.3.

$$R_{seg} = \frac{N_{cr}(10^8)}{(AADT)N_{yr}\left(365\,\frac{days}{yr}\right)L_{mi}}$$

$$= \frac{(154 + 161 + 173 + 162)(10^8)}{\left(\begin{array}{c}38{,}800\,\frac{veh}{day} + 40{,}000\,\frac{veh}{day}\\ +41{,}200\,\frac{veh}{day} + 42{,}400\,\frac{veh}{day}\end{array}\right)}$$

$$\times (1\text{ yr})\left(365\,\frac{days}{yr}\right)(5\text{ mi})$$

$$= 219 \text{ HMVM} \quad (220 \text{ HMVM})$$

The answer is (C).

(b) Test the data for statistical significance. Choose a value of K that corresponds to the level of confidence desired (see Table 6.1). Calculate the threshold rate for each interval and compare to the crash frequency rate using Eq. 6.4.

For a 90% confidence interval,

$$R_{th} = R_M + Ks = 106 \text{ HMVM} + (1.645)(56 \text{ HMVM})$$
$$= 198 \text{ HMVM}$$

For a 95% confidence interval,

$$R_{th} = R_M + Ks = 106 \text{ HMVM} + (1.960)(56 \text{ HMVM})$$
$$= 216 \text{ HMVM}$$

For a 99% confidence interval,

$$R_{th} = R_M + Ks = 106 \text{ HMVM} + (2.58)(56 \text{ HMVM})$$
$$= 250 \text{ HMVM}$$

The crash rate exceeds the threshold rate at confidence intervals of 90% and 95%, but not at 99%.

The answer is (B).

2. (a) Determine the total life of project average annual vehicle kilometers using the projected traffic growth.

$$\text{average traffic} = (\text{AADT})\left(365\ \frac{\text{d}}{\text{yr}}\right)(F/A,\ i\%,\ n)$$
$$\times \left(\frac{\text{distance traveled}}{\text{study years}}\right)$$
$$= \left(10\,000\ \frac{\text{veh}}{\text{d}}\right)\left(365\ \frac{\text{d}}{\text{yr}}\right)$$
$$\times \left(\frac{(1+0.025)^{15}-1}{0.025}\right)\left(\frac{16\ \text{km}}{15\ \text{yr}}\right)$$
$$= 69\,814\,000\ \text{veh}\cdot\text{km/yr}$$

Determine the projected number of crashes for the current roadway and for the proposed alternative, and the annual cost of each.

For the current roadway, the average fatalities are

$$\text{average fatalities} = \left(69\,814\,000\ \frac{\text{veh}\cdot\text{km}}{\text{yr}}\right)\left(\frac{0.0172\ \text{fatalities}}{10^6\ \text{veh}\cdot\text{km}}\right)$$
$$= 1.20\ \text{fatalities/yr}$$

Using the $(F/A,\ i\%,\ n)$ factor, the average annual cost is

$$\left(1.20\ \frac{\text{fatalities}}{\text{yr}}\right)(\$1,000,000)\left(\frac{(1+0.04)^{15}-1}{0.04}\right)\left(\frac{1}{15\ \text{yr}}\right)$$
$$= \$1,602,000/\text{yr}$$

Extend the data table to determine the average projected crashes and average projected crash cost over the life of the project.

crash severity	current roadway (crashes/yr)	current crash ($/yr)	proposed alternative (crashes/yr)	proposed alternative ($/yr)
K	1.20	1,602,000	1.80	2,404,000
A	3.40	908,000	4.80	1,282,000
B	4.20	70,100	6.00	100,100
C	8.31	41,600	10.47	52,400
PDO2	10.47	8800	12.08	10,100
PDO1	6.90	28,800	8.38	34,900
total	34.49	2,659,300	43.50	3,883,500

The average annual increase in the number of crashes per year for the alternative is

$$43.5\ \frac{\text{crashes}}{\text{yr}} - 34.49\ \frac{\text{crashes}}{\text{yr}} = 9.01\ \text{crashes/yr}$$
$$(9.0\ \text{crashes/yr})$$

The answer is (A).

(b) Using the $(A/P,\ i\%,\ n)$ factor, the annualized cost of construction is

$$C = (\$30,000,000)\left(\frac{(0.04)(1+0.04)^{15}}{(1+0.04)^{15}-1}\right) = \$2,698,000$$

The annualized cost of construction and maintenance is

$$C = \$2,698,000 + (16)(\$150,000)$$
$$= \$5,098,000\quad (\$5,100,000)$$

The answer is (B).

(c) The average projected crash cost is the total crash cost divided by the number of crashes projected per year.

$$C_{\text{ave}} = \frac{C_{\text{cr}}}{N_{\text{cr}}} = \frac{3,883,500\ \frac{\$}{\text{yr}}}{43.5\ \frac{1}{\text{yr}}}$$
$$= \$89,276\quad (\$89,000)$$

The answer is (C).

(d) The additional crash cost of implementing the proposed alternative is

$$C_{\text{cr}} = \$3,883,500 - \$2,659,300 = \$1,224,200$$

The benefit-cost ratio of the proposed alternative is

$$B/C = \frac{\$1,224,200}{\$5,098,000}$$
$$= 0.24{:}1$$

The benefit-cost ratio is less than one, meaning there is more cost in predicted crashes than the cost of constructing the proposed alternative.

The answer is (A).

(e) Benefit-cost analysis is best used when comparing all direct benefits and all direct costs. The analysis does not include the benefit of adding the extra lane to increase capacity of the roadway.

The answer is (D).

3. Alternative 1 proposes lowering the speed limit with strict enforcement. With a peak period flow of 8000 vph, all lanes in both directions are operating at near capacity. Referring to Table 2.9, for a free-flow speed of 65 mph, the minimum speed to maintain the flow rate at LOS E is 52.2 mph. From this observation, the traffic is already operating well below the posted speed limit of 65 mph. Therefore, lowering the speed limit would have a negligible effect on the peak period flow. However, lowering the average travel speed limit would reduce the severity of crashes during off-peak periods.

Alternative 2 places the guiderail close to the edge of the shoulder. This placement would reduce the number of vehicles that cross the drainage ditch. The guiderail may not have a pronounced effect on large vehicle crashes as large vehicles frequently vault guiderails, which are primarily designed for automobile protection. Also, the 12 ft clear zone remaining between the guiderail and the ditch slope could no longer be used as a recovery area. Referring to Fig. 6.3, about 60% of vehicles are able to recover from lateral excursions within 20 ft of an off-road encroachment, as opposed to less than 35% recovering from a 10 ft encroachment. Placing the guiderail 10 ft from the edge-of-traveled-way is not preferred by AASHTO guidelines when a greater offset is available.

Alternative 3 places the guiderail 18 ft from the edge of the traveled way. While this allows a greater proportion of vehicles to recover and would prevent many vehicles from traversing the drainage ditch, it would have a negligible effect on large vehicles for the reason outlined in alternative 2. If there were no other countermeasures available, this alternative would be more effective than alternative 1 or alternative 2.

Alternative 4 alters the clear-zone slopes by flattening the ditch side slopes and slightly increasing the recovery area foreslope. The increased slope could be considered traversable because the roadway is flat and nearly straight. Therefore, most of the lateral excursions can be assumed to be nearly straight or nearly parallel with the lanes of traffic. Based on the problem statement, the steep side slopes of the drainage ditch have caused vehicle upsets. Reducing the ditch side slopes to 6H:1V would reduce the number of vehicle upsets as well as the fatality rate. Alternative 4 would be the most effective countermeasure.

The answer is (D).

4. (a) The equivalent uniform annual cost is the present value of the construction cost amortized over the life of the project, A/P, plus the present annual cost of maintenance and the school guard.

Alternative 1

To calculate the local construction cost, multiply the cost by the percentage that must be paid by local sources.

$$\text{local construction cost} = (1.0)(\$100{,}000)$$
$$= \$100{,}000$$

The A/P factor can be found using economic factor tables.

$$(A/P, i\%, n) = (A/P, 5\%, 3) = 0.3672$$

$$\begin{aligned}\text{EUAC} &= (\$100{,}000)(A/P, i\%, n) \\ &\quad + \text{annual maintenance cost} \\ &\quad + \text{school guard cost} \\ &= (\$100{,}000)(0.3672) + \$50{,}000 + \$30{,}000 \\ &= \$117{,}000\end{aligned}$$

Alternative 2

The equivalent uniform annual cost for alternative 2 is

$$\text{local construction cost} = (0.2)(\$2{,}500{,}000)$$
$$= \$500{,}000$$

$$(A/P, i\%, n) = (A/P, 5\%, 20) = 0.0802$$

$$\begin{aligned}\text{EUAC} &= (\$500{,}000)(A/P, i\%, n) \\ &\quad + \text{annual maintenance cost} \\ &\quad + \text{school guard cost} \\ &= (\$500{,}000)(0.0802) + (0.4)(\$50{,}000) + 0 \\ &= \$60{,}100\end{aligned}$$

Alternative 3

The equivalent uniform annual cost for alternative 3 is

$$\text{local construction cost} = (0.2)(\$650{,}000)$$
$$= \$130{,}000$$

$$(A/P, i\%, n) = (A/P, 5\%, 12) = 0.1128$$

$$\begin{aligned}\text{EUAC} &= (\$130{,}000)(A/P, i\%, n) \\ &\quad + \text{annual maintenance cost} \\ &\quad + \text{school guard cost} \\ &= (\$130{,}000)(0.1128) + (0.8)(\$45{,}000) + \$30{,}000 \\ &= \$80{,}700\end{aligned}$$

Alternative 4

The equivalent uniform annual cost for alternative 4 is

$$\text{local construction cost} = (0.2)(\$450{,}000)$$
$$= \$90{,}000$$

$$(A/P, i\%, n) = (A/P, 5\%, 7) = 0.1728$$

$$\text{EUAC} = (\$90{,}000)(A/P, i\%, n)$$
$$+ \text{annual maintenance cost}$$
$$+ \text{school guard cost}$$
$$= (\$90{,}000)(0.1728) + (0.8)(\$30{,}000) + \$45{,}000$$
$$= \$84{,}600$$

Alternative 2 has the lowest equivalent uniform annual cost.

The answer is (B).

(b) The format to determine annual crash cost, ACC, is

$$\text{ACC}_i = \left(\frac{1}{\text{no. crash data years}}\right)$$
$$\times \begin{pmatrix} C_{cr,K} N_{cr} \\ + C_{cr,A} N_{cr} \\ + C_{cr,PDO1} N_{cr} \end{pmatrix}$$

For each alternative,

$$\text{ACC}_1 = \left(\frac{1}{5}\right)\begin{pmatrix}(\$1{,}000{,}000)(5) + (\$200{,}000)(45) \\ +(\$3000)(160)\end{pmatrix}$$
$$= \$2{,}896{,}000$$

$$\text{ACC}_2 = \left(\frac{1}{2}\right)\begin{pmatrix}(\$1{,}000{,}000)(0) + (\$200{,}000)(4) \\ +(\$3000)(14)\end{pmatrix}$$
$$= \$421{,}000$$

$$\text{ACC}_3 = \left(\frac{1}{2}\right)\begin{pmatrix}(\$1{,}000{,}000)(0) + (\$200{,}000)(12) \\ +(\$3000)(28)\end{pmatrix}$$
$$= \$1{,}242{,}000$$

$$\text{ACC}_4 = \left(\frac{1}{2}\right)\begin{pmatrix}(\$1{,}000{,}000)(0) + (\$200{,}000)(16) \\ +(\$3000)(40)\end{pmatrix}$$
$$= \$1{,}660{,}000$$

The EUAC of construction for each alternative is

$$\text{EUAC}_2 = (\text{construction cost})(A/P, i\%, n)$$
$$= (\text{construction cost})(A/P, 5\%, 20)$$
$$= (\$2{,}500{,}000)(0.0802)$$
$$= \$201{,}000$$

$$\text{EUAC}_3 = (\text{construction cost})(A/P, i\%, n)$$
$$= (\text{construction cost})(A/P, 5\%, 12)$$
$$= (\$650{,}000)(0.1128)$$
$$= \$73{,}300$$

$$\text{EUAC}_4 = (\text{construction cost})(A/P, i\%, n)$$
$$= (\text{construction cost})(A/P, 5\%, 7)$$
$$= (\$450{,}000)(0.1728)$$
$$= \$77{,}800$$

The benefit-cost ratio, B/C, is found by

$$B/C_i = \frac{\text{ACC}_1 - \text{ACC}_i}{\text{EUAC}_i}$$

$$B/C_2 = \frac{\$2{,}896{,}000 - \$421{,}000}{\$201{,}000} = 12.3 \quad (12{:}1)$$

$$B/C_3 = \frac{\$2{,}896{,}000 - \$1{,}242{,}000}{\$73{,}300} = 22.6 \quad (23{:}1)$$

$$B/C_4 = \frac{\$2{,}896{,}000 - \$1{,}660{,}000}{\$77{,}800} = 15.9 \quad (16{:}1)$$

The answer is (B).

Topic VII: Support Material

Appendices

2.A Travel Time, Speed, and Delay Study Field Sheet Using the Moving Vehicle Method..... A-1
2.B Warrant 1, Eight-Hour Vehicular Volume A-2
2.C Warrant 2 .... A-3
2.D Warrant 3 .... A-4
2.E Warrant 4 .... A-5
2.F Warrant 9 .... A-7
4.A Design Superelevation Rates A-9
4.B Minimum Radii for Design Superelevation Rates and Design Speeds A-14
4.C Superelevation, Radius, and Design Speed for Low-Speed Urban Streets* A-23
4.D Minimum Radii and Superelevation for Low-Speed Urban Streets[a,b,c] A-25
4.E Minimum Superelevation Runoff for Horizontal Curves A-29
4.F Acceleration Distances for Passenger Cars, Level Conditions A-31
4.G Deceleration Distances for Passenger Vehicles Approaching Intersections........... A-32
4.H AASHTO Design Vehicle Dimensions A-33
4.I Minimum Turning Radii of Design Vehicles ...A-35
4.J Minimum Turning Path for Passenger Car (P) Design Vehicle[a,b] A-37
4.K Turning Characteristics of a Typical Tractor/Semitrailer Combination Truck A-38
4.L Minimum Turning Path for Single-Unit Truck (SU-12 [SU-40]) Design Vehicle[a,b] ...A-39
4.M Minimum Turning Path for Intermediate Semitrailer WB-40 (WB-12) Design Vehicle[a,b,c] A-40
4.N Minimum Turning Path for Interstate Semitrailer WB-67 (WB-20) Design Vehicle[a,b,c] .A-41
4.O Minimum Turning Path for Double-Trailer Combination (WB-20D [WB-67D]) Design Vehicle A-42
4.P Minimum Turning Path for City Transit Bus (CITY-BUS) Design Vehicle[a,b] A-43
4.Q Minimum Turning Path for Conventional School Bus (S-BUS36 [S-BUS11]) Design Vehicle[a,b,c] A-44
4.R Minimum Traveled Way, Passenger (P) Design Vehicle Path A-45
4.S Minimum Traveled Way Designs, Single-Unit Trucks and City Transit Buses A-47
4.T Edge-of-Traveled-Way for Turns at Intersections A-49
4.U Effect of Curb Radii on Right-Turning Paths of Various Vehicles A-57
4.V Cross Street Widths Occupied by Turning Vehicles* A-58
4.W Crosswalk Length Variations with Different Curb Radii and Width of Borders A-60
4.X Corner Setbacks with Different Curb Radii and Width of Borders A-61
5.A AASHTO Soil Classification System A-62
5.B Unified Soil Classification System A-63
5.C USDA Soil Triangle A-64
5.D Performance-Graded Asphalt Binder Specification... A-65
5.E Superpave Mix Design Procedural Outline A-67

Index I-1

APPENDIX 2.A
Travel Time, Speed, and Delay Study Field Sheet Using the Moving Vehicle Method

**Travel Time and Delay Study
Moving Vehicle Method
Field Sheet**

route _____ date _____
start point _____ end point _____
weather _____

run	start time	finish time	travel time	vehicles met	vehicles overtaking	vehicles passed
__bound						
1						
2						
3						
4						
5						
6						
7						
8						
total						
average						
__bound						
1						
2						
3						
4						
5						
6						
7						
8						
total						
average						

comments _____
recorder(s) _____

APPENDIX 2.B
Warrant 1, Eight-Hour Vehicular Volume

condition A—minimum vehicular volume

number of lanes for moving traffic on each approach		vehicles per hour on major street (total of both approaches)				vehicles per hour on higher-volume minor-street approach (one direction only)			
major street	minor street	100%[a]	80%[b]	70%[c]	56%[d]	100%[a]	80%[b]	70%[c]	56%[d]
1	1	500	400	350	280	150	120	105	84
2 or more	1	600	480	420	336	150	120	105	84
2 or more	2 or more	600	480	420	336	200	160	140	112
1	2 or more	500	400	350	280	200	160	140	112

condition B—interruption of continuous traffic

number of lanes for moving traffic on each approach		vehicles per hour on major street (total of both approaches)				vehicles per hour on higher-volume minor-street approach (one direction only)			
major street	minor street	100%[a]	80%[b]	70%[c]	56%[d]	100%[a]	80%[b]	70%[c]	56%[d]
1	1	750	600	525	420	75	60	53	42
2 or more	1	900	720	630	504	75	60	53	42
2 or more	2 or more	900	720	630	504	100	80	70	56
1	2 or more	750	600	525	420	100	80	70	56

[a] basic minimum hourly volume
[b] used for combination of conditions A and B after adequate trial of other remedial measures
[c] may be used when the major-street speed exceeds 40 mph (65 kph) or in an isolated community with a population of less than 10,000
[d] may be used for combination of conditions A and B after adequate trial of other remedial measures when the major-street speed exceeds 40 mph (65 kph) or in an isolated community with a population of less than 10,000

Reprinted from the *Manual on Uniform Traffic Control Devices*, 2009 ed., Table 4C-1, U.S. Department of Transportation, Federal Highway Administration, 2009.

APPENDIX 2.C
Warrant 2

four-hour vehicular volume

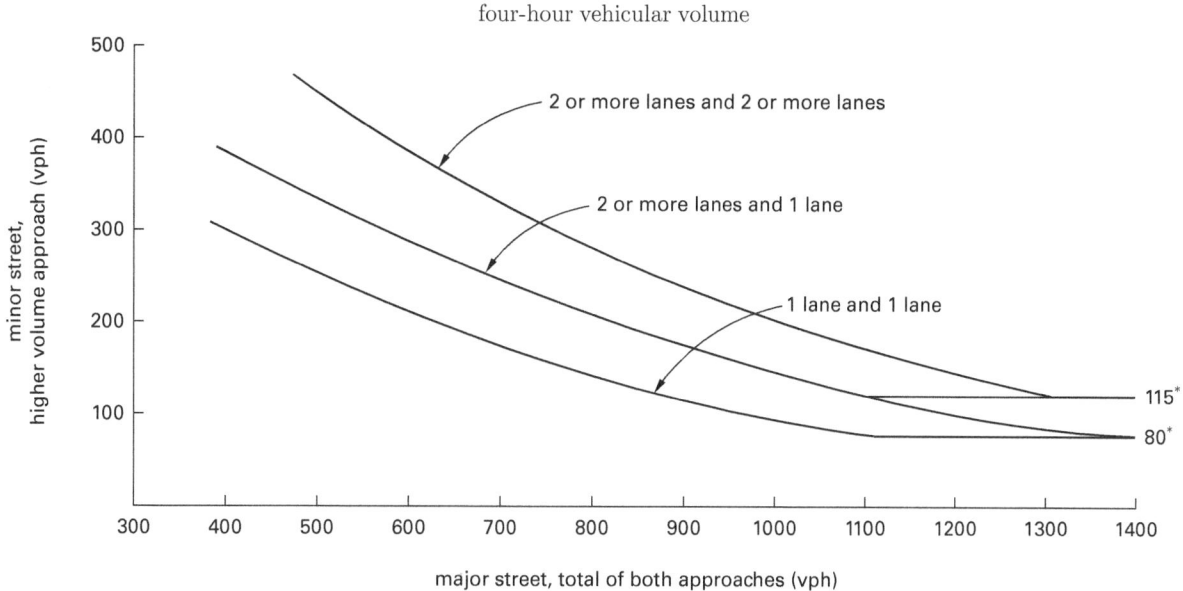

*Note: 115 vph applies as the lower threshold volume for a minor-street approach with two or more lanes, and 80 vph applies as the lower threshold volume for a minor-street approach with one lane.

Reprinted from the *Manual on Uniform Traffic Control Devices*, 2009 ed., Fig. 4C-1, U.S. Department of Transportation, Federal Highway Administration, 2009.

four-hour vehicular volume (70% factor)
(community less than 10,000 people or above 40 mph (65 kph) on major street)

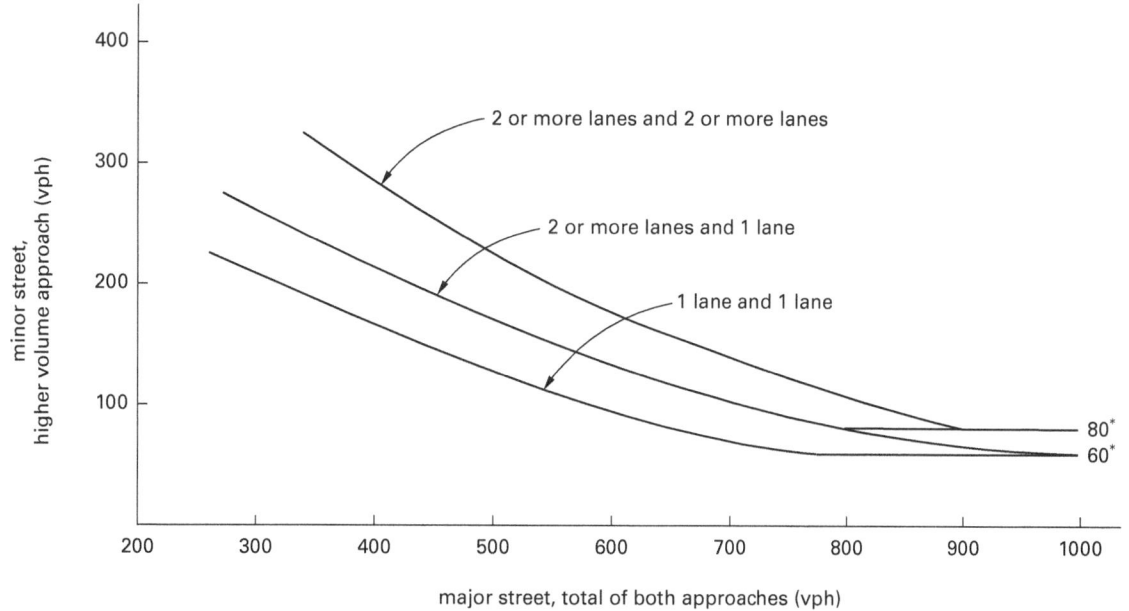

*Note: 80 vph applies as the lower threshold volume for a minor-street approach with two or more lanes, and 60 vph applies as the lower threshold volume for a minor-street approach with one lane.

Reprinted from the *Manual on Uniform Traffic Control Devices*, 2009 ed., Fig. 4C-2, U.S. Department of Transportation, Federal Highway Administration, 2009.

APPENDIX 2.D
Warrant 3

peak hour

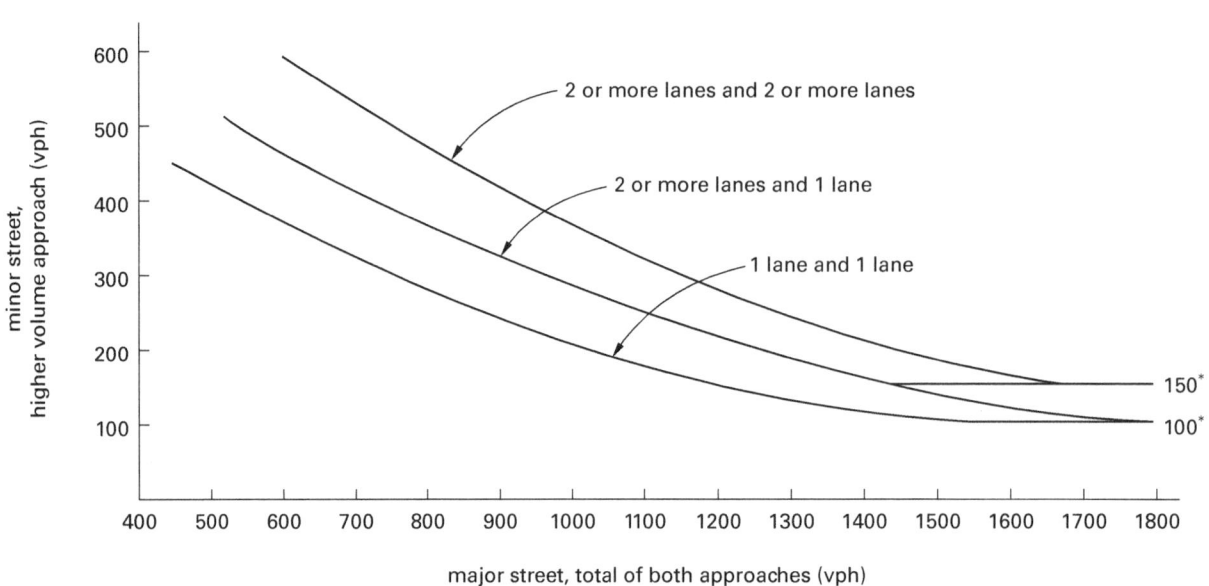

*Note: 150 vph applies as the lower threshold volume for a minor-street approach with two or more lanes, and 100 vph applies as the lower threshold volume for a minor-street approach with one lane.

Reprinted from the *Manual on Uniform Traffic Control Devices*, 2009 ed., Fig. 4C-3, U.S. Department of Transportation, Federal Highway Administration, 2009.

peak hour (70% factor)
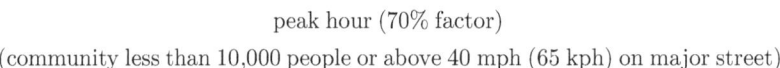
(community less than 10,000 people or above 40 mph (65 kph) on major street)

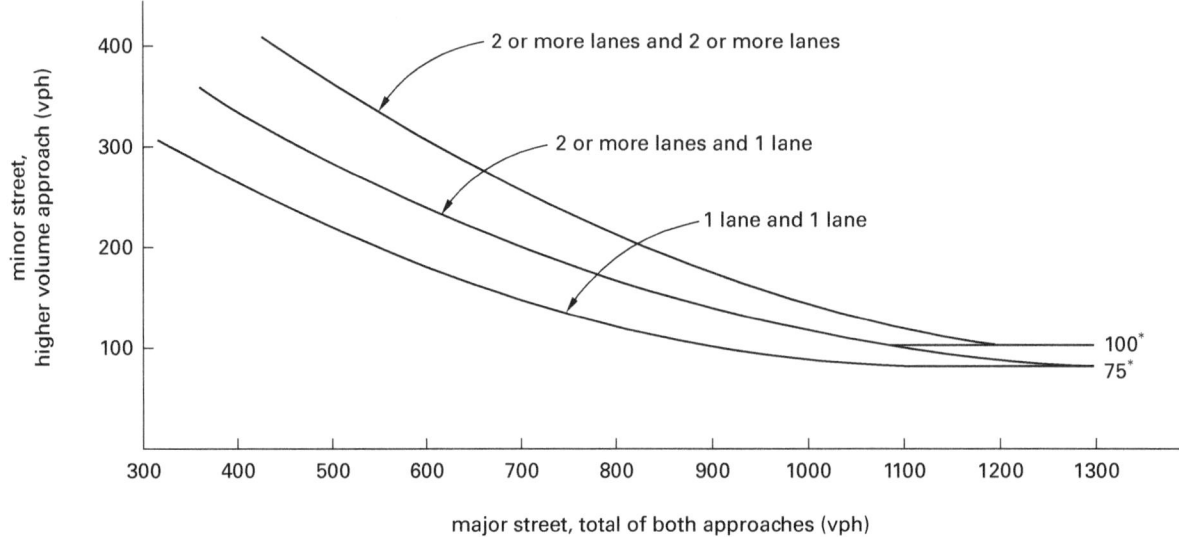

*Note: 100 vph applies as the lower threshold volume for a minor-street approach with two or more lanes, and 75 vph applies as the lower threshold volume for a minor-street approach with one lane.

Reprinted from the *Manual on Uniform Traffic Control Devices*, 2009 ed., Fig. 4C-4, U.S. Department of Transportation, Federal Highway Administration, 2009.

APPENDIX 2.E
Warrant 4

pedestrian four-hour volume

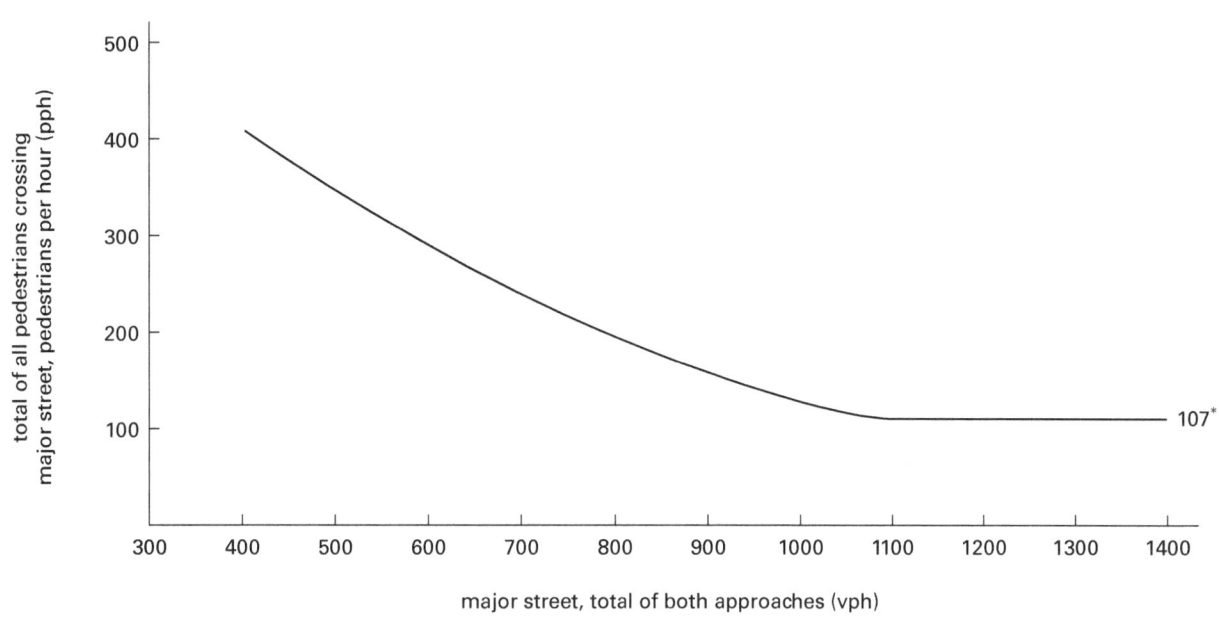

*Note: 107 pph applies as the lower threshold volume.

Reprinted from the *Manual on Uniform Traffic Control Devices*, 2009 ed., Fig. 4C-5, U.S. Department of Transportation, Federal Highway Administration, 2009.

pedestrian four-hour volume (70% factor)

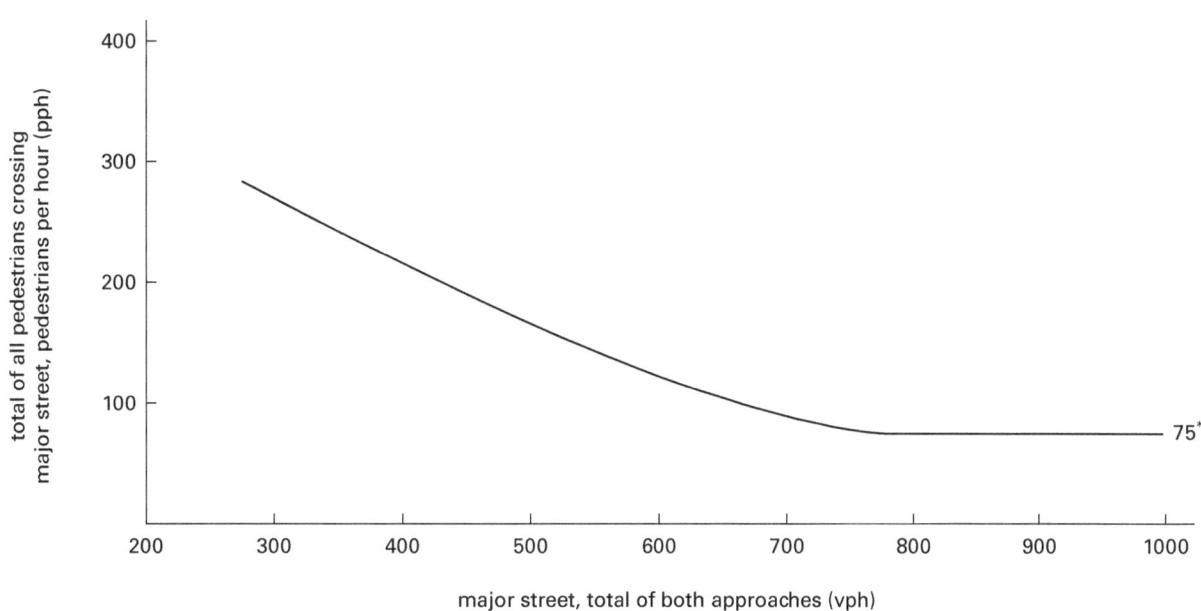

*Note: 75 pph applies as the lower threshold volume.

Reprinted from the *Manual on Uniform Traffic Control Devices*, 2009 ed., Fig. 4C-6, U.S. Department of Transportation, Federal Highway Administration, 2009.

APPENDIX 2.E (continued)
Warrant 4

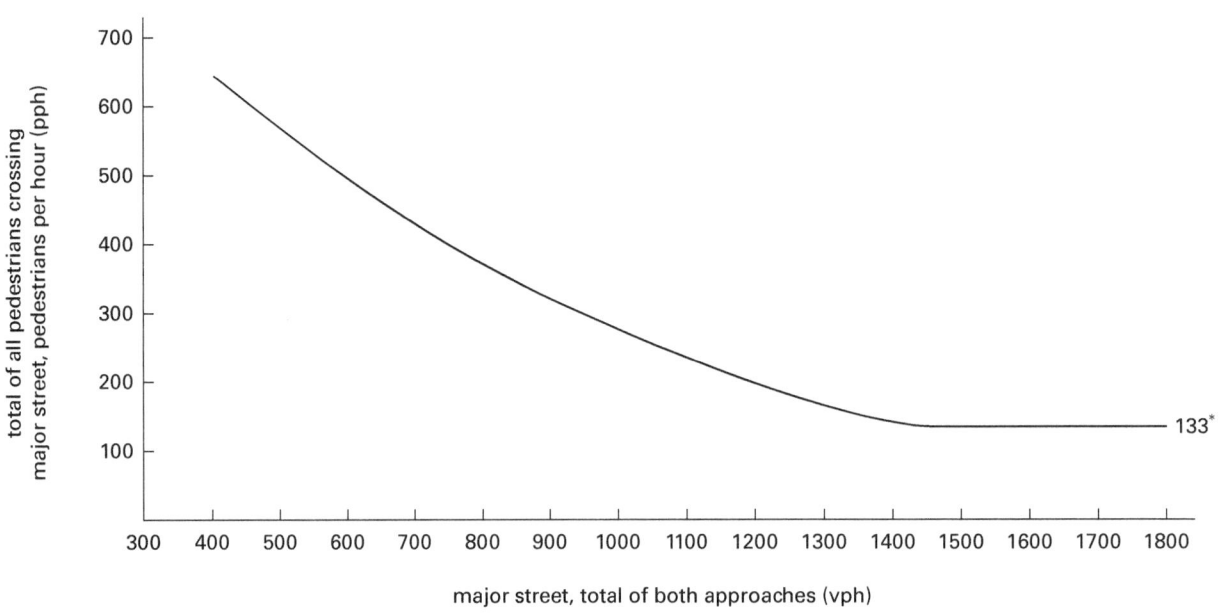

*Note: 133 pph applies as the lower threshold volume.

Reprinted from the *Manual on Uniform Traffic Control Devices*, 2009 ed., Fig. 4C-7, U.S. Department of Transportation, Federal Highway Administration, 2009.

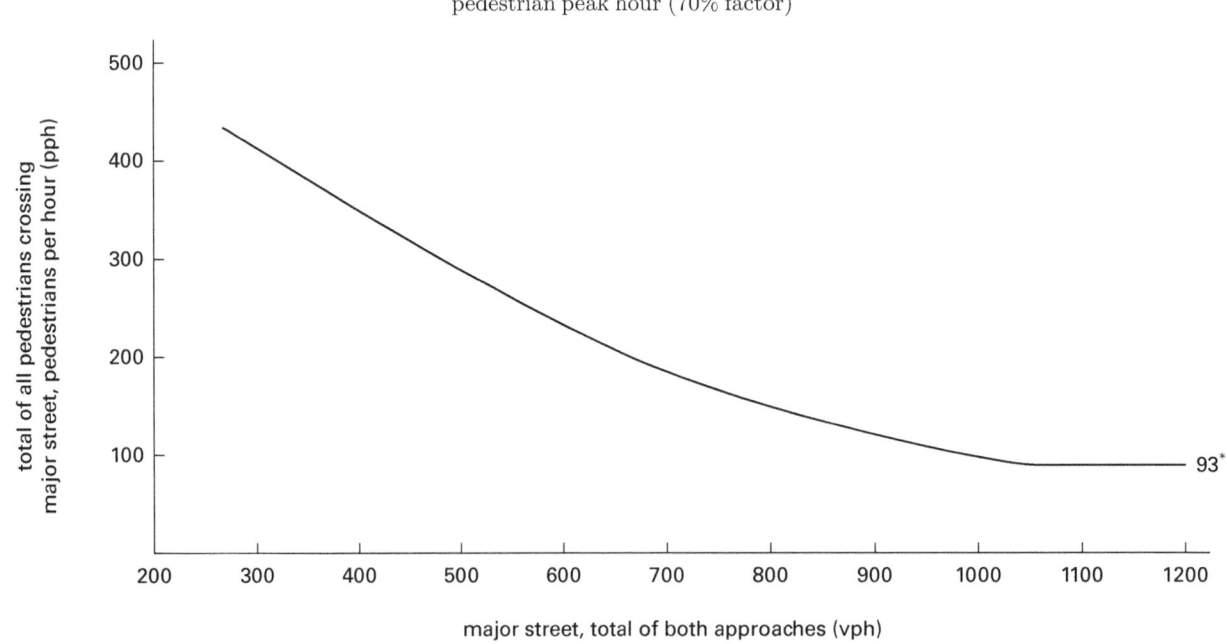

*Note: 93 pph applies as the lower threshold volume.

Reprinted from the *Manual on Uniform Traffic Control Devices*, 2009 ed., Fig. 4C-8, U.S. Department of Transportation, Federal Highway Administration, 2009.

APPENDIX 2.F
Warrant 9

intersection near a grade crossing
(one approach lane at the track crossing)

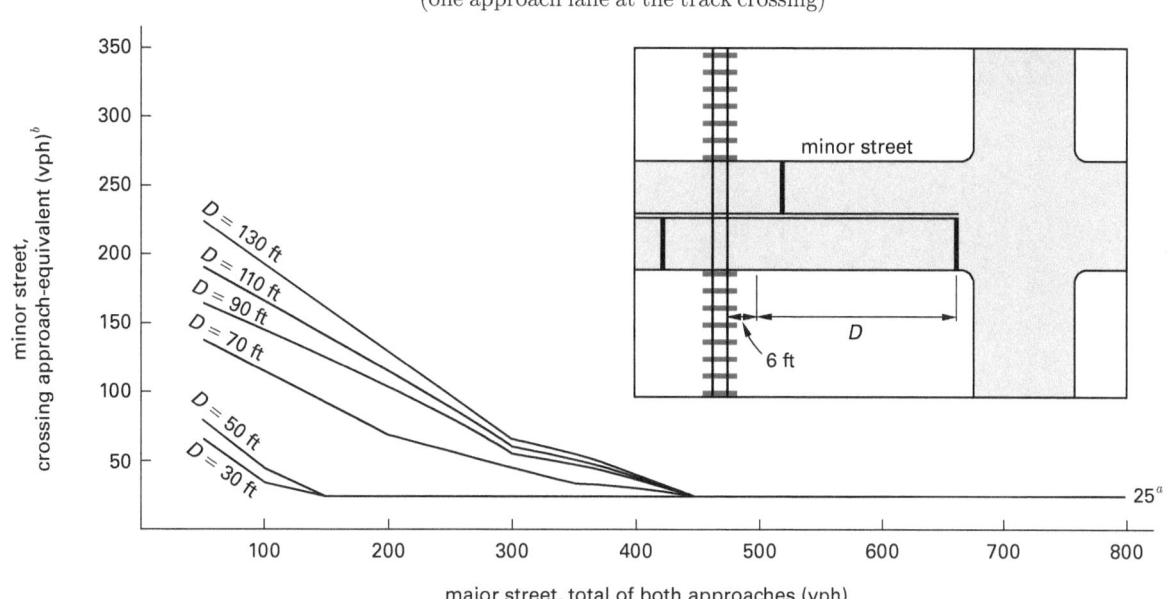

[a]25 vph applies as the lower threshold volume.
[b]vph after applying the adjustment factors, if appropriate

Reprinted from the *Manual on Uniform Traffic Control Devices*, 2009 ed., Fig. 4C-9, U.S. Department of Transportation, Federal Highway Administration, 2009.

intersection near a grade crossing
(two or more approach lanes at the track crossing)

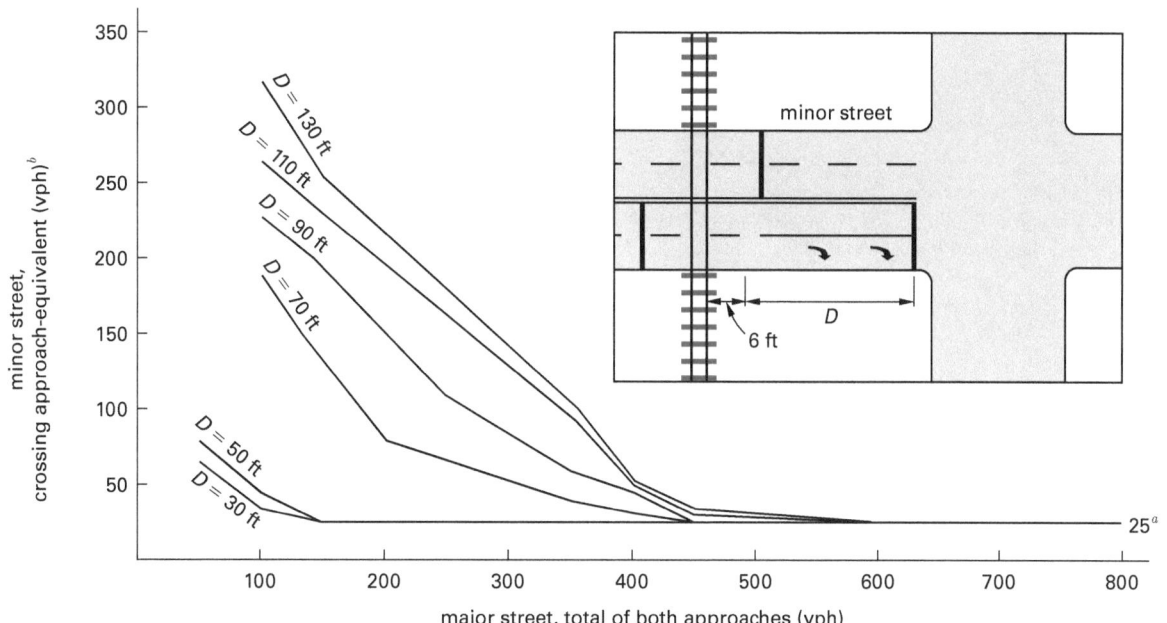

[a]25 vph applies as the lower threshold volume.
[b]vph after applying the adjustment factors, if appropriate

Reprinted from the *Manual on Uniform Traffic Control Devices*, 2009 ed., Fig. 4C-10, U.S. Department of Transportation, Federal Highway Administration, 2009.

APPENDIX 2.F *(continued)*
Warrant 9

adjustment factor for daily frequency of rail traffic

rail traffic per day	adjustment factor
1	0.67
2	0.91
3 to 5	1.00
6 to 8	1.18
9 to 11	1.25
12 or more	1.33

adjustment factor for percentage of high-occupancy buses

high-occupancy buses* on minor-street approach	adjustment factor
0%	1.00
2%	1.09
4%	1.19
6% or more	1.32

adjustment factor for percentage of tractor-trailer trucks

tractor-trailer trucks on minor-street approach	adjustment factor clear storage distance, $D < 70$ ft	adjustment factor clear storage distance, $D \geq 70$ ft
0% to 2.5%	0.50	0.50
2.6% to 7.5%	0.75	0.75
7.6% to 12.5%	1.00	1.00
12.6% to 17.5%	2.30	1.15
17.6% to 22.5%	2.70	1.35
22.6% to 27.5%	3.28	1.64
more than 27.5%	4.18	2.09

*A high-occupancy bus is defined as a bus occupied by at least 20 people.

Reprinted from the *Manual on Uniform Traffic Control Devices*, 2009 ed., Table 4C-2, Table 4C-3, and Table 4C-4, U.S. Department of Transportation, Federal Highway Administration, 2009.

APPENDIX 4.A
Design Superelevation Rates

4% maximum (SI units)

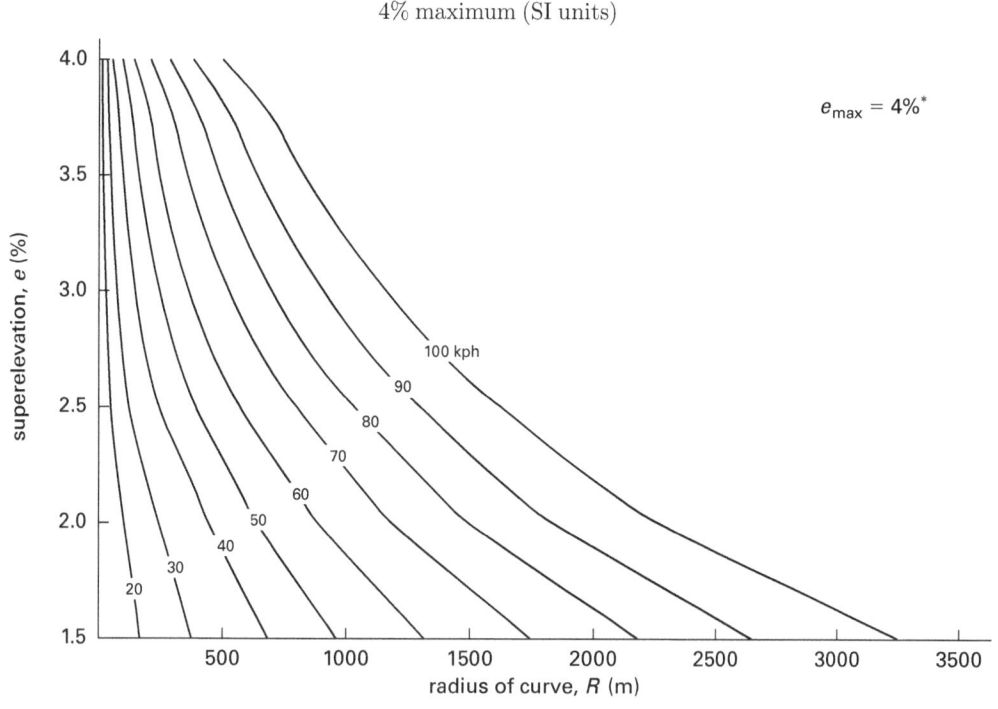

4% maximum (customary U.S. units)

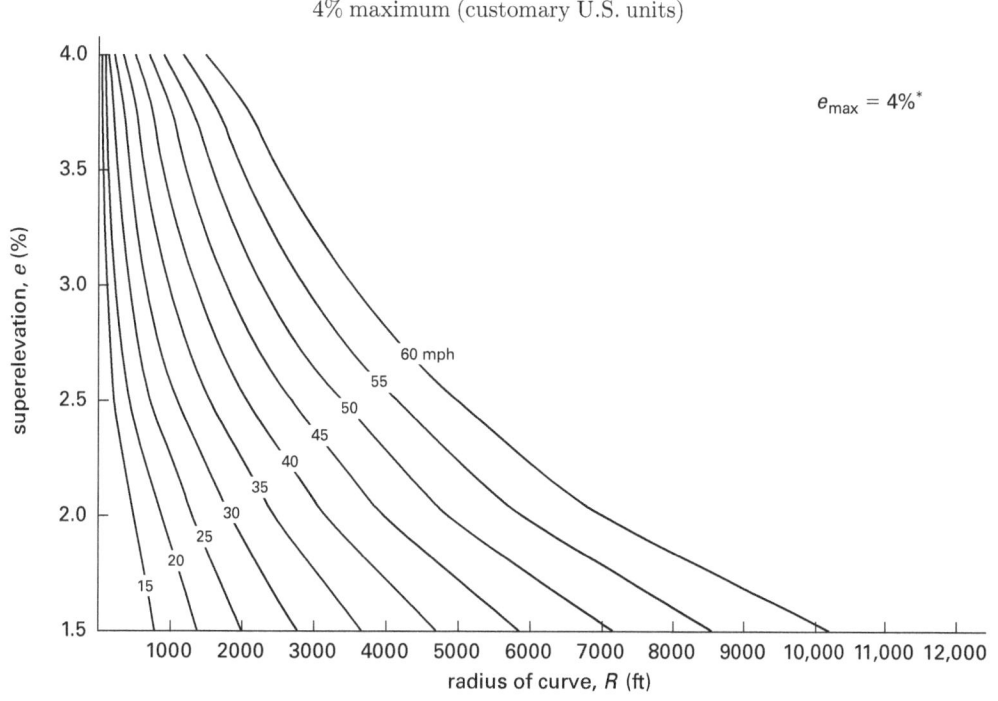

*In recognition of safety considerations, $e_{max} = 4\%$ should be limited to urban conditions.

From *A Policy on Geometric Design of Highways and Streets*, 2011, by the American Association of State Highway and Transportation Officials, Washington, D.C. Figure 3-9. Used by permission.

APPENDIX 4.A *(continued)*
Design Superelevation Rates

6% maximum (SI units)

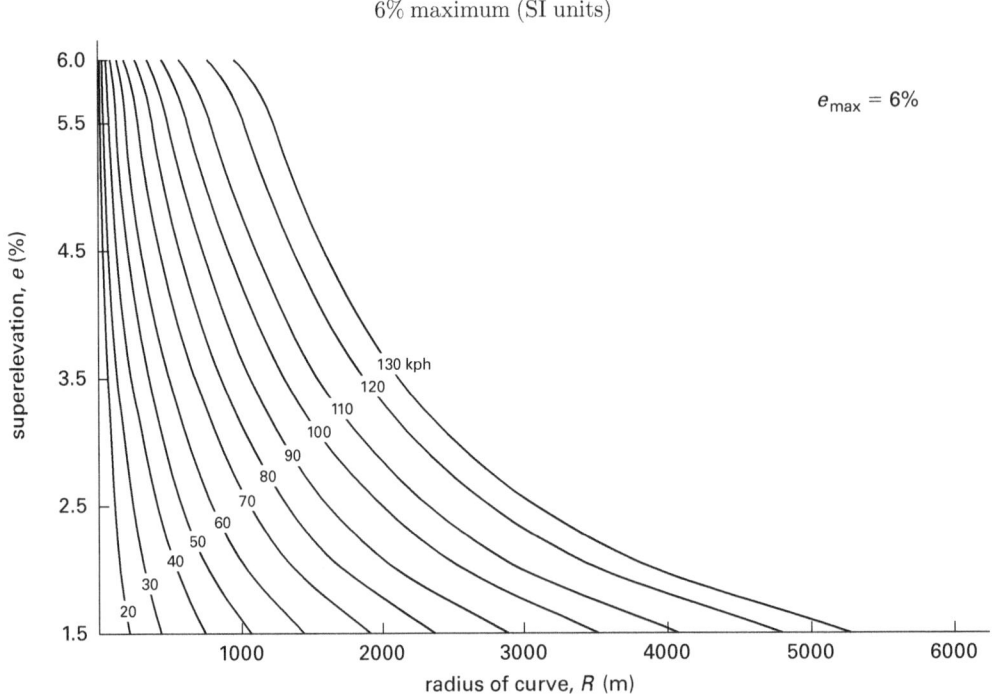

From *A Policy on Geometric Design of Highways and Streets*, 2011, by the American Association of State Highway and Transportation Officials, Washington, D.C. Figure 3-10. Used by permission.

6% maximum (customary U.S. units)

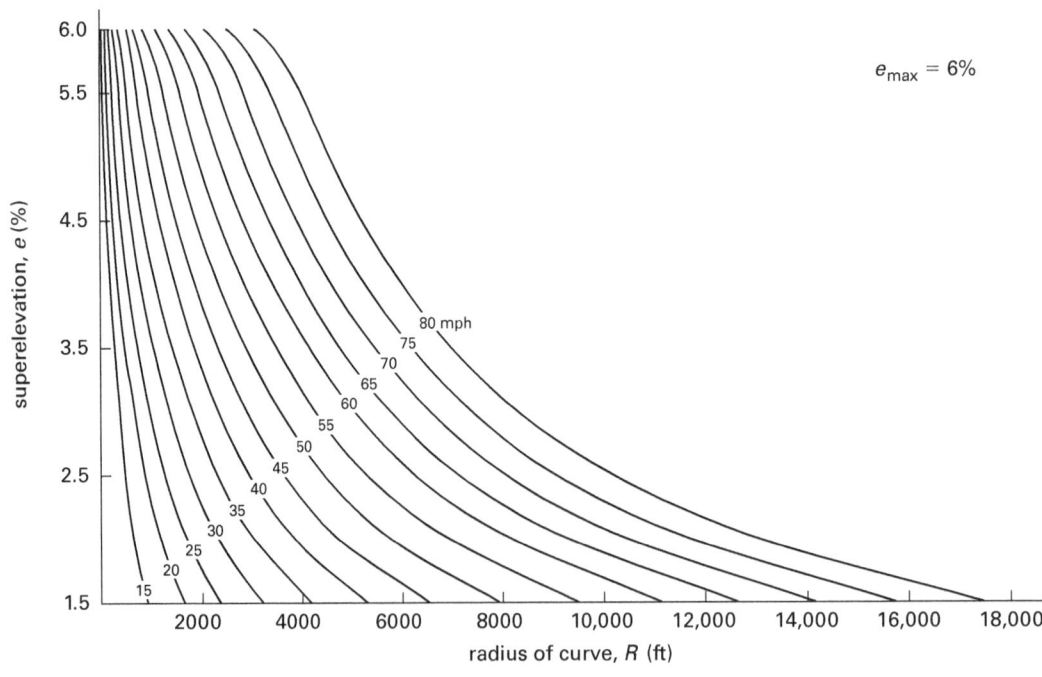

From *A Policy on Geometric Design of Highways and Streets*, 2011, by the American Association of State Highway and Transportation Officials, Washington, D.C. Figure 3-10. Used by permission.

APPENDIX 4.A *(continued)*
Design Superelevation Rates

8% maximum (SI units)

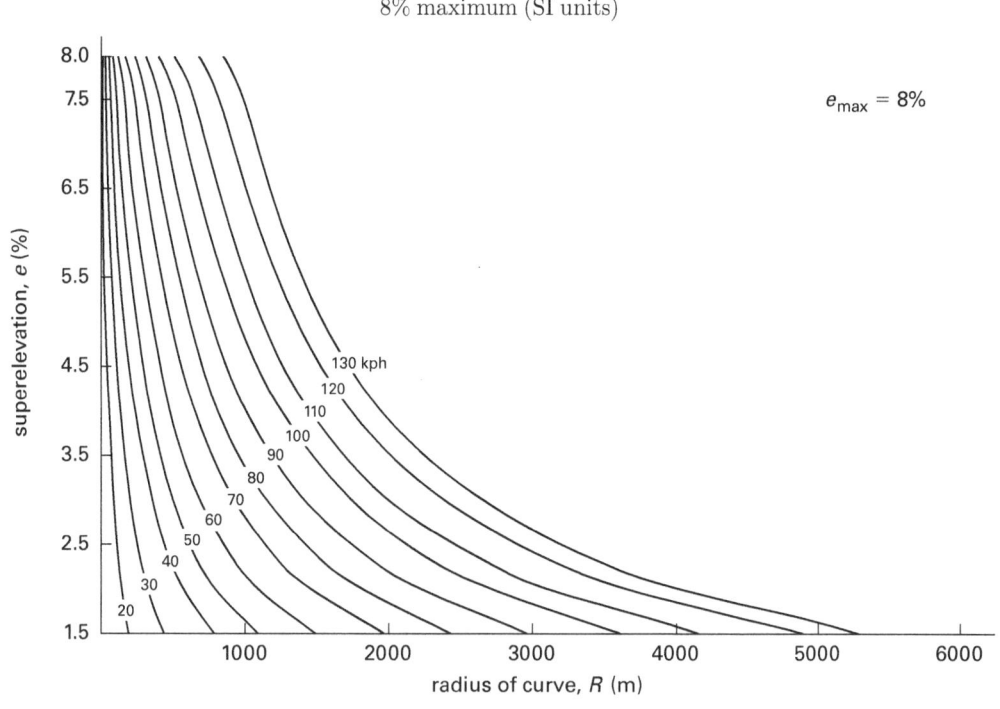

8% maximum (customary U.S. units)

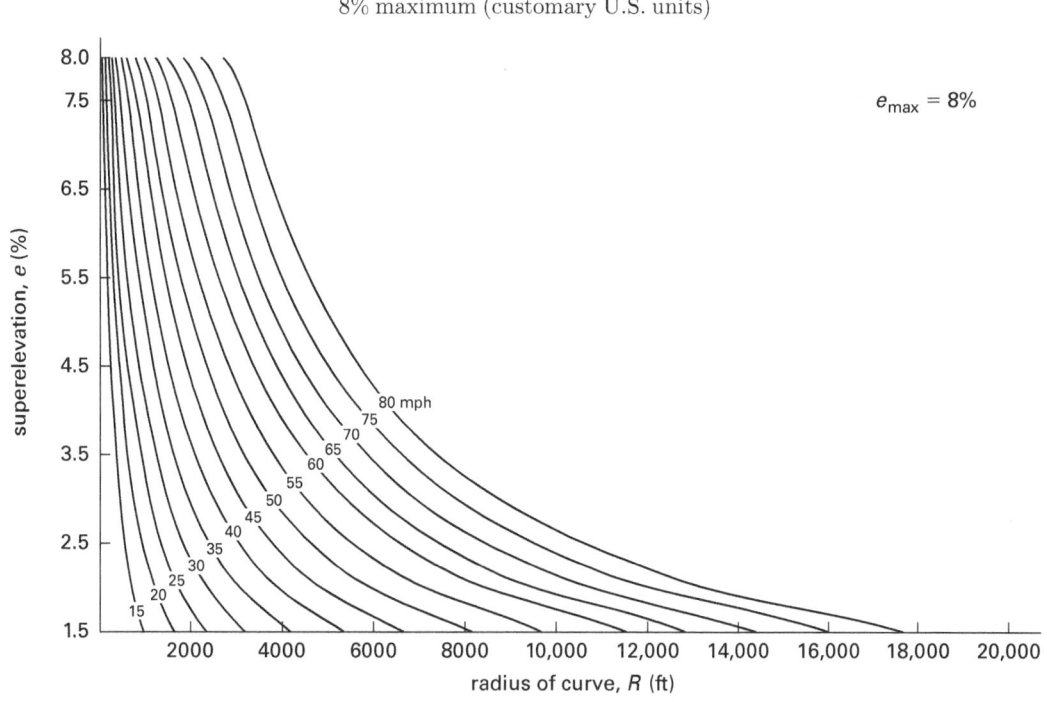

From *A Policy on Geometric Design of Highways and Streets*, 2011, by the American Association of State Highway and Transportation Officials, Washington, D.C. Figure 3-11. Used by permission.

APPENDIX 4.A *(continued)*
Design Superelevation Rates

10% maximum (SI units)

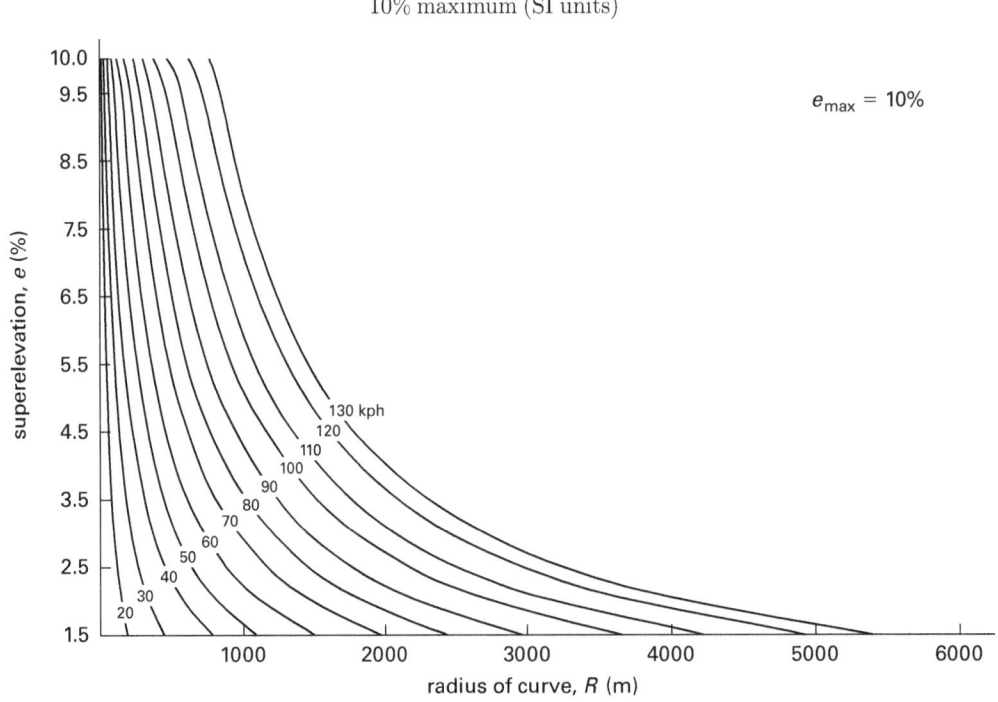

10% maximum (customary U.S. units)

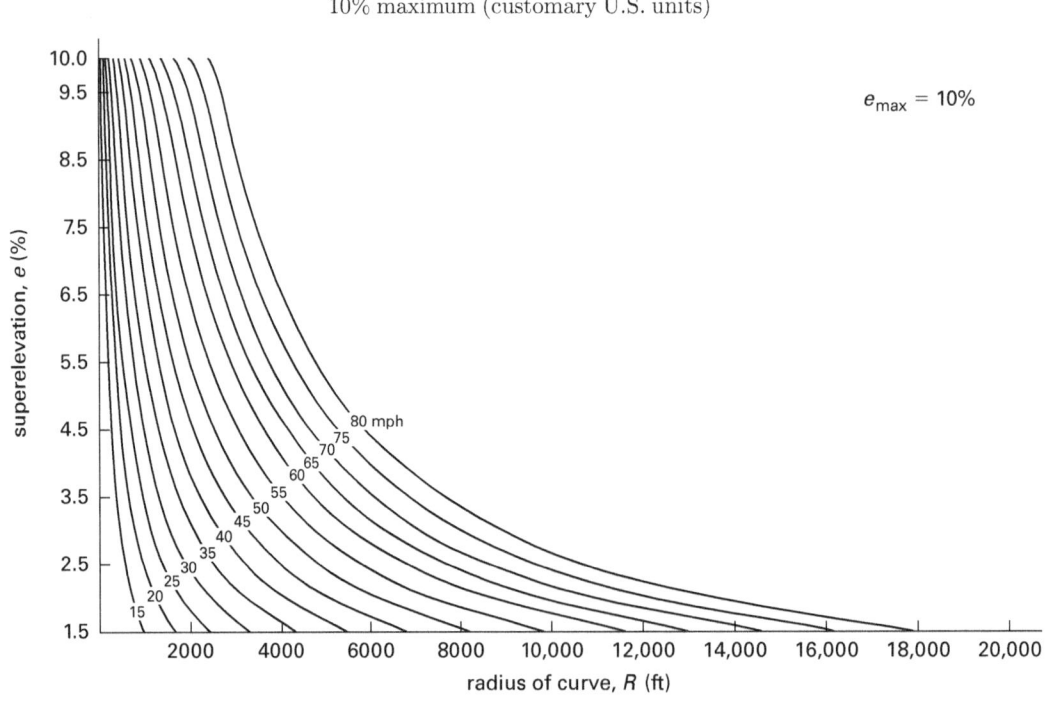

From *A Policy on Geometric Design of Highways and Streets*, 2011, by the American Association of State Highway and Transportation Officials, Washington, D.C. Figure 3-12. Used by permission.

APPENDIX 4.A *(continued)*
Design Superelevation Rates

12% maximum (SI units)

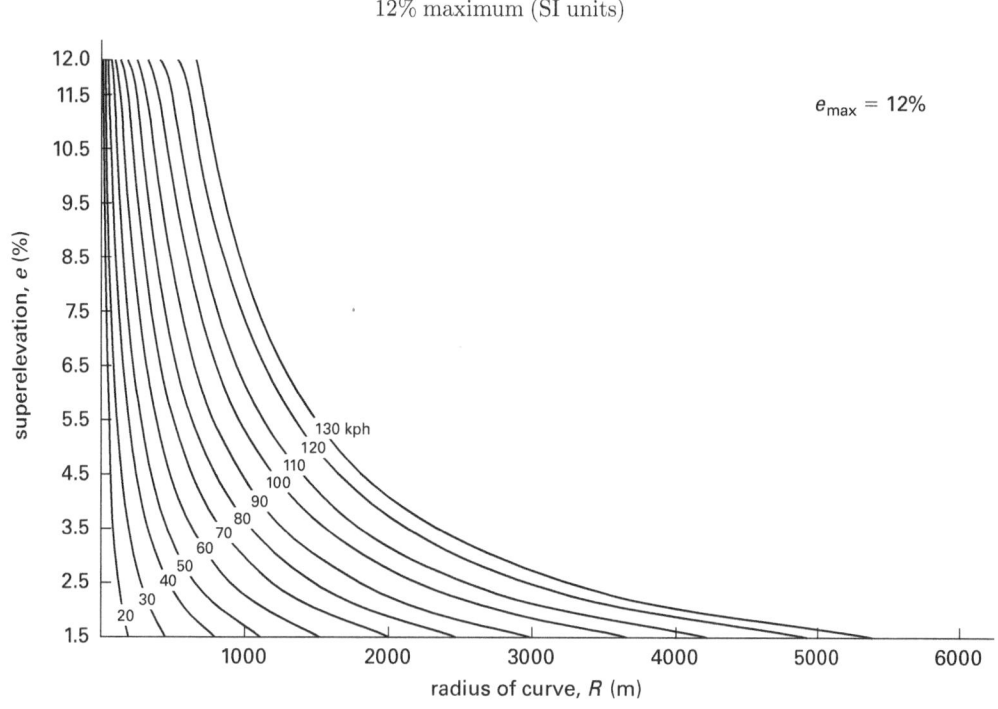

12% maximum (customary U.S. units)

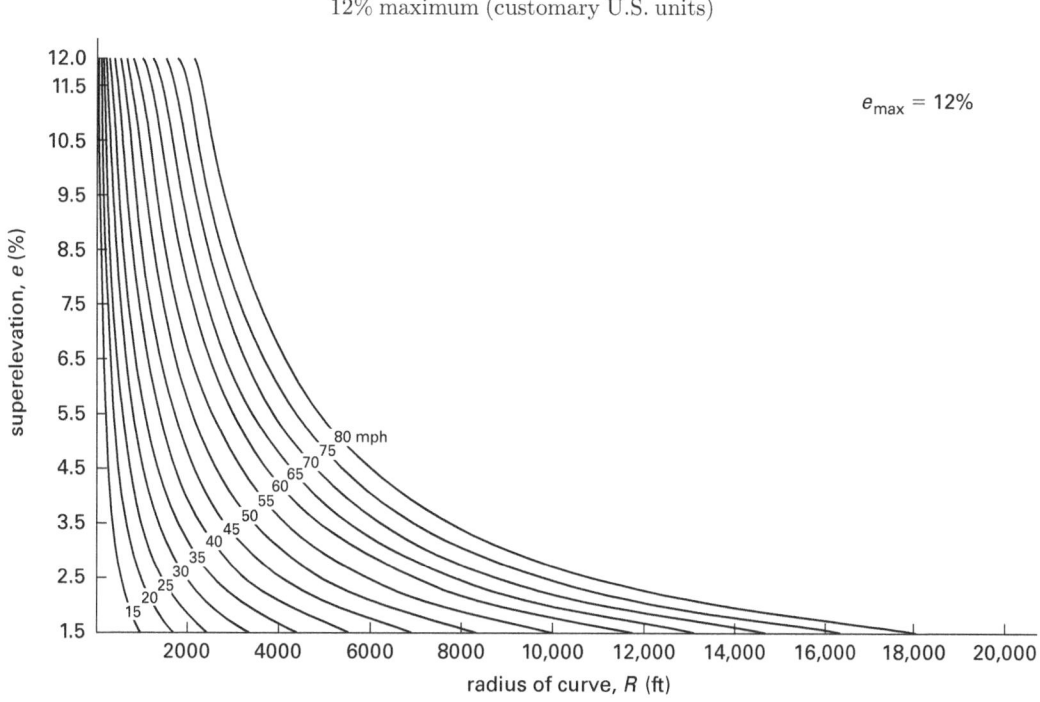

From *A Policy on Geometric Design of Highways and Streets*, 2011, by the American Association of State Highway and Transportation Officials, Washington, D.C. Figure 3-13. Used by permission.

APPENDIX 4.B
Minimum Radii for Design Superelevation Rates and Design Speeds

4% maximum (SI units)*

e (%)	v_d (kph)								
	20	30	40	50	60	70	80	90	100
	R (m)								
NC	163	371	679	951	1310	1740	2170	2640	3250
RC	102	237	441	632	877	1180	1490	1830	2260
2.2	75	187	363	534	749	1020	1290	1590	1980
2.4	51	132	273	435	626	865	1110	1390	1730
2.6	38	99	209	345	508	720	944	1200	1510
2.8	30	79	167	283	422	605	802	1030	1320
3.0	24	64	137	236	356	516	690	893	1150
3.2	20	54	114	199	303	443	597	779	1010
3.4	17	45	96	170	260	382	518	680	879
3.6	14	38	81	144	222	329	448	591	767
3.8	12	31	67	121	187	278	381	505	658
4.0	8	22	47	86	135	203	280	375	492

4% maximum (customary U.S. units)*

e (%)	v_d (mph)									
	15	20	25	30	35	40	45	50	55	60
	R (ft)									
NC	796	1410	2050	2830	3730	4770	5930	7220	8650	10,300
RC	506	902	1340	1680	2490	3220	4040	4940	5950	7080
2.2	399	723	1110	1580	2120	2760	3480	4280	5180	6190
2.4	271	513	836	1270	1760	2340	2980	3690	4500	5410
2.6	201	388	650	1000	1420	1930	2490	3130	3870	4700
2.8	157	308	524	817	1170	1620	2100	2660	3310	4060
3.0	127	251	433	681	982	1370	1800	2290	2860	3530
3.2	105	209	363	576	835	1180	1550	1980	2490	3090
3.4	88	175	307	490	714	1010	1340	1720	2170	2700
3.6	73	147	259	416	610	865	1150	1480	1880	2350
3.8	61	122	215	348	512	730	970	1260	1600	2010
4.0	42	86	154	250	371	533	711	926	1190	1500

*Use of $e_{max} = 4\%$ should be limited to urban conditions.

APPENDIX 4.B *(continued)*
Minimum Radii for Design Superelevation Rates and Design Speeds

6% maximum (SI units)

e (%)	v_d (kph)											
	20	30	40	50	60	70	80	90	100	110	120	130
	R (m)											
NC	194	421	738	1050	1440	1910	2360	2880	3510	4060	4770	5240
RC	138	299	525	750	1030	1380	1710	2090	2560	2970	3510	3880
2.2	122	265	465	668	919	1230	1530	1880	2300	2670	3160	3500
2.4	109	236	415	599	825	1110	1380	1700	2080	2420	2870	3190
2.6	97	212	372	540	746	1000	1260	1540	1890	2210	2630	2930
2.8	87	190	334	488	676	910	1150	1410	1730	2020	2420	2700
3.0	78	170	300	443	615	831	1050	1290	1590	1870	2240	2510
3.2	70	152	269	402	561	761	959	1190	1470	1730	2080	2330
3.4	61	133	239	364	511	697	882	1100	1360	1600	1940	2180
3.6	51	113	206	329	465	640	813	1020	1260	1490	1810	2050
3.8	42	96	177	294	422	586	749	939	1170	1390	1700	1930
4.0	36	82	155	261	380	535	690	870	1090	1300	1590	1820
4.2	31	72	136	234	343	488	635	806	1010	1220	1500	1720
4.4	27	63	121	210	311	446	584	746	938	1140	1410	1630
4.6	24	56	108	190	283	408	538	692	873	1070	1330	1540
4.8	21	50	97	172	258	374	496	641	812	997	1260	1470
5.0	19	45	88	156	235	343	457	594	755	933	1190	1400
5.2	17	40	79	142	214	315	421	549	701	871	1120	1330
5.4	15	36	71	128	195	287	386	506	648	810	1060	1260
5.6	13	32	63	115	176	260	351	463	594	747	980	1190
5.8	11	28	56	102	156	232	315	416	537	679	900	1110
6.0	8	21	43	79	123	184	252	336	437	560	756	951

From *A Policy on Geometric Design of Highways and Streets*, 2011, by the American Association of State Highway and Transportation Officials, Washington, D.C. Table 3–8. Used by permission.

APPENDIX 4.B *(continued)*
Minimum Radii for Design Superelevation Rates and Design Speeds

6% maximum (customary U.S. units)

e (%)	v_d (mph)													
	15	20	25	30	35	40	45	50	55	60	65	70	75	80
	R (ft)													
NC	868	1580	2290	3130	4100	5230	6480	7870	9410	11,100	12,600	14,100	15,700	17,400
RC	614	1120	1630	2240	2950	3770	4680	5700	6820	8060	9130	10,300	11,500	12,900
2.2	543	991	1450	2000	2630	3370	4190	5100	6110	7230	8200	9240	10,400	11,600
2.4	482	884	1300	1790	2360	3030	3770	4600	5520	6540	7430	8380	9420	10,600
2.6	430	791	1170	1610	2130	2740	3420	4170	5020	5950	6770	7660	8620	9670
2.8	384	709	1050	1460	1930	2490	3110	3800	4580	5440	6200	7030	7930	8910
3.0	341	635	944	1320	1760	2270	2840	3480	4200	4990	5710	6490	7330	8260
3.2	300	566	850	1200	1600	2080	2600	3200	3860	4600	5280	6010	6810	7680
3.4	256	498	761	1080	1460	1900	2390	2940	3560	4250	4890	5580	6340	7180
3.6	209	422	673	972	1320	1740	2190	2710	3290	3940	4540	5210	5930	6720
3.8	176	358	583	864	1190	1590	2010	2490	3040	3650	4230	4860	5560	6320
4.0	151	309	511	766	1070	1440	1840	2300	2810	3390	3950	4550	5220	5950
4.2	131	270	452	684	960	1310	1680	2110	2590	3140	3680	4270	4910	5620
4.4	116	238	402	615	868	1190	1540	1940	2400	2920	3440	4010	4630	5320
4.6	102	212	360	555	788	1090	1410	1780	2210	2710	3220	3770	4380	5040
4.8	91	189	324	502	718	995	1300	1640	2050	2510	3000	3550	4140	4790
5.0	82	169	292	456	654	911	1190	1510	1890	2330	2800	3330	3910	4550
5.2	73	152	264	413	595	833	1090	1390	1750	2160	2610	3120	3690	4320
5.4	65	136	237	373	540	759	995	1280	1610	1990	2420	2910	3460	4090
5.6	58	121	212	335	487	687	903	1160	1470	1830	2230	2700	3230	3840
5.8	51	106	186	296	431	611	806	1040	1320	1650	2020	2460	2970	3560
6.0	39	81	144	231	340	485	643	833	1060	1330	1660	2040	2500	3050

APPENDIX 4.B (continued)
Minimum Radii for Design Superelevation Rates and Design Speeds

8% maximum (SI units)

e (%)	\multicolumn{12}{c}{v_d (kph)}
	20	30	40	50	60	70	80	90	100	110	120	130
	\multicolumn{12}{c}{R (m)}
1.5	184	443	784	1090	1490	1970	2440	2970	3630	4180	4900	5360
2.0	133	322	571	791	1090	1450	1790	2190	2680	3090	3640	4000
2.2	119	288	512	711	976	1300	1620	1960	2420	2790	3290	3620
2.4	107	261	463	644	885	1190	1470	1800	2200	2550	3010	3310
2.6	97	237	421	587	808	1080	1350	1650	2020	2340	2760	3050
2.8	88	216	385	539	742	992	1240	1520	1860	2160	2550	2830
3.0	81	199	354	496	684	916	1150	1410	1730	2000	2370	2630
3.2	74	183	326	458	633	849	1060	1310	1610	1870	2220	2460
3.4	68	169	302	425	588	790	988	1220	1500	1740	2080	2310
3.6	62	156	279	395	548	738	924	1140	1410	1640	1950	2180
3.8	57	144	259	368	512	690	866	1070	1320	1540	1840	2060
4.0	52	134	241	344	479	648	813	1010	1240	1450	1740	1950
4.2	48	124	224	321	449	608	766	948	1180	1380	1650	1850
4.4	43	115	208	301	421	573	722	895	1110	1300	1570	1760
4.6	38	106	192	281	395	540	682	847	1050	1240	1490	1680
4.8	33	96	178	263	371	509	645	803	996	1180	1420	1610
5.0	30	87	163	246	349	480	611	762	947	1120	1360	1540
5.2	27	78	148	229	328	454	579	724	901	1070	1300	1480
5.4	24	71	136	213	307	429	549	689	859	1020	1250	1420
5.6	22	65	125	198	288	405	521	656	819	975	1200	1360
5.8	20	59	115	185	270	382	494	625	781	933	1150	1310
6.0	19	55	106	172	253	360	469	595	746	894	1100	1260
6.2	17	50	98	161	238	340	445	567	713	857	1060	1220
6.4	16	46	91	151	224	322	422	540	681	823	1020	1180
6.6	15	43	85	141	210	304	400	514	651	789	982	1140
6.8	14	40	79	132	198	287	379	489	620	757	948	1100
7.0	13	37	73	123	185	270	358	464	591	724	914	1070
7.2	12	34	68	115	174	254	338	440	561	691	879	1040
7.4	11	31	62	107	162	237	318	415	531	657	842	998
7.6	10	29	57	99	150	221	296	389	499	621	803	962
7.8	9	26	52	90	137	202	273	359	462	579	757	919
8.0	7	20	41	73	113	168	229	304	394	501	667	832

From *A Policy on Geometric Design of Highways and Streets*, 2011, by the American Association of State Highway and Transportation Officials, Washington, D.C. Table 3–9. Used by permission.

APPENDIX 4.B *(continued)*
Minimum Radii for Design Superelevation Rates and Design Speeds

8% maximum (customary U.S. units)

e (%)	v_d (mph) 15	20	25	30	35	40	45 R (ft)	50	55	60	65	70	75	80
NC	932	1640	2370	3240	4260	5410	6710	8150	9720	11,500	12,900	14,500	16,100	17,800
RC	676	1190	1720	2370	3120	3970	4930	5990	7150	8440	9510	10,700	12,000	13,300
2.2	605	1070	1550	2130	2800	3570	4440	5400	6450	7620	8600	9660	10,800	12,000
2.4	546	959	1400	1930	2540	3240	4030	4910	5870	6930	7830	8810	9850	11,000
2.6	496	872	1280	1760	2320	2960	3690	4490	5370	6350	7180	8090	9050	10,100
2.8	453	796	1170	1610	2130	2720	3390	4130	4950	5850	6630	7470	8370	9340
3.0	415	730	1070	1480	1960	2510	3130	3820	4580	5420	6140	6930	7780	8700
3.2	382	672	985	1370	1820	2330	2900	3550	4250	5040	5720	6460	7260	8130
3.4	352	620	911	1270	1690	2170	2700	3300	3970	4700	5350	6050	6800	7620
3.6	324	572	845	1180	1570	2020	2520	3090	3710	4400	5010	5680	6400	7180
3.8	300	530	784	1100	1470	1890	2360	2890	3480	4140	4710	5350	6030	6780
4.0	277	490	729	1030	1370	1770	2220	2720	3270	3890	4450	5050	5710	6420
4.2	255	453	678	955	1280	1660	2080	2560	3080	3670	4200	4780	5410	6090
4.4	235	418	630	893	1200	1560	1960	2410	2910	3470	3980	4540	5140	5800
4.6	215	384	585	834	1130	1470	1850	2280	2750	3290	3770	4310	4890	5530
4.8	193	349	542	779	1060	1390	1750	2160	2610	3120	3590	4100	4670	5280
5.0	172	314	499	727	991	1310	1650	2040	2470	2960	3410	3910	4460	5050
5.2	154	284	457	676	929	1230	1560	1930	2350	2820	3250	3740	4260	4840
5.4	139	258	420	627	870	1160	1480	1830	2230	2680	3110	3570	4090	4640
5.6	126	236	387	582	813	1090	1390	1740	2120	2550	2970	3420	3920	4460
5.8	115	216	358	542	761	1030	1320	1650	2010	2430	2840	3280	3760	4290
6.0	105	199	332	506	713	965	1250	1560	1920	2320	2710	3150	3620	4140
6.2	97	184	308	472	669	909	1180	1480	1820	2210	2600	3020	3480	3990
6.4	89	170	287	442	628	857	1110	1400	1730	2110	2490	2910	3360	3850
6.6	82	157	267	413	590	808	1050	1330	1650	2010	2380	2790	3240	3720
6.8	76	146	248	386	553	761	990	1260	1560	1910	2280	2690	3120	3600
7.0	70	135	231	360	518	716	933	1190	1480	1820	2180	2580	3010	3480
7.2	64	125	214	336	485	672	878	1120	1400	1720	2070	2470	2900	3370
7.4	59	115	198	312	451	628	822	1060	1320	1630	1970	2350	2780	3250
7.6	54	105	182	287	417	583	765	980	1230	1530	1850	2230	2650	3120
7.8	48	94	164	261	380	533	701	901	1140	1410	1720	2090	2500	2970
8.0	38	76	134	214	314	444	587	758	960	1200	1480	1810	2210	2670

APPENDIX 4.B (continued)
Minimum Radii for Design Superelevation Rates and Design Speeds

10% maximum (SI units)

e (%)	v_d (kph)											
	20	30	40	50	60	70	80	90	100	110	120	130
	R (m)											
NC	197	454	790	1110	1520	2000	2480	3010	3690	4250	4960	5410
RC	145	333	580	815	1120	1480	1840	2230	2740	3160	3700	4050
2.2	130	300	522	735	1020	1340	1660	2020	2480	2860	3360	3680
2.4	118	272	474	669	920	1220	1520	1840	2260	2620	3070	3370
2.6	108	249	434	612	844	1120	1390	1700	2080	2410	2830	3110
2.8	99	229	399	564	778	1030	1290	1570	1920	2230	2620	2880
3.0	91	211	368	522	720	952	1190	1460	1790	2070	2440	2690
3.2	85	196	342	485	670	887	1110	1360	1670	1940	2280	2520
3.4	79	182	318	453	626	829	1040	1270	1560	1820	2140	2370
3.6	73	170	297	424	586	777	974	1200	1470	1710	2020	2230
3.8	68	159	278	398	551	731	917	1130	1390	1610	1910	2120
4.0	64	149	261	374	519	690	866	1060	1310	1530	1810	2010
4.2	60	140	245	353	490	652	820	1010	1240	1450	1720	1910
4.4	56	132	231	333	464	617	777	953	1180	1380	1640	1820
4.6	53	124	218	315	439	586	738	907	1120	1310	1560	1740
4.8	50	117	206	299	417	557	703	864	1070	1250	1490	1670
5.0	47	111	194	283	396	530	670	824	1020	1200	1430	1600
5.2	44	104	184	269	377	505	640	788	975	1150	1370	1540
5.4	41	98	174	256	359	482	611	754	934	1100	1320	1480
5.6	39	93	164	243	343	461	585	723	896	1060	1270	1420
5.8	36	88	155	232	327	441	561	693	860	1020	1220	1370
6.0	33	82	146	221	312	422	538	666	827	976	1180	1330
6.2	31	77	138	210	298	404	516	640	795	941	1140	1280
6.4	28	72	130	200	285	387	496	616	766	907	1100	1240
6.6	26	67	121	191	273	372	476	593	738	876	1060	1200
6.8	24	62	114	181	261	357	458	571	712	846	1030	1170
7.0	22	58	107	172	249	342	441	551	688	819	993	1130
7.2	21	55	101	164	238	329	425	532	664	792	963	1100
7.4	20	51	95	156	228	315	409	513	642	767	934	1070
7.6	18	48	90	148	218	303	394	496	621	743	907	1040
7.8	17	45	85	141	208	291	380	479	601	721	882	1010
8.0	16	43	80	135	199	279	366	463	582	699	857	981
8.2	15	40	76	128	190	268	353	448	564	679	834	956
8.4	14	38	72	122	182	257	339	432	546	660	812	932
8.6	14	36	68	116	174	246	326	417	528	641	790	910
8.8	13	34	64	110	166	236	313	402	509	621	770	888
9.0	12	32	61	105	158	225	300	386	491	602	751	867
9.2	11	30	57	99	150	215	287	371	472	582	731	847
9.4	11	28	54	94	142	204	274	354	453	560	709	828
9.6	10	26	50	88	133	192	259	337	432	537	685	809
9.8	9	24	46	81	124	179	242	316	407	509	656	786
10.0	7	19	38	68	105	154	210	277	358	454	597	739

From *A Policy on Geometric Design of Highways and Streets*, 2011, by the American Association of State Highway and Transportation Officials, Washington, D.C. Table 3–10a. Used by permission.

APPENDIX 4.B *(continued)*
Minimum Radii for Design Superelevation Rates and Design Speeds

10% maximum (customary U.S. units)

	v_d (mph)													
	15	20	25	30	35	40	45	50	55	60	65	70	75	80
e (%)							R (ft)							
NC	947	1680	2420	3320	4350	5520	6830	8280	9890	11,700	13,100	14,700	16,300	18,000
RC	694	1230	1780	2440	3210	4080	5050	6130	7330	8630	9720	10,900	12,200	13,500
2.2	625	1110	1600	2200	2900	3680	4570	5540	6630	7810	8800	9860	11,000	12,200
2.4	567	1010	1460	2000	2640	3350	4160	5050	6050	7130	8040	9010	10,100	11,200
2.6	517	916	1330	1840	2420	3080	3820	4640	5550	6550	7390	8290	9260	10,300
2.8	475	841	1230	1690	2230	2840	3520	4280	5130	6050	6840	7680	8580	9550
3.0	438	777	1140	1570	2060	2630	3270	3970	4760	5620	6360	7140	7990	8900
3.2	406	720	1050	1450	1920	2450	3040	3700	4440	5250	5930	6680	7480	8330
3.4	377	670	978	1360	1790	2290	2850	3470	4160	4910	5560	6260	7020	7830
3.6	352	625	913	1270	1680	2150	2670	3250	3900	4620	5230	5900	6620	7390
3.8	329	584	856	1190	1580	2020	2510	3060	3680	4350	4940	5570	6260	6990
4.0	308	547	804	1120	1490	1900	2370	2890	3470	4110	4670	5270	5930	6630
4.2	289	514	756	1060	1400	1800	2240	2740	3290	3900	4430	5010	5630	6300
4.4	271	483	713	994	1330	1700	2120	2590	3120	3700	4210	4760	5370	6010
4.6	255	455	673	940	1260	1610	2020	2460	2970	3520	4010	4540	5120	5740
4.8	240	429	636	890	1190	1530	1920	2340	2830	3360	3830	4340	4900	5490
5.0	226	404	601	844	1130	1460	1830	2240	2700	3200	3660	4150	4690	5270
5.2	213	381	569	802	1080	1390	1740	2130	2580	3060	3500	3980	4500	5060
5.4	200	359	539	762	1030	1330	1660	2040	2460	2930	3360	3820	4320	4860
5.6	188	339	511	724	974	1270	1590	1950	2360	2810	3220	3670	4160	4680
5.8	176	319	484	689	929	1210	1520	1870	2260	2700	3090	3530	4000	4510
6.0	164	299	458	656	886	1160	1460	1790	2170	2590	2980	3400	3860	4360
6.2	152	280	433	624	846	1110	1400	1720	2090	2490	2870	3280	3730	4210
6.4	140	260	409	594	808	1060	1340	1650	2010	2400	2760	3160	3600	4070
6.6	130	242	386	564	772	1020	1290	1590	1930	2310	2670	3060	3480	3940
6.8	120	226	363	536	737	971	1230	1530	1860	2230	2570	2960	3370	3820
7.0	112	212	343	509	704	931	1190	1470	1790	2150	2490	2860	3270	3710
7.2	105	199	324	483	671	892	1140	1410	1730	2070	2410	2770	3170	3600
7.4	98	187	306	460	641	855	1100	1360	1670	2000	2330	2680	3070	3500
7.6	92	176	290	437	612	820	1050	1310	1610	1940	2250	2600	2990	3400
7.8	86	165	274	416	585	786	1010	1260	1550	1870	2180	2530	2900	3310
8.0	81	156	260	396	558	754	968	1220	1500	1810	2120	2450	2820	3220
8.2	76	147	246	377	533	722	930	1170	1440	1750	2050	2380	2750	3140
8.4	72	139	234	359	509	692	893	1130	1390	1690	1990	2320	2670	3060
8.6	68	131	221	341	486	662	856	1080	1340	1630	1930	2250	2600	2980
8.8	64	124	209	324	463	633	820	1040	1290	1570	1870	2190	2540	2910
9.0	60	116	198	307	440	604	784	992	1240	1520	1810	2130	2470	2840
9.2	56	109	186	291	418	574	748	948	1190	1460	1740	2060	2410	2770
9.4	52	102	175	274	395	545	710	903	1130	1390	1670	1990	2340	2710
9.6	48	95	163	256	370	513	671	854	1080	1320	1600	1910	2260	2640
9.8	44	87	150	236	343	477	625	798	1010	1250	1510	1820	2160	2550
10.0	36	72	126	200	292	410	540	694	877	1090	1340	1630	1970	2370

APPENDIX 4.B (continued)
Minimum Radii for Design Superelevation Rates and Design Speeds

12% maximum (SI units)

v_d (kph)

e (%)	20	30	40	50	60	70	80	90	100	110	120	130
						R (m)						
NC	210	459	804	1130	1540	2030	2510	3040	3720	4280	4990	5440
RC	155	338	594	835	1150	1510	1870	2270	2770	3190	3740	4080
2.2	139	306	536	755	1040	1360	1690	2050	2510	2900	3390	3710
2.4	127	278	488	688	942	1250	1550	1880	2300	2650	3110	3400
2.6	116	255	448	631	865	1140	1420	1730	2110	2440	2860	3140
2.8	107	235	413	583	799	1060	1320	1600	1960	2260	2660	2910
3.0	99	218	382	541	742	980	1220	1490	1820	2110	2480	2720
3.2	92	202	356	504	692	914	1140	1390	1700	1970	2320	2550
3.4	86	189	332	472	648	856	1070	1300	1600	1850	2180	2400
3.6	81	177	312	443	609	805	1010	1230	1510	1750	2060	2270
3.8	76	166	293	417	573	759	947	1160	1420	1650	1950	2150
4.0	71	157	276	393	542	718	896	1100	1350	1560	1850	2040
4.2	67	148	261	372	513	680	850	1040	1280	1490	1760	1940
4.4	64	140	247	353	487	646	808	988	1220	1420	1680	1850
4.6	60	132	234	335	463	615	770	941	1160	1350	1600	1770
4.8	57	126	222	319	441	586	734	899	1110	1290	1530	1700
5.0	54	119	211	304	421	560	702	860	1060	1240	1470	1630
5.2	52	114	201	290	402	535	672	824	1020	1190	1410	1570
5.4	49	108	192	277	384	513	644	790	973	1140	1360	1510
5.6	47	103	183	265	368	492	618	759	936	1100	1310	1460
5.8	45	98	175	254	353	472	594	730	900	1060	1260	1410
6.0	43	94	167	244	339	454	572	703	867	1020	1220	1360
6.2	41	90	159	234	326	436	551	678	837	981	1180	1310
6.4	39	86	153	225	313	420	531	654	808	948	1140	1270
6.6	37	82	146	216	302	405	512	632	781	917	1100	1230
6.8	35	78	140	208	290	391	494	611	755	888	1070	1200
7.0	34	75	134	200	280	377	478	591	731	860	1040	1160
7.2	32	71	128	192	270	364	462	572	708	834	1010	1130
7.4	30	68	122	185	260	352	447	554	686	810	974	1100
7.6	29	65	117	178	251	340	433	537	666	786	947	1070
7.8	27	61	112	172	243	329	420	521	646	764	921	1040
8.0	26	58	107	165	235	319	407	506	628	743	897	1020
8.2	24	55	102	159	227	309	395	491	610	723	874	989
8.4	23	52	97	154	219	299	383	477	593	704	852	965
8.6	22	50	93	148	212	290	372	464	577	686	831	942
8.8	20	47	88	142	205	281	361	451	562	668	811	921
9.0	19	45	85	137	198	273	351	439	547	652	792	900
9.2	18	43	81	132	191	264	341	428	533	636	774	880
9.4	18	41	77	127	185	256	332	416	520	621	756	861
9.6	17	39	74	123	179	249	323	406	507	606	739	843
9.8	16	37	71	118	173	241	314	395	494	592	723	826
10.0	15	36	68	114	167	234	305	385	482	579	708	809
10.2	14	34	65	110	161	226	296	375	471	566	693	793
10.4	14	33	62	105	155	219	288	365	459	553	679	778
10.6	13	31	59	101	150	212	279	355	448	541	665	763
10.8	12	30	57	97	144	204	270	345	436	529	652	749
10.0	12	28	54	93	139	197	261	335	423	516	639	735
11.2	11	27	51	89	133	189	252	324	411	503	626	722
11.4	11	25	49	85	127	182	242	312	397	488	613	709
11.6	10	24	46	80	120	173	232	300	382	472	598	697
11.8	9	22	43	75	113	163	219	285	364	453	579	685
12.0	7	18	36	64	98	143	194	255	328	414	540	665

From *A Policy on Geometric Design of Highways and Streets*, 2011, by the American Association of State Highway and Transportation Officials, Washington, D.C. Table 3–11a and Table 3–11b. Used by permission.

APPENDIX 4.B (continued)
Minimum Radii for Design Superelevation Rates and Design Speeds

12% maximum (customary U.S. units)

e (%)	v_d (mph) 15	20	25	30	35	40	45 R (ft)	50	55	60	65	70	75	80
NC	950	1690	2460	3370	4390	5580	6910	8370	9990	11,800	13,200	14,800	16,400	18,100
RC	700	1250	1820	2490	3260	4140	5130	6220	7430	8740	9840	11,000	12,300	13,600
2.2	631	1130	1640	2250	2950	3750	4640	5640	6730	7930	8920	9980	11,200	12,400
2.4	574	1030	1500	2060	2690	3420	4240	5150	6150	7240	8160	9130	10,200	11,300
2.6	526	936	1370	1890	2470	3140	3900	4730	5660	6670	7510	8420	9380	10,500
2.8	484	863	1270	1740	2280	2910	3600	4380	5240	6170	6960	7800	8700	9660
3.0	448	799	1170	1620	2120	2700	3350	4070	4870	5740	6480	7270	8110	9010
3.2	417	743	1090	1510	1970	2520	3130	3800	4550	5370	6060	6800	7600	8440
3.4	389	693	1020	1410	1850	2360	2930	3560	4270	5030	5690	6390	7140	7940
3.6	364	649	953	1320	1730	2220	2750	3350	4020	4740	5360	6020	6740	7500
3.8	341	610	896	1250	1630	2090	2600	3160	3790	4470	5060	5700	6380	7100
4.0	321	574	845	1180	1540	1980	2460	2990	3590	4240	4800	5400	6050	6740
4.2	303	542	798	1110	1460	1870	2330	2840	3400	4020	4560	5130	5750	6420
4.4	286	512	756	1050	1390	1780	2210	2700	3240	3830	4340	4890	5490	6120
4.6	271	485	717	997	1320	1690	2110	2570	3080	3650	4140	4670	5240	5850
4.8	257	460	681	948	1260	1610	2010	2450	2940	3480	3960	4470	5020	5610
5.0	243	437	648	904	1200	1540	1920	2340	2810	3330	3790	4280	4810	5380
5.2	231	415	618	862	1140	1470	1840	2240	2700	3190	3630	4110	4620	5170
5.4	220	395	589	824	1090	1410	1760	2150	2590	3060	3490	3950	4440	4980
5.6	209	377	563	788	1050	1350	1690	2060	2480	2940	3360	3800	4280	4800
5.8	199	359	538	754	1000	1300	1620	1980	2390	2830	3230	3660	4130	4630
6.0	190	343	514	723	960	1250	1560	1910	2300	2730	3110	3530	3990	4470
6.2	181	327	492	694	922	1200	1500	1840	2210	2630	3010	3410	3850	4330
6.4	172	312	471	666	886	1150	1440	1770	2140	2540	2900	3300	3730	4190
6.6	164	298	452	639	852	1110	1390	1710	2060	2450	2810	3190	3610	4060
6.8	156	284	433	615	820	1070	1340	1650	1990	2370	2720	3090	3500	3940
7.0	148	271	415	591	790	1030	1300	1590	1930	2290	2630	3000	3400	3820
7.2	140	258	398	568	762	994	1250	1540	1860	2220	2550	2910	3300	3720
7.4	133	246	382	547	734	960	1210	1490	1810	2150	2470	2820	3200	3610
7.6	125	234	366	527	708	928	1170	1440	1750	2090	2400	2740	3120	3520
7.8	118	222	351	507	684	897	1130	1400	1700	2020	2330	2670	3030	3430
8.0	111	210	336	488	660	868	1100	1360	1650	1970	2270	2600	2950	3340
8.2	105	199	321	470	637	840	1070	1320	1600	1910	2210	2530	2880	3260
8.4	100	190	307	452	615	813	1030	1280	1550	1860	2150	2460	2800	3180
8.6	95	180	294	435	594	787	997	1240	1510	1810	2090	2400	2740	3100
8.8	90	172	281	418	574	762	967	1200	1470	1760	2040	2340	2670	3030
9.0	85	164	270	403	554	738	938	1170	1430	1710	1980	2280	2610	2960
9.2	81	156	259	388	535	715	910	1140	1390	1660	1940	2230	2550	2890
9.4	77	149	248	373	516	693	883	1100	1350	1620	1890	2180	2490	2830
9.6	74	142	238	359	499	671	857	1070	1310	1580	1840	2130	2440	2770
9.8	70	136	228	346	481	650	832	1040	1280	1540	1800	2080	2380	2710
10.0	67	130	219	333	465	629	806	1010	1250	1500	1760	2030	2330	2660
10.2	64	124	210	320	448	608	781	980	1210	1460	1720	1990	2280	2600
10.4	61	118	201	308	432	588	757	951	1180	1430	1680	1940	2240	2550
10.6	58	113	192	296	416	568	732	922	1140	1390	1640	1900	2190	2500
10.8	55	108	184	284	400	548	707	892	1110	1350	1600	1860	2150	2460
11.0	52	102	175	272	384	527	682	862	1070	1310	1560	1820	2110	2410
11.2	49	97	167	259	368	506	656	831	1040	1270	1510	1780	2070	2370
11.4	47	92	158	247	351	485	629	799	995	1220	1470	1730	2020	2320
11.6	44	86	149	233	333	461	600	763	953	1170	1410	1680	1970	2280
11.8	40	80	139	218	312	434	566	722	904	1120	1350	1620	1910	2230
12.0	34	68	119	188	272	381	500	641	807	1000	1220	1480	1790	2130

From *A Policy on Geometric Design of Highways and Streets*, 2011, by the American Association of State Highway and Transportation Officials, Washington, D.C. Table 3–12a and Table 3–12b. Used by permission.

APPENDIX 4.C
Superelevation, Radius, and Design Speed for Low-Speed Urban Streets*

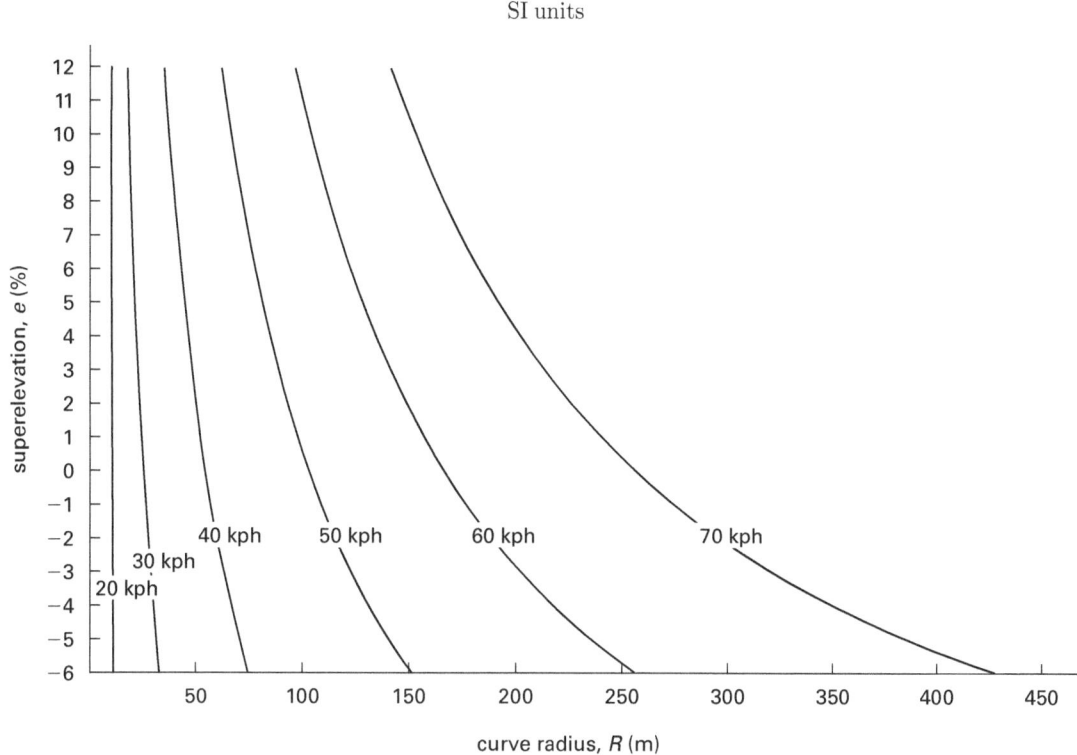

SI units

*Negative superelevation values beyond −2.0% should be used for low type surfaces such as gravel, crushed stone, and earth. However, areas with intense rainfall may use normal cross slopes of −2.5% on high type surfaces.

APPENDIX 4.C *(continued)*
Superelevation, Radius, and Design Speed for Low-Speed Urban Streets*

customary U.S. units

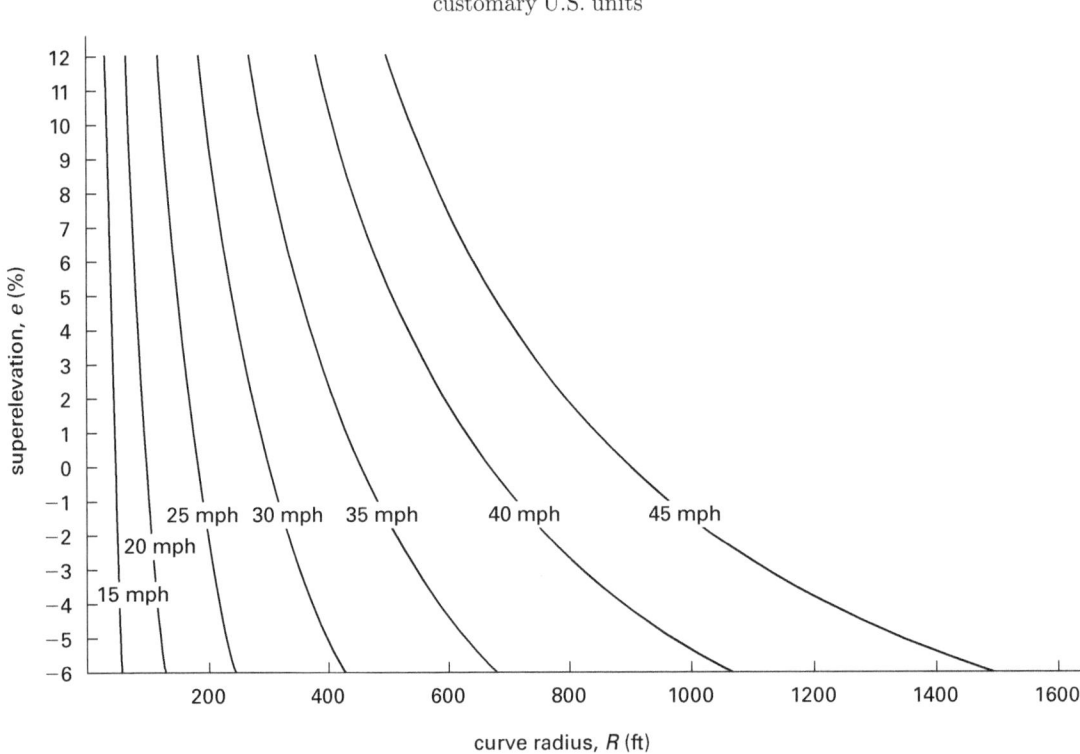

*Negative superelevation values beyond −2.0% should be used for low type surfaces such as gravel, crushed stone, and earth. However, areas with intense rainfall may use normal cross slopes of −2.5% on high type surfaces.

From *A Policy on Geometric Design of Highways and Streets*, 2011, by the American Association of State Highway and Transportation Officials, Washington, D.C. Figure 3-14. Used by permission.

APPENDIX 4.D
Minimum Radii and Superelevation for Low-Speed Urban Streets[a,b,c]

	SI units					
	v_d (kph)					
	20	30	40	50	60	70
e (%)	R (m)					
−6.0	11	32	74	151	258	429
−5.0	10	31	70	141	236	386
−4.0	10	30	66	131	218	351
−3.0	10	28	63	123	202	322
−2.8	10	28	62	122	200	316
−2.6	10	28	62	120	197	311
−2.4	10	28	61	119	194	306
−2.2	10	27	61	117	192	301
−2.0	10	27	60	116	189	297
−1.5	9	27	59	113	183	286
0	9	25	55	104	167	257
1.5	9	24	51	96	153	234
2.0	9	24	50	94	149	227
2.2	8	23	50	93	148	224
2.4	8	23	50	92	146	222
2.6	8	23	49	91	145	219
2.8	8	23	49	90	143	217
3.0	8	23	48	89	142	214
3.2	8	23	48	89	140	212
3.4	8	23	48	88	139	210
3.6	8	22	47	87	138	207
3.8	8	22	47	86	136	205
4.0	8	22	47	86	135	203
4.2	8	22	46	85	134	201
4.4	8	22	46	84	132	199
4.6	8	22	46	83	131	197
4.8	8	22	45	83	130	195
5.0	8	21	45	82	129	193
5.2	8	21	45	81	128	191
5.4	8	21	44	81	127	189
5.6	8	21	44	80	125	187
5.8	8	21	44	79	124	185
6.0	8	21	43	79	123	184
6.2	8	21	43	78	122	182
6.4	8	21	43	78	121	180
6.6	8	20	43	77	120	179
6.8	8	20	42	76	119	177
7.0	7	20	42	76	118	175
7.2	7	20	42	75	117	174
7.4	7	20	41	75	116	172
7.6	7	20	41	74	115	171
7.8	7	20	41	73	114	169
8.0	7	20	41	73	113	168
8.2	7	20	40	72	112	166
8.4	7	19	40	72	112	165
8.6	7	19	40	71	111	163
8.8	7	19	40	71	110	162
9.0	7	19	39	70	109	161
9.2	7	19	39	70	108	159
9.4	7	19	39	69	107	158
9.6	7	19	39	69	107	157
9.8	7	19	38	68	106	156
10.0	7	19	38	68	105	154

APPENDIX 4.D *(continued)*
Minimum Radii and Superelevation for Low-Speed Urban Streets[a,b,c]

	SI units					
	v_d (kph)					
	20	30	40	50	60	70
e (%)			R (m)			
10.2	7	19	38	67	104	153
10.4	7	18	38	67	103	152
10.6	7	18	37	67	103	151
10.8	7	18	37	66	102	150
11.0	7	18	37	66	101	148
11.2	7	18	37	65	101	147
11.4	7	18	37	65	100	146
11.6	7	18	36	64	99	145
11.8	7	18	36	64	98	144
12.0	7	18	36	64	98	143

From *A Policy on Geometric Design of Highways and Streets*, 2011, by the American Association of State Highway and Transportation Officials, Washington, D.C. Table 3–13a. Used by permission.

APPENDIX 4.D (continued)
Minimum Radii and Superelevation for Low-Speed Urban Streets[a,b,c]

customary U.S. units

$e\%$	$v_d = 15$ mph	$v_d = 20$ mph	$v_d = 25$ mph	$v_d = 30$ mph	$v_d = 35$ mph	$v_d = 40$ mph	$v_d = 45$ mph
				R (ft)			
−6.0	58	127	245	429	681	1067	1500
−5.0	56	121	231	400	628	970	1350
−4.0	54	116	219	375	583	889	1227
−3.0	52	111	208	353	544	821	1125
−2.8	51	110	206	349	537	808	1107
−2.6	51	109	204	345	530	796	1089
−2.4	51	108	202	341	524	784	1071
−2.2	50	108	200	337	517	773	1055
−2.0	50	107	198	333	510	762	1039
−1.5	49	105	194	324	495	736	1000
0	47	99	181	300	454	667	900
1.5	45	94	170	279	419	610	818
2.0	44	92	167	273	408	593	794
2.2	44	91	165	270	404	586	785
2.4	44	91	164	268	400	580	776
2.6	43	90	163	265	396	573	767
2.8	43	89	161	263	393	567	758
3.0	43	89	160	261	389	561	750
3.2	43	88	159	259	385	556	742
3.4	42	88	158	256	382	550	734
3.6	42	87	157	254	378	544	726
3.8	42	87	155	252	375	539	718
4.0	42	86	154	250	371	533	711
4.2	41	85	153	248	368	528	703
4.4	41	85	152	246	365	523	696
4.6	41	84	151	244	361	518	689
4.8	41	84	150	242	358	513	682
5.0	41	83	149	240	355	508	675
5.2	40	83	148	238	352	503	668
5.4	40	82	147	236	349	498	662
5.6	40	82	146	234	346	494	655
5.8	40	81	145	233	343	489	649
6.0	39	81	144	231	340	485	643
6.2	39	80	143	229	337	480	637
6.4	39	80	142	227	335	476	631
6.6	39	79	141	226	332	472	625
6.8	39	79	140	224	329	468	619
7.0	38	78	139	222	327	464	614
7.2	38	78	138	221	324	460	608
7.4	38	78	137	219	322	456	603
7.6	38	77	136	217	319	452	597
7.8	38	77	135	216	317	448	592
8.0	38	76	134	214	314	444	587
8.2	37	76	134	213	312	441	582
8.4	37	75	133	211	309	437	577
8.6	37	75	132	210	307	434	572
8.8	37	74	131	208	305	430	567
9.0	37	74	130	207	302	427	563
9.2	36	74	129	205	300	423	558
9.4	36	73	129	204	298	420	553
9.6	36	73	128	203	296	417	549
9.8	36	72	127	201	294	413	544
10.0	36	72	126	200	292	410	540

APPENDIX 4.D *(continued)*
Minimum Radii and Superelevation for Low-Speed Urban Streets[a,b,c]

customary U.S. units

$e\%$	$v_d = 15$ mph	$v_d = 20$ mph	$v_d = 25$ mph	$v_d = 30$ mph	$v_d = 35$ mph	$v_d = 40$ mph	$v_d = 45$ mph
				R (ft)			
10.2	36	72	126	199	290	407	536
10.4	35	71	125	197	288	404	531
10.6	35	71	124	196	286	401	527
10.8	35	71	123	195	284	398	523
11.0	35	70	123	194	282	395	519
11.2	35	70	122	192	280	392	515
11.4	35	69	121	191	278	389	511
11.6	34	69	120	190	276	386	508
11.8	34	69	120	189	274	384	504
12.0	34	68	119	188	272	381	500

[a]Calculated using Superelevation Distribution Method 2.

[b]Superelevation may be optional on low-speed urban streets.

[c]Negative superelevation values beyond –2.0 percent should be used for unpaved surfaces such as gravel, crushed stone, and earth. However, a normal cross slope of –2.5 percent may be used on paved surfaces in areas with intense rainfall.

From *A Policy on Geometric Design of Highways and Streets*, 2011, by the American Association of State Highway and Transportation Officials, Washington, D.C. Table 3–13b. Used by permission.

APPENDIX 4.E
Minimum Superelevation Runoff for Horizontal Curves

SI units

	v_d (kph)																								
	20		30		40		50		60		70		80		90		100		110		120		130	
	no. of lanes rotated, L_R (m)*																								
e (%)	1	2	1	2	1	2	1	2	1	2	1	2	1	2	1	2	1	2	1	2	1	2	1	2
1.5	7	10	7	11	8	12	8	13	9	14	10	15	11	16	12	17	12	18	13	20	14	21	15	23
2.0	9	14	10	14	10	15	11	17	12	18	13	20	14	22	15	23	16	25	18	26	19	28	21	31
2.2	10	15	11	16	11	17	12	18	13	20	14	22	16	24	17	25	18	27	19	29	21	31	23	34
2.4	11	18	12	17	12	19	13	20	14	22	16	24	17	26	18	28	20	29	21	32	23	34	25	37
2.6	12	18	12	19	13	20	14	22	16	23	17	26	19	28	20	30	21	32	23	34	25	37	27	40
2.8	13	19	13	20	14	22	16	23	17	25	18	27	20	30	21	32	23	34	25	37	27	40	29	43
3.0	14	20	14	22	15	23	17	25	18	27	20	29	22	32	23	34	25	37	26	40	28	43	31	46
3.2	14	22	15	23	16	25	18	27	19	29	21	31	23	35	25	37	26	39	28	42	30	45	33	49
3.4	15	23	16	24	17	26	19	28	20	31	22	33	24	37	26	39	28	42	30	45	32	48	35	52
3.6	16	24	17	26	19	26	20	30	22	32	24	35	26	39	28	41	29	44	32	47	34	51	37	56
3.8	17	26	18	27	20	29	21	32	23	34	25	37	27	41	29	44	31	47	33	50	36	54	39	59
4.0	18	27	19	29	21	31	22	33	24	36	26	39	29	43	31	46	33	49	35	53	38	57	41	62
4.2	19	28	20	30	22	32	23	35	25	38	27	41	30	45	32	48	34	52	37	55	40	60	43	65
4.4	20	30	21	32	23	34	24	37	26	40	29	43	32	48	34	51	36	54	39	58	42	63	45	68
4.6	21	31	22	33	24	35	25	38	28	41	30	45	33	50	35	53	38	56	40	61	44	65	47	71
4.8	22	32	23	35	25	37	27	40	29	43	31	47	35	52	37	55	39	59	42	63	45	68	49	74
5.0	23	34	24	36	26	39	28	42	30	45	33	49	36	54	38	57	41	61	44	66	47	71	51	77
5.2	23	35	25	37	27	40	29	43	31	47	34	51	37	56	40	60	43	64	46	68	49	74	53	80
5.4	24	36	26	39	28	42	30	45	32	49	35	53	39	58	41	62	44	66	47	71	51	77	56	83
5.6	25	38	27	40	29	43	31	47	34	50	37	55	40	60	43	64	46	69	49	74	53	80	58	86
5.8	26	39	28	42	30	45	32	48	35	52	38	57	42	63	44	67	47	71	51	76	55	82	60	89
6.0	27	41	29	43	31	46	33	50	36	54	39	59	43	65	46	69	49	74	53	79	57	85	62	93
6.2	28	42	30	45	32	48	34	52	37	56	41	61	45	67	47	71	51	76	54	82	59	88	64	96
6.4	29	43	31	46	33	49	35	53	38	58	42	63	46	69	49	74	52	79	56	84	61	91	66	99
6.6	30	45	32	48	34	51	37	55	40	59	43	65	48	71	51	76	54	81	58	87	63	94	68	102
6.8	31	46	33	49	35	52	38	56	41	61	45	67	49	73	52	78	56	83	60	90	64	97	70	105
7.0	31	47	34	50	36	54	39	58	42	63	46	69	50	76	54	80	57	86	61	92	66	99	72	108
7.2	32	49	35	52	37	56	40	60	43	65	47	71	52	78	55	83	59	88	63	95	68	102	74	111
7.4	33	50	36	53	38	57	41	61	44	67	48	73	53	80	57	85	61	91	65	97	70	105	76	114
7.6	34	51	36	55	39	59	42	63	46	68	50	75	55	82	58	87	62	93	67	100	72	108	78	117
7.8	35	53	37	56	40	60	43	65	47	70	51	77	56	84	60	90	64	96	68	103	74	111	80	120
8.0	36	54	38	58	41	62	44	65	48	72	52	79	58	86	61	92	65	98	70	105	76	114	82	123
8.2	37	55	39	59	42	63	45	68	49	74	54	81	59	89	63	94	67	101	72	108	78	117	84	127
8.4	38	57	40	60	43	65	47	70	50	76	55	82	60	91	64	97	69	103	74	111	80	119	86	130
8.6	39	58	41	62	44	66	48	71	52	77	56	84	62	93	66	99	70	106	76	113	81	122	88	133
8.8	40	59	42	63	45	68	49	73	53	79	58	86	63	95	67	101	72	108	77	116	83	125	91	136
9.0	40	61	43	65	48	69	50	75	54	81	59	88	65	97	69	103	74	110	79	119	85	128	93	139
9.2	41	62	44	66	47	71	51	76	55	83	60	90	66	99	70	106	75	113	81	121	87	131	95	142
9.4	42	63	45	68	48	73	52	78	56	85	62	92	68	102	72	108	77	115	83	124	89	134	97	145
9.6	43	65	46	69	49	74	53	80	58	86	63	94	69	104	74	110	79	118	84	126	91	136	99	148
9.8	44	66	47	71	50	76	54	81	59	88	64	96	71	106	75	113	80	120	86	129	93	139	101	151
10.0	45	68	48	72	51	77	55	83	60	90	65	98	72	108	77	115	82	123	88	132	95	142	103	154
10.2	46	69	49	73	52	79	56	85	61	92	67	100	73	110	78	117	83	125	90	134	97	145	105	157
10.4	47	70	50	75	53	80	58	86	62	94	68	102	75	112	80	119	85	128	91	137	99	148	107	160
10.6	48	72	51	76	55	82	59	88	64	95	89	104	76	114	81	122	87	130	93	140	100	151	109	164
10.8	49	73	52	78	56	83	60	90	65	97	71	106	78	117	83	124	88	133	95	142	102	153	111	167
11.0	50	74	53	79	57	85	61	91	66	99	72	108	79	119	84	126	90	135	97	145	104	158	113	170
11.2	50	76	54	81	58	86	62	93	67	101	73	110	81	121	86	129	92	137	98	148	106	159	115	173
11.4	51	77	55	82	59	88	63	95	68	103	75	112	82	123	87	131	93	140	100	150	108	162	117	176
11.6	52	78	56	84	60	89	64	96	70	104	76	114	84	125	89	133	95	142	102	153	110	165	119	179
11.8	53	80	57	85	61	91	65	98	71	106	77	116	85	127	90	136	97	145	104	155	112	168	121	182
12.0	54	81	58	86	62	93	66	100	72	108	79	118	86	130	92	138	98	147	105	158	114	171	123	185

*One lane rotated is typical for a two-lane highway, two lanes rotated for a four-lane highway, and so on. (See Fig. 4.29.)

APPENDIX 4.E *(continued)*
Minimum Superelevation Runoff for Horizontal Curves

(customary U.S. units)

	v_d (mph)																									
	15		20		25		30		35		40		45		50		55		60		65		70		75		80
	no. of lanes rotated, L_R (ft)*																									
e (%)	1	2	1	2	1	2	1	2	1	2	1	2	1	2	1	2	1	2	1	2	1	2	1	2	1	2
1.5	23	35	24	37	26	39	27	41	29	44	31	47	33	50	36	54	38	58	40	60	42	63	45	68	47	71	52	77
2.0	31	46	32	49	34	51	36	55	39	58	41	62	44	67	48	72	51	77	53	80	56	84	60	90	63	95	59	103
2.2	34	51	36	54	38	57	40	60	43	64	46	68	49	73	53	79	56	84	59	88	61	92	66	99	69	104	75	113
2.4	37	55	39	58	41	62	44	65	46	70	50	74	53	80	58	86	61	92	64	96	67	100	72	108	76	114	82	123
2.6	40	60	42	63	45	67	47	71	50	75	54	81	58	87	62	94	66	100	69	104	73	109	78	117	82	123	89	134
2.8	43	65	45	68	48	72	51	76	54	81	58	87	62	93	67	101	71	107	75	112	78	117	84	126	88	133	96	144
3.0	46	69	49	73	51	77	55	82	58	87	62	93	67	100	72	108	77	115	80	120	84	126	90	135	95	142	103	154
3.2	49	74	52	78	55	82	58	87	62	93	66	99	71	107	77	115	82	123	85	128	89	134	96	144	101	152	110	165
3.4	52	78	55	83	58	87	62	93	66	99	70	106	76	113	82	122	87	130	91	138	95	142	102	153	107	161	117	175
3.6	55	83	58	88	62	93	65	98	70	105	74	112	80	120	86	130	92	138	96	144	100	151	108	162	114	171	123	185
3.8	58	88	62	92	65	98	69	104	74	110	79	118	84	127	91	137	97	146	101	152	106	159	114	171	120	180	130	195
4.0	62	92	65	97	69	103	73	109	77	116	83	124	89	133	96	144	102	153	107	160	112	167	120	180	126	189	137	206
4.2	65	97	68	102	72	108	76	115	81	122	87	130	93	140	101	151	107	161	112	168	117	176	126	189	133	199	144	216
4.4	68	102	71	107	75	113	80	120	85	128	91	137	98	147	106	158	112	169	117	176	123	184	132	198	139	208	151	226
4.6	71	106	75	112	79	118	84	125	89	134	95	143	102	153	110	166	117	176	123	184	128	193	138	207	145	218	158	237
4.8	74	111	78	117	82	123	87	131	93	139	99	149	107	160	115	173	123	184	128	192	134	201	144	216	152	227	165	247
5.0	77	115	81	122	86	129	91	136	97	145	103	155	111	167	120	180	128	191	133	200	140	209	150	225	158	237	171	257
5.2	80	120	84	126	89	134	95	142	101	151	108	161	116	173	125	187	133	199	139	208	145	218	156	234	164	246	178	267
5.4	83	125	88	131	93	139	98	147	105	157	112	168	120	180	130	194	138	207	144	216	151	226	162	243	171	256	185	278
5.6	86	129	91	136	96	144	102	153	108	163	116	174	124	187	134	202	143	214	149	224	156	234	168	252	177	265	192	288
5.8	89	134	94	141	99	149	105	158	112	168	120	180	129	193	139	209	146	222	155	232	162	243	174	261	183	275	199	298
6.0	92	138	97	146	103	154	109	164	116	174	124	186	133	200	144	216	153	230	160	240	167	251	180	270	189	284	206	309
6.2	95	143	101	151	106	159	113	169	120	180	128	192	138	207	149	223	158	237	165	248	173	260	186	279	196	294	213	319
6.4	98	148	104	156	110	165	116	175	124	186	132	199	142	213	154	230	163	245	171	256	179	268	192	288	202	303	219	329
6.6	102	152	107	161	113	170	120	180	128	192	137	205	147	220	158	238	169	253	176	264	184	276	196	297	208	313	226	339
6.8	105	157	110	165	117	175	124	185	132	197	141	211	151	227	163	245	174	260	181	272	190	285	204	306	215	322	233	350
7.0	108	162	114	170	120	180	127	191	135	203	145	217	156	233	168	252	179	266	187	280	195	293	210	315	221	332	240	360
7.2	111	166	117	175	123	185	131	196	139	209	149	223	160	240	173	259	184	276	192	288	201	301	216	324	227	341	247	370
7.4	114	171	120	180	127	190	135	202	143	215	153	230	164	247	178	266	189	283	197	296	207	310	222	333	234	351	254	381
7.6	117	175	123	185	130	195	138	207	147	221	157	236	169	253	182	274	194	291	203	304	212	318	228	342	240	350	261	391
7.8	120	180	126	190	134	201	142	213	151	226	161	242	173	260	187	281	199	299	208	312	218	327	234	351	246	369	267	401
8.0	123	185	130	195	137	206	145	218	155	232	166	248	178	267	192	288	204	306	213	320	223	335	240	360	253	379	274	411
8.2	126	189	133	199	141	211	149	224	159	238	170	254	182	273	197	295	209	314	219	328	229	343	246	369	259	388	281	422
8.4	129	194	136	204	144	216	153	229	163	244	174	261	187	280	202	302	214	322	224	336	234	352	252	378	265	398	288	432
8.5	132	198	139	209	147	221	156	235	166	250	178	267	191	287	206	310	220	329	229	344	240	360	258	387	272	407	295	442
8.8	135	203	143	214	151	226	160	240	170	255	182	273	196	293	211	317	225	337	235	352	246	368	264	396	278	417	302	453
9.0	138	208	146	219	154	231	164	245	174	261	186	279	200	300	216	324	230	345	240	360	251	377	270	405	284	426	309	463
9.2	142	212	149	224	158	237	167	251	178	267	190	286	204	307	221	331	235	352	245	368	257	385	276	414	291	436	315	473
9.4	145	217	152	229	161	242	171	256	182	273	194	292	209	313	226	338	240	360	251	376	262	393	282	423	297	445	322	483
9.6	148	222	156	234	165	247	175	262	186	279	199	298	213	320	230	346	245	368	256	384	268	402	288	432	303	455	329	494
9.8	151	226	159	238	168	252	178	267	190	285	203	304	218	327	235	353	250	375	261	392	273	410	294	441	309	464	336	504
10.0	154	231	162	243	171	257	182	273	194	290	207	310	222	333	240	360	255	383	267	400	279	419	300	450	316	474	343	514
10.2	157	235	165	248	175	262	185	278	197	296	211	317	227	340	245	367	260	391	272	408	285	427	306	459	322	483	350	525
10.4	160	240	169	253	178	267	189	284	201	302	215	323	231	347	250	374	266	398	277	415	290	435	312	468	328	493	357	535
10.6	163	245	172	258	182	273	193	289	205	308	219	329	236	353	254	382	271	406	283	424	296	444	318	477	335	502	363	545
10.8	166	249	175	263	185	278	196	295	209	314	223	335	240	360	259	389	276	414	288	432	301	452	324	486	341	512	370	555
11.0	169	254	178	268	189	283	200	300	213	319	228	341	244	367	264	396	281	421	293	440	307	450	330	495	347	521	377	566
11.2	172	258	182	272	192	288	204	305	217	325	232	348	249	373	269	403	286	429	299	448	313	469	336	504	354	531	384	576
11.4	175	263	185	277	195	293	207	311	221	331	236	354	253	380	274	410	291	437	304	456	318	477	342	513	360	540	391	586
11.6	178	268	188	282	199	298	211	316	225	337	240	360	258	387	278	418	296	444	309	464	324	486	348	522	366	549	398	597
11.8	182	272	191	287	202	303	215	322	228	343	244	366	262	393	283	425	301	452	315	472	329	494	354	531	373	559	405	607
12.0	185	277	195	292	208	309	218	327	232	348	248	372	267	400	288	432	306	460	320	480	335	502	360	540	379	569	411	617

*One lane rotated is typical for a two-lane highway, two lanes rotated for a four-lane highway, and so on. (See Fig. 4.29.)

From *A Policy on Geometric Design of Highways and Streets*, 2011, by the American Association of State Highway and Transportation Officials, Washington, D.C. Table 3–17a and Table 3–17b. Used by permission.

APPENDIX 4.F
Acceleration Distances for Passenger Cars, Level Conditions

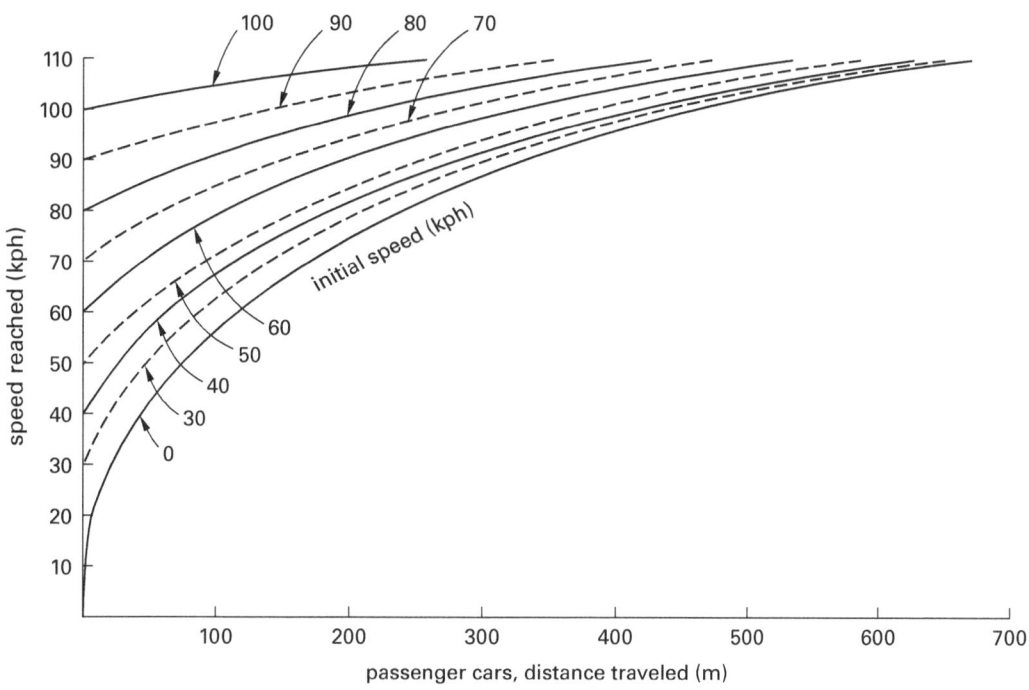

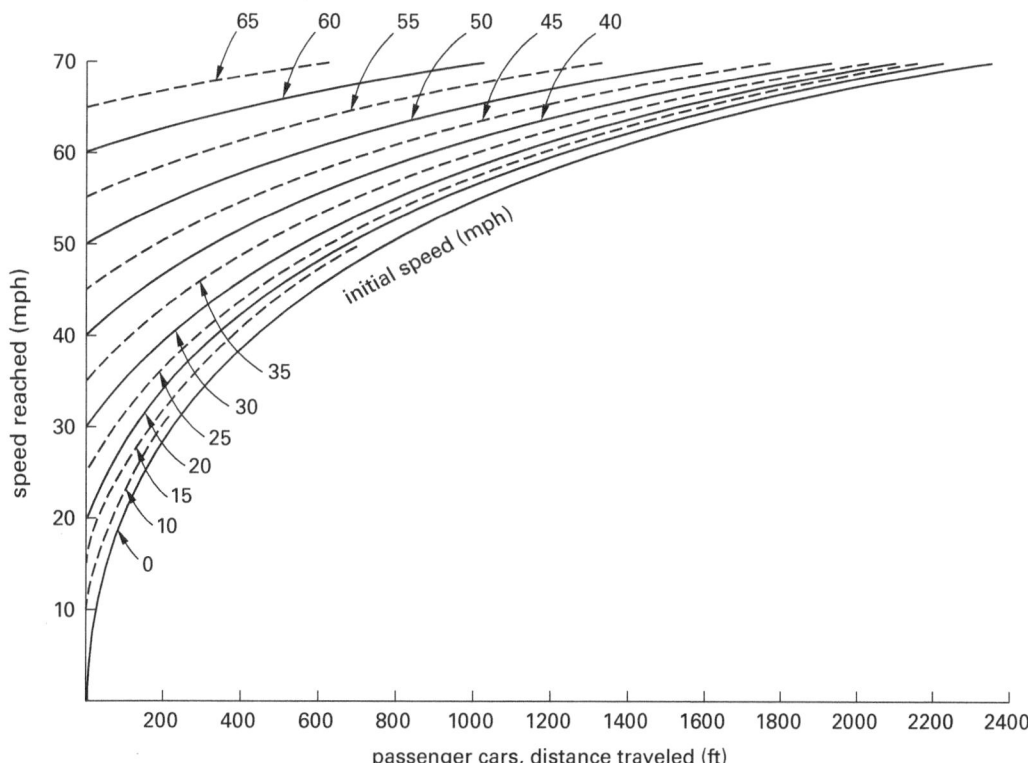

From *A Policy on Geometric Design of Highways and Streets*, 2011, by the American Association of State Highway and Transportation Officials, Washington, D.C. Figure 2-24. Used by permission.

APPENDIX 4.G
Deceleration Distances for Passenger Vehicles Approaching Intersections

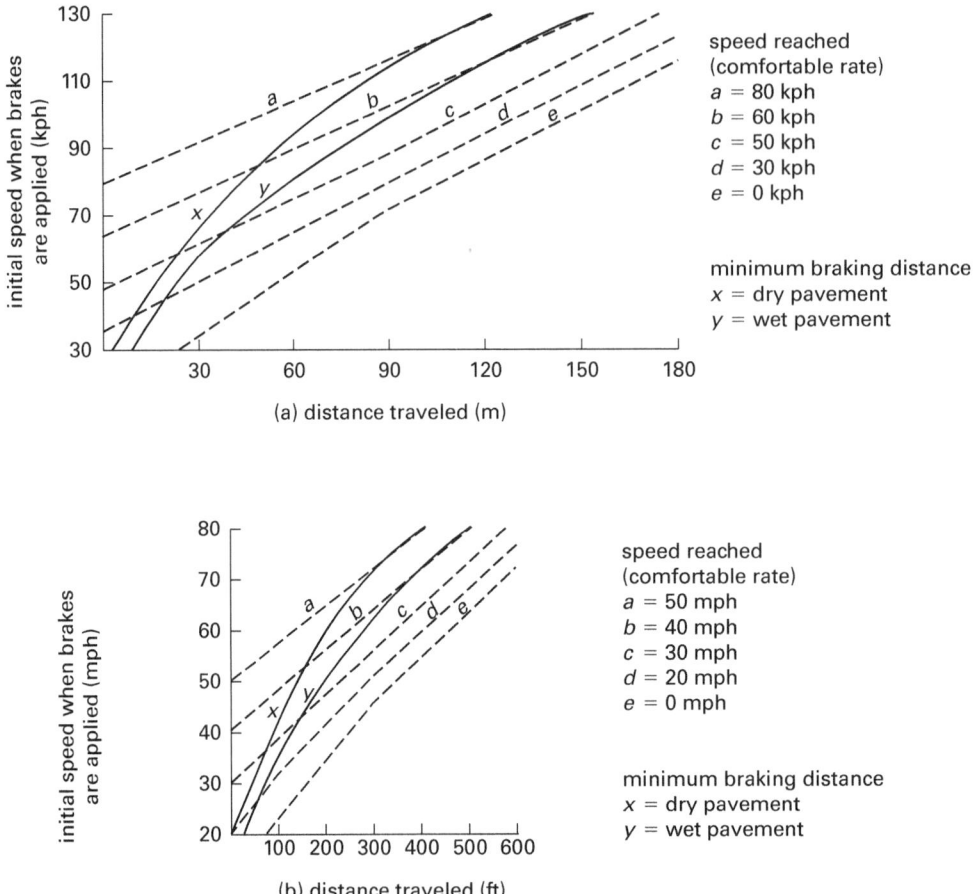

From *A Policy on Geometric Design of Highways and Streets*, 2011, by the American Association of State Highway and Transportation Officials, Washington, D.C. Figure 2-25. Used by permission.

APPENDIX 4.H
AASHTO Design Vehicle Dimensions

SI units[a]

		dimensions (m)										
		overall			overhang							typical kingpin to center of rear tandem axle
design vehicle type	symbol	height	width	length	front	rear	WB_1	WB_2	S	T	WB_3	WB_4	
passenger car	P	1.30	2.13	5.79	0.91	1.52	3.35	–	–	–	–	–	–
single-unit truck	SU-9	3.35–4.11	2.44	9.14	1.22	1.83	6.10	–	–	–	–	–	–
single-unit truck (three axle)	SU-12	3.35–4.11	2.44	12.04	1.22	3.20	7.62	–	–	–	–	–	–
buses													
intercity bus (motor coaches)	BUS-12	3.66	2.59	12.36	1.93	2.73[d]	7.70	–	–	–	–	–	–
	BUS-14	3.66	2.59	13.86	1.89	2.73[d]	8.69	–	–	–	–	–	–
city transit bus	CITY-BUS	3.20	2.59	12.19	2.13	2.44	7.62	–	–	–	–	–	–
conventional school bus (65 pass.)	S-BUS 11	3.20	2.44	10.91	0.79	3.66	6.49	–	–	–	–	–	–
large school bus (84 pass.)	S-BUS 12	3.20	2.44	12.19	2.13	3.96	6.10	–	–	–	–	–	–
articulated bus	A-BUS	3.35	2.59	18.29	2.62	3.05	6.71	5.91	1.89[e]	4.02[e]	–	–	–
combination trucks													
intermediate semitrailer	WB-12	4.11	2.44	13.87	0.91	1.37[d]	3.81	7.77	–	–	–	–	7.77
interstate semitrailer	WB-19[b]	4.11	2.59	21.03	1.22	1.37[d]	5.94	12.50	–	–	–	–	12.50
interstate semitrailer	WB-20[c]	4.11	2.59	22.40	1.22	1.37[d]	5.94	13.87	–	–	–	–	13.87
"double-bottom" semitrailer/trailer	WB-20D	4.11	2.59	22.04	0.71	0.91	3.35	7.01	0.91[f]	2.13[f]	6.86	–	7.01
Rocky Mountain semitrailer/trailer	WB-28D	4.11	2.59	29.67	0.71	0.91	5.33	12.19	1.37	2.13	6.86	–	12.34
triple semitrailer/trailer	WB-30T	4.11	2.59	31.94	0.71	0.91	3.35	6.86	0.91[g]	2.13[g]	6.86	6.86	7.01
turnpike double semitrailer/trailer	WB-33D[b]	4.11	2.59	34.75	0.71	1.37[a]	3.72	12.19	1.37[h]	3.05[h]	12.19	–	12.34
recreational vehicles													
motor home	MH	3.66	2.44	9.14	1.22	1.83	6.10	–	–	–	–	–	–
car and camper trailer	P/T	3.05	2.44	14.84	0.91	3.66	3.35	–	1.52	5.39	–	–	–
car and boat trailer	P/B	–	2.44	12.80	0.91	2.44	3.35	–	1.52	4.57	–	–	–
motor home and boat trailer	MH/B	3.66	2.44	16.15	1.22	2.44	6.10	–	1.83	4.57	–	–	–

[a]Since vehicles are manufactured in customary U.S. dimensions to provide only one physical size for each design vehicle, the values shown in the design drawings have been soft converted from numbers listed in feet, and then the numbers in this table have been rounded to the nearest hundreth of a meter.
[b]Design vehicle with 14.63 m trailer as adopted in the 1982 Surface Transportation Assistance Act (STAA).
[c]Design vehicle with 16.15 m trailer as grandfathered in with the 1982 Surface Transportation Assistance Act (STAA).
[d]This is the length of the overhang from the back axle of the tandem axle assembly.
[e]Combined dimension is 5.91 m and articulating section is 1.22 m wide.
[f]Combined dimension is typically 3.05 m.
[g]Combined dimension is typically 3.05 m.
[h]Combined dimension is typically 3.81 m.
[i]Dimensions are for a 150–200 hp tractor excluding any wagon length.
[j]To obtain the total length of tractor and one wagon, add 5.64 m to the tractor length. Wagon length is measured from front of drawbar to rear of wagon, and drawbar is 1.98 m long.

- WB_1, WB_2, and WB_4 are the effective vehicle wheelbases, or distances between axle groups, starting at the front and working toward the back of each unit.
- S is the distance from the rear effective axle to the hitch point or point of articulation.
- T is the distance from the hitch point of articulation measured back to the center of the next axle or center of the tandem axle assembly.

APPENDIX 4.H (continued)
AASHTO Design Vehicle Dimensions

customary U.S. units

design vehicle type	symbol	overall height	overall width	overall length	overhang front	overhang rear	WB_1	WB_2	S	T	WB_3	WB_4	typical kingpin to center of rear tandem axle
passenger car	P	4.3	7.0	19.0	3.0	5.0	11.0	–	–	–	–	–	–
single-unit truck	SU-30	11.0–13.5	8.0	30.0	4.0	6.0	20.0	–	–	–	–	–	–
single-unit truck (three axle)	SU-40	11.0–13.5	8.0	39.5	4.0	10.5	25.0	–	–	–	–	–	–
buses													
intercity bus (motor coaches)	BUS-40	12.0	8.5	40.5	6.3	9.0[c]	25.3	–	–	–	–	–	–
	BUS-45	12.0	8.5	45.5	6.2	9.0[c]	28.5	–	–	–	–	–	–
city transit bus	CITY-BUS	10.5	8.5	40.0	7.0	8.0	25.0	–	–	–	–	–	–
conventional school bus (65 pass.)	S-BUS 36	10.5	8.0	35.8	2.5	12.0	21.3	–	–	–	–	–	–
large school bus (84 pass.)	S-BUS 40	10.5	8.0	40.0	7.0	13.0	20.0	–	–	–	–	–	–
articulated bus	A-BUS	11.0	8.5	60.0	8.6	10.0	22.0	19.4	6.2[d]	13.2[d]	–	–	–
combination trucks													
intermediate semitrailer	WB-40	13.5	8.0	45.5	3.0	4.5[c]	12.5	25.5	–	–	–	–	25.5
interstate semitrailer	WB-62[a]	13.5	8.5	69.0	4.0	4.5[c]	19.5	41.0	–	–	–	–	41.0
interstate semitrailer	WB-67[b] or WB-67	13.5	8.5	73.5	4.0	4.5[c]	19.5	45.5	–	–	–	–	45.5
"double-bottom" semitrailer/trailer	WB-67D	13.5	8.5	72.3	2.3	3.0	11.0	23.0	3.0[e]	7.0[e]	22.5	–	23.0
Rocky Mountain semitrailer/trailer	WB-92D	13.5	8.5	97.3	2.3	3.0	17.5	40.0	4.5	7.0	22.5	–	40.5
triple semitrailer/trailer	WB-100T	13.5	8.5	104.8	2.3	3.0	11.0	22.5	3.0[f]	7.0[f]	22.5	22.5	23.0
turnpike double semitrailer/trailer	WB-109D[a]	13.5	8.5	114.0	2.3	4.5[g]	12.2	40.0	4.5[g]	10.0[g]	40.0	–	40.5
recreational vehicles													
motor home	MH	12	8.0	30.0	4.0	6.0	20.0	–	–	–	–	–	–
car and camper trailer	P/T	10	8.0	48.7	3.0	12.0	11.0	–	5.0	17.7	–	–	–
car and boat trailer	P/B	–	8.0	42.0	3.0	8.0	11.0	–	5.0	15.0	–	–	–
motor home and boat trailer	MH/B	12	8.0	53.0	4.0	8.0	20.0	–	6.0	15.0	–	–	–

[a] Design vehicle with 48.0 ft trailer as adopted in the 1982 Surface Transportation Assistance Act (STAA).

[b] Design vehicle with 53.0 ft trailer as grandfathered in with the 1982 Surface Transportation Assistance Act (STAA).

[c] This is the length of the overhang from the back axle of the tandem axle assembly.

[d] Combined dimension is 19.4 ft and articulating section is 4 ft wide.

[e] Combined dimension is typically 10.0 ft.

[f] Combined dimension is typically 10.0 ft.

[g] Combined dimension is typically 12.5 ft.

[h] Dimensions are for a 150–200 hp tractor excluding any wagon length.

[i] To obtain the total length of tractor and one wagon, add 18.5 ft to the tractor length. Wagon length is measured from front of drawbar to rear of wagon, and drawbar is 6.5 ft long.

- WB_1, WB_2, and WB_4 are the effective vehicle wheelbases, or distances between axle groups, starting at the front and working toward the back of each unit.
- S is the distance from the rear effective axle to the hitch point or point of articulation.
- T is the distance from the hitch point of articulation measured back to the center of the next axle or center of the tandem axle assembly.

From *A Policy on Geometric Design of Highways and Streets*, 2011, by the American Association of State Highway and Transportation Officials, Washington, D.C. Table 2–1b. Used by permission.

APPENDIX 4.I
Minimum Turning Radii of Design Vehicles

SI units[a]

design vehicle type	passenger car	single-unit truck	single-unit truck (three axle)	intercity bus (motor coach)		city transit bus	conventional school bus (65 pass.)	large[e] school bus (84 pass.)	articulated bus	intermediate semitrailer
symbol	P	SU	SU-12	BUS-12	BUS-14	CITY-BUS	S-BUS11	S-BUS12	A-BUS	WB-12
minimum design turning radius (m)	7.26	12.73	15.60	12.70	13.40	12.80	11.75	11.92	12.0	12.16
centerline[d] turning radius, CTR (m)	6.40	11.58	14.46	11.53	12.25	11.52	10.64	10.79	10.82	10.97
minimum inside radius (m)	4.39	8.64	11.09	7.41	7.54	7.54	7.25	7.71	6.49	5.88

design vehicle type	interstate semitrailer	"double-bottom" combination	Rocky Mountain double	triple semitrailer/trailer	turnpike double semitrailer/trailer	motor home	car and camper trailer	car and boat trailer	motor home and boat trailer
symbol	WB-19[b]	WB-20[c]	WB-20D	WB-28D	WB-30T	WB-33D[b]	MH	P/T	P/B	MH/B
minimum design turning radius (m)	13.66	13.66	13.66	24.98	13.66	18.25	12.11	10.03	7.26	15.19
centerline[d] turning radius, CTR (m)	12.50	12.50	12.47	23.77	12.47	17.04	10.97	9.14	6.40	14.02
minimum inside radius (m)	2.25	0.59	5.83	16.94	2.96	4.19	7.92	5.58	2.44	10.67

[a]Numbers in table have been rounded to the nearest hundreth of a meter.

[b]Design vehicle with 14.63 m trailer as adopted in the 1982 Surface Transportation Assistance Act (STAA).

[c]Design vehicle with 16.15 m trailer as grandfathered in with the 1982 Surface Transportation Assistance Act (STAA).

[d]The turning radius assumed by a designer when investigating possible turning paths and is set at the centerline of the front axle of the vehicle. If the minimum turning path is assumed, the CTR approximately equals the minimum design turning radius minus one-half the front width of the vehicle.

[e]School buses are manufactured from 42 passenger to 84 passenger sizes. This corresponds to wheelbase lengths of 3.35 m to 6.1 m, respectively. For these different sizes, the minimum design turning radii vary from 8.78 m to 12.01 m, and the minimum inside radii vary from 4.27 m to 7.74 m.

APPENDIX 4.I (continued)
Minimum Turning Radii of Design Vehicles

customary U.S. units

design vehicle type	passenger car	single-unit truck	single-unit truck (three axle)	intercity bus (motor coach)		city transit bus	conventional school bus (65 pass.)	large[d] school bus (84 pass.)	articulated bus	intermediate semitrailer
symbol	P	SU-30	SU-40	BUS-40	BUS-45	CITY-BUS	S-BUS36	S-BUS40	A-BUS	WB-40
minimum design turning radius (ft)	23.8	41.8	51.2	41.7	44.0	41.6	38.6	39.1	39.4	39.9
centerline[c] turning radius, CTR (ft)	21.0	38.0	47.4	37.8	40.2	37.8	34.9	35.4	35.5	36.0
minimum inside radius (ft)	14.4	28.4	36.4	24.3	24.7	24.5	23.8	25.3	21.3	19.3

design vehicle type	interstate semitrailer		"double-bottom" combination	Rocky Mountain double	triple semi-trailer/ trailer	turnpike double semi-trailer/ trailer	motor home	car and camper trailer	car and boat trailer	motor home and boat trailer
symbol	WB-62[a]	WB-67[b]	WB-67D	WB-92D	WB-100T	WB-109D[a]	MH	P/T	P/B	MH/B
minimum design turning radius (ft)	44.8	44.8	44.8	82.0	44.8	59.9	39.7	32.9	23.8	49.8
centerline[c] turning radius, CTR (ft)	41.0	41.0	40.9	78.0	40.9	55.9	36.0	30.0	21.0	46.0
minimum inside radius (ft)	7.4	1.9	19.1	55.6	9.7	13.8	26.0	18.3	8.0	35.0

[a]Design vehicle with 48 ft trailer as adopted in the 1982 Surface Transportation Assistance Act (STAA).

[b]Design vehicle with 53 ft trailer as grandfathered in with the 1982 Surface Transportation Assistance Act (STAA).

[c]The turning radius assumed by a designer when investigating possible turning paths and is set at the centerline of the front axle of the vehicle. If the minimum turning path is assumed, the CTR approximately equals the minimum design turning radius minus one-half the front width of the vehicle.

[d]School buses are manufactured from 42 passenger to 84 passenger sizes. This corresponds to wheelbase lengths of 11.0 ft to 20.0 ft, respectively. For these different sizes, the minimum design turning radii vary from 28.1 ft to 39.1 ft, and the minimum inside radii vary from 17.7 ft to 25.3 ft.

From *A Policy on Geometric Design of Highways and Streets*, 2011, by the American Association of State Highway and Transportation Officials, Washington, D.C. Table 2-2a and Table 2-2b. Used by permission.

APPENDICES **A-37**

APPENDIX 4.J
Minimum Turning Path for Passenger Car (P) Design Vehicle[a,b]

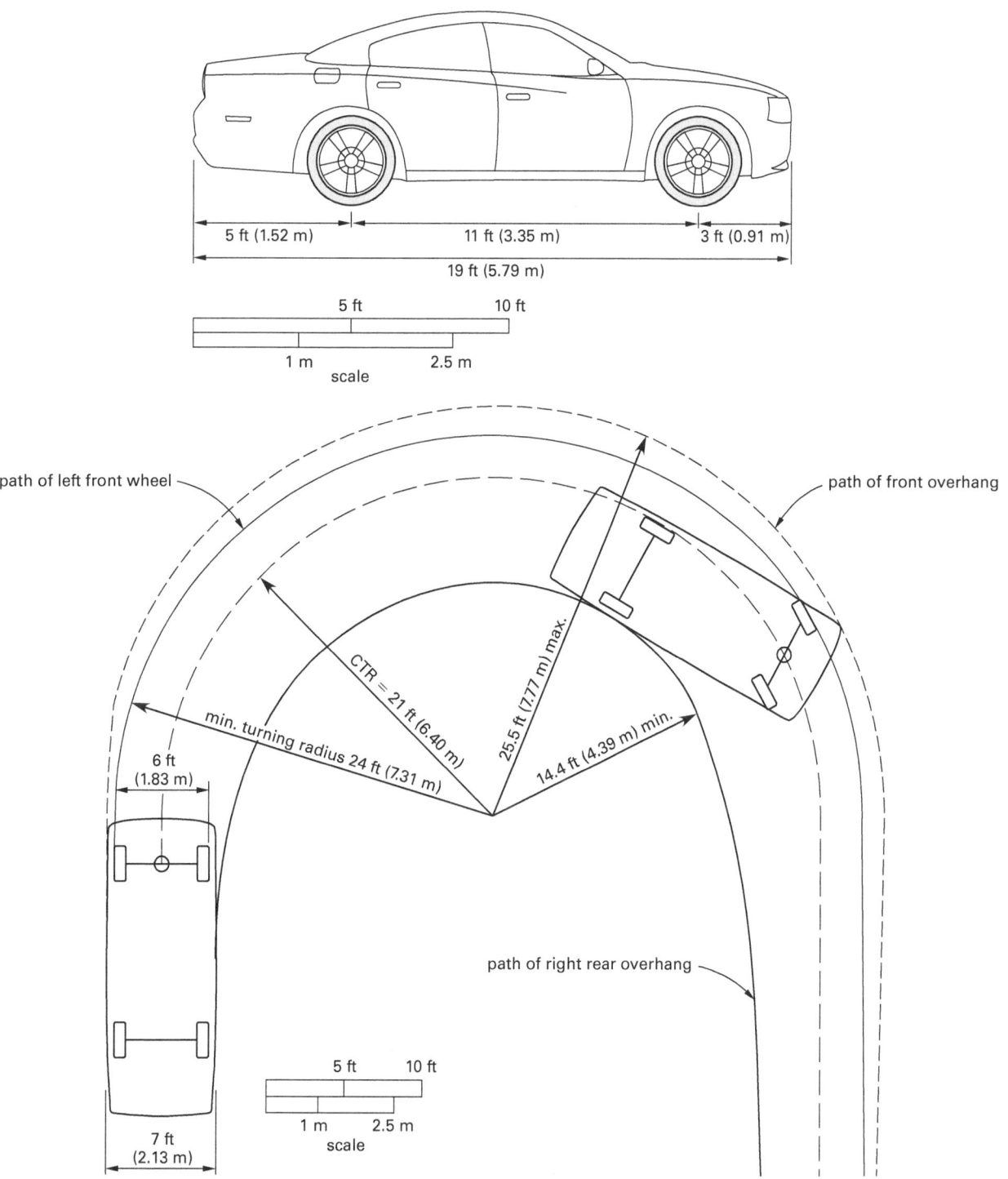

[a]assumed steering angle is 31.6°
[b]CTR = centerline turning radius at front axle

From *A Policy on Geometric Design of Highways and Streets*, 2011, by the American Association of State Highway and Transportation Officials, Washington, D.C. Figure 2-1. Used by permission.

APPENDIX 4.K
Turning Characteristics of a Typical Tractor/Semitrailer Combination Truck

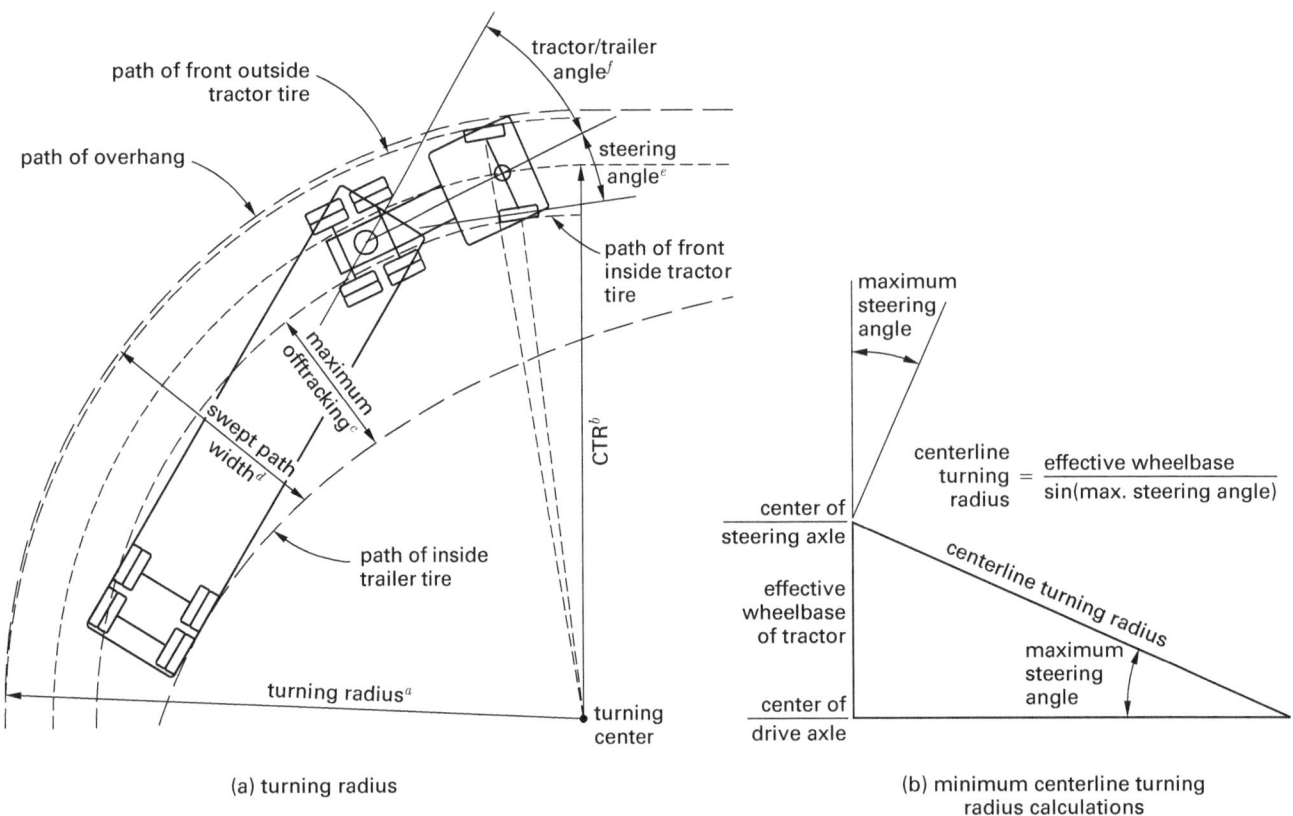

(a) turning radius

(b) minimum centerline turning radius calculations

[a]*Turning radius:* The circular arc formed by the turning path radius of the front outside tire of a vehicle. This radius is also described by vehicle manufacturers as the *turning curb radius*.

[b]*CTR:* The turning radius of the centerline of the front axle of a vehicle.

[c]*Offtracking:* The difference in the paths of the front and rear wheels of a tractor/semitrailer as it negotiates a turn. The path of the rear tires of the turning truck does not coincide with that of the front tires, and this effect is shown in the drawing above.

[d]*Swept path width:* The amount of roadway width that a truck covers in negotiating a turn. The swept path width is equal to the amount of offtracking plus the width of the tractor unit. The most significant dimension affecting the swept path width of a tractor/semitrailer is the distance from the kingpin to the rear trailer axle or axles. The greater this distance is, the greater the swept path width.

[e]*Steering angle:* The maximum angle of turn built into the steering mechanism of the front wheels of a vehicle. This maximum angle controls the minimum turning radius of the vehicle.

[f]*Tractor/trailer angle:* The angle between adjoining units of a tractor/semitrailer when the combination unit is placed into a turn. This angle is measured between the longitudinal axes of the tractor and trailer as the vehicle turns. The maximum tractor/trailer angle occurs when a vehicle makes a 180° turn at the minimum turning radius. This angle is reached slightly beyond the point where the maximum swept path width is achieved.

From *A Policy on Geometric Design of Highways and Streets*, 2011, by the American Association of State Highway and Transportation Officials, Washington, D.C. Figure 2-10 and Fig. 2-11. Used by permission.

APPENDIX 4.L
Minimum Turning Path for Single-Unit Truck (SU-12 [SU-40]) Design Vehicle[a,b]

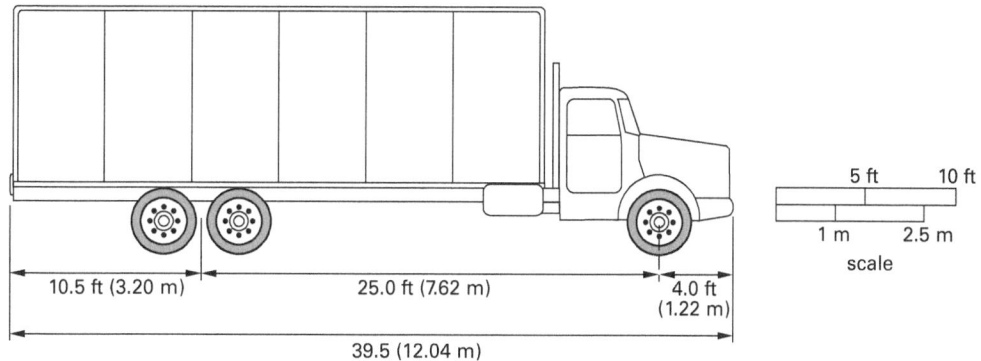

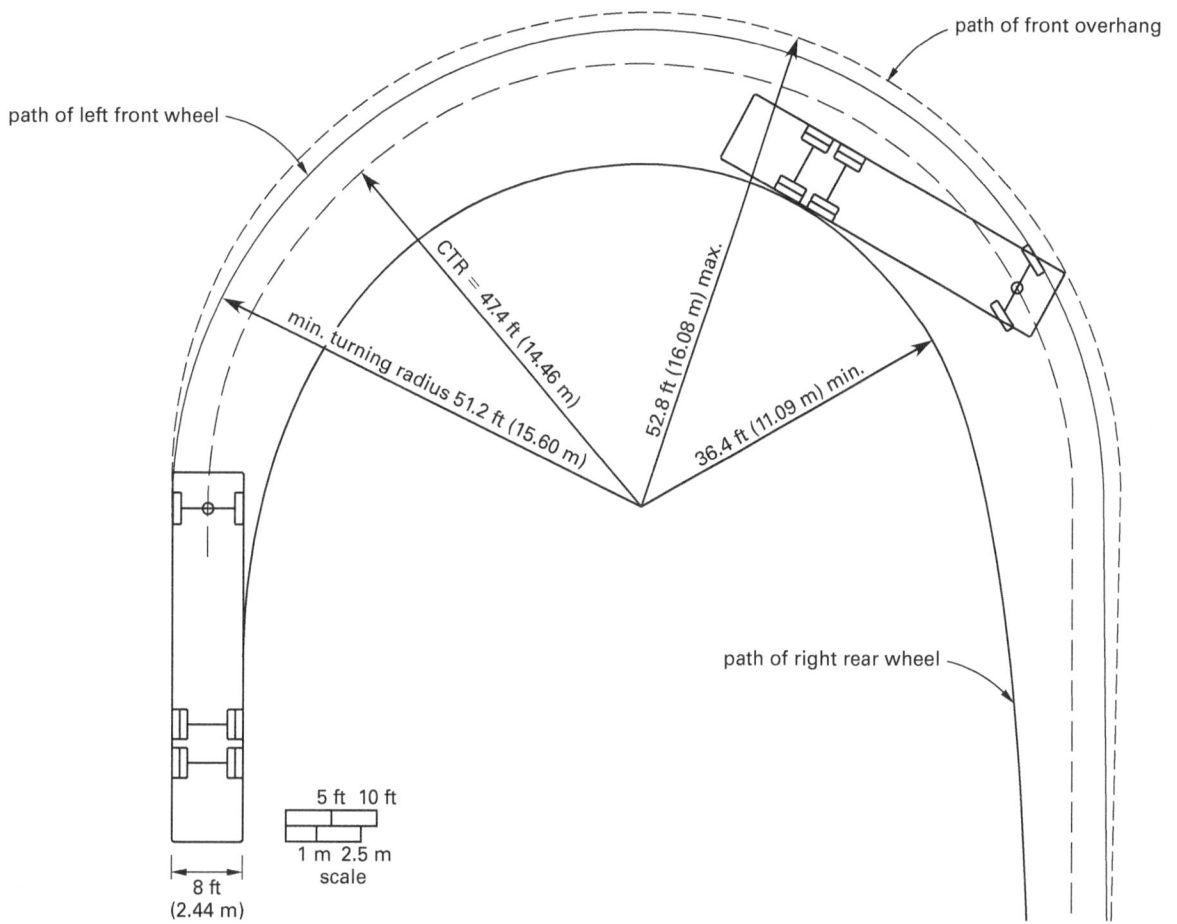

[a] maximum steering angle is 31.8°
[b] CTR = centerline turning radius at front axle

From *A Policy on Geometric Design of Highways and Streets*, 2011, by the American Association of State Highway and Transportation Officials, Washington, D.C. Figure 2-3. Used by permission.

APPENDIX 4.M
Minimum Turning Path for Intermediate Semitrailer WB-40 (WB-12) Design Vehicle[a,b,c]

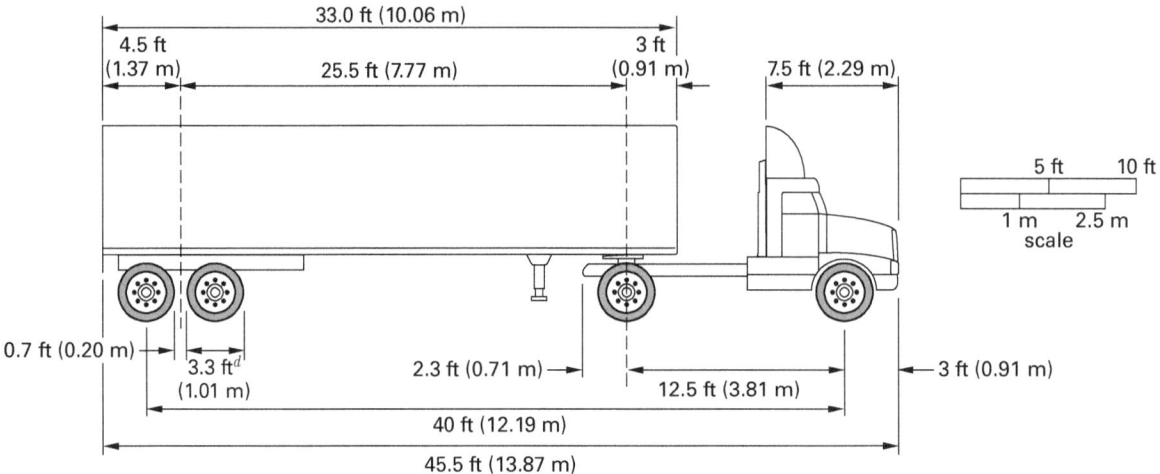

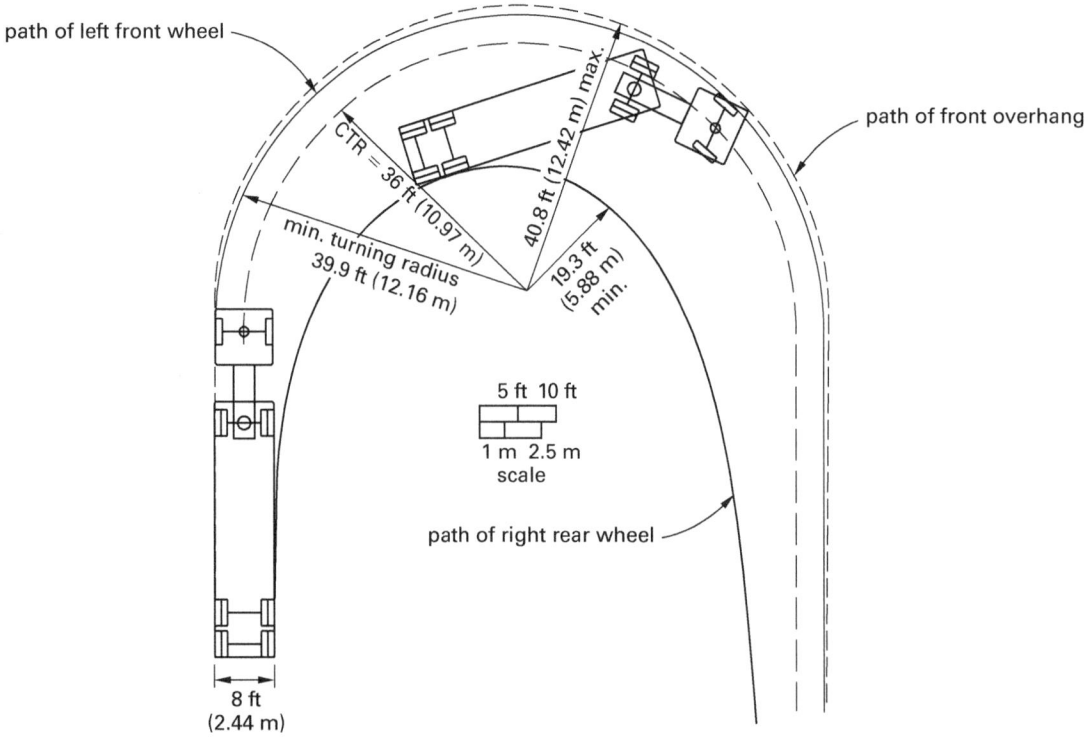

[a]Maximum steering angle is 20.3°.
[b]AA1 is 46°.
[c]CTR = centerline turning radius at front axle
[d]Typical tire size and space between tires applies to all trailers.

From *A Policy on Geometric Design of Highways and Streets*, 2011, by the American Association of State Highway and Transportation Officials, Washington, D.C. Figure 2-13. Used by permission.

APPENDIX 4.N
Minimum Turning Path for Interstate Semitrailer WB-67 (WB-20) Design Vehicle[a,b,c]

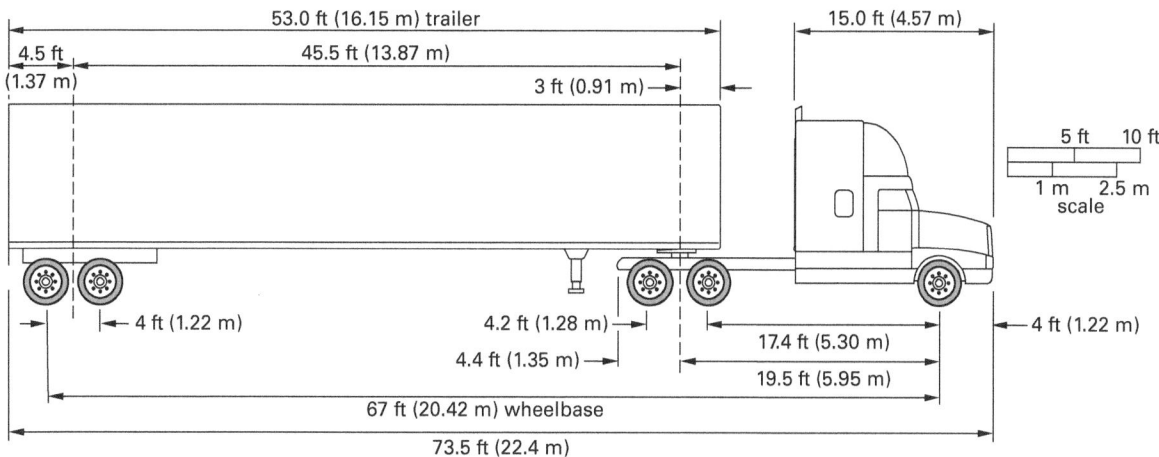

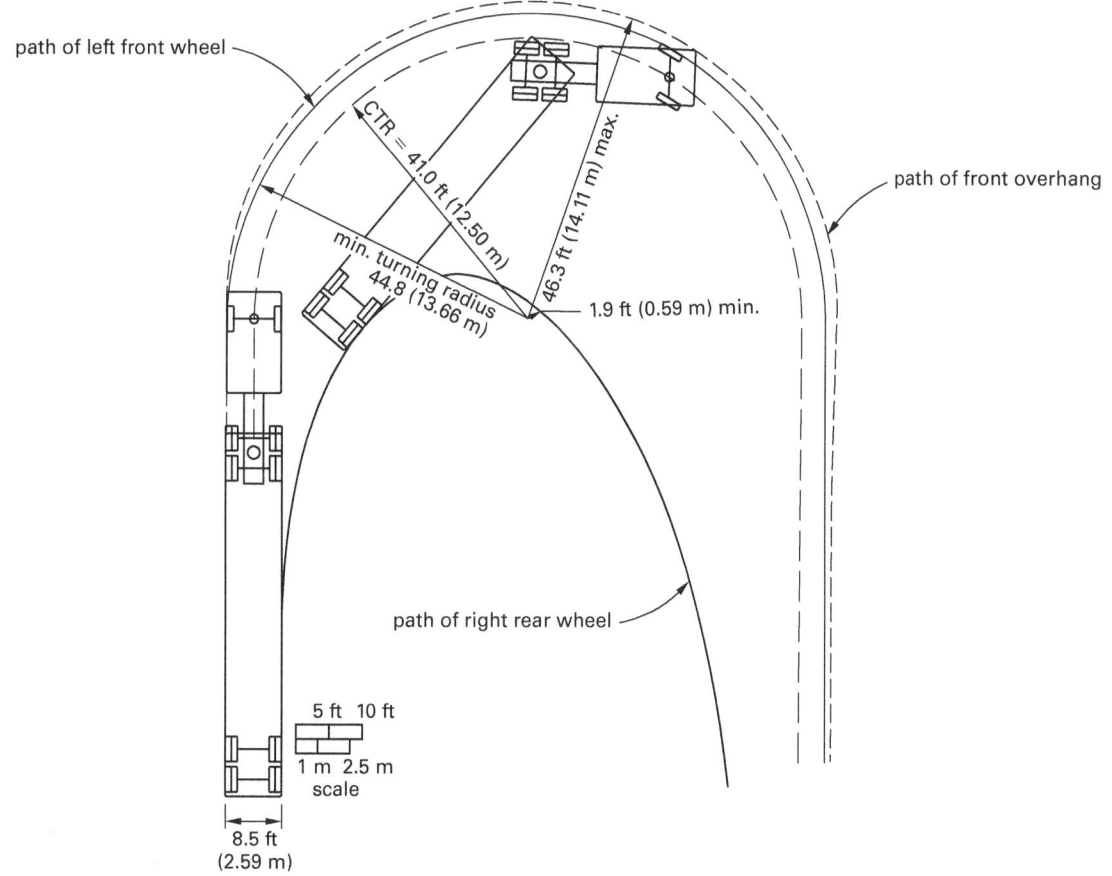

[a]Maximum steering angle is 28.4°.
[b]Assumed tractor/trailer angle is 68.5°.
[c]CTR = centerline turning radius at front axle

From *A Policy on Geometric Design of Highways and Streets*, 2011, by the American Association of State Highway and Transportation Officials, Washington, D.C. Figure 2-15. Used by permission.

APPENDIX 4.O
Minimum Turning Path for Double-Trailer Combination (WB-20D [WB-67D]) Design Vehicle

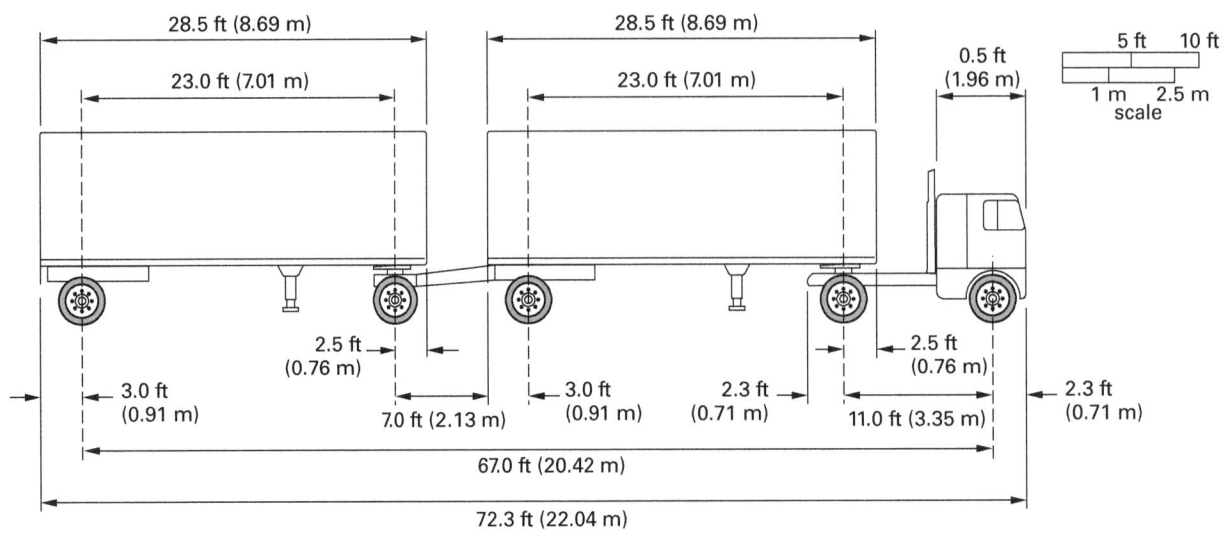

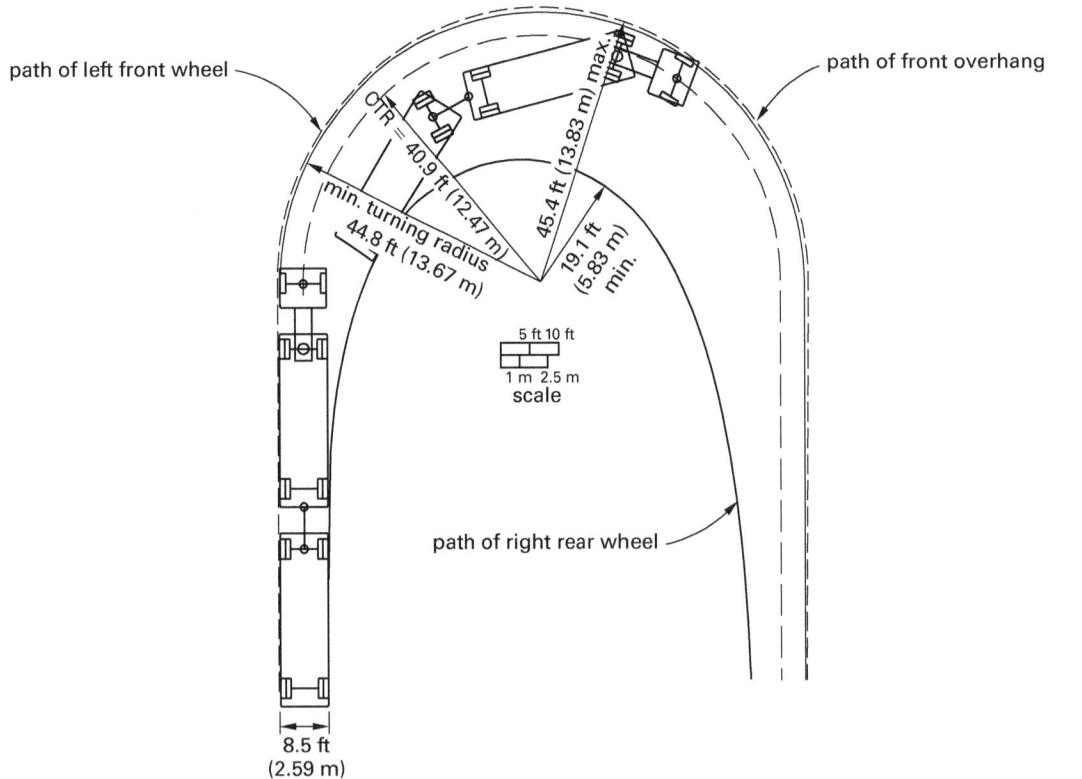

CTR = centerline turning radius at front axle

From *A Policy on Geometric Design of Highways and Streets*, 2011, by the American Association of State Highway and Transportation Officials, Washington, D.C. Figure 2-16. Used by permission.

APPENDIX 4.P
Minimum Turning Path for City Transit Bus (CITY-BUS) Design Vehicle[a,b]

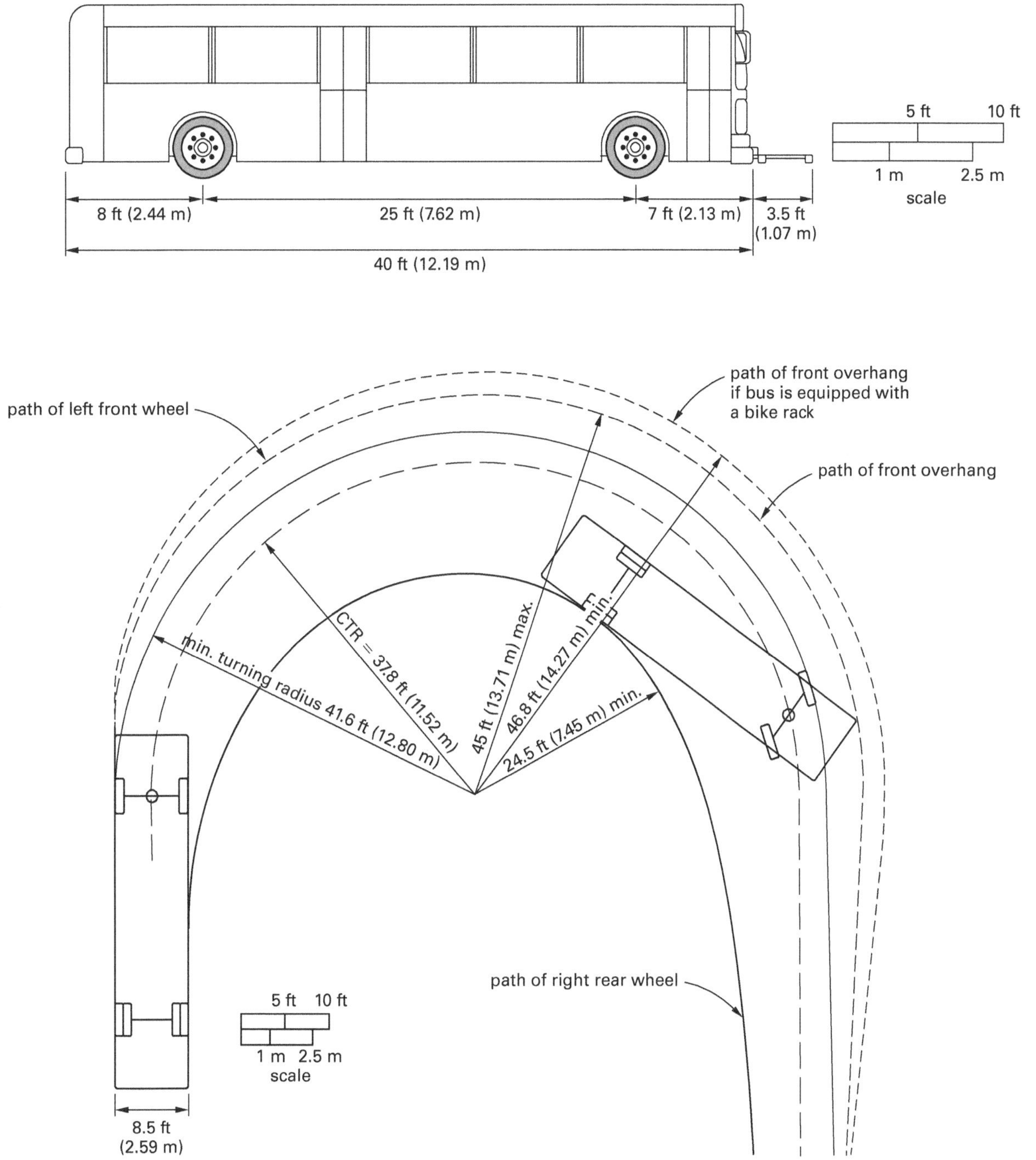

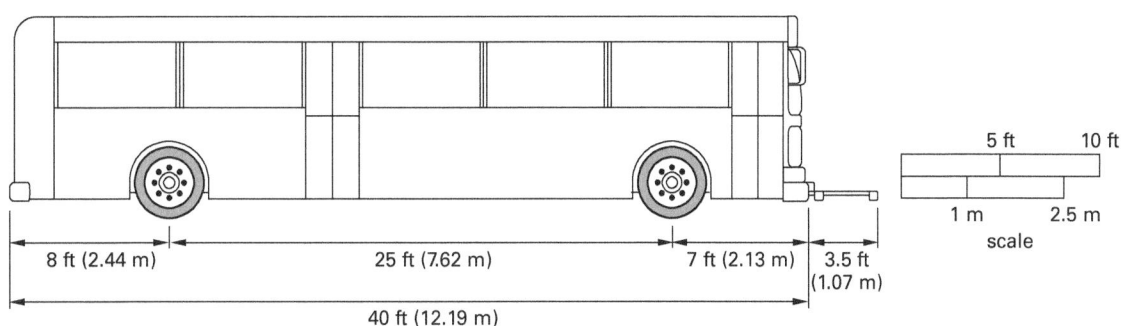

[a] assumed steering angle is 41.4°
[b] CTR = centerline turning radius at front axle

From *A Policy on Geometric Design of Highways and Streets*, 2011, by the American Association of State Highway and Transportation Officials, Washington, D. C. Figure 2-6. Used by permission.

APPENDIX 4.Q
Minimum Turning Path for Conventional School Bus (S-BUS36 [S-BUS11]) Design Vehicle[a,b,c]

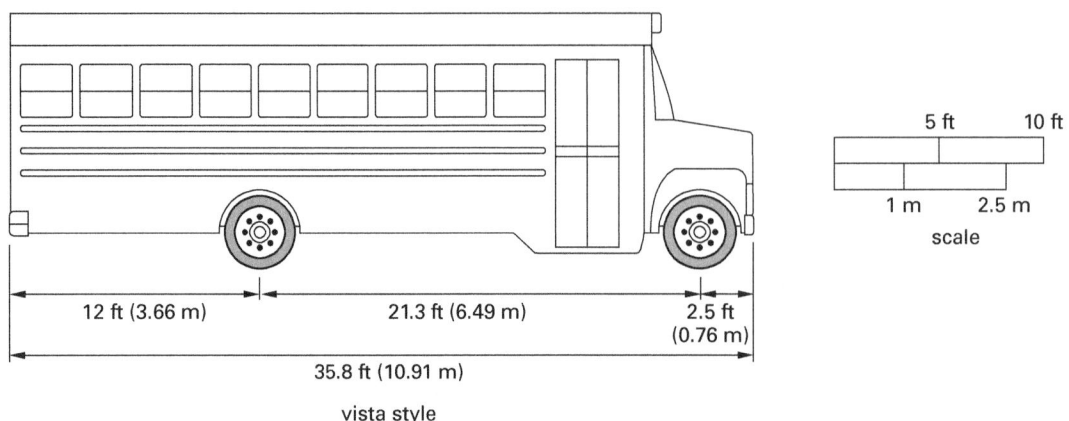

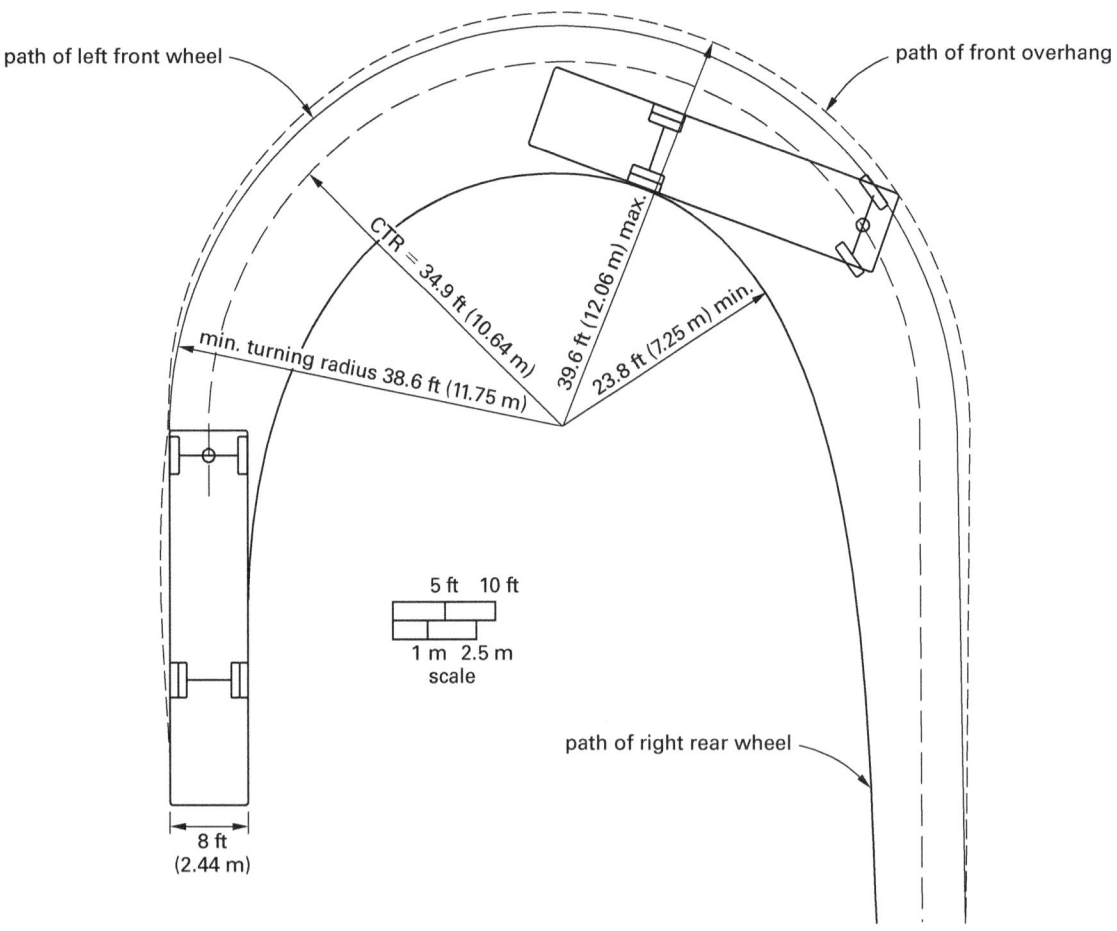

[a]Maximum steering angle is 37.6°.
[b]CTR = centerline turning radius at front axle

From *A Policy on Geometric Design of Highways and Streets*, 2011, by the American Association of State Highway and Transportation Officials, Washington, D.C. Figure 2-7. Used by permission.

APPENDIX 4.R
Minimum Traveled Way, Passenger (P) Design Vehicle Path

SI units

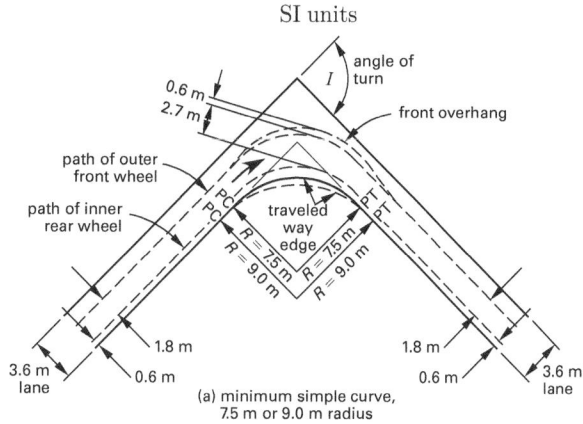

(a) minimum simple curve, 7.5 m or 9.0 m radius

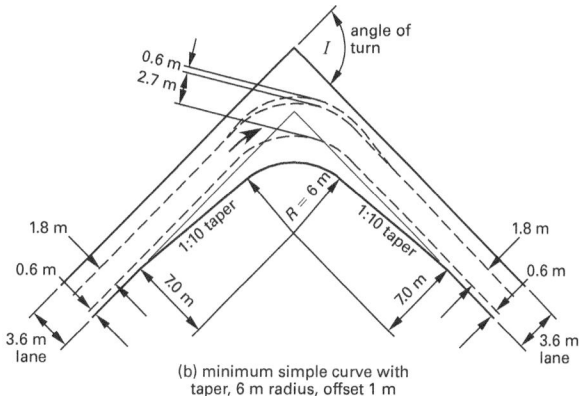

(b) minimum simple curve with taper, 6 m radius, offset 1 m

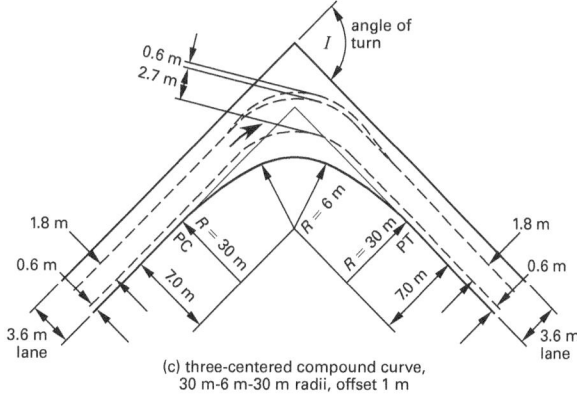

(c) three-centered compound curve, 30 m-6 m-30 m radii, offset 1 m

APPENDIX 4.R *(continued)*
Minimum Traveled Way, Passenger (P) Design Vehicle Path

customary U.S. units

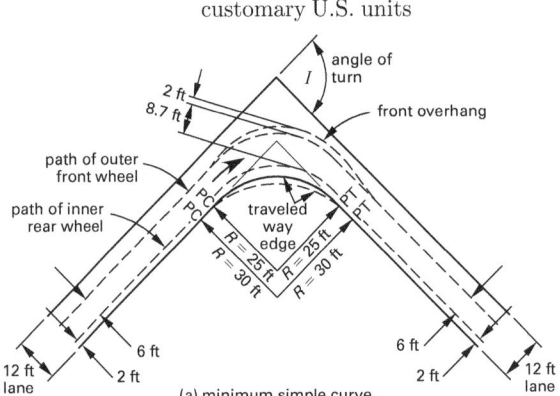

(a) minimum simple curve, 25 ft or 30 ft radius

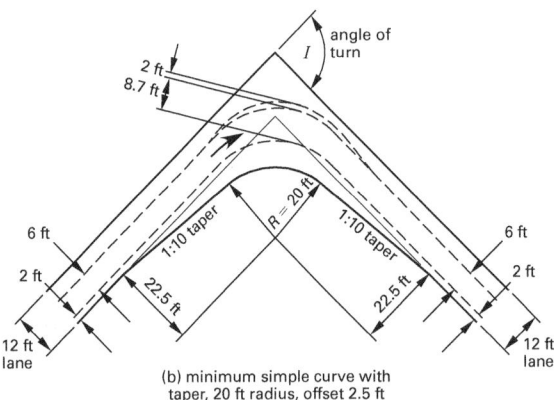

(b) minimum simple curve with taper, 20 ft radius, offset 2.5 ft

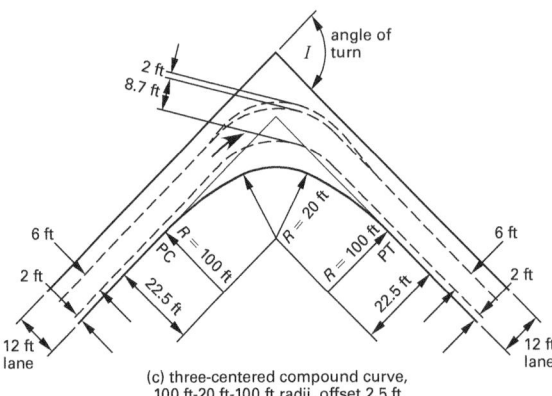

(c) three-centered compound curve, 100 ft-20 ft-100 ft radii, offset 2.5 ft

From *A Policy on Geometric Design of Highways and Streets*, 2011, by the American Association of State Highway and Transportation Officials, Washington, D.C. Figure 9-23. Used by permission.

APPENDIX 4.S
Minimum Traveled Way Designs, Single-Unit Trucks and City Transit Buses

SI units

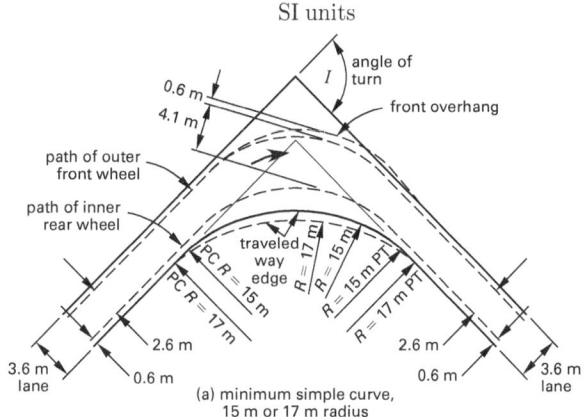

(a) minimum simple curve, 15 m or 17 m radius

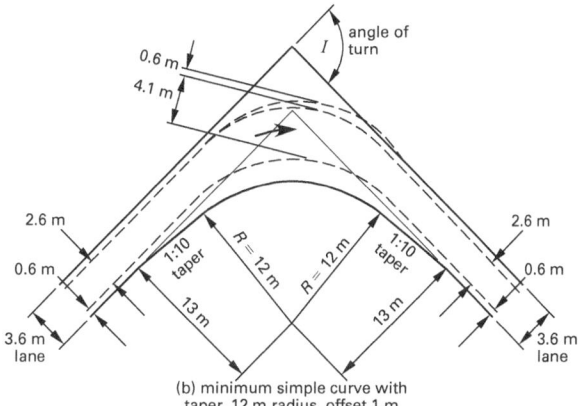

(b) minimum simple curve with taper, 12 m radius, offset 1 m

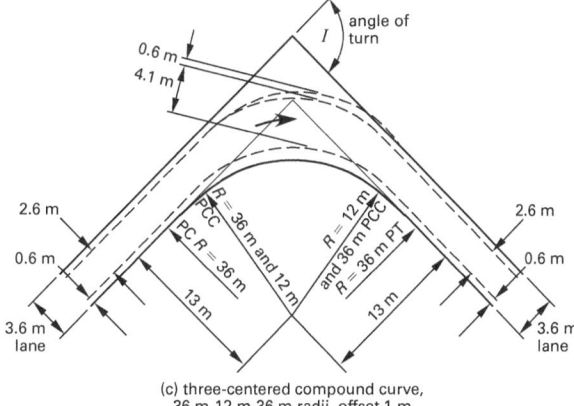

(c) three-centered compound curve, 36 m-12 m-36 m radii, offset 1 m

APPENDIX 4.S *(continued)*
Minimum Traveled Way Designs, Single-Unit Trucks and City Transit Buses

customary U.S. units

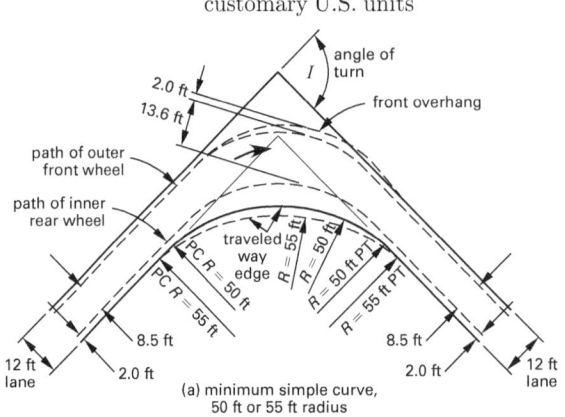

(a) minimum simple curve, 50 ft or 55 ft radius

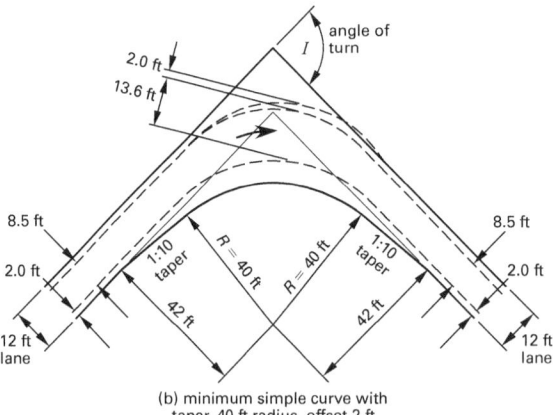

(b) minimum simple curve with taper, 40 ft radius, offset 2 ft

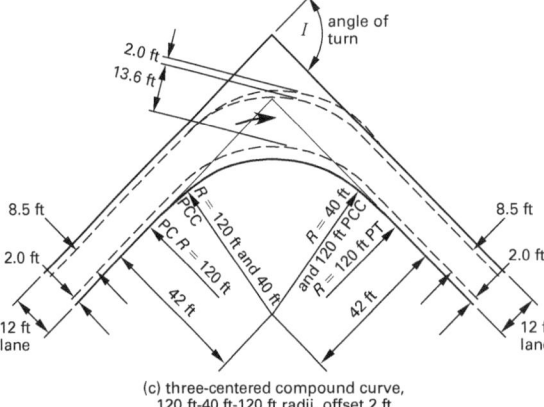

(c) three-centered compound curve, 120 ft-40 ft-120 ft radii, offset 2 ft

From *A Policy on Geometric Design of Highways and Streets*, 2011, by the American Association of State Highway and Transportation Officials, Washington, D.C. Figure 9-24. Used by permission.

APPENDIX 4.T
Edge-of-Traveled-Way for Turns at Intersections

SI units

angle of turn	design vehicle	simple curve radius (m)	simple curve radius with taper		
			radius (m)	offset (m)	taper L:T
30°	P	18	–	–	–
	SU-9	30	–	–	–
	SU-12	–	–	–	–
	WB-12	45	–	–	–
	WB-19	110	67	1.0	15:1
	WB-20	116	67	1.0	15:1
	WB-28D	111	58	1.0	15:1
	WB-30T	77	37	1.0	15:1
	WB-33D	145	77	1.1	20:1
45°	P	15	–	–	–
	SU-9	23	–	–	–
	SU-12	35	–	–	–
	WB-12	36	–	–	–
	WB-19	70	43	1.2	15:1
	WB-20	76	43	1.3	15:1
	WB-28D	82	44	1.2	15:1
	WB-30T	60	35	0.8	15:1
	WB-33D	–	60	1.3	20:1
60°	P	12	–	–	–
	SU-9	18	–	–	–
	SU-12	30	–	–	–
	WB-12	28	–	–	–
	WB-19	50	43	1.2	15:1
	WB-20	60	43	1.3	15:1
	WB-28D	70	37	1.5	15:1
	WB-30T	46	29	0.8	15:1
	WB-33D	–	54	1.3	20:1
75°	P	11	8	0.6	10:1
	SU-9	17	14	0.6	10:1
	SU-12	27	18	0.6	10:1
	WB-12	–	18	0.6	15:1
	WB-19	–	43	1.2	20:1
	WB-20	–	43	1.3	20:1
	WB-28D	–	34	1.5	15:1
	WB-30T	–	26	1.0	15:1
	WB-33D	–	42	1.7	20:1
90°	P	9	6	0.8	10:1
	SU-9	15	12	0.6	10:1
	SU-12	24	14	1.2	10:1
	WB-12	–	14	1.2	10:1
	WB-19	–	36	1.3	30:1
	WB-20	–	37	1.3	30:1
	WB-28D	–	30	1.8	10:1
	WB-30T	–	25	0.8	15:1
	WB-33D	–	35	0.9	15:1
105°	P	–	6	0.8	8:1
	SU-9	–	11	1.0	10:1
	SU-12	–	14	1.2	10:1
	WB-12	–	12	1.2	10:1
	WB-19	–	35	1.0	15:1

APPENDIX 4.T *(continued)*
Edge-of-Traveled-Way for Turns at Intersections

SI units

angle of turn	design vehicle	simple curve radius (m)	simple curve radius with taper		
			radius (m)	offset (m)	taper L:T
	WB-20	–	35	1.0	15:1
	WB-28D	–	24	2.4	10:1
	WB-30T	–	22	1.0	15:1
	WB-33D	–	28	2.8	20:1
120°	P	–	6	0.6	10:1
	SU-9	–	9	1.0	10:1
	SU-12	–	11	1.8	8:1
	WB-12	–	11	1.5	8:1
	WB-19	–	30	1.5	15:1
	WB-20	–	31	1.6	15:1
	WB-28D	–	24	2.1	10:1
	WB-30T	–	20	1.1	15:1
	WB-33D	–	26	2.8	20:1
135°	P	–	6	0.5	10:1
	SU-9	–	9	1.2	10:1
	SU-12	–	12	1.2	8:1
	WB-12	–	9	2.5	15:1
	WB-19	–	24	1.5	20:1
	WB-20	–	25	1.6	20:1
	WB-28D	–	23	2.2	10:1
	WB-30T	–	19	1.7	15:1
	WB-33D	–	25	2.6	20:1
150°	P	–	6	0.6	10:1
	SU-9	–	9	1.2	8:1
	SU-12	–	11	2.1	8:1
	WB-12	–	9	2.0	8:1
	WB-19	–	18	3.0	10:1
	WB-20	–	19	3.1	10:1
	WB-28D	–	20	3.4	10:1
	WB-30T	–	19	2.2	10:1
	WB-33D	–	20	4.6	10:1
180°	P	–	5	0.2	20:1
	SU-9	–	9	0.5	10:1
	SU-12	–	11	2.0	10:1
	WB-12	–	6	3.0	5:1
	WB-19	–	17	3.0	15:1
	WB-20	–	16	4.2	10:1
	WB-28D	–	17	5.1	10:1
	WB-30T	–	17	3.1	10:1
	WB-33D	–	17	6.1	10:1

From *A Policy on Geometric Design of Highways and Streets*, 2011, by the American Association of State Highway and Transportation Officials, Washington, D.C. Table 9-15. Used by permission.

APPENDIX 4.T *(continued)*
Edge-of-Traveled-Way for Turns at Intersections

customary U.S. units

angle of turn	design vehicle	simple curve radius (ft)	simple curve radius with taper		
			radius (ft)	offset (ft)	taper L:T
30°	P	60	–	–	–
	SU-30	100	–	–	–
	SU-40	140	–	–	–
	WB-40	150	–	–	–
	WB-62	360	220	3.0	15:1
	WB-67	380	220	3.0	15:1
	WB-92D	365	190	3.0	15:1
	WB-100T	260	125	3.0	15:1
	WB-109D	475	260	3.5	20:1
45°	P	50	–	–	–
	SU-30	75	–	–	–
	SU-40	115	–	–	–
	WB-40	120	–	–	–
	WB-62	230	145	4.0	15:1
	WB-67	250	145	4.5	15:1
	WB-92D	270	145	4.0	15:1
	WB-100T	200	115	2.5	15:1
	WB-109D	–	200	4.5	20:1
60°	P	40	–	–	–
	SU-30	60	–	–	–
	SU-40	100	–	–	–
	WB-40	90	–	–	–
	WB-62	170	140	4.0	15:1
	WB-67	200	140	4.5	15:1
	WB-92D	230	120	5.0	15:1
	WB-100T	150	95	2.5	15:1
	WB-109D	–	180	4.5	20:1
75°	P	35	25	2.0	10:1
	SU-30	55	45	2.0	10:1
	SU-40	90	60	2.0	10:1
	WB-40	–	60	2.0	15:1
	WB-62	–	145	4.0	20:1
	WB-67	–	145	4.5	20:1
	WB-92D	–	110	5.0	15:1
	WB-100T	–	85	3.0	15:1
	WB-109D	–	140	5.5	20:1
90°	P	30	20	2.5	10:1
	SU-30	50	40	2.0	10:1
	SU-40	80	45	4.0	10:1
	WB-40	–	45	4.0	10:1
	WB-62	–	120	4.5	30:1
	WB-67	–	125	4.5	30:1
	WB-92D	–	95	6.0	10:1
	WB-100T	–	85	2.5	15:1
	WB-109D	–	115	2.9	15:1
105°	P	–	20	2.5	8:1
	SU-30	–	35	3.0	10:1
	SU-40	–	45	4.0	10:1
	WB-40	–	40	4.0	10:1
	WB-62	–	115	3.0	15:1

APPENDIX 4.T *(continued)*
Edge-of-Traveled-Way for Turns at Intersections

customary U.S. units

angle of turn	design vehicle	simple curve radius (ft)	simple curve radius with taper		
			radius (ft)	offset (ft)	taper $L{:}T$
	WB-67	–	115	3.0	15:1
	WB-92D	–	80	8.0	10:1
	WB-100T	–	75	3.0	15:1
	WB-109D	–	90	9.2	20:1
120°	P	–	20	2.0	10:1
	SU-30	–	30	3.0	10:1
	SU-40	–	35	6.0	8:1
	WB-40	–	35	5.0	8:1
	WB-62	–	100	5.0	15:1
	WB-67	–	105	5.2	15:1
	WB-92D	–	80	7.0	10:1
	WB-100T	–	65	3.5	15:1
	WB-109D	–	85	9.2	20:1
135°	P	–	20	1.5	10:1
	SU-30	–	30	4.0	10:1
	SU-40	–	40	4.0	8:1
	WB-40	–	30	8.0	15:1
	WB-62	–	80	5.0	20:1
	WB-67	–	85	5.2	20:1
	WB-92D	–	75	7.3	10:1
	WB-100T	–	65	5.5	15:1
	WB-109D	–	85	8.5	20:1
150°	P	–	18	2.0	10:1
	SU-30	–	30	4.0	8:1
	SU-40	–	35	7.0	8:1
	WB-40	–	30	6.0	8:1
	WB-62	–	60	10.0	10:1
	WB-67	–	65	10.2	10:1
	WB-92D	–	65	11.0	10:1
	WB-100T	–	65	7.3	10:1
	WB-109D	–	65	15.1	10:1
180°	P	–	15	0.5	20:1
	SU-30	–	30	1.5	10:1
	SU-40	–	35	6.4	10:1
	WB-40	–	20	9.5	5:1
	WB-62	–	55	10.0	15:1
	WB-67	–	55	13.8	10:1
	WB-92D	–	55	16.8	10:1
	WB-100T	–	55	10.2	10:1
	WB-109D	–	55	20.0	10:1

From *A Policy on Geometric Design of Highways and Streets*, 2011, by the American Association of State Highway and Transportation Officials, Washington, D.C. Table 9-15. Used by permission.

APPENDIX 4.T *(continued)*
Edge-of-Traveled-Way for Turns at Intersections

SI units

angle of turn	design vehicle	three-centered compound curve radii (m)	symmetric offset (m)	three-centered compound curve radii (m)	asymmetric offset (m)
30°	P	–	–	–	–
	SU-9	–	–	–	–
	SU-12	–	–	–	–
	WB-12	–	–	–	–
	WB-19	–	–	–	–
	WB-20	140-53-140	1.2	91-53-168	0.6–1.4
	WB-28D	168–16–168	1.2	61–46–168	0.6–1.8
	WB-30T	67-24-67	1.4	61-24-91	0.8–1.5
	WB-33D	168-76-168	1.5	76-61-198	0.5–2.1
45°	P	–	–	–	–
	SU-9	–	–	–	–
	SU-12	–	–	–	–
	WB-12	–	–	–	–
	WB-19	140-72-140	0.6	36-43-150	1.0–2.6
	WB-20	140-53-140	1.2	76-38-183	0.3–1.8
	WB-28D	160–47–160	1.5	61–43–152	0.5–1.8
	WB-30T	76-24-76	1.4	67-24-91	0.8–1.7
	WB-33D	168-61-168	1.5	61-52-198	0.5–2.1
60°	P	–	–	–	–
	SU-9	–	–	–	–
	SU-12	–	–	–	–
	WB-12	–	–	–	–
	WB-19	120-30-120	4.5	34-30-67	3.0–3.7
	WB-20	122-30-122	2.4	76-38-183	0.3–1.8
	WB-28D	146–34–146	1.8	46–34–152	0.9–2.7
	WB-30T	76-24-76	1.4	61-24-91	0.6–1.7
	WB-33D	198-46-198	1.7	61-43-183	0.5–2.4
75°	P	30-8-30	0.6	–	–
	SU-9	36-14-36	0.6	–	–
	SU-12	61–11–61	1.5	18–14–61	0.3–1.4
	WB-12	36-14-36	1.5	36-14-60	0.6–2.0
	WB-19	134-23-134	4.5	43-30-165	1.5–3.6
	WB-20	128-23-128	3.0	61-24-183	0.3–0.3
	WB-28D	152–29–152	2.1	46–30–152	0.3–2.4
	WB-30T	76-24-76	1.4	30-24-91	0.5–1.5
	WB-33D	213-38-213	2.0	46-34-168	0.5–3.5
90°	P	30-6-30	0.8	–	–
	SU-9	36-12-36	0.6	–	–
	SU-12	61–9–61	2.1	18–14–61	0.3–1.4
	WB-12	36-12-36	1.5	36-12-60	0.6–2.0
	WB-19	120-21-120	3.0	48-21-110	2.0–3.0
	WB-20	134-20-134	3.0	61-21-183	0.3–3.4
	WB-28D	143–23–143	3.0	46–27–152	0.5–2.6
	WB-30T	76-21-76	1.4	61-21-91	0.3–1.5
	WB-33D	213-34-213	2.0	30-29-168	0.6–3.5
105°	P	30-6-30	0.8	–	–
	SU-9	30-11-30	1.0	–	–
	SU-12	61–11–61	1.8	18–12–58	0.5–1.8
	WB-12	30-11-30	1.5	30-17-60	0.6–2.5

APPENDIX 4.T *(continued)*
Edge-of-Traveled-Way for Turns at Intersections

		SI units			
		three-centered compound		three-centered compound	
angle of turn	design vehicle	curve radii (m)	symmetric offset (m)	curve radii (m)	asymmetric offset (m)
	WB-19	160-15-160	4.5	110-23-180	1.2–3.2
	WB-20	152-15-152	4.0	61-20-183	0.3–3.4
	WB-28D	154–24–152	2.4	46–24–152	0.6–3.0
	WB-30T	76-18-76	1.5	30-18-91	0.5–1.8
	WB-33D	213-29-213	2.4	46-24-152	0.9–4.6
120°	P	30-6-30	0.6	–	–
	SU-9	30-9-30	1.0	–	–
	SU-12	61–11–61	1.8	18–12–58	0.5–1.5
	WB-12	36-9-36	2.0	30-9-55	0.6–2.7
	WB-19	160-21-160	3.0	24-17-160	5.2–7.3
	WB-20	168-14-168	4.6	61-18-183	0.6–3.8
	WB-28D	152–21–152	3.0	46–21–137	0.9–3.2
	WB-30T	76-18-76	1.5	30-18-91	0.5–1.8
	WB-33D	213-26-213	2.7	46-21-152	2.0–5.3
135°	P	30-6-30	0.5	–	–
	SU-9	30-9-30	1.2	–	–
	SU-12	61–12–61	1.2	18–12–55	0.5–1.5
	WB-12	36-9-36	2.0	30-8-55	1.0–4.0
	WB-19	180-18-180	3.6	30-18-195	2.1–4.3
	WB-20	168-14-168	5.0	61-18-183	0.6–3.8
	WB-28D	137–21–137	2.7	46–20–137	2.1–4.1
	WB-30T	76-18-76	1.7	30-18-91	0.8–2.0
	WB-33D	213-21-213	3.8	46-20-152	2.1–5.6
150°	P	23-6-23	0.6	–	–
	SU-9	30-9-30	1.2	–	–
	SU-12	61–11–61	2.0	18–21–61	0.3–1.4
	WB-12	30-9-30	2.0	28-8-48	0.3–3.6
	WB-19	145-17-145	4.5	43-18-170	2.4–3.0
	WB-20	168-14-168	5.8	61-17-183	2.0–5.0
	WB-28D	107–18–107	4.6	37–20–137	1.8–4.0
	WB-30T	76-18-76	2.1	30-18-91	1.5–2.4
	WB-33D	213-20-213	4.6	61-20-152	2.7–5.6
180°	P	15-5-15	0.2	–	–
	SU-9	30-9-30	0.5	–	–
	SU-12	46–11–46	1.9	15–11–40	1.7–2.1
	WB-12	30-6-30	3.0	26-6-45	2.0–4.0
	WB-19	245-14-245	6.0	30-17-275	4.5–4.5
	WB-20	183-14-183	6.2	30-17-122	1.8–4.6
	WB-28D	122–17–122	5.1	37–18–122	2.7–4.4
	WB-30T	76-17-76	2.9	30-17-91	2.6–3.2
	WB-33D	213-17-213	6.1	61-18-152	3.0–6.4

From *A Policy on Geometric Design of Highways and Streets*, 2011, by the American Association of State Highway and Transportation Officials, Washington, D.C. Table 9-16. Used by permission.

APPENDIX 4.T *(continued)*
Edge-of-Traveled-Way for Turns at Intersections

customary U.S. units

angle of turn	design vehicle	three-centered compound curve radii (ft)	symmetric offset (ft)	three-centered compound curve radii (ft)	asymmetric offset (ft)
30°	P	–	–	–	–
	SU-30	–	–	–	–
	SU-40	–	–	–	–
	WB-40	–	–	–	–
	WB-62	–	–	–	–
	WB-67	460-175-460	4.0	300-175-550	2.0–4.5
	WB-92D	550–155–550	4.0	200–150–500	2.0–6.0
	WB-100T	220-80-220	4.5	200-80-300	2.5–5.0
	WB-109D	550-250-550	5.0	250-200-650	1.5–7.0
45°	P	–	–	–	–
	SU-30	–	–	–	–
	SU-40	–	–	–	–
	WB-40	–	–	–	–
	WB-62	460-240-460	2.0	120-140-500	3.0–8.5
	WB-67	460-175-460	4.0	250-125-600	1.0–6.0
	WB-92D	525–155–525	5.0	200–140–500	1.5–6.0
	WB-100T	250-80-250	4.5	200-80-300	2.5–5.5
	WB-109D	550-200-550	5.0	200-170-650	1.5–7.0
60°	P	–	–	–	–
	SU-30	–	–	–	–
	SU-40	–	–	–	–
	WB-40	–	–	–	–
	WB-62	400-100-400	15.0	110-100-220	10.5–12.5
	WB-67	400-100-400	8.0	250-125-600	1.0–6.0
	WB-92D	480-110-480	6.0	150-110-500	3.0–9.0
	WB-100T	250-80-250	4.5	200-80-300	2.0–5.5
	WB-109D	650-150-650	5.5	200-140-600	1.5–8.0
75°	P	100-25-100	2.0	–	–
	SU-30	120-45-120	2.0	–	–
	SU-40	200-35-200	5.0	60-45-200	1.0–4.5
	WB-40	120-45-120	5.0	120-45-195	2.0–6.5
	WB-62	440-75-440	15.0	140-100-540	5.0–12.0
	WB-67	420-75-420	10.0	200-80-600	1.0–10.0
	WB-92D	500-95-500	7.0	150-100-500	1.0–8.0
	WB-100T	250-80-250	4.5	100-80-300	1.5–5.0
	WB-109D	700-125-700	6.5	150-110-550	1.5–11.5
90°	P	100-20-100	2.5	–	–
	SU-30	120-40-120	2.0	–	–
	SU-40	200-30-200	7.0	60-45-200	1.0–4.5
	WB-40	120-40-120	5.0	120-40-200	2.0–6.5
	WB-62	400-70-400	10.0	160-70-360	6.0–10.0
	WB-67	440-65-440	10.0	200-70-600	1.0–11.0
	WB-92D	470-75-470	10.0	150-90-500	1.5–8.5
	WB-100T	250-70-250	4.5	200-70-300	1.0–5.0
	WB-109D	700-110-700	6.5	100-95-550	2.0–11.5
105°	P	100-20-100	2.5	–	–
	SU-30	100-35-100	3.0	–	–
	SU-40	200-35-200	6.0	60-40-190	1.5–6.0
	WB-40	100-35-100	5.0	100-55-200	2.0–8.0

APPENDIX 4.T (continued)
Edge-of-Traveled-Way for Turns at Intersections

customary U.S. units

angle of turn	design vehicle	three-centered compound curve radii (ft)	symmetric offset (ft)	three-centered compound curve radii (ft)	asymmetric offset (ft)
	WB-62	520-50-520	15.0	360-75-600	4.0–10.5
	WB-67	500-50-500	13.0	200-65-600	1.0–11.0
	WB-92D	500-80-500	8.0	150-80-500	2.0–10.0
	WB-100T	250-60-250	5.0	100-60-300	1.5–6.0
	WB-109D	700-95-700	8.0	150-80-500	3.0–15.0
120°	P	100-20-100	2.0	–	–
	SU-30	100-30-100	3.0	–	–
	SU-40	200-35-200	6.0	60-40-190	1.5–5.0
	WB-40	120-30-120	6.0	100-30-180	2.0–9.0
	WB-62	520-70-520	10.0	80-55-520	24.0–17.0
	WB-67	550-45-550	15.0	200-60-600	2.0–12.5
	WB-92D	500-70-500	10.0	150-70-450	3.0–10.5
	WB-100T	250-60-250	5.0	100-60-300	1.5–6.0
	WB-109D	700-85-700	9.0	150-70-500	7.0–17.4
135°	P	100-20-100	1.5	–	–
	SU-30	100-30-100	4.0	–	–
	SU-40	200-40-200	4.0	60-40-180	1.5–5.0
	WB-40	120-30-120	6.5	100-25-180	3.0–13.0
	WB-62	600-60-600	12.0	100-60-640	14.0–7.0
	WB-67	550-45-550	16.0	200-60-600	2.0–12.5
	WB-92D	450-70-450	9.0	150-65-450	7.0–13.5
	WB-100T	250-60-250	5.5	100-60-300	2.5–7.0
	WB-109D	700-70-700	12.5	150-65-500	14.0–18.4
150°	P	75-20-75	2.0	–	–
	SU-30	100-30-100	4.0	–	–
	SU-40	200-35-200	6.5	60-40-200	1.0–4.5
	WB-40	100-30-100	6.0	90-20-160	1.0–12.0
	WB-62	480-55-480	15.0	140-60-560	8.0–10.0
	WB-67	550-45-550	19.0	200-55-600	7.0–16.4
	WB-92D	350-60-350	15.0	120-65-450	6.0–13.0
	WB-100T	250-60-250	7.0	100-60-300	5.0–8.0
	WB-109D	700-65-700	15.0	200-65-500	9.0–18.4
180°	P	50-15-50	0.5	–	–
	SU-30	100-30-100	1.5	–	–
	SU-40	150–35–150	6.2	50–35–130	5.5–7.0
	WB-40	100-20-100	9.5	85-20-150	6.0–13.0
	WB-62	800-45-800	20.0	100-55-900	15.0–15.0
	WB-67	600-45-600	20.5	100-55-400	6.0–15.0
	WB-92D	400–55–400	16.8	120–60–400	9.0–14.5
	WB-100T	250-55-250	9.5	100-55-300	8.5–10.5
	WB-109D	700-55-700	20.0	200-60-500	10.0–21.0

From *A Policy on Geometric Design of Highways and Streets*, 2011, by the American Association of State Highway and Transportation Officials, Washington, D.C. Table 9-16. Used by permission.

APPENDIX 4.U
Effect of Curb Radii on Right-Turning Paths of Various Vehicles

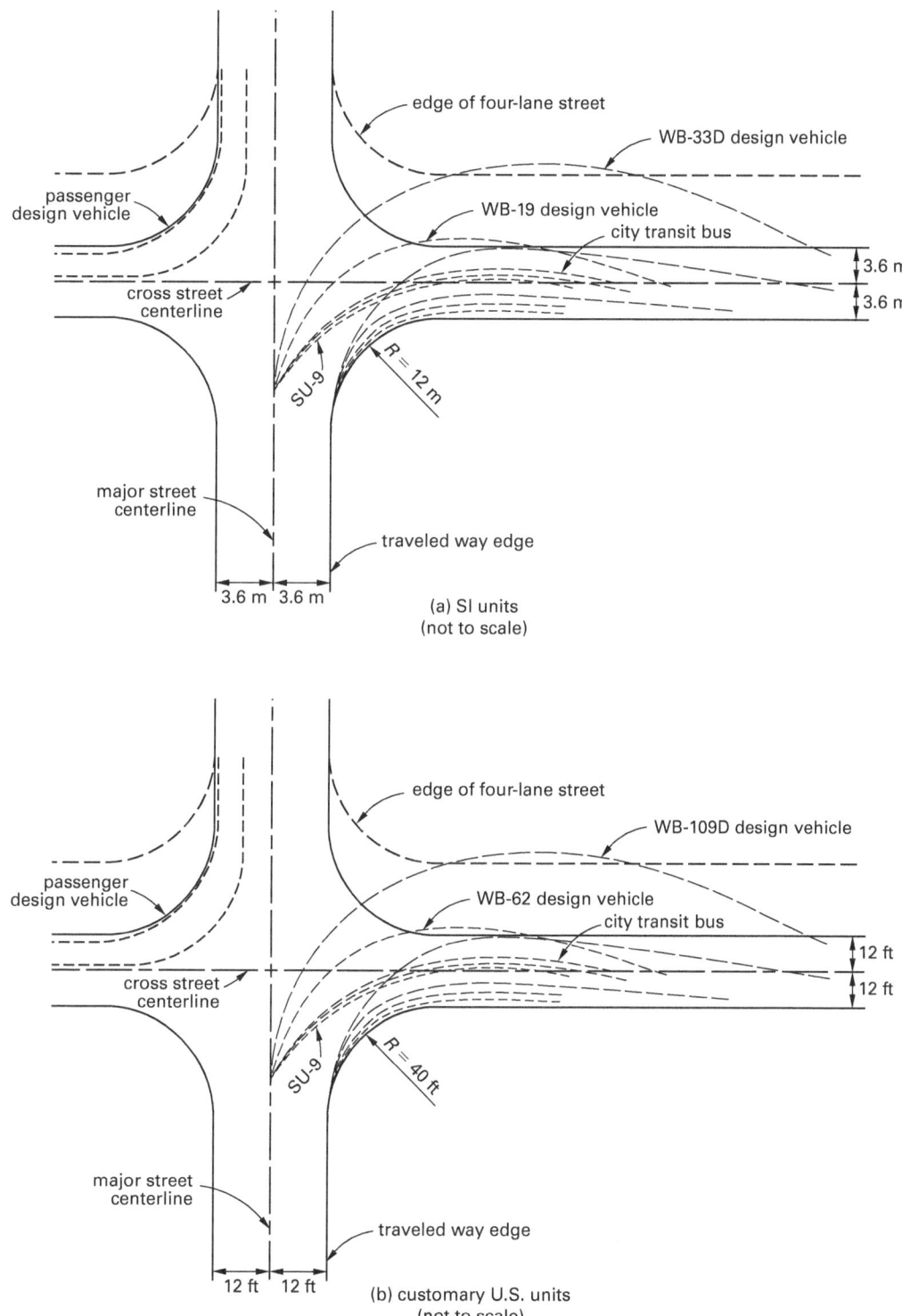

From *A Policy on Geometric Design of Highways and Streets*, 2011, by the American Association of State Highway and Transportation Officials, Washington, D.C. Figure 9-32. Used by permission.

APPENDIX 4.V
Cross Street Widths Occupied by Turning Vehicles*

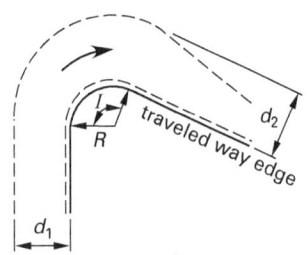

case A
Vehicle turns from proper lane
and swings wide on cross street.*
d_1 is 12 ft (3.6 m), and d_2 is variable.

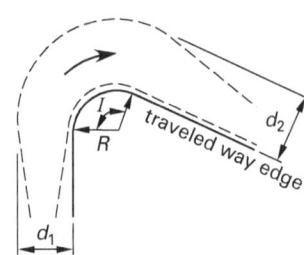

case B
Turning vehicle swings equally
wide on both streets.*
d_1 and d_2 are both variable.

SI units

angle of intersection (I)	design vehicle	d_2 for cases A and B where									
		$R = 4.5$ m		$R = 6$ m		$R = 7.5$ m		$R = 9$ m		$R = 12$ m	
		A (m)	B (m)	A (m)	B (m)	A (m)	B (m)	A (m)	B (m)	A (m)	B (m)
30°	SU-9	4.3	4.0	4.3	4.0	4.0	4.0	4.0	4.0	4.0	4.0
	SU-12	4.9	4.6	4.6	4.3	4.6	4.3	4.6	4.3	4.3	4.0
	BUS	6.7	5.2	5.8	5.2	5.8	5.2	5.8	5.2	5.5	5.2
	WB-12	4.3	4.3	4.3	4.3	4.3	4.3	4.3	4.3	4.3	4.3
	WB-19	–	–	–	–	–	–	–	–	8.2	5.2
	WB-20	–	–	–	–	–	–	–	–	8.5	5.5
60°	SU-9	5.8	4.9	5.8	4.9	5.2	4.6	4.9	4.6	4.3	4.3
	SU-12	7.3	5.8	5.8	5.5	5.8	5.2	5.5	4.9	5.2	4.6
	BUS	8.5	6.4	7.9	6.1	7.3	6.1	7.0	5.8	6.7	5.5
	WB-12	7.3	5.8	6.7	5.8	6.4	5.8	5.8	5.5	5.2	4.9
	WB-19	–	–	–	–	–	–	–	–	9.1	6.7
	WB-20	–	–	–	–	–	–	–	–	11.3	7.3
90°	SU-9	7.9	6.1	7.0	5.5	5.8	4.9	5.2	4.6	4.0	4.0
	SU-12	8.8	6.4	7.9	5.8	6.7	5.8	5.8	5.2	4.6	4.6
	BUS	11.6	7.0	10.0	6.7	9.1	6.7	7.6	6.4	6.7	5.5
	WB-12	9.4	6.7	8.2	6.4	7.0	6.4	5.8	5.5	5.2	4.9
	WB-19	–	–	–	–	–	–	–	–	11.9	7.0
	WB-20	–	–	–	–	–	–	–	–	11.9	7.6
120°	SU-9	10.4	6.7	8.2	5.8	6.4	5.5	5.2	4.9	4.0	4.0
	SU-12	12.2	7.6	10.4	7.0	8.2	6.1	6.7	5.5	4.9	4.6
	BUS	14.0	8.5	12.2	7.6	9.8	7.0	7.9	5.8	5.8	5.5
	WB-12	11.3	7.0	8.8	6.7	7.3	6.7	5.8	5.5	5.2	4.9
	WB-19	–	–	–	–	–	–	–	–	7.9	6.7
	WB-20	–	–	–	–	–	–	–	–	9.1	7.0
150°	SU-9	12.2	7.6	9.8	6.4	6.7	5.8	5.2	4.9	3.6	3.6
	SU-12	15.2	9.1	12.5	7.9	10.1	7.0	7.6	5.8	2.9	4.6
	BUS	14.6	8.5	12.2	7.6	9.8	7.0	6.7	5.5	5.2	4.9
	WB-12	11.9	7.3	8.8	6.7	7.0	6.7	5.8	5.5	5.2	4.9
	WB-19	–	–	–	–	–	–	–	–	6.1	5.5
	WB-20	–	–	–	–	–	–	–	–	8.2	5.5

*P design vehicle turns within 12 ft (3.6 m) where $R = 15$ ft (4.5 m) or more. No parking on either street.

APPENDIX 4.V (continued)
Cross Street Widths Occupied by Turning Vehicles*

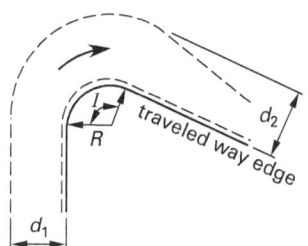

case A
Vehicle turns from proper lane
and swings wide on cross street.*
d_1 is 12 ft (3.6 m), and d_2 is variable.

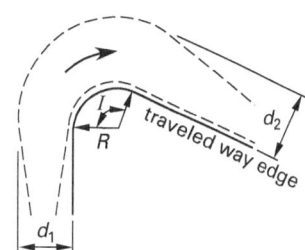

case B
Turning vehicle swings equally
wide on both streets.*
d_1 and d_2 are both variable.

customary U.S. units

angle of intersection (I)	design vehicle	d_2 for cases A and B where									
		$R = 15$ ft		$R = 20$ ft		$R = 25$ ft		$R = 30$ ft		$R = 40$ ft	
		A (ft)	B (ft)	A (ft)	B (ft)	A (ft)	B (ft)	A (ft)	B (ft)	A (ft)	B (ft)
30°	SU-30	14	13	14	13	13	13	13	13	13	13
	SU-40	16	15	15	14	15	14	15	14	14	13
	BUS	22	17	19	17	19	17	19	17	18	17
	WB-40	14	14	14	14	14	14	14	14	14	14
	WB-62	–	–	–	–	–	–	–	–	27	17
	WB-67	–	–	–	–	–	–	–	–	28	18
60°	SU-30	19	16	19	16	17	15	16	15	14	14
	SU-40	24	19	19	18	19	17	18	16	17	15
	BUS	28	21	26	20	24	20	23	19	22	18
	WB-40	24	19	22	19	21	19	19	18	17	16
	WB-62	–	–	–	–	–	–	–	–	30	22
	WB-67	–	–	–	–	–	–	–	–	37	24
90°	SU-30	26	20	23	18	19	16	17	15	13	13
	SU-40	29	21	26	19	22	19	19	17	15	15
	BUS	38	23	33	22	30	22	25	21	22	18
	WB-40	31	22	27	21	23	21	19	18	17	16
	WB-62	–	–	–	–	–	–	–	–	39	23
	WB-67	–	–	–	–	–	–	–	–	39	25
120°	SU-30	34	22	27	19	21	18	17	16	13	13
	SU-40	40	25	34	23	27	20	22	18	16	15
	BUS	46	28	40	25	32	23	26	19	19	18
	WB-40	37	23	29	22	24	22	19	18	17	16
	WB-62	–	–	–	–	–	–	–	–	26	22
	WB-67	–	–	–	–	–	–	–	–	30	23
150°	SU-30	40	25	32	21	22	19	17	16	12	12
	SU-40	50	30	41	26	33	23	25	19	16	15
	BUS	48	28	40	25	32	23	22	18	17	16
	WB-40	39	24	29	22	23	22	19	18	17	16
	WB-62	–	–	–	–	–	–	–	–	20	18
	WB-67	–	–	–	–	–	–	–	–	27	18

*P design vehicle turns within 12 ft (3.6 m) where $R = 15$ ft (4.5 m) or more. No parking on either street.

From *A Policy on Geometric Design of Highways and Streets*, 2011, by the American Association of State Highway and Transportation Officials, Washington, D.C. Table 9-17. Used by permission.

APPENDIX 4.W
Crosswalk Length Variations with Different Curb Radii and Width of Borders

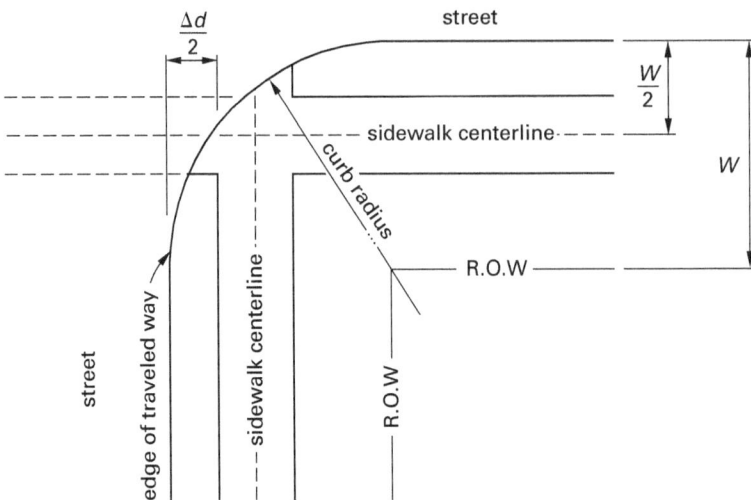

Δd = increased walking distance between curbs resulting from curved curb return at intersections
W = width of R.O.W. measured from edge of traveled way
R = curb radius

From *A Policy on Geometric Design of Highways and Streets*, 2011, by the American Association of State Highway and Transportation Officials, Washington, D.C. Figure 9-34. Used by permission.

APPENDIX 4.X
Corner Setbacks with Different Curb Radii and Width of Borders

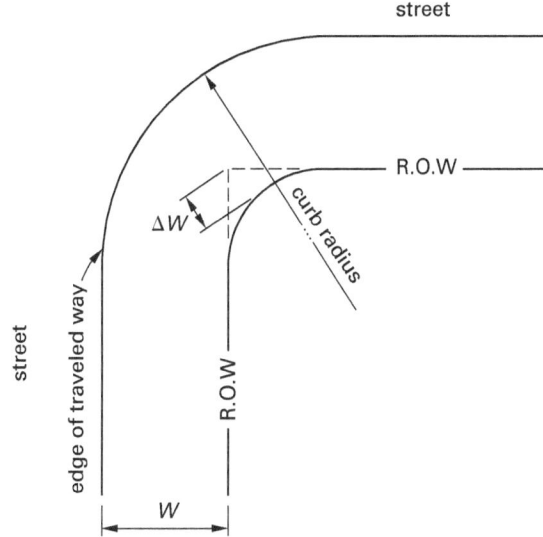

ΔW = additional corner setback along 45° radial
W = width of R.O.W. measured from edge of traveled way
R = curb radius

From *A Policy on Geometric Design of Highways and Streets*, 2011, by the American Association of State Highway and Transportation Officials, Washington, D.C. Figure 9-35. Used by permission.

APPENDIX 5.A
AASHTO Soil Classification System

	granular materials (35% or less passing no. 200 sieve)							silt-clay materials (more than 35% passing no. 200 sieve)				
	A-1		A-3	A-2				A-4	A-5	A-6	A-7	A-8
	A-1-a	A-1-b		A-2-4	A-2-5	A-2-6	A-2-7				A-7-5 or A-7-6	
sieve analysis: % passing no. 10 no. 40 no. 200	50 max 30 max 15 max	50 max 25 max	51 min 10 max	35 max	35 max	35 max	35 max	36 min	36 min	36 min	36 min	
characteristics of fraction passing no. 40: LL: liquid limit PI: plasticity index	6 max		NP	40 max 10 max	41 min 10 max	40 max 11 min	41 min 11 min	40 max 10 max	41 min 10 max	40 max 11 min	41 min 11 min	
usual types of significant constituents	stone fragments gravel and sand		fine sand	silty or clayey gravel and sand				silty soils		clayey soils		peat, highly organic soils
general subgrade rating	excellent to good							fair to poor				unsatisfactory

APPENDIX 5.B
Unified Soil Classification System

major division		group symbol	laboratory classification criteria		soil description
			finer than 200 sieve (%)	supplementary requirements	
coarse-grained (over 50% by weight coarser than no. 200 sieve)	gravelly soils (over half of coarse fraction larger than no. 4)	GW GP GM GC	0–5[a] 0–5[a] 12 or more[a] 12 or more[a]	D_{60}/D_{10} greater than 4 $D_{30}^2/(D_{60}D_{10})$ between 1 and 3 not meeting above gradation for GW PI less than 4 or below A-line PI over 7 and above A-line	well-graded gravels, sandy gravels gap-graded or uniform gravels, sandy gravels silty gravels, silty sandy gravels clayey gravels, clayey sandy gravels
	sandy soils (over half of coarse fraction finer than no. 4)	SW SP SM SC	0–5[a] 0–5[a] 12 or more[a] 12 or more[a]	D_{60}/D_{10} greater than 6 $D_{30}^2/(D_{60}D_{10})$ between 1 and 3 not meeting above gradation requirements PI less than 4 or below A-line PI over 7 and above A-line	well-graded, gravelly sands gap-graded or uniform sands, gravelly sands silty sand clayey sands, clayey gravelly sands
fine-grained (over 50% by weight finer than no. 200 sieve)	low compressibility (liquid limit less than 50)	ML CL OL	plasticity chart plasticity chart plasticity chart, organic odor or color		silts, very fine sands, silty or clayey fine sands, micaceous silts low plasticity clays, sandy or silty clays organic silts and clays of high plasticity
	high compressibility (liquid limit 50 or more)	MH CH OH	plasticity chart plasticity chart plasticity chart, organic odor or color		micaceous silts, diatomaceous silts, volcanic ash highly plastic clays and sandy clays organic silts and clays of high plasticity
soils with fibrous organic matter[1]		Pt	fibrous organic matter; will char, burn, or glow		peat, sandy peats, and clayey peat

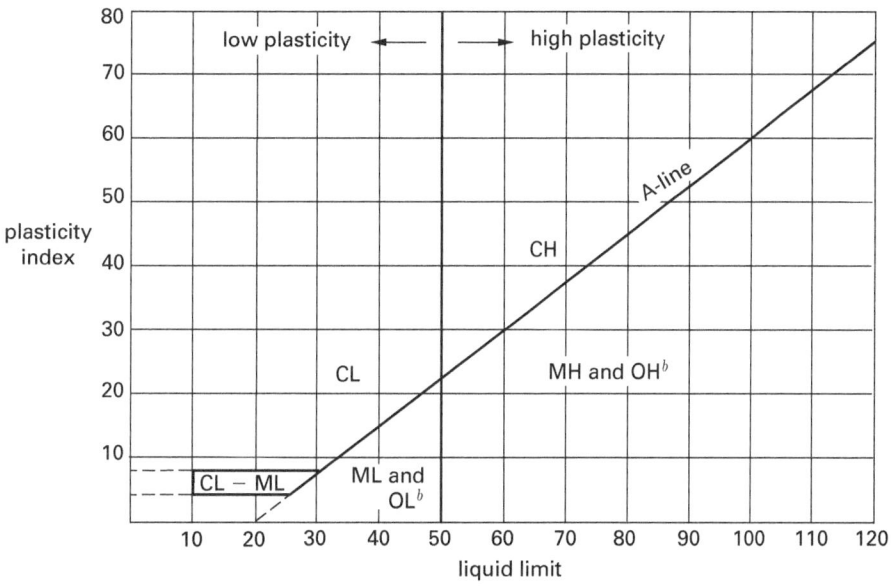

Note: Plasticity chart for the classification of fine-grained soils. Tests made on fraction finer than no. 40 sieve, 0.425 mm.

---

[1][a]For soils having 5–12% passing the no. 200 seive, use a dual symbol such as GW-GC.
[b]Distinguishing between M and O classifications requires identifying organic components by observation, odor, or other testing.

APPENDIX 5.C
USDA Soil Triangle

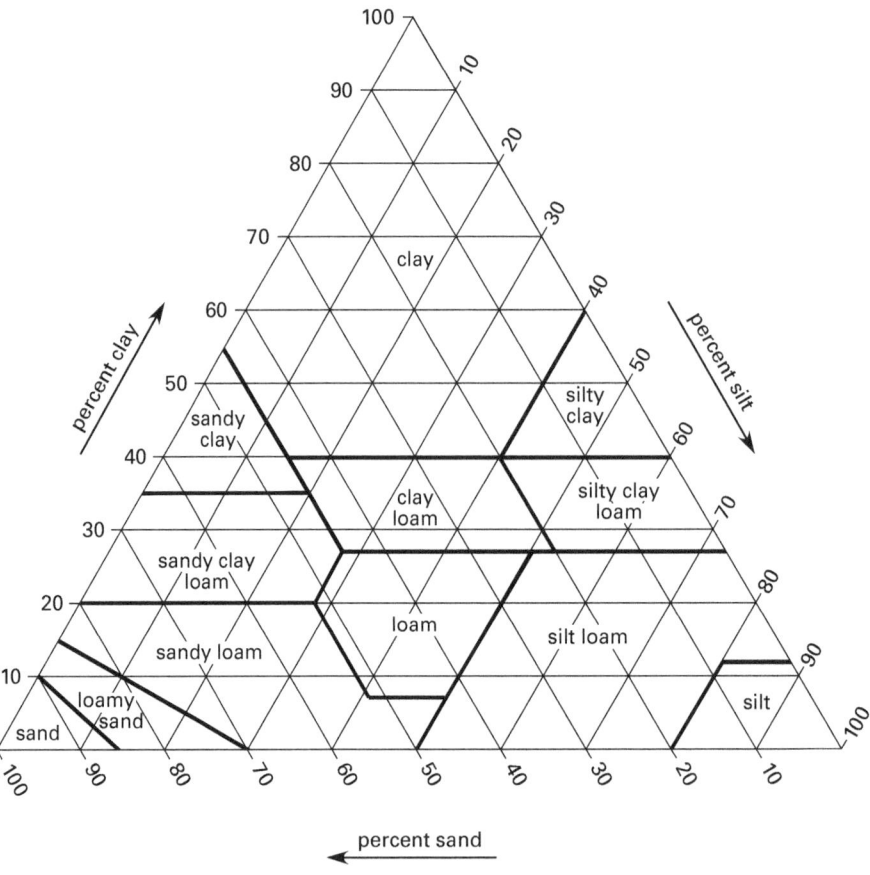

Reprinted from *Soil Survey Manual*, USDA Handbook 18.

APPENDIX 5.D
Performance-Graded Asphalt Binder Specification

performance grade	PG 46			PG 52							PG 58					PG 64					
	34	40	46	10	16	22	28	34	40	46	16	22	28	34	40	10	16	22	28	34	40
average 7 day max. pavement design temp. (°C)	<46			<52							<58					<64					
min. pavement design temp. (°C)	−34	−40	−46	−10	−16	−22	−28	−34	−40	−46	−16	−22	−28	−34	−40	−10	−16	−22	−28	−34	−40
original binder																					
flash point temp., T 48, min. (°C)	230																				
viscosity, T 316: max. 3 Pa·s, test temp. (°C)	135																				
dynamic shear, T 315: $G^*/\sin\delta$, min. 1.10 kPa test temp. at 10 rad/s (°C)	46			52							58					64					
rolling thin-film oven residue (T 240)																					
mass change, max. (%)	1.00																				
dynamic shear, T 315: $G^*/\sin\delta$, min. 2.20 kPa test temp. at 10 rad/s (°C)	46			52							58					64					
pressure aging vessel (R 28)																					
PAV aging temp. (°C)	90			90							100					100					
dynamic shear, T 315: $G^*/\sin\delta$, max. 5000 kPa test temp. at 10 rad/s (°C)	10	7	4	25	22	19	16	13	10	7	25	22	19	16	13	31	28	25	22	19	16
creep stiffness, T 313: S, max. 300 MPa m-value, min. 0.300 test temp. at 60 s (°C)	−24	−30	−36	0	−6	−12	−18	−24	−30	−36	−6	−12	−18	−24	−30	0	−6	−12	−18	−24	−30
direct tension, T 314: failure strain, min. 0.300 test temp. at 60 s (°C)	−24	−30	−36	0	−6	−12	−18	−24	−30	−36	−6	−12	−18	−24	−30	0	−6	−12	−18	−24	−30
critical low cracking temp., PP 42: critical low cracking temp. determined by PP 42, test temp. (°C)	−24	−30	−36	0	−6	−12	−18	−24	−30	−36	−6	−12	−18	−24	−30	0	−6	−12	−18	−24	−30

APPENDIX 5.D *(continued)*
Performance-Graded Asphalt Binder Specification

performance grade	PG 70						PG 76					PG 82				
	10	16	22	28	34	40	10	16	22	28	34	10	16	22	28	34
average 7 day max. pavement design temp. (°C)	<70						<76					<82				
min. pavement design temp. (°C)	−10	−16	−22	−28	−34	−40	−10	−16	−22	−28	−34	−10	−16	−22	−28	−34
original binder																
flash point temp., T 48, min. (°C)	230															
viscosity, ASTM D 4402: max. 3 Pa·s, test temp. (°C)	135															
dynamic shear, TP 5: $G^*/\sin\delta$, min. 1.10 kPa test temp. at 10 rad/s (°C)	70						76					82				
rolling thin film oven residue (T 240)																
mass change, max. (%)	1.00															
dynamic shear, TP 5: $G^*/\sin\delta$, min. 2.20 kPa test temp. at 10 rad/s (°C)	70						76					82				
pressure aging vessel residue (PP 1)																
PAV aging temp. (°C)	100						100					100				
dynamic shear, TP 5: $G^*/\sin\delta$, max. 5000 kPa test temp. at 10 rad/s (°C)	34	31	28	25	22	19	37	34	31	28	25	40	37	34	31	28
creep stiffness, T 313: S, max. 300 MPa m-value, min. 0.300 test temp. at 60 s (°C)	0	−6	−12	−18	−24	−30	0	−6	−12	−18	−24	0	−6	−12	−18	−24
direction tension, T 314: failure strain, min. 0.300 test temp. at 60 s (°C)	0	−6	−12	−18	−24	−30	0	−6	−12	−18	−24	0	−6	−12	−18	−24
critical low cracking temp., PP 42: critical cracking temp. determined by PP 42, test temp. (°C)	0	−6	−12	−18	−24	−30	0	−6	−12	−18	−24	0	−6	−12	−18	−24

From *Standard Specification for Performance-Grade Asphalt Binder*, 2010, by the American Association of State Highway and Transportation Officials, Washington, D.C. Used by permission.

APPENDIX 5.E
Superpave Mix Design Procedural Outline

I. Selection of Materials
    A. Select asphalt binder.
        1. Determine project weather conditions using weather database.
        2. Select reliability.
        3. Determine design temperatures.
        4. Verify asphalt binder grade.
        5. Determine temperature-viscosity relationship for lab mixing and compaction.
    B. Select aggregates.
        1. consensus properties
            a. combined gradation
            b. coarse aggregate angularity
            c. fine aggregate angularity
            d. flat and elongated particles
            e. clay content
        2. agency and other properties
            a. specific gravity
            b. toughness
            c. soundness
            d. deleterious materials
            e. other
    C. Select modifiers.

II. Selection of Design Aggregate Structure
    A. Establish trial blends.
        1. Develop three blends.
        2. Evaluate combined aggregate properties.
    B. Compact trial blend specimens.
        1. Establish trial asphalt binder content.
            a. Superpave method
            b. engineering judgment method
        2. Establish trial blend specimen size.
        3. Determine $N_{initial}$ and N_{design}.
        4. Batch trial blend specimens.
        5. Compact specimens and generate densification tables.
        6. Determine mixture properties (G_{mm} and G_{mb}).
    C. Evaluate trial blends.
        1. Determine percentage of G_{mm} at $N_{initial}$ and N_{design}.
        2. Determine percentage of air voids and voids in mineral aggregate.
        3. Estimate asphalt binder content to achieve 4% air voids.
        4. Estimate mix properties at estimated asphalt binder content.
        5. Determine dust-asphalt ratio.
        6. Compare mixture properties to criteria.
    D. Select most promising design aggregate structure for further analysis.

APPENDIX 5.E *(continued)*
Superpave Mix Design Procedural Outline

III. Selection of Design Asphalt Binder Content
- A. Compact design aggregate structure specimens at multiple binder contents.
    1. Batch design aggregate structure specimens.
    2. Compact specimens and generate densification tables.
- B. Determine mixture properties versus asphalt binder content.
    1. Determine percentage of G_{mm} at $N_{initial}$, N_{design}, and N_{max}.
    2. Determine volumetric properties.
    3. Determine dust-asphalt ratio.
    4. Graph mixture properties versus asphalt binder content.
- C. Select design asphalt binder content.
    1. Determine asphalt binder content at 4% air voids.
    2. Determine mixture properties at selected asphalt binder content.
    3. Compare mixture properties to criteria.

IV. Evaluation of Moisture Sensitivity of Design Asphalt Mixture Using AASHTO T238

Reprinted with permission from the Asphalt Institute, *Superpave Mix Design (SP-2)*, 2001, App. C.

Index

0.45 power line, 5-13
20/20 vision, 6-22
85th percentile driver behavior, 6-21

A

A
   Soil Primer, 5-9
   -weighted scale, 1-16
AADT, 6-2, 6-5
AASHTO, 4-2
Abrasion test, Los Angeles, 5-23
Absolute noise level, 1-16
Absorption asphalt, 5-16
Acceleration
   centripetal, 4-30
   lateral, 4-30
   limit, transit, 4-50
   radial, 4-30
   rate, 4-42
   vertical, train, 4-51
Access
   -point density, 2-14
   -point density adjustment factor, two-lane highway (tbl), 2-21
   points per mile, 2-12
Accessible parking space, 2-35
Accident, 6-4, 6-16
   data, 6-16
Act, Americans with Disabilities, 2-35
Action recourse, 1-13
Activity, full movement, 1-8
Acuity
   recognition, 6-21
   visual, 6-21
   visual, center cone, 6-21
ADA, 2-35
Additional curve elements (fig), 4-11
Adjusted volume, 1-9
Adjustment
   factor, access-point density, two-lane highway (tbl), 2-21
   factor, lane and shoulder width (tbl), 2-21
   factors, 2-20
   walkway width, 3-7
ADT, 6-2
Advance warning area, 6-26
Adverse crown, 4-35
Aerial
   mapping, 4-3
   mapping contour, 4-3
   mapping scale, 4-3
Aggregate, 5-12
   angularity, coarse, 5-23
   angularity, fine, 5-23
   material, 5-9
   mixture gradation, Superpave (tbl), 5-14
   natural, 5-12
   processed, 5-12
   property, consensus, 5-23
   selection, design, 5-26
   size for asphalt pavement (tbl), 5-10
   specification, Superpave, 5-13
   synthetic, 5-12
   uncompacted void of fine, 5-23
   voids in mineral, 5-15
Ahead
   stationing, 4-7
   tangent, 4-11

Aid, mobility, 3-3
Air
   quality study, 1-16
   quality study, mesoscale, 1-17
   quality study, microscale, 1-17
   temperature, 5-24
   void, 5-16
   voids, percentage, 5-16
Alignment, 4-2
   field, 4-10
   final, 4-10
   layout, 4-10
   layout, route (fig), 4-23
   preliminary, 4-10
   subtangent, 4-11
Alligator cracking, 5-12
Alphabet, federal, 6-22
Amber
   interval, 1-10
   phase length, PRT, 6-21
Ambient sound, 1-16
American
   Association of State Highways and Transportation Officials (AASHTO), 4-2
   Railway Engineering and Maintenance Association (AREMA), 4-2
Americans with Disabilities Act, 2-35
Analysis
   benefit-cost, 1-15, 6-20
   bicycle facility, 3-20
   capacity, 2-9
   capacity planning, 1-10
   compacted paving mixture, 5-14
   conflict, iii, 6-1, 6-27
   cost-benefit, highway safety, 6-19
   crash, iii, 6-1, 6-4, 6-5
   crash data, 6-5
   data, 2-9
   economic, iii, 6-1, 6-19
   environmental impact, 4-3
   life-cycle cost, 1-13
   off-street facility (tbl), 3-13
   operational, 1-12
   planning, 1-12
   roadside clearance, iii, 6-1, 6-5
   site impact, 1-15
   traffic capacity, 2-9
Angle
   bearing (fig), 4-4
   central, 4-5, 4-6
   circular curve deflection (fig), 4-13
   conversion, azimuth-bearing, 4-4
   deflection, 4-4, 4-5, 4-7, 4-12
   encroachment, 6-8
   intersection, 4-5
   negative deflection, 4-11
   positive deflection, 4-11
   slope, 4-30
   steering, A-38
   subtended, 4-12
   to the right, 4-4
   tractor/trailer, A-38
   vehicle orientation, 6-8
Angularity
   coarse aggregate, 5-23
   fine aggregate, 5-23
Annual daily traffic, average, 6-5
Apparent specific gravity, 5-15

Approach
   maneuver, 4-25
   structural, 3-19
   tangent length, 4-28
Arc
   circular, 4-5
   definition, 4-6
   definition curve, 4-6
   length, 4-7
Area
   circulation, 3-12
   clear runout, 6-8
   cross-flow, 3-10
   geographic, 5-24
   nonattainment, 1-17
   study, 1-5
   temporary holding, 3-12
AREMA, 4-2
Arterial, 1-11
ASDT, 6-2
Aspect, 2-26
Asphalt, 5-10
   absorption, 5-16
   binder content (tbl), 5-27
   content, effective, 5-15
   content, optimum, 5-20
   hot mix, 5-11
   pavement, 5-10
   pavement behavior, 5-12
   pavement failure, 5-11
   pavement, aggregate size (tbl), 5-10
   pavement, reclaimed, 5-12
   rutting, weak, 5-12
   surfacing layer (fig), 5-10
   target optimum content, 5-18
   voids filled with, 5-16
   volumetric properties, 5-14
Assessment risk, 1-14
Association, National Highway Traffic Safety, 6-5
Assumed meridian, 4-3
ATS, 2-23
   adjustment factor, 2-22
   heavy vehicle adjustment factor, 2-21
Attainable target, 1-2
Attenuator crash, 6-23
Automatic
   counting method, 1-9
   data collection, 2-5
Automobile
   mode, 2-2
   signalized intersection analysis (fig), 2-29
Available
   net corner time-space, 3-14
   total time-space, 3-13
Average
   annual daily traffic (AADT), 1-10, 6-2, 6-5
   end area method, 5-2
   flow, 3-11
   running speed, 4-32
   running speed (tbl), 4-32
   seasonal daily traffic, 6-2
   speed method, 2-5
   start-up time, 3-3
   travel time, 2-6
   walking speed, 3-3

Avoidance
crash, 6-3
maneuver A, 4-26
maneuver B, 4-26
maneuver C, 4-26
maneuver D, 4-26
maneuver E, 4-26

Axis
rotation, 4-35
vision, 6-21

Axle count, 1-9

Azimuth, 4-3, 4-4
-bearing angle conversion, 4-4
from north, 4-4
from south, 4-4
from the back line, 4-4
north meridian, 4-4
south meridian, 4-5

B

Back
stationing, 4-7
tangent, 4-11

Backslope, 6-8, 6-10
Backslopes, 6-12

Balance
line, 5-7
line (fig), 5-7
point, 5-7
point (fig), 5-7

Balancing speed, 3-28
Ballast shoulder, 5-4
Bank volume, 5-4
Banking curve, 4-30
Barn roof section, 6-10
Barrier Flare Rates, 6-15

Base
condition, roadway, 6-16
free-flow speed, 2-10, 2-12

Basic freeway segment, input data (tbl), 2-13
Bayes adjustment, empirical, 6-17

Bearing, 4-4
angle (fig), 4-4
angle tangent, 4-4
equivalent (fig), 4-4
ratio, California, 5-9

Beginning
of curve, 4-5, 4-7
of curve point, 4-7

Behavior
85th percentile driver, 6-21
asphalt pavement, 5-12
driver, 4-27, 6-21
risk-taking, 3-11

Benefit
-cost analysis, 1-15, 6-20
-cost determination, 6-20
road-user, 4-45

Berths, 3-30
Between letter space, 6-22
BFFS, 2-10, 2-12

Bicycle, 2-2
facility analysis, 3-20
friction coefficient, 4-48
mode, 2-9
perception-reaction time, 4-48
stopping distance, 4-48
stopping distance (tbl), 4-49

Bicycle/pedestrian, 3-20

Bicyclist
eye height, 4-49
lateral movement, 4-48
minimum operating space, 4-48

Bike
lane, sight distance, 4-49
lane, vertical curve, 4-49
lanes, 3-20

Bikeway
extended, maximum grade (tbl), 4-50
horizontal curve minimum radius, 4-49
horizontal curve sight distance, 4-49
side friction, 4-49
sight distance, 4-49
stopping distance, 4-49
superelevation rate, 4-49
superelevation, minimum, 4-49
vertical curve, 4-49
width, minimum, 4-48

Binder, 5-14, 5-24
grade selection method, 5-24
high temperature grade selection (tbl), 5-25
temperature grade, 5-24

Bituminous pavement, 5-10
Bleeding (ftn), 5-22

Blindness
color, 6-22
glare, 6-22

Board
electronic counting, 1-9
mechanical counting, 1-9

Book, Green, 6-8

Borrow, 5-4
volume, 5-5

Bottleneck, 4-45
relief, 4-45

Brake
reaction distance, 4-23
reaction time, 4-23

Braking distance, 4-24
Break period, 1-9
Broad traffic study, 1-2

BRT, 3-36
roadway design, 4-50

Buffer space, 6-27
Bulk specific gravity, 5-15
Bumping grade, 5-25

Bus
blockage, 2-30
flow, 3-32
intercity, 3-32
rapid transit, 3-36
stops, 3-32
urban transit, 3-32

Busway
design, 4-50
station design, 3-31

C

Calculating material problem, 5-11
Calibration factor, 6-17

California
bearing ratio, 5-9
kneading compactor, 5-21

Capacity, 2-9
analysis, 2-9
analysis method, 2-9
hauling, 5-5
junction, 3-27
line haul, 3-27
parking, 2-36
planning analysis, 1-10
walkway, 3-7

Captive riders, 3-27
Car equivalent, passenger, 2-10
Category crash, 6-4
Center cone, visual acuity, 6-21

Centerline
geometry, 4-10
radius, 4-42
roadway, 4-18

Central
angle, 4-5, 4-6
curve radius, 4-16

Centrifugal force, 4-30, 4-31
Centrifuge kerosene equivalent, 5-21
Centripetal acceleration, 4-30
Channelization taper, 6-26
Chart, Snellen, 6-22
Checking closure, 4-22
Choice riders, 3-27

Chord
circular arc, 4-42
definition, 4-6
definition curve, 4-6, 4-47
deflection (fig), 4-12
length, 4-6
long, 4-6, 4-42

Circular
arc, 4-5
arc chord, 4-42
curve, 4-5
curve deflection (fig), 4-12
curve deflection angle (fig), 4-13
curve roadway profile (fig), 4-36
curve, fully spiraled (fig), 4-17

Circulation area, 3-12

Class
I two-lane highway, 2-19
II two-lane highway, 2-19
III two-lane highway, 2-19

Classification
soil, Unified (tbl), v, 6-35, A-63
soil, USCS (tbl), v, 6-35, A-63

Clay content, 5-23
Clean Air Act, 1-1

Clear
runout area, 6-8
zone, 4-42, 6-3, 6-8
-zone correction factor, 6-10
-zone distance, 6-8
-zone distance, recommended, 6-8
zone, object in, 6-3, 6-10

Clearance
analysis, roadside, iii, 6-1, 6-5
from obstruction, horizontal, 4-42
from overhead obstruction, vertical, 4-41
minimum vertical, 4-41, 4-48
on horizontal curve, 4-42
on vertical curve, 4-39
to obstruction, 6-3

Clearinghouse, National Work Zone Safety Information, 6-25

Closed
alignment traverse (fig), 4-10
traverse, 4-22

Closure checking, 4-22
Cloverleaf, 4-44
Clutter, surrounding, 6-3
Coarse aggregate angularity, 5-23
Code, International Building, 2-35

Coefficient
bicycle friction, 4-48
correlation, 1-4
of determination, 1-3, 1-4

Cohesiometer, Hveem, 5-22

Collection
automatic data, 2-5
data, 2-5
manual data, 2-5

Collector
-distributor lane, 4-44
systems, 2-2

Collision, 6-16
diagram, 6-18
diagram (fig), 6-18

Color
blindness, 6-22
traffic control device, 6-22
traffic sign, 6-22

Comfort factor, 4-31
Community need, 1-14
Compacted paving mixture analysis, 5-14

Compactor
kneading, 5-21
Superpave design gyratory (tbl), 5-27
Superpave gyratory, 5-26

Concrete
pavement, 5-10
slab thickness, 5-9

Condition
crossing signal phase, 3-12
one, 3-12
signal phase, 3-12
signal phase, one, 3-12
signal phase, two, 3-12
two, 3-12

Confidence level, 2-3, 6-5
Configuration ramp, 4-44
Conflict, within pedestrian cross-flow, 3-8
Conflict, 6-27
    analysis, iii, 6-1, 6-27
    probability, 3-8
Consensus aggregate property, 5-23
Constant deceleration rate, 4-24
Constraint, 1-13
    threshold, 1-13
Construction
    of transportation facilities, 5-1
    transportation, air quality impact, 1-17
Content
    asphalt binder (tbl), 5-27
    clay, 5-23
    effective asphalt, 5-15
Contour, aerial mapping, 4-3
Control
    delay, 2-32
    delay and LOS, 2-33
    delays, 3-27
    environmental, 1-14
    signal, 2-27
    traverse, 4-23
    zone, temporary traffic, iii, 6-1, 6-26
Conversion, azimuth-bearing angle, 4-4
Coordinated signal system, 2-27
Cordon survey, 1-6
Correlation coefficient, 1-4
Corridor, 4-2
    alignment study, 4-2
    survey, 1-6
Cost
    -benefit analysis, highway safety, 6-19
    crash, 6-20
    crash (tbl), 6-20
    delivery, 5-6
    -effectiveness index, 6-19
    -effectiveness method, 6-19
    encroachment, 6-20
    equipment, 5-6
    expected crash, 6-20
    labor, 5-6
Count
    axle, 1-9
    intersection, 1-10
    location, 1-10
    pathway mid-block segment, 1-10
    period, 1-7
    roadway mid-block segment, 1-10
    vehicle, 1-10
Counter
    permanent automatic, 1-9
    portable automatic, 1-9
Countermeasure, 6-2, 6-17
    highway safety, 6-16
Counting
    method, 1-8
    method, automatic, 1-9
    method, manual, 1-8
Crack shrinkage, 5-12
Cracking
    alligator, 5-12
    fatigue, 5-12
    low-temperature, 5-12
    reduction, fatigue, 5-12
Crash, 6-4, 6-16
    analysis, iii, 6-1, 6-4, 6-5
    attenuator, 6-23
    avoidance, 6-3
    barrier impact, 6-23
    barrier impact, deceleration during, 6-23
    category, 6-4
    cost, 6-20
    cost (tbl), 6-20
    cost estimate, societal (tbl), 6-19
    cost, expected, 6-20
    cushion, 6-23
    data, 6-5, 6-16
    data analysis, 6-5
    estimation, 6-16

    experience, 2-28
    exposure, 6-4
    fixed-object, 6-6
    fixed-object (tbl), 6-6
    frequency, 6-16
    frequency (ftn), 6-16
    frequency data (fig), 6-19
    frequency rate, 6-4
    modification factor, 6-17
    prediction, 6-7
    probability, 6-8
    rate, 6-4, 6-16
    rate (ftn), 6-16
    severity, 6-16
    site condition, 6-18
    statistics, 6-19
    threshold, minimum reportable, 6-4
    type, 6-4, 6-16
Crest
    vertical curve, 4-18, 4-49
    vertical curve (fig), 4-19
    vertical curve (tbl), 4-52
    vertical curve sight distance, 4-49
Criteria
    noise abatement, 1-16
    noise abatement (tbl), 1-17
    optimization, 1-14
    pollutant, 1-17
    schedule and phasing, 1-14
Critical
    clearance distance, 4-39, 4-41
    foreslopes, 6-12
Cross
    -flow areas, 3-10
    -flow walkway interference, 3-8
    section element, 6-3
    section, railroad, 5-3, 5-4
    section, roadway, 5-2
    slope, 4-30
Crossing signal phase condition (fig), 3-12, 3-14
Crosswalk time-space available, 3-14
Crosswalks, 3-3
Crown
    adverse, 4-35
    normal, 4-35
Crst vertical curve (tbl), 4-53
CTR, A-38
Curb, 5-3
Curvature point of, 4-5, 4-7
Curve, 4-2
    arc definition, 4-6
    banking, 4-30
    beginning of, 4-5, 4-7
    bike lane vertical, 4-49
    bikeway vertical, 4-49
    chord definition, 4-6, 4-47
    circular, 4-5
    clearance on horizontal, 4-42
    clearance on vertical, 4-39
    crest vertical, 4-18, 4-49
    crest vertical (fig), 4-19
    crest vertical (tbl), 4-52
    degree of, 4-6, 4-16
    degree of (fig), 4-6
    elements, additional (fig), 4-11
    end of, 4-5, 4-7
    fully spiraled circular (fig), 4-17
    grade, 4-19
    horizontal, 4-5
    horizontal circular (fig), 4-5
    layout, 4-6, 4-7
    length of, 4-6
    midpoint of, 4-11
    parabolic, 4-18, 4-19
    parabolic (fig), 4-19
    point of continuing (PCC), 4-47
    point, beginning of, 4-7
    radius, 4-30, 4-42, 6-3
    sag vertical, 4-18, 4-28
    sag vertical (fig), 4-19
    sag vertical (tbl), 4-52
    simple, 4-5

    spiral, 4-16
    spiral (tbl), 4-17
    symmetrical vertical, 4-20
    tangent, 4-11
    tangent length, 4-7
    tangent to, 4-5
    unsymmetrical vertical, 4-20
    vertical, 4-18
    widening, 6-3
Cushion, crash, 6-23
Cut, 5-2
    and fill quantity, 5-4
    and fill section, 5-2
    and fill section (fig), 5-3
    and fill slope, 5-2
    slope, 6-8, 6-10

D

Daily traffic, average, 6-2
Damage
    property, 6-4
    scale, KABCO, 6-16
    threshold of hearing, 1-16
Data
    accident, 6-16
    analysis, 2-9
    analysis, crash, 6-5
    collection, 2-5
    collection, automatic, 2-5
    collection, manual, 2-5
    crash, 6-5, 6-16
    crash frequency (fig), 6-19
    facility, 6-16
    input, for basic freeway segments (tbl), 2-13
    input, for multilane segments (tbl), 2-13
    reduction, 1-9
    toll collection system, 2-8
    traffic volume, 6-16
Deceleration
    during crash barrier impact, 6-23
    limit, transit, 4-50
    rate, 4-43
    rate, typical vehicle, 4-48
Decision
    sight distance, 4-23, 4-25, 4-26
    sight distance (tbl), 4-26
Declination, 4-3
    east, 4-3
    magnetic, 4-3
    minus, 4-3
    plus, 4-3
    west, 4-3
Definition
    arc, 4-6
    chord, 4-6
    curve, arc, 4-6
Deflection, 4-11
    angle, 4-4, 4-5, 4-7, 4-12
    angle method, 4-9, 4-12
    angle, circular curve (fig), 4-13
    chord (fig), 4-12
    circular curve (fig), 4-12
    left, 4-5
    negative, 4-5
    positive, 4-5
    right, 4-5
    total curve (fig), 4-12
Deformation, permanent, 5-12
Degree
    of curve, 4-6, 4-7, 4-16
    of curve (fig), 4-6
Delay, 1-12, 2-3
    parking terminal LOS (tbl), 2-37
    pedestrian intersection, 3-11
Deleterious material, 5-24
Delineation, 6-3
Delivery cost, 5-6
Demand, 2-9
    volume, 2-12

Density, 2-10
    pedestrian, 3-6
Depth perception, 6-22
Design
    analysis, 2-10
    BRT roadway, 4-50
    busway, 4-50
    busway station, 3-31
    designation, 4-44
    driver, 6-21
    final, 4-3
    friction factor, 4-31
    Guide, Roadside, 6-8
    pavement, 5-9
    pavement thickness, 5-9
    safe roadway, 6-2
    selection, aggregate, 5-26
    speed, 3-3, 4-9, 4-23, 4-30, 4-31, 4-32, 4-48, 6-5
    speed (tbl), 4-32
    Standards for Accessible, 2-35
    stopping distance, 6-21
    superelevation rates, v, 6-35, A-9, A-10, A-11, A-12, A-13
    variable-slope, 6-10
    vehicle, 2-36
Designation, design, 4-44
Desire line, 1-10
Destination, trip, 1-2
Determination
    benefit-cost, 6-20
    coefficient of, 1-3, 1-4
Device, portable recording, 1-9
Devices, Manual on Uniform Traffic Control, 6-22
Diagnosis
    procedure, transportation network, 6-18
    transportation network, 6-18
Diagram
    collision, 6-18
    collision (fig), 6-18
    free-body (fig), 4-31
    mass, iii, 5-1, 5-7
    mass-haul, 5-7
    traffic link and node, 1-10
    traffic network, 1-10
    travel time, speed, and delay, 2-9
Diamond interchange, 4-44
Direct observation method, 2-8
Direction, 4-4
Directional
    interchange, 4-44
    split, unbalanced, 3-7
    volume, 2-6
Discomfort, threshold of, 1-16
Discontinuous objects, 3-17
Distance
    between stations (fig), 3-29
    bicycle stopping, 4-48
    bicycle stopping (tbl), 4-49
    brake reaction, 4-23
    braking, 4-24
    clear-zone, 6-8
    crest vertical curve sight, 4-49
    critical clearance, 4-39, 4-41
    decision sight, 4-23, 4-25, 4-26
    decision sight (tbl), 4-26
    design stopping, 6-21
    diversion, 3-18
    external, 4-5, 4-19
    external (fig), 4-19
    freehaul, 5-7
    haul (fig), 5-7
    headlight sight, 4-28
    horizontal sightline (fig), 4-42
    passing sight, 4-23, 4-26
    passing sight (tbl), 4-27
    perception-reaction, 4-23
    PIEV, 4-23
    PRT sight, 6-21
    radial offset, 4-42
    recommended clear-zone, 6-8
    route, 4-7
    shy, 3-7, 6-10
    shy (tbl), 3-8
    sight, 4-23, 4-28, 6-3
    stopping sight, 4-23, 4-28, 4-42
    stopping sight (tbl), 4-24
    tangent, 4-5
    total stopping, 4-43
    traffic sign viewing, 6-22
Distraction, 6-3
Distribution
    Gaussian, 2-4
    normal, 2-4
    Student's t-, 2-4
Ditch, 6-3
Diversion
    distance, 3-18
    distance component (fig), 3-18
Downgrade
    specific, 2-17
    specific, passenger car equivalents (tbl), 2-23
Downstream taper, 6-26, 6-27
Draft environmental impact statement, 1-2
Drain grate, 6-3
Drainage
    ditches, 6-13
    roadway, 4-28, 4-36
    structures, 6-12
    system, 5-3
    system, railroad, 5-3
Drawing, profile, 5-7
Driver
    age, PRT, 6-21
    behavior, 4-27, 6-21
    behavior, 85th percentile, 6-21
    design, 6-21
    education, 6-21
    error, 6-23
    eye height, 4-26, 4-27, 4-42
    noncompliance, 6-21
    performance, 6-21
    reaction time (fig), 4-25
    reasonable worst case, 6-21
Dumping, 5-5
Dust
    proportion, 5-16
    -to-binder ratio, 5-16
Dwell time, 3-28, 3-33
Dynamic programming, 1-13

E

Earthwork, 5-2
    limit of, 5-2
    mass-haul quantities, 5-4
    side slope, 6-3
Easement, 4-16
East declination, 4-3
Economic
    analysis, iii, 6-1, 6-19
    analysis process, 6-19
    appraisal, HSM, 6-19
    overhaul limit, 5-7
Edge-of-traveled-way, v, 6-35, A-49, A-50, A-51, A-52, A-53, A-54, A-55, A-56
Education, driver, 6-21
Effective
    asphalt content, 5-15
    available time-space, pedestrian, 3-15
    green time, 3-11, 3-15
    length, walkway obstruction, 3-7
    specific gravity, 5-15
    walkway width, 3-5, 3-7
    width reduction, 3-7
Effectiveness, 6-5
    cost-, method, 6-19
    index, cost-, 6-19
Eight-hour vehicular volume, 2-27
Elderly persons, 3-3
Electronic counting board, 1-9
Element
    additional curve (fig), 4-11
    cross section, 6-3
    roadway network, 6-16
Element, roadway, 6-16
Elevation turning point, 4-19
Elliptical pedestrian space (fig), 3-5
Elongated particle, flat and, 5-23
Emergency stopping rate, 4-24
Emotion, 6-21
Empirical Bayes adjustment, 6-17
Encroachment
    angle, 6-8
    cost, 6-20
    path, 6-8
Encroachments, 6-7
End
    area method, average, 5-2
    of curve, 4-5, 4-7
Endwall, 6-3
Envelope, impact, 6-8
Environmental
    control, 1-14
    impact analysis, 4-3
Environments, 3-3
Equation
    fitted curve, 1-3
    regression, 1-3
Equipment cost, 5-6
Equivalent
    bearing (fig), 4-4
    centrifuge kerosene, 5-21
    passenger car, 1-10, 2-10, 2-25
    single-axle loads, 5-22
Equivalents, PTSF passenger car (tbl), 2-23
Error
    driver, 6-23
    permitted, 2-3
ESALs, 5-22
Estimation, crash, 6-16
Evaluation, link-based, 3-16
Excavating and loading, 5-5
Excavation, unsuitable, 5-5
Excess
    expected average crash frequency, 6-17
    value, 6-18
Expected
    average crash frequency, 6-18
    crash cost, 6-20
Exposure
    crash, 6-4
    highway safety, 6-16
Extended bikeway, maximum grade (tbl), 4-50
External
    distance, 4-5, 4-19
    distance (fig), 4-19
    -external trip, 1-5
    -internal trip, 1-5
    trip, 1-5
Eye
    height, bicyclist, 4-49
    height, driver, 4-26, 4-27
    visual axis, 6-21

F

Facility, 2-9, 3-16
    data, 6-16
    interrupted flow, 2-9
    parking, 2-35
    pedestrian, 3-3
    transportation, 6-16
    uninterrupted flow, 2-9, 3-9
Factor
    calibration, 6-17
    clear-zone correction, 6-10
    comfort, 4-31
    design friction, 4-31
    heavy vehicle adjustment, 2-25
    shrinkage, 5-5
    side friction, 4-31, 4-32

INDEX I-5

superelevation transition adjustment (fig), 4-38
swell, 5-5
Failure asphalt pavement, 5-11
FAST, 1-1
Fatality, 6-4
Fatigue
   cracking, 5-12
   cracking reduction, 5-12
FDW interval, 3-11
Feature, topographic, 4-2
Federal
   -Aid Highway Act, 1-1
   alphabet, 6-22
FFS, 2-10, 2-12
   Adjustment for Right-Side Lateral Clearance, 2-14
FHWA, 6-24
Field
   alignment, 4-10
   condition, crash, 6-18
   condition, transportation, 6-16
   layout, 4-12
   layout method, 4-9
   maximum audible, 1-16
   measured free-flow speed, 2-12
   staking, 4-10
Fill, 5-2
   and cut slope, 5-2
   quantity, cut and, 5-4
   slope, 6-8, 6-10
Final
   alignment, 4-10
   design, 4-3
   location survey, 4-3
Fine, 5-9
   aggregate angularity, 5-23
Fire walkway, 3-10
Fitted curve equation, 1-3
Fixed
   delay, 2-3
   gauge, 4-46
   object (fig), 3-8
   -object crash, 6-6
   -object crash (tbl), 6-6
   routing, 3-32
Fixing America's Surface Transportation Act (FAST), 1-1
Flashing don't walk interval, 3-11
Flat and elongated particle, 5-23
Flexible pavement, 5-10
Floating car method, 2-5
Flow
   average, 3-11
   bus, 3-32
   facility, interrupted, 2-9
   facility, uninterrupted, 2-9
   free, 6-2
   friction, 2-10
   interrupted, 3-11
   pedestrian, 3-3
   quality index, 3-9
   rate, 2-10
   rate, maximum, 3-6
   rate, unit pedestrian, 3-9
   traffic, 1-2
   two-directional, 3-7
   typical design-day traffic, 1-8
   unidirectional, 3-7
   uniform pedestrian, 3-9
   value, asphalt mix, 5-19
Flushing, 5-22
Flushing (ftn), 5-22
Force
   centrifugal, 4-30, 4-31
   lateral, 4-30
   maximum g-, 6-23
   stopping, 6-23
Foreslope, 6-8, 6-10
Foreslopes, 6-12
Forward stationing, 4-7
Foundation, 5-9
Four-hour vehicular volume, 2-27

Free
   -flow speed, base, 2-10
   -body diagram (fig), 4-31
   -flow speed, 2-10, 2-12
   flow, 6-2
Freehaul distance, 5-7
Freeway, 1-11, 2-9, 2-10, 2-12
   methodology (fig), 2-11
Frequency
   crash, 6-16
   crash (ftn), 6-16
   rate, crash, 6-4
Friction
   coefficient, bicycle, 4-48
   factor, design, 4-31
   flow, 2-10
   maximum side (tbl), 4-34
Full movement activity, 1-8
Fully
   or semi-actuated signal control (tbl), 2-46
   spiraled circular curve (fig), 4-17
Function
   objective, 1-12
   optimization, 1-12
   safety performance, 6-16

G

g-force, maximum, 6-23
Gauge, 4-46
   fixed, 4-46
Gaussian distribution, 2-4
GDHS, 4-2
General
   terrain, 2-16
   terrain segment, 2-17
Generalized
   DSV for rural multilane highway (tbl), 2-39
   DSV for urban multilane highways (tbl), 2-40
Generation, trip, 1-3, 2-35
Generator, trip, 1-2
Geographic area, 5-24
Geometric
   data, 2-12
   design, 4-2
   Design of Highways and Streets (*GDHS*), 4-2
Geometry
   centerline, 4-10
   horizontal, 4-11
   roadside, 6-20
Glare, 6-22
   blindness, 6-22
   recovery, 6-22
Gradation, 5-13
   limit, 5-13
   maximum density (fig), 5-13
   Superpave aggregate mixture (tbl), 5-14
Grade
   binder temperature, 5-24
   bumping, 5-25
   change, rate per station, 4-19
   curve, 4-19
   extended bikeway, maximum (tbl), 4-50
   point of, 4-19
   railroad, 4-47
   rate, change in, 4-28
   separation, 4-44
   separation warrant, 4-44
Gradient
   maximum relative, 4-36
   maximum relative (tbl), 4-37
Grate, drain, 6-3
Gravity
   apparent specific, 5-15
   bulk specific, 5-15
   effective specific, 5-15
   maximum specific, 5-15
Green Book, 4-2, 6-8

Grid meridian, 4-3
Guide, Roadside Design, 6-8
Guiderail, 6-3

H

Haul
   capacity, line, 3-27
   diagram, mass-, 5-7
   distance (fig), 5-7
Hauler return, 5-6
Hauling, 5-5
   capacity, 5-5
   mass, 5-5
   volume, 5-5
HCM, 1-2, 1-10, 3-36
Headlight sight distance, 4-28
Heavy
   vehicle adjustment factor, 2-25
   vehicle adjustments, 2-16
   vehicle PCEs (tbl), 2-17
Height
   bicyclist eye, 4-49
   driver eye, 4-26, 4-27, 4-42
   load, 4-41
   object, 4-26, 4-27, 4-42
Hierarchy of data quality, 1-5
High pavement design temperature, 5-24
Highway, 6-16
   Capacity Manual (*HCM*), 1-2, 1-10
   curve, 4-7
   mode, 2-9
   multilane, 1-11, 2-9
   pavement thickness, 5-9
   segment, 6-4
   two-lane, 1-11, 2-9
Highways, 2-12
Holding area waiting time, 3-14
Home-interview origin-destination survey, 1-5
Horizontal
   acceleration rates, 3-29
   circular curve (fig), 4-5
   clearance from obstruction, 4-42
   control line, 4-18
   curve, 4-5
   curve layout, iii, 4-1, 4-9
   curve minimum radius, bikeway, 4-49
   curve sight distance, bikeway, 4-49
   curve, clearance on, 4-42
   deceleration rates, 3-29
   geometry, 4-11
   load, 5-9
   sightline distance (fig), 4-42
   sightline offset, 4-42
   sightline offset (HSO), 4-11
Hot mix asphalt, 5-11
Hourly volume, 2-6
HSM crash estimation method, 6-16
HSO, 4-42
Hveem
   cohesiometer, 5-22
   mix design method, 5-21
   stabilometer, 5-21
Hveem's resistance value, 5-9

I

IBC, 2-35
Identification, 6-21
Impact
   crash barrier, 6-23
   envelope, 6-8
   neighborhood, 1-14
   traffic noise, 1-16
Impaired mobility, 3-3
Improvement, safety, 4-45
In situ volume, 5-4
Inaccessible PI (fig), 4-15
Incremental delay, 2-33
Index, cost-effectiveness, 6-19
Initial queue delay, 2-33
Injury, 6-4

Input
- data for basic freeway segments (tbl), 2-13
- data for multilane highway segments (tbl), 2-13

Inside
- lane, 4-42
- loop ramp, 4-44

Integer programming, 1-13
Interchange configuration (fig), 4-45
Intercity buses, 3-32
Interference
- cross-flow walkway, 3-8
- surface traffic, 3-27

Intermodal Surface Transportation Efficiency (ISTEA), 1-1
Internal
- -external trip, 1-5
- -internal trip, 1-5

International Building Code, 2-35
Interrupted
- flow, 3-11
- flow facility, 2-9

Intersection, 1-10, 1-11, 6-3, 6-4
- angle, 4-5
- count, 1-10
- delay, pedestrian, 3-11
- LOS, 2-33
- LOS procedure for signalized (fig), 3-12
- point of, 4-5
- roadway, 4-43
- unsignalized, 3-5

Interstate highway system, 4-41
Interval
- amber, 1-10
- FDW, 3-11
- flashing don't walk, 3-11
- walk, 3-11

Interview
- method, 2-8
- roadside, 1-6

Irreducible value, 1-2
Isogonic line, 4-3
ISTEA, 1-1

J
Junction capacity, 3-27

K
K
- -value, 4-28
- -value (tbl), 4-52
- -value method, 4-27

KABCO severity scale, 6-16
Kerosene equivalent, centrifuge, 5-21
Kneading compactor, 5-21

L
Labor cost, 5-6
Landings, 3-19
Lane
- and shoulder width adjustment factor (tbl), 2-21
- bike, 3-20
- group, 2-33
- groups, 2-28
- inside, 4-42
- width, 2-12

Lateral
- acceleration, 4-30
- clearance, 2-12
- force, 4-30
- movement, bicyclist, 4-48
- obstruction, 4-48

Layer, asphalt surfacing (fig), 5-10
Layout
- alignment, 4-10
- curve, 4-6, 4-7
- field, 4-12
- horizontal curve, iii, 4-1, 4-9

LCCA, 1-13

Left
- -hand exit ramp, 4-44
- deflection, 4-5

Length
- approach tangent, 4-28
- arc, 4-7
- chord, 4-6
- curve tangent, 4-7
- minimum runoff, 4-37
- minimum vertical curve (tbl), 4-29
- of curve, 4-6
- PRT amber phase, 6-21
- runoff, 4-36
- superelevation transition, 4-35
- taper, 6-27
- vertical curve, 4-28, 4-29

Letter height, sign, 6-22
Level
- absolute noise, 1-16
- confidence, 2-3, 6-5
- -of-service rating, 1-12
- of service, 1-10, 3-9
- of service (LOS), 2-10
- reference, 1-16
- significance, 2-3
- sound, 1-16
- terrain, 2-16, 4-9

License
- plate method, 2-8
- plate study, 1-5

Life-cycle cost analysis, 1-13
Light
- -intensity-contrast ratio, 6-22
- rail transit, 3-36
- rail vehicles, 3-28

Limit
- economic overhaul, 5-7
- gradation, 5-13
- of capacity, 2-11
- of earthwork, 5-2
- of profitable haul, 5-7
- transit acceleration, 4-50

Limited area study, 1-2
Line
- 0.45 power, 5-13
- azimuth from the back, 4-4
- balance, 5-7
- balance (fig), 5-7
- desire, 1-10
- haul capacity, 3-27
- horizontal control, 4-18
- isogonic, 4-3
- maximum density gradation, 5-13
- of sight, 4-42
- of sight, underpass (fig), 4-40
- stops, 3-32

Linear programming, 1-13
Link, 3-16
- -based evaluation, 3-16

Load
- height, 4-41
- horizontal, 5-9
- traffic (tbl), 5-25
- volume, 5-5

Loading, 5-5
- excavating and, 5-5
- pavement, 5-9

Location
- count, 1-10
- error, rounding (fig), 4-23
- preliminary, 4-3
- survey and design, final, 4-2
- survey, final, 4-3
- survey, preliminary, 4-2

Long
- chord, 4-6, 4-42
- -Term Pavement Performance program, 5-25

Los Angeles abrasion test, 5-23
LOS, 1-12
- control delay, 2-33
- criteria (tbl), 2-17
- methodology, 3-36

nominal walkway capacity, 3-9
parking terminal (tbl), 2-47
parking terminal delay (tbl), 2-37
pedestrian, 3-9
pedestrian, signalized intersection (tbl), 3-12
procedure, signalized intersection (fig), 3-12
signalized intersection, 3-11
turning vehicle effect, 3-15
walkway, 3-9

Low
- pavement design temperature, 5-25
- -temperature cracking, 5-12

LRT, 3-36
LRV systems, 3-28
LTPP, 5-25
LTPPBind, 5-25

M
Magnetic
- declination, 4-3
- meridian, 4-3

Mainline route, 4-41
Major
- arterial, 2-10
- collector roads, 2-2

Maneuver
- approach, 4-25
- avoidance, A, 4-26
- avoidance, B, 4-26
- avoidance, C, 4-26
- avoidance, D, 4-26
- avoidance, E, 4-26

Maneuvering and weaving section, 6-3
Manual
- counting method, 1-8
- data collection, 2-5
- *for Assessing Safety Hardware*, 6-3
- *Highway Capacity* (HCM), 1-10
- *on Uniform Traffic Control Devices*, 6-22

Map, traffic flow, 1-7
Mapping
- aerial, 4-3
- contour, aerial, 4-3
- scale, aerial, 4-3

Marshall
- method, mix design, 5-18
- method, stability, 5-18
- mix criteria, 5-20
- mix method, 5-18
- mix test (fig), 5-19
- stability, 5-18

Mass
- diagram, iii, 5-1, 5-7
- -haul diagram, 5-7
- -haul quantities, earthwork, 5-4
- hauling, 5-5
- transit, 3-26

Material
- aggregate, 5-9
- deleterious, 5-24
- problem calculation, 5-11
- procedure, 5-11
- selection, 1-14

Maximum
- audible field, 1-16
- comfortable stopping rate, 4-24
- density gradation (fig), 5-13
- density gradation line, 5-13
- flow rate, 2-17, 3-6
- *g*-force, 6-23
- grade, extended bikeway (tbl), 4-50
- hauling weight, 5-5
- relative gradient, 4-36
- relative gradient (tbl), 4-37
- relative slope, 4-37
- service flow, 2-18
- side friction (tbl), 4-34
- size, 5-13
- size, nominal, 5-13
- specific gravity, 5-15

superelevation rate, 4-33
superelevation rate (tbl), 4-34
Mean travel speed, 2-3
Measure noise reduction, 1-16
Mechanical counting board, 1-9
Median type, 2-12
Merging taper, 6-27
Meridian
   1, 4-3
   assumed, 4-3
   grid, 4-3
   magnetic, 4-3
   north, 4-4
   principal, 4-3
   south, 4-4
Mesoscale study, 1-17
Met vehicle, 2-5, 2-6
Method
   5, superelevation, 4-33
   automatic counting, 1-9
   average end area, 5-2
   average speed, 2-5
   binder grade selection, 5-24
   capacity analysis, 2-9
   counting, 1-8
   crash estimation, HSM, 6-16
   deflection angle, 4-9, 4-12
   direct observation, 2-8
   field layout, 4-9
   floating car, 2-5
   Hveem, 5-21
   interview, 2-8
   K-value, 4-27
   license plate, 2-8
   manual counting, 1-8
   Marshall, 5-18
   modified rate of change, 4-20
   Monte Carlo, 6-7
   moving vehicle, v, 2-5, 6-35, A-1
   peak searching, 6-17
   prismoidal, 5-2
   proportional offset, 4-19
   pyramid, 5-22
   side friction distribution, 4-32
   side friction distribution (fig), 4-33
   simple ranking, 6-17
   sliding window, 6-17
   superelevation, 4-35
   superelevation (fig), 4-35
   superelevation distribution, 4-32
   superelevation distribution (fig), 4-33
   tangent offset, 4-9, 4-20
   tangent offset (fig), 4-15
   test vehicle, 2-5
   toll road card, 2-8
   vehicle intercept, 1-6
Method (*see also type*), Marshall mix, 5-18
Methodology, transit, 3-33
Metropolitan planning organization (MTO), 1-2
Microscale study, 1-17
Middle
   ordinate, 4-6, 4-11, 4-19, 4-42
   ordinate (fig), 4-19
Midpoint of curve, 4-11
Minimum
   bikeway width, 4-48
   curve radius, 4-35
   curve radius (tbl), 4-34
   operating space, bicyclist, 4-48
   radius, v, 6-35, A-14, A-15, A-16, A-17, A-18, A-19, A-20, A-21, A-22
   reportable crash threshold, 6-4
   runoff length, 4-37
   sample size, 2-3
   superelevation, bikeway, 4-49
   vertical clearance, 4-41, 4-48
   vertical curve length (tbl), 4-29
   walking path, 3-5
Minor
   arterials, 2-2
   collector roads, 2-2
Minus declination, 4-3

Mix
   design, Hveem method, 5-21
   design, Marshall method, 5-18
   flow value, 5-19
   stability, 5-19
   traffic flow vehicle, 4-43
Mixture
   gradation, Superpave aggregate (tbl), 5-14
   voids in the total, 5-16
Mobility
   aids, 3-3
   impaired, 3-3
Mode
   bicycle, 2-9
   highway, 2-9
   pedestrian, 2-9
   transit, 2-9
Modification factor, crash, 6-17
Modified rate of change method, 4-20
Module parking, 2-36
Monte Carlo method, 6-7
Mountainous terrain, 2-17, 4-9
Movement group, 2-28
Moving vehicle method, v, 2-5, 6-35, A-1
MPC, 4-11
Multilane
   highway, 1-11, 2-9
   highway maximum service flow rates (tbl), 2-18
   highway segment, input data (tbl), 2-13
   ramp, 4-44
Multiple objective, 1-13
MUTCD, 6-25

N

National
   Ambient Air Quality Standards, 1-17
   Geodetic Survey (NGS), 4-23
   Highway System (NHS), 6-3
   Highway Traffic Safety Association, 6-5
   Mass Transit Assistance Act, 1-2
   Work Zone Safety Information Clearinghouse, 6-25
Natural aggregate, 5-12
NCHRP Report 350, 6-3
Need, community, 1-14
Negative
   deflection, 4-5
   deflection angle, 4-11
Neighborhood impact, 1-14
Net corner time-space available, 3-14
Network
   diagnosis, transportation, 6-18
   node, transportation, 6-16
   screening, transportation, 6-17
   transportation, 6-16
NGS, 4-23
NHTSA, 6-5
No-passing zones, 2-22
Node transportation network, 6-16
Noise, 1-16
   abatement criteria, 1-16
   abatement criteria (tbl), 1-17
   reduction measure, 1-16
   study, 1-16
Nominal
   maximum size, 5-13
   walkway capacity LOS, 3-9
Non-recoverable foreslopes, 6-12
Non-recoverable slope, 6-8
Nonattainment area, 1-17
Noncompliance
   driver, 6-21
   signal, 3-11
Nonmerging taper, 6-27
Normal
   crown, 4-35
   distribution, 2-4

North
   azimuth from, 4-4
   meridian, 4-4
   meridian azimuth, 4-4
Number of lanes, 2-12

O

O-D
   study, 1-5
   survey, 1-5
Object
   discontinuous, 3-17
   fixed (fig), 3-8
   height, 4-26, 4-27, 4-42
   height, sight distance, 4-49
   in the clear zone, 6-3, 6-10
Objective
   function, 1-12
   multiple, 1-13
   single, 1-13
   sublevel, 1-12
   superior, 1-13
Obstruction
   clearance, 6-3
   lateral, 4-48
   vertical clearance from overhead, 4-41
   walkway, 3-7
Occurrence, rate of, 6-4
Off, 3-32
   -street facility analysis (tbl), 3-13
   -street pedestrian walkway, 3-20
Offset
   horizontal sightline, 4-11, 4-42
   tangent (fig), 4-19
Offtracking, A-38
On
   -line stops, 3-34
   -street performance of transit, 3-33
Onboard transit survey, 1-6
One-quadrant, 4-44
Open
   alignment traverse (fig), 4-10
   traverse, 4-22
Operation, railroad, 3-36
Operational
   analysis, 1-12
   delay, 2-3
Optimization, 1-12
   criteria, 1-14
   function, 1-12
   productivity, 1-14
   unknown, 1-13
   variable, 1-13
Order, 4-23
Ordinate
   middle, 4-6, 4-11, 4-19, 4-42
   middle (fig), 4-19
Origin-destination study, 1-5
Outside lane, 1-11
Overhaul
   limit, economic, 5-7
   paid, 5-7
Overhead obstruction, vertical clearance from, 4-41
Overpass, roadway, 4-46
Oversaturated, 2-10
Overtaking vehicle, 2-5, 2-6
Oxidation, 5-14

P

Pacing zone, 3-5
Paid overhaul, 5-7
Pain, threshold of, 1-16
Parabolic
   curve, 4-18, 4-19
   curve (fig), 4-19
Pareto's 80/20 rule, 5-27
Parking
   capacity, 2-36
   facility, 2-35
   module, 2-36
   space, 2-36
   space, accessible, 2-35

stall, 2-36
terminal delay LOS (tbl), 2-37
terminal LOS, recommended design parameter (tbl), 2-47
volume, peak hour (tbl), 2-36
Particle, flat and elongated, 5-23
Passed vehicle, 2-5
Passenger
car equivalent, 1-10, 2-10, 2-25
car equivalents (PCEs), 2-22
car equivalents for level terrain (tbl), 2-23
car equivalents for rolling terrain (tbl), 2-23
car equivalents for specific downgrades (tbl), 2-23
Passing
capacity, 2-24
demand, 2-24
maneuver, 4-27
sight distance, 4-23, 4-26, 4-52
sight distance (tbl), 4-27
zone length (tbl), 4-28
Path
encroachment, 6-8
minimum walking, 3-5
shared-use, 3-20
Pathway mid-block segment count, 1-10
Pattern, wave, 3-3
Pavement
asphalt, 5-10
behavior, asphalt, 5-12
bituminous, 5-10
concrete, 5-10
design, 5-9
flexible, 5-10
loading, 5-9
reclaimed asphalt, 5-12
surface, 5-3
temperature, 5-24
thickness design, 5-9
thickness, highway, 5-9
Paving mixture analysis, compacted, 5-14
PCC, 4-47
PCEs, 2-17
for SUT and TT, 2-38
Peak
arrival rate, 2-35
departure rate, 2-35
hour, 2-27
hour flow rate, 1-8
hour parking volume (tbl), 2-36
-hour volume, 2-12
searching method, 6-17
Pedestrian, 2-2
cross flow, conflict within (fig), 3-8
density, 3-6
effective available time-space, 3-15
facilities, 3-3
flow, 3-3
flow rate, 1-13, 3-3, 3-17
flow, platoon, 3-10
flow, uniform, 3-9
intersection delay, 3-11
LOS, 3-9
LOS, signalized intersection (tbl), 3-12
-mode LOS criteria (tbl), 3-10
mode, 2-9
service volume, 3-9
space, 3-6, 3-16
space requirement (fig), 3-5
space, elliptical (fig), 3-5
speed, 3-6
speed and flow (fig), 3-7
start-up time, 3-3
travel flow, 3-19
trip purpose, 3-3
unit flow rate, 3-6, 3-9
volume, 2-27
walkway, off-street, 3-20
Per
-hour basis, 5-6
-vehicle-mile basis, 5-6

Percentage of air voids, 5-16
Perception, 6-21
depth, 6-22
-reaction response, 6-21
-reaction distance, 4-23
-reaction subprocess, 6-21
-reaction time, 6-21
-reaction time, bicycle, 4-48
Performance
driver, 6-21
function, safety, 6-16
Period
break, 1-9
count, 1-7
Peripheral
vision, 6-21
zone, 6-21
Permanent
automatic counter, 1-9
deformation, 5-12
Permitted error, 2-3
Person
-trip, 1-7
-trip end, 1-7
PFFS, 2-23
PI inaccessible (fig), 4-15
PIEV, 6-21
distance, 4-23
time, 4-23
Planning
analysis, 1-12, 2-10
study, 1-14
Platform locations, 3-30
Platoon, 3-10
flow, 3-11
pedestrian flow, 3-10
ratio, 2-31
Plus
declination, 4-3
stationing, 4-7
Pneumatic road tube, 1-9
Point, 2-9, 3-16
balance, 5-7
balance (fig), 5-7
beginning of curve, 4-7
of continuing curve (PCC), 4-47
of curvature, 4-5, 4-7
of grade, 4-19
of impending skid, 4-31
of intersection, 4-5
of tangency, 4-5, 4-7, 4-19
performance and link performance (fig), 3-16
subtangent, 4-11
tangent, 4-11
turning, 4-18, 4-19
Policy on Geometric Design of Highways and Streets (GDHS), 4-2
Pollutant criteria, 1-17
Ponding, 4-36
Portable
automatic counter, 1-9
recording device, 1-9
Portion tangent runoff (tbl), 4-39
Positive
deflection, 4-5
deflection angle, 4-11
Postcard study, 1-6
POT, 4-11
Pre-maneuver time, 4-25, 4-26
Predicted average crash frequency, 6-18
Prediction crash, 6-7
Preemption, signal, 3-32
Preemptive width, 3-7
Preliminary
alignment, 4-10
design report, 4-10
engineering, 2-10
location, 4-3
location survey, 4-2
Primer, A Soil, 5-9

Principal
arterial systems, 2-2
meridian, 4-3
Prismoidal method, 5-2
Probability
conflict, 3-8
crash, 6-8
Problem calculation, material, 5-11
Procedure
diagnosis, transportation network, 6-18
material, 5-11
Superpave mix design, 5-23
Process
economic analysis, 6-19
screening, 6-4
Processed aggregate, 5-12
Productivity optimization, 1-14
Profile
circular curve roadway (fig), 4-36
drawing, 5-7
Program
Long-Term Pavement Performance, 5-25
Roadside Safety Analysis, 6-20
Programming
dynamic, 1-13
integer, 1-13
linear, 1-13
Project
retrofit, 1-16
transportation, 6-16
type I, 1-16
type II, 1-16
Property, 5-23
asphalt volumetric, 5-14
consensus aggregate, 5-23
damage, 6-4
source-specific, 5-23
volumetric, 5-14
Proportion dust, 5-16
Proportional offset method, 4-19
PRT, 6-21
amber phase length, 6-21
driver age, 6-21
sight distance, 6-21
PTSF, 2-23
grade adjustment factor, 2-44
passenger car equivalents (tbl), 2-23
Public transportation, 1-1, 3-26
Pull-off stops, 3-32
Purpose, pedestrian trip, 3-3
Pyramid method, 5-22

Q

Quality
flow index, 3-9
of service ratings, 3-33
Quantity, cut and fill, 5-4
Queue
backup, 2-12
discharge, 2-12
red-phase, 3-12
storage ratio, 2-33
wave pattern, 3-3

R

R-value, 5-9
Radial
acceleration, 4-30
offset distance, 4-42
Radius, 4-6
bikeway horizontal curve, minimum, 4-49
centerline, 4-42
central curve, 4-16
curve, 4-30, 4-42, 6-3
minimum, v, 6-35, A-14, A-15, A-16, A-17, A-18, A-19, A-20, A-21, A-22
minimum curve, 4-35
minimum curve (tbl), 4-34
turning, A-38
turning curb, A-38

Railroad
    cross section, 5-3, 5-4
    drainage system, 5-3
    grade, 4-47
    operations, 3-36
    subgrade, 5-3
Ramp, 6-3
    configuration, 4-44
Ranking, simple, 6-17
RAP, 5-12
Rapid transit, bus, 3-36
Rate
    acceleration, 4-42
    change in grade, 4-28
    constant deceleration, 4-24
    crash, 6-4, 6-16
    crash (ftn), 6-16
    crash frequency, 6-4
    deceleration, 4-43
    emergency stopping, 4-24
    flow, 2-10
    horizontal acceleration, 3-29
    horizontal deceleration, 3-29
    maximum comfortable stopping, 4-24
    maximum flow, 3-6
    maximum superelevation, 4-33
    maximum superelevation (tbl), 4-34
    of grade change per station, 4-19
    of occurrence, 6-4
    peak arrival, 2-35
    peak departure, 2-35
    peak hour flow, 1-8
    pedestrian flow, 3-3
    pedestrian unit flow, 3-6
    superelevation, 4-32
    typical vehicle deceleration, 4-48
    weighted average trip generation, 1-4
Rating, quality of service, 3-33
Ratio
    California bearing, 5-9
    dust-to-binder, 5-16
    light-intensity-contrast, 6-22
    water-cement, 5-11
Reaction, 6-21
Reasonable worst case driver, 6-21
Reclaimed, asphalt pavement, 5-12
Recognition acuity, 6-21
Recommended
    clear zone distance, 6-8
    design parameter, parking terminal LOS (tbl), 2-47
Reconnaisance, 4-2
Recourse action, 1-13
Recoverable
    backslopes, 6-12
    foreslopes, 6-12
    slope, 6-8
Recovery glare, 6-22
Red-phase queue, 3-12
Reduced lateral clearance, 2-12
Reduction
    data, 1-9
    effective width, 3-7
    fatigue cracking, 5-12
Reference level, 1-16
Regression equations, 1-3
Rehabilitation, 5-1
Relationship, volume and speed, 1-10
Relative slope, maximum, 4-37
Reliability, 5-25
Relief
    bottleneck, 4-45
    spot congestion, 4-45
Renewal, 5-1
Report, preliminary design, 4-10
Reportable crash threshold, minimum, 6-4
Requirement, volumetric design, 5-28
Residual value, 1-15
Resistance value, Hveem's, 5-9
Response, perception-reaction, 6-21
Restricted zone, 5-13
Restriction, site, 1-14
Retrofit project, 1-16

Return of hauler, 5-6
Rider
    captive, 3-27
    choice, 3-27
Right
    angle to the, 4-4
    deflection, 4-5
    -side clearance, 2-12
Risk
    assessment, 1-14
    -taking behavior, 3-11
Road-user benefit, 4-45
Roadside
    clearance, 2-12
    clearance analysis, iii, 6-1, 6-5
    Design Guide (*RSDG*), 4-2, 6-3, 6-8
    geometry, 6-20
    interview, 1-6
    interview sample size, 1-6
    safety analysis program, 6-20
Roadway, 6-16
    centerline, 4-18
    cross section, 5-2
    density, 2-11
    design, safe, 6-2
    drainage, 4-28, 4-36
    intersection, 4-43
    mid-block segment count, 1-10
    network, 2-28
    network element, 6-16
    overpass, 4-46
    template, 5-2
    terrain, 2-16
    type, 1-11
    underpass, 4-46
Rolling terrain, 2-17, 4-9
Roof section, barn, 6-10
Rotation axis, 4-35
Roundabout, 2-34
Rounding, location error (fig), 4-23
Route
    alignment, 4-2
    alignment layout (fig), 4-23
    alignment tangent, 4-22
    distance, 4-7
Routing, fixed, 3-32
RSAP, 6-20
RSDG, 4-2, 6-5
Running
    speed, 2-8
    speed, average, 4-32
    speed, average (tbl), 4-32
    time, 2-3, 3-29, 3-33
Runoff
    flow rate, 6-13
    length, 4-36
    superelevation, v, 6-35, A-29, A-30
Runout
    length, 6-14
    lengths, tangent, v, 6-35, A-29, A-30
    tangent, 4-35
Rural
    areas, 2-3
    local roads, 2-2
Rush load, 3-32
Rutting, 5-12
    weak asphalt, 5-12
    weak subgrade, 5-12

S
Safe
    Accountable, Flexible, Efficient Transportation Equity Act: A Legacy for Users (SAFETEA-LU), 1-1
    roadway design, 6-2
SAFETEA-LU, 1-1
Safety
    improvement, 4-45
    performance function, 6-16
    traffic, 6-2
    work zone, iii, 6-1, 6-24

Sag
    vertical curve, 4-18, 4-28
    vertical curve (fig), 4-19
    vertical curve (tbl), 4-52, 4-53
Salvage value, 1-15
Sample size, roadside interview, 1-6
Sampling short period, 1-8
Scale
    A-weighted, 1-16
    aerial mapping, 4-3
Schedule and phasing criteria, 1-14
School crossing, 2-27
Screening
    process, 6-4
    transportation network, 6-17
Seasonal daily traffic, average, 6-2
Secondary source, 1-6
Section
    barn roof, 6-10
    cut and fill, 5-2
    cut and fill (fig), 5-3
    railroad cross, 5-4
    weaving and maneuvering, 6-3
Segment, 2-9, 3-16
    basic freeway, input data (tbl), 2-13
    highway, 6-4
    multilane highway, input data (tbl), 2-13
    roadway, 6-16
    study, 2-5
Selection
    binder high temperature grade (tbl), 5-25
    design, aggregate, 5-26
    material, 1-14
    method, binder grade, 5-24
Sensory zone, 3-5
Separation grade, 4-44
Service
    flow rate, 2-11
    level, 3-9
    level of, 1-10
    volume, pedestrian, 3-9
Severity crash, 6-16
Shadow vehicle, 6-27
Shared
    -use, 3-20
    -use paths, 3-21
Shear strength, 5-13
Sheet
    tally, 1-9
    travel time and delay study, 2-6
Shifting taper, 6-27
Short
    period sampling, 1-8
    -term flow, 3-11
Shoulder, 5-3
    ballast, 5-4
    taper, 6-27
    width, 2-20
Shrinkage, 5-4
    crack, 5-12
    factor, 5-5
Shy
    distance, 3-7, 6-10
    distance (tbl), 3-8
    line, 6-14
Side
    friction distribution method, 4-32
    friction distribution method (fig), 4-33
    friction factor, 4-31, 4-32
    friction, bikeway, 4-49
    slope, earthwork, 6-3
Sidewalk, 3-3
    width, 3-16
Sight
    distance, 4-23, 4-27, 4-28, 6-3
    distance, bike lane, 4-49
    distance, bikeway, 4-49
    distance, crest vertical curve, 4-49
    distance, horizontal curve, bikeway, 4-49
    distance, object height, 4-49

distance, PRT, 6-21
distance, undercrossing (fig), 4-40
line of, 4-42
Sightline offset, horizontal, 4-11, 4-42
Sign letter height, 6-22
Signal
    noncompliance, 3-11
    phase condition, 3-12
    phase condition one, 3-12
    phase condition two, 3-12
    phase condition, crossing (fig), 3-12, 3-14
    phase duration, 2-32
    phase, timing, 3-32
    preemption, 3-32
    traffic, 2-26, 3-32
Signalized
    intersection LOS, 3-11
    intersection LOS, pedestrian (tbl), 3-12
    intersection, LOS procedure (fig), 3-12
Significance level, 2-3
Simple
    curve, 4-5
    ranking method, 6-17
Single
    -axle loads, equivalent, 5-22
    objective, 1-13
Site
    condition, crash, 6-18
    condition, transportation, 6-16
    impact analysis, 1-15
    -impact studies, 1-2
    restriction, 1-14
    topography, 4-45
    transportation network, 6-16
Size
    asphalt pavement aggregate (tbl), 5-10
    maximum, 5-13
    minimum sample, 2-3
    nominal maximum, 5-13
Skid, point of impending, 4-31
Slab thickness, 5-9
Sliding window method, 6-17
Slope
    angle, 4-30
    cross, 4-30
    cut, 6-8, 6-10
    cut and fill, 5-2
    earthwork side, 6-3
    fill, 6-8, 6-10
    maximum relative, 4-37
    non-recoverable, 6-8
    recoverable, 6-8
    traversable, 6-8
Snellen chart, 6-22
Snow and ice, 3-19
Soil
    Classification System, Unified (tbl), v, 6-35, A-63
    classification, USCS (tbl), v, 6-35, A-63
Sound, 1-16
    ambient, 1-16
    level, 1-16
Soundness, 5-23
Source
    secondary, 1-6
    -specific property, 5-23
South
    azimuth from, 4-4
    meridian, 4-4
    meridian azimuth, 4-5
Space
    accessible parking, 2-35
    between letter, 6-22
    bicyclist minimum operating, 4-48
    buffer, 6-27
    -mean speed, 2-3
    mean speed, 2-3, 2-7
    parking, 2-36
    pedestrian, 3-6
    requirement, pedestrian (fig), 3-5
    traffic, 6-27
    within letter, 6-22
    work, traffic, 6-27
Spacing, vehicle arrival, 3-28
Specific
    downgrade, 2-17
    property, source-, 5-23
    upgrade, 2-17
Specification, Superpave aggregate, 5-13
Speed
    average running, 4-32
    average running (tbl), 4-32
    average walking, 3-3
    base free-flow, 2-12
    curves, 2-15
    design, 3-3, 4-9, 4-23, 4-30, 4-31, 4-32, 4-48, 6-5
    design (tbl), 4-32
    -determining methodologies, 2-19
    -flow curves (fig), 2-15
    free-flow, 2-10, 2-12
    free-flow, base, 2-10
    pedestrian, 3-6
    running, 2-8
    space mean, 2-3, 2-7
    traffic (tbl), 5-25
    travel, 4-48
    walking, 3-3
Spiral
    curve, 4-16
    curve (tbl), 4-17
    superelevation transition, 4-39
    transition (fig), 4-17
Split, unbalanced directional, 3-7
Spot congestion relief, 4-45
Stability
    asphalt mix, 5-19
    correlation factor, 5-18
    Hveem, 5-21
    Marshall, 5-18
Stabilometer
    Hveem, 5-21
    value, 5-21
Stair width, 3-19
Stairway, 3-19
    up-direction, 3-10
Stakeout, 4-12
Staking field, 4-10
Stall parking, 2-36
Standards
    for Accessible Design, 2-35
    National Ambient Air Quality, 1-17
Start
    -up time, average, 3-3
    -up time, pedestrian, 3-3
Statement, draft environmental impact, 1-2
Station, 4-7
    access capacity, 3-27
    delay, 3-28
    -meter, 5-7
    platform, 3-27
    rate of grade change per, 4-19
    -yard, 5-7
    zero plus zero zero, 4-7
Stationing, 4-8
    ahead, 4-7
    back, 4-7
    forward, 4-7
    plus, 4-7
Stations, distance between (fig), 3-29
Statistics, crash, 6-19
Steering angle, A-38
Stop
    bus, 3-32
    line, 3-32
    on-line, 3-34
    pull-off, 3-32
Stopped-time delay, 2-3
Stopping
    distance, bicycle, 4-48
    distance, bicycle (tbl), 4-49
    distance, bikeway, 4-49
    distance, design, 6-21
    force, 6-23
    sight distance, 4-23, 4-28, 4-42
    sight distance (tbl), 4-24, 4-52
Strategic Highway Research Program (SHRP), 5-11
Street, urban, 2-9
Strength, shear, 5-13
Structural approach, 3-19
Student's t-distribution, 2-4
Study
    air quality, 1-16
    area, 1-5
    broad traffic, 1-2
    corridor alignment, 4-2
    license plate, 1-5
    limited area, 1-2
    mesoscale, 1-17
    microscale, 1-17
    noise, 1-16
    O-D, 1-5
    origin-destination, 1-5
    planning, 1-14
    postcard, 1-6
    segment, 2-5
    site-impact, 1-2
    tag-on vehicle, 1-6
    traffic flow, 1-2
    traffic impact, 1-2
    traffic volume, 1-7
    trip attraction, 1-2
    trip generation, 1-2
Subarc, 4-12
Subbase, 5-3
Subgrade, 5-3
    railroad, 5-3
    rutting, weak, 5-12
Sublevel objective, 1-12
Subprocess, perception-reaction, 6-21
Subtangent
    alignment, 4-11
    point, 4-11
Subtended angle, 4-12
Superelevation, 4-16, 4-30, 4-32, 6-3
    distribution method, 4-32
    distribution method (fig), 4-33
    method, 4-35
    method (fig), 4-35
    method 5, 4-33
    minimum bikeway, 4-49
    rate, 4-32
    rate, bikeway, 4-49
    rate, maximum, 4-33
    rate, maximum (tbl), 4-34
    rates, design, v, 6-35, A-9, A-10, A-11, A-12, A-13
    runoff, v, 6-35, A-29, A-30
    transition, 4-35, 4-47
    transition adjustment factor (fig), 4-38
    transition length, 4-35
    transition, spiral, 4-39
Superior objective, 1-13
Superpave, 5-9
    aggregate mixture gradation (tbl), 5-14
    aggregate specification, 5-13
    design gyratory compactor (tbl), 5-27
    gyratory compactor, 5-26
    mix design procedure, 5-23
Surface
    pavement, 5-3
    traffic interference, 3-27
Surprise, 6-21
Surrounding clutter, 6-3
Survey
    and design, final location, 4-2
    cordon, 1-6
    corridor, 1-6
    final location, 4-3
    home-interview origin-destination, 1-5
    National Geodetic (NGS), 4-23
    O-D, 1-5
    onboard transit, 1-6
    preliminary location, 4-2, 4-3
Susceptibility temperature, 5-14

Swell, 5-4
    factor, 5-5
Swept path width, A-38
Symmetrical vertical curve, 4-20
Synthetic aggregate, 5-12
System
    capacity, 3-27
    drainage, 5-3
    LRV, 3-28
    railroad drainage, 5-3
    transit, 3-27

T

t-distribution, Student's, 2-4
Tag-on vehicle study, 1-6
Tally sheet, 1-9
Tangency, point of, 4-5, 4-7, 4-19
Tangent, 4-2
    ahead, 4-11
    back, 4-11
    bearing angle, 4-4
    curve, 4-11
    distance, 4-5
    length, approach, 4-28
    offset (fig), 4-19
    offset method, 4-9, 4-20
    offset method (fig), 4-15
    point, 4-11
    route alignment, 4-22
    runoff portion (tbl), 4-39
    runout, 4-35
    runout lengths, v, 6-35, A-29, A-30
    superelevation transition, 4-39
    to curve, 4-5
Taper, 6-26
    channelization, 6-26
    downstream, 6-26
    length, 6-27
    length criteria (tbl), 6-27
    merging, 6-27
    nonmerging, 6-27
    shifting, 6-27
    shoulder, 6-27
    two-way traffic, 6-27
Target
    attainable, 1-2
    optimum asphalt content, 5-18
TEA-21, 1-1
Temperature
    air, 5-24
    cracking, low-, 5-12
    high-design pavement, 5-24
    low-design pavement, 5-25
    pavement, 5-24
    susceptibility, 5-14
Temporary
    holding area, 3-12
    traffic control, 6-25
    traffic control zone section (fig), 6-26
Terminal, 3-3
Terrain
    general, 2-16
    level, 4-9
    level, passenger car equivalents (tbl), 2-23
    mountainous, 4-9
    rolling, 4-9
    rolling, passenger car equivalents (tbl), 2-23
Test
    Los Angeles abrasion, 5-23
    Marshall mix (fig), 5-19
    vehicle, 2-6
    vehicle method, 2-5
Thickness
    concrete slab, 5-9
    design, pavement, 5-9
    highway pavement, 5-9
    slab, 5-9
Three
    -leg directional interchange, 4-44
    -leg interchange, 4-44
    -level, 4-44

Threshold
    constraint, 1-13
    minimum reportable crash, 6-4
    of discomfort, 1-16
    of hearing damage, 1-16
    of pain, 1-16
Time
    average travel, 2-6
    bicycle perception-reaction, 4-48
    brake reaction, 4-23
    driver reaction (fig), 4-25
    dwell, 3-28
    effective green, 3-15
    holding area waiting, 3-14
    -in-queue delay, 2-3
    pedestrian start-up, 3-3
    perception-reaction, 6-21
    PIEV, 4-23
    pre-maneuver, 4-25, 4-26
    running, 3-29
    -space available, crosswalk, 3-14
    -space, effective available pedestrian, 3-15
    -space, turning vehicle interference, 3-15
    start-up, 3-3
    total crossing, 3-15
    total crosswalk occupancy, 3-15
    travel, 2-3
    -velocity, 3-28
Timing signal phases, 3-32
Toll
    collection system data, 2-8
    road card method, 2-8
Top ballast, 5-3
Topographic
    conditions, 4-2
    feature, 4-2
Topography, 4-9
    site, 4-45
Total
    crossing time, 3-15
    crosswalk occupancy time, 3-15
    curve deflection (fig), 4-12
    delay, 2-3
    ramp density, 2-12
    stopping distance, 4-43
    time-space available, 3-13
    trip end, 1-7
Toughness, 5-23
Tractor/trailer angle, A-38
Traffic
    average annual daily, 6-2, 6-5
    average annual daily (AADT), 1-10
    average daily, 6-2
    average seasonal daily, 6-2
    calming, 2-34
    capacity analysis, 2-9
    circles, 2-34
    composition, 2-12
    control device color, 6-22
    Control Devices, Manual on Uniform, 6-22
    flow, 1-2
    flow map, 1-7
    flow study, 1-2
    flow vehicle mix, 4-43
    impact study, 1-2
    link and node diagram, 1-10
    load (tbl), 5-25
    network diagram, 1-10
    noise impact, 1-16
    safety, 6-2
    sign color, 6-22
    sign viewing distance, 6-22
    signal, 2-26
    signal warrant, 2-27
    signal, warrant 1, v, 6-35, A-2
    signal, warrant 2, v, 6-35, A-3
    signal, warrant 3, v, 6-35, A-4
    signal, warrant 4, v, 6-35, A-5, A-6
    signal, warrant 9, v, 6-35, A-7, A-8
    signals, 2-27, 3-32

    space, 6-27
    speed (tbl), 5-25
    volume, 4-46
    volume data, 6-16
    volume study, 1-7
Train
    vertical acceleration, 4-51
    vertical curve length, 4-50
Transit
    acceleration limit, 4-50
    deceleration limit, 4-50
    light rail, 3-36
    mass, 3-26
    methodology (fig), 3-33
    mode, 2-2, 2-9, 3-31, 3-32
    on-street performance, 3-33
    rapid, bus, 3-36
    systems, 3-27
    urban scheduled, 3-31
Transition
    spiral (fig), 4-17
    superelevation, 4-35, 4-47
    tangent superelevation, 4-39
Transportation
    construction, air quality impact, 1-17
    Equity Act for the 21st Century (TEA-21), 1-1
    network, 6-16
    network screening, 6-17
    network, diagnosis, 6-18
    project, 6-16
Transverse slopes, 6-13
Travel
    speed, 2-3, 4-48
    -time delay, 2-3
    time, 2-3
    time and delay studies, 2-2
    time and delay study field sheet, 2-6
    time, speed, and delay diagram, 2-9
Traversable slope, 6-8
Traverse, 4-22
    closed, 4-22
    closed alignment (fig), 4-10
    control, 4-23
    open, 4-22
    open alignment (fig), 4-10
Treatment, highway safety, 6-16
Triangle, USDA, 5-9
Trip
    attraction, 1-5
    attraction study, 1-2
    destination, 1-2
    end, person-, 1-7
    end, total, 1-7
    external, 1-5
    external-external, 1-5
    external-internal, 1-5
    generation, 1-2
    Generation, 2-35
    generation data, 1-5
    generation study, 1-2
    generator, 1-2, 1-5
    internal-external, 1-5
    internal-internal, 1-5
    person-, 1-7
True meridian, 4-3
Trumpet shape, 4-44
TTC zones, 6-25
Tube pneumatic road, 1-9
Turning
    curb radius, A-38
    point, 4-18, 4-19
    point elevation, 4-19
    radius, A-38
    vehicle interference, time-space, 3-15
    vehicle, LOS effect, 3-15
Turnover, 2-36
Two
    -directional flow, 3-7
    -lane highway, 1-11, 2-9
    -lane highway methodology (fig), 2-20
    -lane highway segment analysis, 2-19

-lane highway, access point density
    adjustment (tbl), 2-21
-lane highway, 2-10
-way left-turn lanes, 2-12
-way traffic taper, 6-27
Type
    crash, 6-4
    I project, 1-16
    II project, 1-16
    of traffic, 2-12
    roadway, 1-11
Typical
    design-day traffic flow, 1-8
    vehicle deceleration rate, 4-48

U

Unbalanced directional split, 3-7
Uncompacted void of fine aggregate, 5-23
Uncontrolled access points, 2-14
Undercrossing sight distance (fig), 4-40
Underpass, 4-41
    line of sight (fig), 4-40
    roadway, 4-46
Unidirectional flow, 3-7
Unified Soil Classification System (tbl), v, 6-35, A-63
Uniform
    delay, 2-32
    pedestrian flow, 3-9
    segment, 1-10
Uninterrupted
    -flow facility, 2-10
    flow facility, 2-9, 3-9
    flow methodology, 2-19
Unit flow rate, pedestrian, 3-6
Unknown optimization, 1-13
Unsaturated, 2-10
Unsignalized intersection (fig), 3-5
Unsuitable excavation, 5-5
Unsymmetrical vertical curve, 4-20
Upgrade, specific, 2-17
Upgrades, 2-17
Urban
    areas, 2-2
    local streets, 2-2
    Mass Transportation Assistance Act, 1-1
    scheduled transit, 3-31
    street, 2-9
    transit buses, 3-32
USCS (tbl), v, 6-35, A-63
USDA, triangle, 5-9

V

Value
    engineering (VE), 5-27
    excess, 6-18
    Hveem's resistance, 5-9
    irreducible, 1-2
    K-, 4-28
    K- (tbl), 4-52
    methodology, 5-27
    R-, 5-9
    residual, 1-15
    salvage, 1-15
Variable
    optimization, 1-13
    -slope design, 6-10
Variation, 4-3
Vehicle
    arrival spacing, 3-28
    count, 1-10
    design, 2-36
    intercept method, 1-6
    met, 2-5, 2-6
    orientation angle, 6-8
    overtaking, 2-5, 2-6
    passed, 2-5
    shadow, 6-27
    test, 2-6
Vehicular flow rate, 1-13
Velocity, time-, 3-28
Vertex, 4-5

Vertical
    acceleration, train, 4-51
    clearance from overhead obstruction, 4-41
    clearance, minimum, 4-41, 4-48
    curve, 4-18
    curve length, 4-27, 4-28, 4-29
    curve length, train, 4-50
    curve, bike lane, 4-49
    curve, bikeway, 4-49
    curve, clearance on, 4-39
    curve, crest, 4-18, 4-49
Viewing distance, traffic sign, 6-22
Viscoelastic, 5-11
Vision
    20/20, 6-22
    axis, 6-21
    peripheral, 6-21
Visual
    acuity, 6-21
    acuity, center cone, 6-21
    axis, eye, 6-21
VMA criterion, 5-19
Void
    air, 5-16
    filled with asphalt, 5-16
    in mineral aggregate, 5-15
    in the total mixture, 5-16
    percentage of air, 5-16
Volition, 6-21
Volume, 2-10
    adjusted, 1-9
    and speed relationship, 1-10
    bank, 5-4
    borrow, 5-5
    directional, 2-6
    hauling, 5-5
    hourly, 2-6
    in situ, 5-4
    peak hour parking (tbl), 2-36
    traffic, 4-46
Volumetric
    design requirement, 5-28
    property, 5-14

W

Walk interval, 3-11
Walking
    path, minimum, 3-5
    speed, 3-3, 3-16
    speed, average, 3-3
Walkway
    capacity, 3-7
    capacity, nominal LOS, 3-9
    fire, 3-10
    interference, cross-flow, 3-8
    LOS, 3-9
    obstruction, 3-7
    obstruction effective length, 3-7
    off-street pedestrian, 3-20
    width adjustment, 3-7
    width, effective, 3-5, 3-7
Walkways, 3-3, 3-10
Warning area, advanced, 6-26
Warrant
    1, 2-27
    2, 2-27
    3, 2-27
    4, 2-27
    5, 2-27
    6, 2-27
    7, 2-28
    8, 2-28
    grade separation, 4-44
    traffic signal, 2-27
Warrant 1 traffic signal, v, 6-35, A-2
Warrant 2 traffic signal, v, 6-35, A-3
Warrant 3 traffic signal, v, 6-35, A-4
Warrant 4 traffic signal, v, 6-35, A-5, A-6
Warrant 9 traffic signal, v, 6-35, A-7, A-8
Waste, 5-4

Water
    -cement ratio, 5-11
    ponding, 4-36
Wave
    pattern, 3-3
    pattern queue, 3-3
Weak
    asphalt rutting, 5-12
    subgrade rutting, 5-12
Weaving and maneuvering section, 6-3
Weighted average trip generation rate, 1-4
West declination, 4-3
Wheelchairs, 3-10
Widening curve, 6-3
Width
    adjustment, walkway, 3-7
    effective walkway, 3-7
    minimum bikeway, 4-48
    preemptive, 3-7
    sidewalk, 3-16
    stair, 3-19
    swept path, A-38
Within letter spaces, 6-22
Work
    space, traffic, 6-27
    zone, 6-24
    zone safety, iii, 6-1, 6-24, 6-25

Y

Yard, station-, 5-7

Z

Zero plus zero zero, station, 4-7
Zone
    clear, 4-42, 6-8
    clear, object in, 6-3, 6-10
    length, passing (tbl), 4-28
    pacing, 3-5
    peripheral, 6-21
    restricted, 5-13
    sensory, 3-5
    work, 6-24